PROCEEDINGS OF THE

THIRD SEMINAR ON QUANTUM GRAVITY

PROCEEDINGS OF THE

THIRD SEMINAR ON QUANTUM GRAVITY

OCTOBER 23-25 1984, MOSCOW, USSR

EDITORS: **M. A. MARKOV**
Academy of Sciences of the USSR
Moscow, USSR

V. A. BEREZIN
Institute for Nuclear Research
Moscow, USSR

V. P. FROLOV
P. N. Lebedev Physical Institute
Moscow, USSR

World Scientific

Published by

World Scientific Publishing Co Pte Ltd.
P. O. Box 128, Farrer Road, Singapore 9128
242, Cherry Street, Philadelphia PA 19106-1906, USA

Library of Congress Cataloging in Publication Data

Seminar on Quantum Gravity (3rd : 1984 : Moscow, R. S. F. S. R.)
Proceedings of the Third Seminar on Quantum Gravity.

Seminar held Oct. 23–25, 1984 in Moscow, USSR.
1. Quantum Gravity — Congresses. 2. Supergravity — Congresses. 3. Black Holes (Astronomy) — Congresses. 4. Cosmology — Congresses. I. Markov, M. A. (Moisei Aleksandrowich), 1908– . II. Berezin, V. A. III. Frolov, V. P. IV. Title.
QC178.S455 1984 530.1 8514497
ISBN 9971-978-90-3

Printed in Singapore by Kim Hup Lee Printing Co. Pte. Ltd.

PREFACE

The Third Moscow Seminar on Quantum Gravity was held in October 23-25, 1984 by the Nuclear Physics Department of the Academy of Sciences of the USSR and the Institute for Nuclear Research of the Academy of Sciences of the USSR. The aim of the Seminar was to discuss the most important problems of the modern Quantum Gravity, namely: i) Quantum Gravity: the state of art; ii) Quantum effects in Cosmology and Problems of the Inflationary Universe; iii) Quantum black-hole physics; iv) the recent development in Supergravity and Kaluza-Klein theories. The seminar gathered 180 participants.

The Editors express their sincere gratitude to all contributors to these Proceedings for their cooperation in respect of time limitation, and accurate and patient preparation of their manuscripts in the camera ready form. The talks in the Proceedings are arranged in sections in accordance with their presentation at the Seminar.

Editors: M.A. Markov
V.A. Berezin
V.P. Frolov

CONTENTS

SECTION I

QUANTUM GRAVITY : GENERAL

ENTROPY IN AN OSCILLATING UNIVERSE IN THE ASSUMPTION OF UNIVERSE "SPLITTING" INTO NUMEROUS SMALLER "DAUGHTER" UNIVERSE

M.A.Markov
Institute for Nuclear Research of the USSR Academy of Sciences, 60th October anniversary pr. 7a, Moscow, 117312, USSR

ABSTRACT

We consider the possibility that in the collapse of a closed universe "daughter" universes (with smaller mass and entropy) may form in the regions of the mother universe where gravitational mass defect due to fluctuations reaches certain values. The model describes transformation of the classical Friedmann universe into the classical De Sitter universe in the assumption that the gravitational constant is a decreasing function of the matter density (ρ) and ρ does not exceed some arbitrary density ρ_0. At $\rho_0 \to \rho_{Pl} = c^5/\hbar G^2 = 10^{94}$ gr/cm³ the De Sitter universe is no longer classical: its "classical" radius (r_{cl}) becomes of Planck's length practically independent of the mass (M_0) of the Friedmann universe: $r_{cl} = l_{Pl} + l_{Pl}(m_{Pl}/M_0)$; $M_0 \gg m_{Pl}$ Possible consequences of indistinguishability of universes in this state for subsequent oscillations (i.e. subsequent mass ans entropy values) are discussed

Special theory of relativity unified what seem to be so different essences ino one space-time essence. Space, time and matter are in a sense already united by general theory of relativity. This process is evidently far from being completed in many aspects. The properties of the matter itself, the variety of its forms are still far from having an adequate theoretical description. We mean attractive ideas of organic unity of all forms of interactions, the search for symmetries which will provide an insight into the variety of particles, the rebirth of the ideas by Kaluza-Klein that space may appear to be multidimensional.

Cosmological problems, especially the problem of early universe, now attracts great attention.

The heterofore abstract idea concerning the possible existence of a variety of universes gradually becomes a subject of concrete discussions not only in the framework of general cosmological concepts, such as, for example, the chaotic appearance of different universes[1)], but also in what concerns possible matter penetration through various barriers that separate universes. The idea of various universes is discussed below in connection with the problem of entropy increase in scenarios of oscillating universes. There also arises the possibility of a peculiar splitting of universes, for example in a collapse, into universes with smaller amount of particles and entropy.

The interpretation of quantum theory itself seems to be unfinished too. Here we mean also the conception of the Everett type but the one which gives quite a specific interpretation to the existence of various universes. This conception or, more precisely, the interpretation of quantum theory is often reffered to (starting from the papers by Everett, Wheeler-Graham, De Witt[2)]) in connection with the description of a closed universe. The absence of the classical outside observer typical of the Copenhagen interpretation of a closed universe (i.e. just an oscillating universe) has always been the reason for recalling this peculiar case which gives rise to certain questions in the framework of the Copenhagen interpretation of quantum theory.

As distinguished from open universes, in the description of oscillating universes the initial state can be chosen either at the moment of its maximum (R_{max}) or minimum (R_{min}) radius. In our classical (non-quantum) consideration the initial state of an oscillating universe is taken at the moment of maximum expansion where the validity of the classical /i.e. non-quantum) description is extended to a smaller universes where it is necessary to estimate the applicability limits of the classically obtained results. Let it be a universe of the type of our Universe which at the present time has the density, say $\rho \sim 10^{-29}$gr, the total mass $M_0 \sim 10^{55}$gr, and $R_{max} \sim 10^{28}$cm.

1. FORMATION OF "DAUGHTER" UNIVERSES

First we consider macroscopical classical universes. We speak of such matter fluctuations due to which a given part of a universe can form its own universe. So, in principle, inside an expanding Friedman universe there may appear a closed collapsing universe if for a certain amount of particles a fluctuation is possible due to which the gravitational mass defect of these particles will turn out to be equal to the entire bare mass of this ensemble of particles.

In an expanding univrese such a daughter universe could in principle appear in an extremely rare black hole matter fluctuation.

Favourable possibilities for daughter universe formation may evidently appear in a collapse of oscillating universes. In the process of collapse even of an ideally homogeneous universe there initially develop fluctuational inhomogeneties. At some stage of a collapsing universe, matter density becomes close to the critical one,

$\rho \to \rho_{pl} = c^5/\hbar G^2 \sim 10^{94}$ gr/cm³. So, for a dust-like matter of the universe with bare mass $M_0 \sim 10^{55}$ gr, this limiting density is reached in our scenarios for a radius of the order of 10^{-13} cm. For a corresponding universe with a pressure typical of radiation, the critical density is reached at a radius of 10^{-3} cm. If in a very small part of the universe there occurs such a fluctuation of N particles in which gravitational defect will turn out equal to the entire particle mass of this ensemble, the ensemble forms a closed universe (we call it a daughter universe). The latter differs from its mother universe only by smallness of its maximum dimensions and smallness of the number of particles, the number of baryons or, generally speaking, of matter and correspondingly by small-ness of the global entropy inherited from the mother.

Further, according to our scenarios, in this small (daughter) universe the vavuum-like term gradually increases and the universe takes the De Sitter's form.

It is natural to assume that if such fluctuations do occur, daughter universes must at first be of small mass, i.e. have a small M_0 during further expansion.

But in subsequent oscillations their mass increases due to specific entropy increase, i.e. specific universe heating.

In this sense our Universe, if closed, may be regarded as a daughter universe which has splitted from its mother universe. If we knew the mass increase per one universe oscillation cycle then, assuming that our Universe appeared with minimal parameters, we would calculate the maximum time necessary for a universe similar to ours to appear as a result of subsequent oscillations.

Generally speaking, in a daughter universe there may exist its own closed universes, etc.

Such a peculiar hierarchic structure of the Universe as a whole may be called micro-macrosymmetric with a peculiar relativization of the concepts "macro" and "micro"[11].

The idea that daughter universe can be formed does not contradict the law of entropy increase, but entropy increase in oscillations does not contradict the finite dimensions and the lifetime of our Universe.

2. CLASSICAL MODEL OF FRIEDMAN-DE SITTER UNIVERSE.

We consider the set of Einstein's equations which describes evolution of the classical Friedmann universe in the process of its collapse into the classical (macroscopic) De Sitter universe. Such an evolution develops in the assumption that the gravitational constant is a decreasing function of the energy density. At given parameters of the Friedmann universe the limiting energy density determines the dimension of the classical (macroscopic) De Sitter universe if (ρ) is limited to the condition

$$\rho_0 \ll \rho_{Pl} = 10^{94} gr/cm^3.$$

Such a version of the classical Einstein theory could have arisen immediately after the appearance of the known Friedmann's work, i.e. six years ago, but the tendency of the limiting energy density to the Planck's value ($\rho \to \rho_{Pl}$) leads to the De Sitter universe of Planck's dimensions. In this case the universe cannot already be considered as an object of the classical theory. But in this case the corresponding classical characteristics, as is seen from what follows, are of heuristic interest for the discussion of the quantum character of further evolution already of a De Sitter micro-universe into a Friedmann macro-universe.

In recent years my collaborators and I tried to derive a set of equations of General Relativity to describe the universe evolution in such a way thet in a collapse the Friedmann universe would automatically become of the De Sitter type with Planck's mass and radius.

At first, on a simple model of a dust-like zero-pressure universe[3)4)5)] the idea was illustrated by the following modification of the Einstein equations

$$R_\mu^\nu - \frac{1}{2}R\delta_\mu^\nu = \frac{8\pi G_0}{c^2}\left[\rho\left(1-\frac{\rho^2}{\rho_{pl}^2}\right)^n \delta_0^\nu \delta_\mu^0 + \Lambda'\left(\frac{\rho}{\rho_{pl}}\right)^{2m} \delta_\mu^\nu\right] = K_\mu^\nu \quad , \tag{1}$$

where $T_\mu^\nu = \rho\delta_0^\nu\delta_\mu^0 c^2$; $\Lambda' = \Lambda c^2/8\pi G_0$ has the dimension of density $\rho < \rho_{pl} = c^5/\hbar G_0^2 = 10^{94}$ gr/cm³. It is of interest that in the left-hand side here the role of gravitating energy density is played by the expression

$$\Theta_\mu^\nu = T_\mu^\nu \left(1 - \frac{\rho^2}{\rho_{pl}^2}\right) \quad . \tag{2}$$

Since the covariant divergence in the left-hand side vanishes, the divergence in the right-hand side

$$K_{\mu}{}^{\nu}{}_{;\nu} = 0 \tag{3}$$

At $\rho^2 \ll \rho_{pl}^2$ the Λ -like term tends to zero and $\Theta_\mu^\nu \to T_\mu^\nu$, i.e. in the region $\rho^2 \ll \rho_{pl}^2$ the equations coincide with the classical Einstein equations which describe in the particular case the isotropic Friedmann universe with the scale factor a(t).

At $\rho^2 \to \rho_{pl}^2$ the tensor Θ_μ^ν gradually vanishes and the Λ -like term tends to the constant Λ , i.e. a Friedmann universe becomes a De Sitter universe.

The choice of the constants n,m,Λ gives an automatic fulfilment of the condition

$$\rho < \rho_{pl} \tag{4}$$

but we interpret as a new law of nature[5)] analogous to the laws $v \leq c$: $S \geq \hbar/2$.

The rate of radius variation a(t) gives for $\dot{a} = 0$ two values of the universe radius. The radius of maximum expansion and the minimum radius of the universe at the moment of collapse cessation (again $\dot{a} = 0$ is the moment of bounce) for n=m=1, $\Lambda'/\rho_{pl} = 2$

$$a_{max} \sim G_0 M_0/c^2, \quad a_{min} \sim l_{pl} \tag{5}$$

At $\Lambda'/\rho_{pl} = 2$ the density ρ_{max} is very close to ρ_{pl}, but never reaches it

$$\rho_{max} \sim \frac{\rho_{pl} M_o}{8\rho_{pl} \ell_{pl}^3 + M_o} = \frac{\rho_{pl} M_o}{8 m_{pl} + M_o} \sim \rho_{pl}\left(1 - \frac{m_{pl}}{M_o}\right). \qquad (6)$$

That is why the function Θ in (I) never becomes negative. According to (6) the maximum density ρ_{max} for all universes with masses from several Planck's mass to infinitely large masses differs little from the Planck density to a large accuracy.

It is also of importance that the "classical" bounce radius (the minimum radius of the universe) for all universes with masses $m_{pl} < M_o \leq \infty$ lies in the limits

$$a_{min} \sim \ell_{pl} + \ell_{pl}\left(1 - \frac{m_{pl}}{M_o}\right) \qquad (7)$$

Here m_{pl} is the Planck's mass. M_o is the total mass of a dust-like universe.

In other words, in our scenario <u>all</u> the collapsing universes with particle mass M_o

$$m_{pl} \leq M_o \leq \infty \qquad (8)$$

in the bounce state have the classical radius (a_{min})

$$\ell_{pl} < a_{min} < 2\ell_{pl}$$

But the theory we are dealing with is classical, i.e. not quantum and is inapplicable for these values of the scale factor. This circumstance may turn out to be lucky also for the solution of the entropy problem in an oscillating universe.

The latter statement now sounds as a paradox. But the point is that if the classical value a_{min} had a real meaning in the quantum description of collapse development in time, then by replacing t by -t in a further expansion of the universe we would obtain at best the former value of a_{max}, i.e. the former value of M_o or as a result of entropy increase a new value of $M_o' > M_o$.

But in the framework of a mini-universe with minimum dimensions quantum uncertainty $(\Delta a_{min}/a_{min} \sim 1)$ can just be interpreted as that of the bounce radius which in the classical description is rigidly connected with the mass of the universe at the moment of maximum expansion and, therefore, with the entropy problem. From the above only one thing is now clear: the classical rigid connection between the bounce radius a_{min} and the mass of the universe M_o is violated by Wheeler-De Witt quantization. At this stage we have no quantum description of the bounce phenomenon and of the consequence which will arise in this situation (surely different from the classical one) in the universe at its new expansion state that follows the bounce.

First, the question may arise whether the wave function has here the usual, i.e. the Copenhagen interpretation which is the solution to the corresponding Wheeler-De Witt's equation. May be the main difficulty in understanding the situation is that an essential point in the classical description of the "bounce" phenomenon is time: $a(t_1)$ before bounce, $a(t_2)$ after bounce, and $a(t)$ as $t_1 \rightarrow t_2$. Is it accidental that in the quantum description which is at present at our disposal (Wheeler-De Witt) the time is absent? Or, may be we cannot introduce ityet? And what if the time description of the bounce is in principle impossible? If the concept of time is meaningless in this case is there any hope that in such scenario (where matter is transformed into a vacuum state of the metric) the universe must appear only in the form and amount which it had before, only with the same quantitative parameters?

To answer this question, one should evidently consider three hypothetic possibilities.

1. Further it will be possible to show that in our scenario, too, the universe after bounce will undergo collapse exactly in the inverse order.

In this case the entropy problem is solved only by the

above-mentioned possibility of daughter universe formation.

II. There exists a mere chance absolute (God really dices)[2], and it is quite by chance that we find ourselves in such a universe where life could appear.

III. In the state of maximum collapse a universe is a set, a multitude of different universes, and it is not by an absolute chance, but by chance as it is understood in the statistical mechanics[2] that we find ourselves in the universe where there were conditions for the appearance of organized matter.

In the two latter cases along with the solution of the entropy problem in the appearance of daughter universes, the problem of entropy, of the origin of the universe, of its finite historical past can also be solved in a way specific for us, the inhabitants of this universe.

As concerns the development of classical mathematical methods of this theory and understanding of its place among the existing theories, we can say the following. Aman and the present author[6] have considered the version of a universe with a nonzero pressure. They have also shown that there exist such values, n = 1, m = 3/2, Λ'/ρ_{Pl} = 2/3, for which in the case of p = 1/3ε our Universe bounces at a^{-3} cm, i.e. the Planck region plays no role here, but the relative smearing of the bounce is, of course, by many orders of magnitude larger than the Wheeler's one for a_{max}[4].

In the paper[7] modified Einstein equations are derived on the basis of the variational principle.

The action function is written, as usual, in the form

$$S = \frac{c^4}{16\pi G_0} \int (R + 2\varkappa\varepsilon)\sqrt{-g}\, d^4x \quad , \qquad (9)$$

where $\varkappa$ is assumed to be some function of the energy density

$$\varkappa = \frac{8\pi G_0}{c^4} \Psi(\varepsilon) \quad . \qquad (10)$$

Variation of S about the metric g_{ik} leads to the modified Einstein equations

$$R^i_k - \frac{1}{2} R \delta^i_k = \left(\varepsilon \frac{\partial \varkappa}{\partial \varepsilon} + \varkappa\right) T^i_k - \varepsilon^2 \frac{\partial \varkappa}{\partial \varepsilon} \delta^i_k \ ,$$

$$\varkappa(\varepsilon) = \frac{1}{2} \int_0^\varepsilon G(\varepsilon)\, d\varepsilon \qquad (11)$$

where $T^i_k = (\varepsilon + p) u^i u_k - p\delta^i_k$; $\quad G(\varepsilon) = \varepsilon \frac{\partial \varkappa}{\partial \varepsilon} + \varkappa$.

It is of interest that in this version of the theory the Λ -term is not introduced a priori into the action function, but a Λ -like term appeares automatically in the Einstein equations if $\varkappa$ depends on the energy density. If such a dependence is absent, no Λ -like term appeares. The $\varkappa$ function is imposed only with the condition of asymptotical freedom. In this scenario there is no particle production or annihilation since in the derivation of eq. (II) a continuity equation $(\rho^* u^i)_{;i} = 0$ is used: figuratively speaking, particles do not vanish during collapse, but each of them becomes gravitationally less and less material due to the appearance of asymptotical freedom, and when the tensor $\Theta^i_k = (\varepsilon \frac{\partial \varkappa}{\partial \varepsilon} + \varkappa)\, T^i_k$ tends to zero as $\varepsilon \to \infty$, "immaterial souls of particles" form vacuum of the De Sitter universe.

Now one writes the expression for the action in an isotropic universe and the corresponding Hamiltonian.

In the case of the Friedmann universe the action

$$S \equiv \frac{3\pi c^4}{4G_0} \int \left[a\dot{a}^2 - a + \frac{a^3}{3} \int_0^\varepsilon G(\varepsilon)\, d\varepsilon\right] dt$$

and the Hamiltonian

$$H \equiv -\frac{G_0}{3\pi c^4} \frac{p^2}{a} + \frac{3\pi c^4 a}{4G_0} \left(-V(a) - 1\right)$$

where

$$p = -3\pi c^4 a\dot{a}/2G_0$$

The potential $V(a) = -\frac{1}{3} a^2 \int_0^{\varepsilon(a)} G(\varepsilon)\, d\varepsilon$

The Einstein equation for the 00 component (H = 0) is

$$\dot{a}^2 + V(a) = -1 \tag{12}$$

at $G(\varepsilon) = \frac{8\pi G_0}{c^4}\left(1 + \frac{\varepsilon}{\varepsilon_{pl}}\right)^{-2}$; $M l_{pl}^3 > m_{pl} a_{min}^3$; $a_{min} \sim l_{pl}$.

In this version it is only the asymptotical freedom that "works", particle production is absent.

In the subsequent papers particle production was taken into account in the action function by the term

$$\lambda[(\varepsilon u^{\nu})_{;\nu} - F(\varepsilon)\,\varepsilon_{,\nu} u^{\nu}]$$

where λ is the Lagrange factor (Berezin, Markov).

In one of the papers[8)] (Markov, Mukhanov) the hydrodynamic energy density ε is replaced by the energy density of a scalar field.

The scalar field Lagrangian

$$S = \frac{1}{2}\int [\varphi^{;l}\varphi_{;l}\, F\left(\frac{\varphi^2 m^2}{B}\right) + m^2\varphi^2 \psi\left(\frac{\varphi^2 m^2}{B}\right)]\sqrt{-g}\, d^4x$$

is used. The conditions imposed on the F and ψ functions make this version of the theory a physical analogue of hydrodynamic versions. For example:

$$F\left(\frac{\varphi^2 m^2}{B}\right) = \left(1 - \frac{\varphi^2 m^2}{B}\right); \quad \psi = \frac{\varphi^2 m^2}{B}$$

where $G_0(1 - \varphi^2 m^2/B) \to 0$ expresses asymptotical freedom, and the expression for the energy density

$$\theta_{\mu}^{\nu} \to 0 \qquad \text{as} \qquad \varphi^2 m^2 \to B$$

as before, is the limiting value already for the density $m^2\varphi^2$. Replacement of hydrodynamic energy density by the energy density of a scalar field illustrates a close similarity of the theory to the so-called inflation versions of the early universe.

The applicability limits of the theory are clear. So, the conditions $\rho < \rho_{pl}$ is equivalent to fulfilment of the condition

$$R_{\alpha\beta\gamma\delta} R^{\alpha\beta\gamma\delta} < 1/\ell_{Pl}^4$$

i.e. the applicability of such a formalism does not require quantization of a gravitational field.

Replacement of the hydrodynamic density by the density of a scalar field automatically leads to a nonlinear equation for the scalar field. In particular case this leads to the presence of the term φ^3 in the equation (φ^4 in the Lagrangian).

The scalar field is taken with a nonzero mass. In the author's opinion (he would like to think so) matter appeared from the vacuum state in the form of scalar particles of Planck's mass, which are elementary black holes (a cold gas at the initial moment). Splitting into all possible existiong particles they form the Big Bang state. And vice versa, in the collapse process (it is desirable that...) all the matter (including gravitational energy) vanishes in the vacuum of the De Sitter universe in the form of elementary black holes.

There is an interesting possibility of the formation of special (limiting) black holes, charged or with spin, which are known to undergo np Hawking's decay. The temperature of these black holes $T = 0$.

In the assumption that $\rho \leq \rho_{Pl}$ there must be evidently very few of such black holes. They must appear only in pairs and only by means of fluctuation mechanism and at rather large distance from one another (otherwise they annihilate). It is not excluded that the existence of a very small amount of such stable elementary black holes does not contradict astrophysical data. What is very important is that the existence of such stable particles does not contradict the known Hawking's theorems[10)] because in a closed space these particles, when produced, then annihilate completely (in pairs).

REFERENCES

1.[a] Linde,A.D. Phys.Lett. 132 B 317-20 (1983).
1.[b] Linde, A.D. "The inflationario. In reports on Progress in Physics", 47 N°8, August 1984.
2. The Many-Worlds Interpretation of Quantum Mechanics, ed. by Bryces, S De Witt and Neil Grahm. Prienceton Series in Physics., Prienston University Press (1973).
3. Markov, M.A. Preprint Inst. for Nucl. Res., p-0227 (1981).
4. Markov, M.A. Annals of Physics. 177, 333 (1984)
5. Markov, M.A. Pis'ma JETP 36 (1982) 214
6. Aman, E.G. and Markov, M.A. Preprint p-0290 Oscillating Universe in State $p \neq 0$(1983).
7. Markov, M.A. and Mukhanov V.F. Preprint "De Sitter-like Initial State of the Universe as a Result of Asymptotical Disappearance of Gravitational Interactions of Matter". Preprint P-0351 (1984).
8. Markov, M.A. and Mukhanov, V.F. Phys.Lett. (1984).
9. Markov, M.A. "On the Nature of Matter" Nauka, Moscow, 1976.
10. Hawking S.W. Phys.Rev. D14 2460 (1978)
Markov, M.A. in "Quantum Gravity" ed. by Markov, M.A. and West, P.C. Plenum Press, New York and London (1984).

ON THE MANY-WORLDS INTERPRETATION OF QUANTUM THEORY

V.F. Mukhanov
Institute for Nuclear Research of the Academy of Sciences of the USSR, 60-th October Anniversary Prospect 7[a], Moscow 117312, USSR

This contribution is a natural continuation of the report by M.A.Markov "The Problem of Entropy in an Oscillating Universe in the Light of the Hypothesis of a Possible "Splitting" of a Universe into Many Smaller Daughter Universes". One of the possibilities pointed out in the report by M.A.Markov is connected with the Everett's interpretation of the wave function of a closed universe. In this connection we think it necessary to suggest our speculations concerning the concepts by Everett, De Witt and other authors who have developed and popularized the Everett's interpretation of quantum theory. Besides, the recent results by Hawking et al. concerning quantization of a universe can be in a natural way interpreted in terms of the Everett's concept. We consider here such a possibility.

The present report is a result of numerous discussions with M.A.Markov.

The authors are grateful to B.De Witt for fruitful discussions.

1. INTRODUCTION

At the present time there exist two most selfconsistent interpretations of quantum theory: one of them, usually called the Copenhagen (orthodoxal) conception was developed by Bohr and his followers, the other, the many-worlds interpretation, was proposed by Everett. The Copenhagen interpretation is considered to be generally accepted, whereas the many-worlds interpretation is little known, and a very small number of papers are devoted to it [1)-7)]. This is possibly due to the fact that the questions exhibiting the advantage of this interpretation have not been widely discussed up to now. We mean, in particular, the problems of early universe, which only now have become an object of intensive studies. We shall not discuss in detail why the interpretation of the physical theory is needed, we shall only mention the fact that the essence of the theory is not limited to the mathematical aspects.

The questions connected with interpretation naturally arise in the attempt to associate the mathematical scheme of quantum theory with measurements.

Mathematics of quantum theory includes:

M.I. Hilbert space. The state of a closed quantum system is in some correspondence with a normalized vector $|\Psi\rangle$ of a corresponding Hilbert space.

M 2. A set of operators $\hat{A}$, $\hat{B}$, ... (not necesserily commuting) acting in this space. The observable quantities are put into correspondence with Hermitian operators.

M 3. The Schroedinger equation:

$$i\hbar \frac{\partial|\Psi\rangle}{\partial t} = H|\Psi\rangle \qquad (I)$$

2. MEASUREMENTS

First, we note that the measurement of a certain physical quantity, to which there corresponds an operator $\hat{A}$, always gives one of the eigenvalues of this operator, i.e. one of the numbers A_e for which the equation

$$\hat{A}\,|\Psi_e\rangle = A_e\,|\Psi_e\rangle \tag{1}$$

is resolvable .

If to a quantum system there corresponds the eigenvector $|\Psi_e\rangle$ of the operator $\hat{A}$, the measurement gives with necessity a corresponding eigenvalue.

Second, in any theory of measurements, under any interpretation of quantum theory the following experimental facts should be reproduced:

R (repeatability). The results of two immediately subsequent measurements of some quantity are in agreement, i.e. are the same. (If the operator of an observable quantity commutes with the Hamiltonian, the experiments can be performed with time intervals).

S (statistics). Let us consider $\mathcal{N}$ equally prepared systems, to each of which there corresponds the vector $|\Psi\rangle$. In each of the systems we measure the observable $\mathcal{A}$ ($\mathcal{A} \leftrightarrow \hat{A}$, $\hat{A}|\Psi_e\rangle = A_e|\Psi_e\rangle$)Then we have $n(A_e)/\mathcal{N} = |\langle\Psi|\Psi_e\rangle|^2$ as $\mathcal{N} \to \infty$, where $n(A_e)$ is the number of the systems for which the measured $\mathcal{A}$ proved to be equal to A_e.

Above we have restricted ourselves only to the statements which must be valid under any interpretation of quantum theory.Let us find out whether measurement can be described only by the Schroedinger equation.

Let us consider a system S interacting with an apparatus $\mathcal{A}$ aimed at measuring an observable $\mathcal{A}$ which characterizes S. Let the vector $|\Psi^S\rangle$ correspond to the system S and the vector $|\Psi^A[\ldots]\rangle$ to the apparatus in the initial state (before measurement) - (since the apparatus consists of a large number of quantum objects, it seems that its state can also be described by the vector of a corresponding Hilbert space). Hilbert space of a complete system $\Sigma = S + A$ is a direct product of the Hilbert spaces of a given system and an apparatus.

First, we consider a particular case where before measurement the system S is in a "state" $|\psi^S_m\rangle$, which is the eigenstate of the "measured operator" $\hat{A}$, i.e. $\hat{A}|\psi^S_m\rangle = A_m|\psi^S_m\rangle$. Let the system-apparatus interaction be such that the system S is in the same state $|\psi^S_m\rangle$ after measurement and the state of the apparatus changes into the following: $|\psi^A[\dots]\rangle \to |\psi^A[A_m]\rangle$.

Suppose that to different initial "states" $|\psi^S_m\rangle$ of the system S there correspond different final (after the interaction) "states" $|\psi^A[A_m]\rangle$ of the apparatus (and, accordingly, "indications of apparatus"). Only in this case can the quantity $\mathcal{A}$ be measured. Measurements by means of apparatus interacting in the indicated way with a given system are called "good measurements" or first-type measurements [8)-9)]. In what follows we consider only such measurements. It has turned out that for concrete physical systems this type of interactions can be described by means of the Schroedinger equation, i.e. there exists an unitary operator U_A, such that

$$U_A(|\psi^S_e\rangle|\psi^A[\dots]\rangle) = |\psi^S_e\rangle|\psi^A[A_e]\rangle \tag{2}$$

Concrete examples can be found in the books by von Neumann [8)] and De Witt [10)] (see also the works by Bohr and Rosenfeld [11)] and De Witt [4)]). Thus, in a particular case, where $|\psi^S\rangle = |\psi^S_e\rangle$ the measurement procedure can be described completely by the Schroedinger equation.

Now let us pass to the general case where the wave vector corresponding to the system S is an arbitrary superposition of the eigenvectors of the "measured operator" $\hat{A}$ i.e.

$$|\psi^S\rangle = \sum_m a_m |\psi^S_m\rangle . \tag{3}$$

From linearity of the Schroedinger equation (I) it follows that as a result of the above type of interaction between the system S and the apparatus the vector $|\psi^\Sigma\rangle = |\psi^S\rangle|\psi^A[\ldots]\rangle$ characterizing the initial state of the quantum system and of the apparatus is transformed as follows

$$|\psi^\Sigma\rangle \rightarrow U_A\left(|\psi^S\rangle|\psi^A[\ldots]\rangle\right) = \tag{4}$$

$$U_A\left(\sum_m a_m|\psi^S_m\rangle|\psi^A[\ldots]\rangle\right) = \sum_m a_m|\psi^S_m\rangle|\psi^A(A_m)\rangle .$$

It is clear that the superposition (4) cannot describe the result of only one single measurement since it includes terms corresponding, in fact, to all possible results. Introduction of a second, third and a larger number of apparatus described by the Schroedinger equation does not help. As a result, one always obtains a superposition of the type (4). The question arises how the result (4) can be interpreted. The attempt to answer this question suggests different interpretations of quantum theory.

3. ORTODOXAL (COPENHAGEN) CONCEPT

In as much as quantum theory is also fit for describing individual objects (but not only ensembles) [12] then, the superposition (4) being unable to correspond to the result of a single measurement, the measurement procedure does not seem to be exhaustively described by the Schroedinger equation. The Bohr's concept suggests that this is actually the case, and to complete the measurement procedure, one should involve additionally an outside "classical" observer who is not described by the Schroedinger equation. Further it is stated that in the "interaction" between a quantum system and a "classical" observer the wave vector of the quantum system collapses, in particular, the superposition (4) reduces to one of its terms by the probability laws, i.e.

$$\text{"classical observer"} + \sum_m a_m |\Psi_m^S\rangle |\Psi^A[A_m]\rangle \begin{array}{l} \nearrow |\Psi_1^S\rangle|\Psi^A[A_1]\rangle \\ \rightarrow |\Psi_2^S\rangle|\Psi^A[A_2]\rangle \\ \quad \cdots \\ \searrow^{P_n=|a_n|^2} |\Psi_n^S\rangle|\Psi^A[A_n]\rangle \\ \quad \cdots \end{array} \quad (5)$$

The probabilities $P_n = |a_n|^2$ are introduced <u>a priori.</u> The meaning of the word "classical" in the orthodoxal concept comes down, in fact, to the reduction (5). Thus, as a result of measurement the wave vector of the system S reduces (undergoes the "quantum jump") to one of the eigenstates of the "measured operator" in accordance with the above probabilities: $|\Psi^S\rangle \xrightarrow{P_m=|a_m|^2} |\Psi_m^S\rangle$. It is clear, therefore, that with such an approach one obtains <u>repeatability</u> (R) and <u>statistics</u> (S) of measurements.

In our opinion, the Copenhagen interpretation is based on the following two points.

1')The idea of probability reduction in measurements. Here "Quantum mechanics is a theory that attempts to describe in mathematical language a world in which chance is not a measure of our ignorance but is absolute" (B.De Witt [5]). From this it follows that there exist two ways of wave vector evolution which are not reduced to each other [8]:

a) quantum jumps under measurements.

b) evolution according to the Schroedinger equation.

2')The wave vector is a <u>mathematical tool</u> for the description of the results of measurements and has no meaning independent of concrete experiments [13)-15)]. It is necessary to accept this point is, first, because the $|\psi\rangle$-vector is associated to the state of an individual object but not of an ensemble of objects and, second, because of reduction. If we abandon this point, it will be difficult to interpret the mental experiments by Schroedinger("Schroedinger cat") [16)] by Einstein-Podolsky-Rosen [17)], by Wigner ("Wigner's fried") [18)] and others [3)].

In the Bohr's interpretation quantum theory should necesserily be supplemented with the postulate of measurements and for substantiation of quantum theory one should introduce the concept of "classicalness" which is not deduced in the framework of the mathematics of quantum theory. The Bohr's interpretation is consistent and satisfactory when applied to the description of laboratory experiments. But in the recent years especially urgent have become the problems connected with the description of early quantum Universe. Is quantization of the entire Universe meaningful from the point of view of the Bohr's concept of quantum theory? The answer to this question is negative. Indeed, in this case the vector has no independent meaning in the absence of observer. Hence, quantization of the Universe (manipulations with the -function) is also meaningless (especially in a closed Universe) because outside the Universe there are no observers and apparatus that could make measurements. Besides, the entire Universe, including "classical observers" that carry out experiments within the Universe cannot of course be described by the deterministic Schroedinger equation.

4. MANY-WORLDS INTERPRETATION

In the orthodox theory reduction was necessary first of all for fixing in the superposition (4) a concrete term which describes the interaction between the system S and the apparatus. This term was further in a natural way put into correspondence with the results of measurements. Reduction should inevitably be postulated if we require that the wave vector $|\Psi\rangle$ be put in a direct correspondence with an individual quantum object. Let us try to abandon this requirement. Suppose the superposition (4)

$$\sum_m a_m |\Psi^S_m\rangle |\Psi^A[A_m]\rangle$$

simultaneously describes the results of an ensemble of experiments and the vector $|\Psi^S\rangle = \sum a_m |\Psi^S_m\rangle$ is put into correspondence with an ensemble of objects. At first glance such an assumption seem unlawful for we know that quantum mechanics also describes the behaviour of individual objects in our Universe [12]. Suppose, however, that there exists many universes and that the superposition (4) describes simultaneously the results of experiments carried out in different universes, and to each concrete universe and to the result of a measurement in it there corresponds one of the terms in the superposition (4). Then, on the one hand, the vector $|\Psi\rangle$ characterizes only an ensemble of objects each of which is situated in its universe and, on the other hand, in a given universe quantum theory describes an individual object whose behaviour is affected by corresponding objects from other universes. The existence of other universes is just what accounts for interference effects. Using this approach, one can interpret quantum mechanics as a theory the very existence of which is due to the existence of many worlds. The assumption concerning the existence of many universes is the essence of the Everett's (many-worlds) interpretation of quantum theory. In the Everett's interpretation of quantum theory the reduction of the wave vector is not needed for the explanation of measurements. It abandons both the postulates, 1') and 2'), which are the basis of the Bohr's interpretation.

1) The wave vector evolves only according to the Schroedinger equation ("Quantum jumps" for which probability is absolute are absent).

2) The $|\Psi\rangle$ -vector is in one-to-one correspondence with the states of objects from an ensemble of systems and is meaningful irrespective of experiments. To the state of an individual object in a concrete universe there corresponds a component in the expansion of the - vector. (One-to-one correspondence may hold here since in this case an arbitrary $|\Psi\rangle$ -vector is in correspondence to an ensemble of objects, and quantum jumps are absent. None of the above-mentioned paradoxes appear).

To demonstrate agreement between the many-worlds interpretation and experiment, it is necessary to show that repeatability (R) and statistics (S) of measurements are reproducible.

Repeatability. Suppose that the observable system S, which was initially in the state $|\Psi^S\rangle = \sum a_m |\Psi^S_m\rangle$ $(\hat{A}|\Psi^S_m\rangle = A_m|\Psi^S_m\rangle)$ interacts with two apparatus I and II to the initial state of which there correspond the vectors $|\Psi^{\mathrm{I}A}[\ldots]\rangle$ and $|\Psi^{\mathrm{II}A}[\ldots]\rangle$. Let the interactions be described by the Schroedinger equation and be such that the observable $\mathcal{A}$ can be measured, i.e. (see (2))

$$|\Psi^S_m\rangle|\Psi^{\mathrm{I}A}[\ldots]\rangle \rightarrow U_{\mathrm{I}A}(|\Psi^S_m\rangle|\Psi^{\mathrm{I}A}[\ldots]\rangle) = |\Psi^S_m\rangle|\Psi^{\mathrm{I}A}[A_m]\rangle \quad (6)$$

$$|\Psi^S_m\rangle|\Psi^{\mathrm{II}A}[\ldots]\rangle \rightarrow U_{\mathrm{II}A}(|\Psi^S_m\rangle|\Psi^{\mathrm{II}A}[\ldots]\rangle) = |\Psi^S_m\rangle|\Psi^{\mathrm{II}A}[A_m]\rangle \quad (7)$$

Then, after a first measurement the wave function of the total system $|\Psi^{\Sigma}\rangle = |\Psi^{S}\rangle|\Psi^{IA}[\ldots]\rangle|\Psi^{IIA}[\ldots]\rangle$ is transformed as follows (this follows from (6) and from the linear Schroedinger equation)

$$|\Psi^{\Sigma}\rangle \rightarrow U_{IA}\left(\sum a_m|\Psi^{S}_m\rangle|\Psi^{IA}[\ldots]\rangle|\Psi^{IIA}[\ldots]\rangle\right) = \sum_m a_m|\Psi^{S}_m\rangle|\Psi^{IA}[A_m]\rangle|\Psi^{IIA}[\ldots]\rangle, \quad (8)$$

and after a second measurement, as follows from (7), we have

$$\sum_m a_m|\Psi^{S}_m\rangle|\Psi^{IA}[A_m]\rangle|\Psi^{IIA}[\ldots]\rangle \rightarrow U_{IIA}\left(\sum_m a_m|\Psi^{S}_m\rangle|\Psi^{IA}[A_m]\rangle|\Psi^{IIA}[\ldots]\rangle\right) = \sum_m a_m|\Psi^{S}_m\rangle|\Psi^{IA}[A_m]\rangle|\Psi^{IIA}[A_m]\rangle. \quad (9)$$

To each concrete universe there corresponds one of the elements of the superposition (9) and in each element the indication of the apparatus are in agreement. (The vectors $|\Psi^{I,II\,A}[A_m]\rangle$ correspond to the same value of the measured quantity). Thus, in each universe the repeatability of measurements takes place.

Statistics. Now let us show that the formula for the probabilities: $P_n = |a_n|^2$ (see S) need not be postulated (see (5)). It follows from some rather general considerations based on the concept of measure in the Hilbert space. Let us consider systems prepared in the same state $|\Psi^{S}\rangle$. Let in each of these systems the observable quantity $\mathfrak{A}$ be measured. Bearing in mind this measurement, we expand $|\Psi^{S}\rangle$ in eingenvectors of the operator $\hat{A}$ $(\mathcal{A} \leftrightarrow \hat{A})$, which in this case form a preferable basis of the Hilbert space of the system

$$|\psi^s\rangle = \sum_m a_m |\psi^s_m\rangle .$$

Not to shade the main idea, we assume for simplicity that the spectrum of the operator $\hat{A}$ is discrete and finite: $A_1, A_2, \ldots, A_n$. Then

$$|\Psi\rangle = \underbrace{|\psi^s\rangle|\psi^s\rangle \ldots |\psi^s\rangle}_{\mathcal{N} \text{ times}} = \tag{9a}$$

$$= \sum_{\ell, m, \kappa, \ldots} a_\ell a_m \ldots a_\kappa |\psi^s_\ell\rangle|\psi^s_m\rangle \ldots |\psi^s_\kappa\rangle .$$

In each element of the sum (9a) the vector $|\psi^s_1\rangle$ comes across p_1 times, the vector $|\psi^s_2\rangle$ - p_2 times, ..., the vector $|\psi^s_n\rangle$ - p_n times, and

$$p_1 + p_2 + \ldots p_n = \mathcal{N} \ , \ 0 \leq p_1, p_2, \ldots, p_n \leq \mathcal{N}$$

Let us represent $|\Psi\rangle$ in the form:

$$|\Psi\rangle = \sum_{p_1, p_2, \ldots p_n} c(p_1, p_2, \ldots p_n) \, |\Psi(p_1, p_2, \ldots p_n)\rangle ,$$

where

$$|\Psi(p_1, p_2, \ldots, p_n)\rangle = \frac{1}{\mathcal{N}or} \sum_{perm} \underbrace{|\psi^s_1\rangle \ldots |\psi^s_1\rangle}_{p_1} \ldots \underbrace{|\psi^s_n\rangle \ldots |\psi^s_n\rangle}_{p_n}$$

Nor is a normalization factor. Each of the $|\Psi(p_1, p_2, \ldots p_n)\rangle$-vectors corresponds to certain frequency rates $p_1, p_2, \ldots p_n$ of the measurable quantity. With an account of further measurement to different universes there correspond different vectors $|\Psi(p_1, p_2, \ldots, p_n)\rangle$

For $c(p_1, p_2, \ldots p_n)$ one can obtain fairly readily the formula

$$|C(p_1,p_2,\dots,p_n)|^2 = \frac{\mathcal{N}!}{p_1!\,p_2!\dots p_n!}\,|a_1|^{2p_1}\dots|a_n|^{2p_n} \quad (10)$$

As $\mathcal{N}\to\infty$ the polynomial distribution (IO) for the squares of moduli of the coefficients $|C(p_1,p_2,\dots p_n)|^2$ goes over to a multidimensional Gaussian distribution with the centre corresponding to $p_1\sim\mathcal{N}|a_1|^2$, $p_2\approx\mathcal{N}|a_2|^2,\dots,p_n\approx\mathcal{N}|a_n|^2$, and the dispersion is proportional to $\sqrt{\mathcal{N}|a_1|\dots|a_n|}$. Hence, as $\mathcal{N}\to\infty$ the main contribution (in the sense of norm) into the complete wave vector $|\Psi\rangle$ is made by those vectors $|\Psi(p_1,p_2,\dots,p_n)\rangle$ for which the frequencies $p_1,p_2,\dots,p_n$ to an accuracy of standard statistical fluctuations $\propto\sqrt{\mathcal{N}}$, correspond to the frequencies required by the measurement statistics S. On the basis of the above conclusion it is easy to prove the theorem that one can introduce the frequency operator which satisfies the property [19)-21)]

$$\lim_{\mathcal{N}\to\infty}\hat{F}^{\mathcal{N}}_{k}|\Psi\rangle = |a_k|^2|\Psi\rangle$$

Thus, we see that the formula for the probability, $P_n=|a_n|^2$ is related to the general properties of the Hilbert space of an ensemble of systems and arises in a natural way in the framework of the mathematical scheme of quantum theory. For its complete substantiation it is undoubtedly necessary to postulate the relation between "typicalness" of events and the measure in the Hilbert space.This postulate is however more natural than the initial one. If we assume that the "typicalness" of a concrete universes is determined by the value of the coefficient before a term corresponding to this universes in the superposition (4), then in "typical" universes the observable statistics of measurements (S) is reproducible. It can be assumed, for example, that the number of a certain type of universes is proportional to the squared modulus of a corresponding coefficient in the expansion (4). Then the concept of "typicalness" can be identified with the concept of the fraction of universes of a given type in the total number of universes. The question of the total number of universes (whether it is finite or infinite) is open and requires further investigation. From the point of view of the many-worlds concept all possible values of the observable quantities are realized in different universes. This means that all the conceivable processes, if the probability of their realization is not exactly equal to zero, take place in reality. In particular, the events of small probability (for example, heat overflow from a cold body to a hot one) are also realized. But in the most "typical" universes (in the majority of worlds) the processes obey, however, the known physical laws (for example, entropy increase). The task of physics is to describe these laws, but not individual events. Such laws are inherent only in the most "typical" universes, and therefore, only such universes are of interest, because only for them can one make certain predictions.

Everett has come to the many-worlds interpretation of quantum theory when he was considering the theory of measurements in quantum mechanics. From the point of view of the description of "good" measurements, the many-worlds theory is at present practically completed except for the problems of language formalization and a selfconsistent definition of the concepts of "system", "apparatus", "state", "observable quantity" [4]. The theory of measurements is however far from exhausing the subject of the many-worlds interpretation of quantum mechanics. To approach the questions which naturally arise in the attempt of a consistent consideration of the many-worlds theory, it is necessary to clarify the role of the apparatus in the previous speculations. We have studied above a much simplified case where each given universe consists, in fact, of an apparatus and an observable system S and all the physical processes in each universe are reduced to measurement, i.e. to a specific interaction between system and apparatus. A complete wave function is in this case a superposition of products of the wave functions of the device and the system. The apparatus was supposed to be a macroscopical object, and to different wave functions of the apparatus $|\psi^A[A_\ell]\rangle$ there correspond macroscopically different states. In order that different indications of the apparatus could differ macroscopically, the apparatus should be organized in a special way (for example, it should be an unstable system). The vectors $|\psi^A[A_\ell]\rangle$ corresponding to different indications of the apparatus (and, accordingly, to different values A_ℓ of a measured quantity) must be orthogonal. This is necessary for realization of measurement and agree with the fact that the vectors $|\psi^A[A_\ell]\rangle$ must be eigenvectors of some Hermitian operator because they correspond to certain values of some observable characteristic of the apparatus. These properties of the apparatus, permit first, measurement by a macroscopic observer who perceives world in terms of classical physics and, second, distinguish in the class of Hermitian operators characterizing the microsystem a certain subclass with eigenvalues characterizing the properties of an observable system in any concrete universe. This is realized because in the interaction with the apparatus the observable characteristics of micro-objects are in one-to-one correspondence with certain macroscopic apparatus characteristics, the reality of which can be judged on the basis of the concepts of the classical physics. In the Everett theory the expansion of complete wave function has an absolute physical meaning. This expansion determines, in fact, an ensemble of really existing universes.

For interpretation of the results it is always necessary to determine some preferable basis of operators and expand a complete wave function in their eigenvectors. These operators correspond to really existing characteristics of objects. In the above case of "good" measurement such a preferable basis (which corresponds to the expansion (4)) has been chosen, in fact, on the basis of the quasi-classical concepts (arm positions), suggested by specific properties of the apparatus. A preferable basis of operators has a physical meaning, for it determines really existing characteristics of objects in various universes. In the general case the question of the choice of such a basis (without associating it with measurements) is now unsolved. To solve this problem, one should evidently study in more detail the general structure of quantum theory (to establish complete sets of commutating operators), the quasi-classical approximation and the theory of measurements. The necessary set of operators, which compose a complete basis and correspond to really existing characteristics of quantum objects must possibly be a commuting one. If this assumption is valid, it entails limitedness of the space-time description of universes, the necessity to reformulate the concept of particle. the necessity to change the statement of the high-energy problems in quantum field theory. It is possible that conceptual problems of quantum gravity, the problem of particle identification in a curved space-time, the problem of measurements in quantum theory of gravity, etc. will be solved on this way [7]. To use either the theory of "good" measurements or quasi-classical analogues is now the only possible way in searching for a correct, at least an approximate form of expansion of the wave function for determining an ensemble of universes, until the problem of the choice of a preferable basis is solved in a closed form.

5 QUANTIZATION OF THE UNIVERSE

In the Everett's interpretation quantum theory is closed. Measuring instruments and an observer himself are only complicated macroscopic quantum systems with special properties. The classical description of macro-objects appears to be the limiting case of the quantum description when an elementary quantum of action can be neglected and the measurement procedure is reduced to specific interaction between macro and micro objects, which can be exhaustively described in terms of the Schroedinger equation. The many-worlds interpretation abandons both principal assumptions of the Bohr's interpretation. The equations of the theory that describe the whole ensemble of universes are fully deterministic and have no probable "quantum jumps". The wave vector completely characterizes the "real" state of an ensemble of objects existing in different universes irrespective of the fact whether they undergo measurements. From what has been said it is clear that in the Everett theory quantization of a closed universe is essential, and the complete wave function of a universe can be naturally interpreted as a characteristic of an ensemble of different universes.

The complete wave function of a universe, after the choice of a corresponding basis in the Hilbert space, can be represented as a functional which depends on the 3-geometry ${}^{(3)}\mathcal{T}$ and on the material fields q (for example, a scalar field $q \equiv \varphi(x_1, x_2, x_3)$ given on corresponding three-dimensional hypersurfaces, i.e.

$$\Psi = \Psi\,({}^{(3)}\tau, q).$$

This wave function Ψ obeys the following Schroedinger equation which disregards time dependence

$$\mathcal{H}\Psi = 0 \qquad \text{(II)}$$

This equation results from canonical quantization of gravity [22]. Equation (II), where $\mathcal{H}$ is a super-Hamiltonian, splits into two equations. The first is the invariance condition for Ψ under coordinate transformations on a three-dimensional hypersurface. The second is the Schroedinger equation proper, also called the Wheeler - De Witt equation

$$H\left({}^{(3)}\tau, \frac{\delta}{\delta\,{}^{(3)}\tau}, q, \frac{\delta}{\delta q}\right)\Psi({}^{(3)}\tau, q) = 0. \qquad (12)$$

It indicates physically admissible vectors in the entire Hilbert space of a given dynamic system. The solution of eq.(12) can formally be represented as a functional integral. For the Wheeler-De Witt equation boundary conditions are necessary. As such conditions Hartle and Hawking [23] have proposed to choose those which in the Euclidean functional integral (it is assumed to be equivalent to the integral over the Minkowski metrics in the sense of contour integration) correspond to summation over all compact closed (without boundaries) four-dimensional manifolds including a given 3-geometry ${}^{(3)}\tau$ for a given matter distribution on it:

$\Psi({}^{(3)}\tau, q) =$

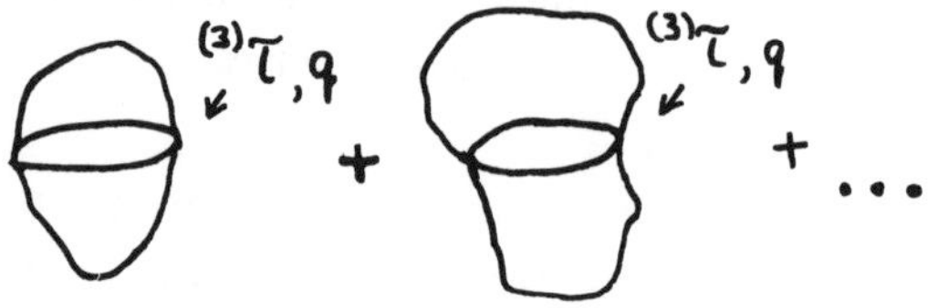

If we proceed from the analogy with an ordinary quantum field theory, the obtained wave function must correspond to the state of gravitational vacuum, i.e. to the state of minimum excitation [23]. The proposed boundary conditions are very natural since the time parameter does not explicitly enter (12).

With given boundary conditions equation (12) cannot be now solved in the general form. To have an idea of the character of its solution, one usually considers "minisuperspace" models [22)-26)]. In such models an infinite number of the degrees of freedom of a gravitational field is frozen and only several degrees of freedom are quantized. The papers [22)-26)] consider homogeneous, isotropic universes filled with matter. In this case the 3-geometry ${}^{(3)}\mathcal{T}$ can be completely characterized by the value of the scale factor a (which corresponds to a physical degree of freedom - time), and the matter can be characterized for example, by the density ε [22)] or by the value of a homogeneous scalar field φ [23)-26)] depending on what the universe is filled with. In the papers by Hawking et al. [24)-26)] the solution of the Wheeler-De Witt equation is considered in the case of "minisuperspace" model for an isotropic universe filled with a uniform non-conformal scalar field . In [27)] an attempt is made to overstep the limits of the "minisuperspace" model. The boundary conditions are chosen there in the form proposed in ref. [23)]. The solution of the equation is obtained analytically in the quasi-classical approximation (the functional integral was estimated by the stationary phase method) and the numerical simulation methods are used which provide a deeper insight into the behaviour of the wave function $\Psi(a,\varphi)$ [26)]. The obtained wave function describes the state of minimum excitation. How should it be interpreted? In the framework of the Everett's concept of quantum theory for the interpretation of the complete wave function of the universe we must find a correct expansion of this function, different terms of which correspond to different universes. As has already been mentioned, the problem of finding such an expansion in the general form cannot be solved until a preferable basis is chosen. We cannot use for this purpose the theory of good measurements. Searching for such an expansion, we can therefore apply only quasi-classical analogues. The very form of the complete wave function practically predetermines the form of the expansion necessary for its interpretation. A certain part of the complete wave function which describes the minimum excitation state in the case of a non-conformal scalar field can approximately be represented in the form of superposition of the wave functions each of which describes quasiclassical universes with different maximum expansion radii a_{max} [24)-25)]. At high energy densities these universes underdo the De Sitter stage of different duration. This is due to non-conformity of the scalar field which immitates the Λ-term under the condition $|\dot{\varphi}|^2 \ll V(\varphi)$ [28)]. Undoubtedly, this only brings nearer the unknown expansion. It is obtained, in fact, in the quasi-classical limit and distinguishes only classically different universes. Since in the many-world interpretation of quantum theory there is no fundamental difference between macroscopic and microscopic objects, in the unknown complete expansion of the wave function we must also distinguish microscopically different universes (the existence of such universes are responsible for the interference effects). How to achieve this in the case of a"good" measurement we know already. But it is unclear how we can continuously pass over from one limiting case (classically different universes) to another (for example, universes differing only by the spin of one electron). To this end, one should solve in a closed form the problem of the choice of a preferable basis.

In the case of a non-conformal scalar field the coefficients in the expansion of the complete wave function before the terms corresponding to universes of the type of ours are essentially nonzero, whereas in the case of a conformal field they are negligibly small.[23)] This is due to the fact that only a nonconformal scalar field leads to the De Sitter stage at which even an initially Planck-length universe expands to macroscopic scales. If "typicalness" of a universe is determined by the value of a corresponding coefficient, the role of the De Sitter stage in the universe evolution is reduced to the fact that due to this stage the universes of the type of our become "typical" in a gravitational vacuum.

Thus, there are evidently some grounds to suppose that the state of minimum excitation of a gravitational field (gravitational vacuum) can be thought of as an ensemble of closed universes with radii of maximal expanding, beginning with the Planck's one and up to very large scales ($> 10^{28}$ cm). Miscroscopically different universes of this ensemble are responsible for ordinary quantum-mechanical interference effects. Universes of the type of ours, which have passed through the De Sitter stage of evolution, are typical in a gravitational vacuum just due to this stage.

The above speculations, the same as the results of the calculations [23)-27)] are, in our opinion, of a rather preliminary character, and considerable work is undoubtedly required for clarifying this point.

6. CONCLUDING REMARKS

In conclusion we would like to list the main

advantages of quantum theory in the many-worlds interpretation as compared with the Bohr's interpretation.

I. Quantum theory in the Everett's interpretation is a closed theory. Classical physics is the limiting case of quantum physics (as $\hbar \to 0$).

2. The equations of the theory are deterministic and measurements are completely described by the Schroedinger equation.

3. The wave function gives a one-to-one description of reality irrespective of experiments.

4. Within the many-worlds interpretation quantization of a closed universe as a whole is meaningful. The wave function is interpreted in a natural way. Proceeding from the Everett's interpretation, one should reconsider many concepts of the modern quantum physics. It is not excluded that on this way one will succeed in solving a number of most fundamental physical problems.

REFERENCES

1. Everett H. III Rev. Mod. Phys., 29, 454 (1957)
2. Wheeler J.A. Rev. Mod. Phys., 29, 463 (1957)
3. De Witt B.S. and Graham N. "The many-worlds interpretation of quantum mechanics" (Princeton University Press, Princeton, 1973)
4. DeWitt B.S. in Proc. of the Int. School of Physics "Enrico Fermi", Course IL : "Foundations of quantum mechanics" ed. B.D'Espagnat (Academic Press, New York, 1971)
5. DeWitt B.S. Phys. Today, 23, 30 (1970)
6. Ballentine L.E. et.al. Phys. Today 24, 36 (1971)
7. Smolin L. in "Quantum theory of gravity" ed. Christensen (London, 1984)
8. Neumann von, J. "Mathematical Foundations of Quantum

Mechanics" (Princeton N.J., 1955)
9. London F. and Bauer E. "La theorie de l'observation en mecanique quantitue (Paris, 1939)
10. DeWitt B.S. "Dynamical theory of groups and Fields" (Gordon and Breach, New York, London, Paris, 1965)
11. Bohr N. and Rosenfeld L. Kgl. Danske Videnskab. Selskab, Mat.-fys. Med., 12, No.8 (1933)
12. See, for example, Jauch J.M. in Proc. of the Int. School of Physics "Enrico Fermi", Course IL: "Foundations of quantum mechanics" ed. B.D.'Espagnat (Academic Press, New York, 1971)
13. Bohr N. Dialectica, 2, 312 (1948)
14. Bohr N. in "A.Einstein, philosopher-scientist" Evanston, 201 (1949)
15. Einstein A. in Albert Einstein, Hedwiga und Max Born. (Munchen, 1968)
16. Schroedinger E. Naturwissenschaften, 23, 812 (1935)
17. Einstein A., Podolsky B. and Rosen N. Phys. Rev. 47, 777 (1935)
18. Wigner E. "Symmetries and Reflections" (Indiana University Press, 1967)
19. Finkelstein D. Transactions of the New York Academy of Sciences, 25, 621 (1963)
20. Graham N. in DeWitt and Graham, eds. op. cit (1973)
21. Hartle J.B. Am. J. Phys. 36, 704 (1968)
22. DeWitt B.S. Phys. Rev. 160, 1113 (1967)
23. Hartle J.B. and Hawking S.W. Phys. Rev. D28, 2960 (1983)
24. Hawking S.W. Nucl. Phys. B239, 257 (1984)
25. Hawking S.W. in "Relativity, Groups and Topology" Les Houches. Session XL, North-Holland (1984)
26. Hawking S.W. and Wu Z.C. "Numerical Calculations of Minisuperspace Models", D.A.M.P.T. preprint (1984)

27. Halliwell J.J. and Hawking S.W. "The origin of Structure in the Universe" D.A.M.P.T. preprint (1984)

28. Linde A.D. Phys. Lett., 119B, 177 (1983)

RELATIVISTIC THEORY OF GRAVITATION

A.A.Logunov, M.A.Mestvirishvili

Institute for High Energy Physics, Serpukhov, Moscow region, USSR

ABSTRACT

In the present paper a relativistic theory of gravitation (RTG) is constructed in a unique way on the basis of the special relativity and geometrization principle. In this, a gravitational field is treated as the Faraday-Maxwell spin-2 and spin-0 physical field possessing energy and momentum. The source of a gravitational field is the total conserved energy-momentum tensor of matter and of a gravitational field in Minkowski space. In the RTG, the conservation laws are strictly fulfilled for the energy-momentum and for the angular momentum of matter and a gravitational field. The theory explains the whole available set of experiments on gravitation. In virtue of the geometrization principle, the Riemannian space in our theory is of field origin, since it appears as an effective force space due to the action of a gravitational field on matter. The RTG leads to an exceptionally strong prediction: the Universe is not closed but just "flat". This suggests that in the Universe a "hidden mass" should exist in some form of matter.

INTRODUCTION

In this paper the relativistic theory of gravitation (RTG) is constructed on the basis of the special relativity, and the ideas by Poincaré, Minkowski, Einstein and Hilbert get their further development. Also the investigations of authors [1,2] are reflected and developed here.

First the principle of relativity was applied to mechanical phenomena only. But then Henri Poincaré formulated it as the universal principle for all physical phenomena [3]: "The laws of physical phenomena should be the same both for an observer at rest and for a one who is in the state of a uniform translation motion. So, we do not and cannot have any means to distinct whether we are in such a motion or not". Even by now people used to think that the essence of the principle of relativity is restricted by the existence of only one class of coordinate systems, the so-called inertial reference frames within which physical processes proceed in the same way. However, as shown in ref. [4], the pseudo-Euclidean space-time geometry discovered by Minkowski allows to formulate the generalized principle of relativity, valid both for the class of inertial and that of noninertial frames. The generalized principle of relativity was formulated in refs. [4]: whichever physical reference system is chosen, inertial or noninertial, one can always find an infinite set of other frames, where physical phenomena are simultaneous with those in the initial reference frame. Thus, we do not and cannot have any experimental means to distinguish in what particular reference frame among this infinite set we are.

The discovery of the pseudo-Euclidean space-time geometry allows to formulate physical laws both in inertial and noninertial reference frames, and thus to disprove the erroneous statements [5] on inapplicability of the special theory of relativity to accelerated reference frames. This means that when describing physical phenomena in Minkowski space we may, depending on a specific physical problem, choose any suitable reference frame adequate for the given problem and, hence, set a corresponding metric tensor γ^{ik} of Minkowski space. According to the ideology of the general relativity (GR), the special principle of relativity cannot be applied for gravitational phenomena. It was that very central point in which almost seventy years ago Einstein and Hilbert turned away from a special theory of relativity when constructing GR. This resulted in giving up the conservation laws for the energy-momentum and angular momentum, as well as in the development of unphysical concepts on the nonlocalizability of gravitational energy, and of many other things, which have nothing to do with gravitation. These two eminent scientists left the surprisingly simple Minkowski space with the maximal, ten-parameter, group of space motion and entered the maze of the Riemannian geometry, which entangled the subsequent generations of physicists engaged in gravitation. Some authors even consider giving up the energy-momentum conservation laws in GR to be the most important principle step of this theory which overthrew the concept of energy. But it would be too thoughtless if we renounced the most important law of nature, i.e. the conservation law of energy-momentum and the angular momentum of a closed system without sound experimental grounds. It was shown in refs. [1] that, since the GR does not and cannot have conservation laws for the energy-momentum of both matter and a gravitational field, then the inert mass defined in Einstein theory has no physical sense, the gravitational radiation flux, as it is defined in the GR, can always be eliminated by the corresponding choice of the admissible reference frame, and hence, the Einstein quadrupole formula for the gravitational field radiation does not follow from GR. The general relativity does not basically suggest that a binary system loses energy in the form of gravitational radiation. The GR does not have the classical Newtonian limit and, consequently, does not satisfy the most fundamental principle of physics, i.e. the correspondence principle.

That is where the absence of energy momentum conservation laws in GR brings us if we reject dogma, seriously consider the essence of the problem and make the analysis close to an elementary one.

All of it testifies to the fact that the GR is not a satisfactory physical theory. Therefore, the problem of constructing a classical theory of gravitation which would satisfy all the requirements imposed on a physical theory, is quite vital.

As opposed to the GR, our theory is based on the special principle of relativity which we, following Poincaré, consider universal and, consequently, applicable to gravitational phenomena as well. Thus, in our approach the conservation laws for the energy-momentum and angular momentum are fulfilled strictly and have a covariant character. Therefore, our theory contains no pseudotensors and as a consequence no unphysical concepts of nonlocalizability of the gravitational energy arise. Figuratively speaking, our problem is to construct, without leaving Minkowski space, an effective field Riemannian space with the help of a tensor gravitation field and the geometrization principle, with the conservation laws for matter being strictly fulfilled. This will allow us to use, if necessary, Riemannian space already equipped with the conservation laws for matter. Note, that Riemannian space constructed in such a way is, literally, of a field origin since the effective force space is generated by a gravitation field of Faraday-

Maxwell type. Thus, in the present paper we shall carry out this program developing the ideas of refs. [6]. In this we manage to preserve with necessity the Hilbert-Einstein equations supplementing them with four new field equations. According to new equations, a gravitation field has, in the general case, only 2 and 0 spins. This theory changes the established concepts of space-time influenced by the GR, takes out of the maze of the Riemannian geometry and is in spirit of the modern theories in elementary particle physics.

The theory developed in this paper is based on the concept of a gravitational field being a physical field of Faraday-Maxwell type and possessing energy-momentum. Thus, a gravitational field, as well as all other physical fields, is characterized by the energy-momentum tensor of the system. We consider a gravitational field as a spin 2 and spin 0 physical field, and a free gravitational field to have spin 2. The space-time geometry for all physical fields is pseudo-Euclidean (Minkowski space). Thus, the conservation laws for the energy-momentum and angular momentum of a closed system are rigorously fulfilled. This is the principal distinction between our theory and the Einstein GR. Another important problem arising in the construction of a theory of gravity, is that concerning the interaction between a gravitational field and matter. We think a gravitational field to be universal, and to act on all forms of matter identically. We construct our theory on the basis of the geometrization principle [1], which says that the equations of motion of matter under the action of the tensor gravitation field ϕ^{ik} in Minkowski space with the metric tensor γ^{ik}, may be identically represented as the equations of motion of matter in the effective Riemannian space-time with the metric tensor g^{ik} depending on the gravitational field ϕ^{ik} and the metric tensor γ^{ik}. In this way we introduce the concept of an effective Riemannian space of field nature. Proceeding from Minkowski space and the geometrization principle, the Lagrangian density has a general form

$$L = L_g(\tilde{\gamma}^{ik}, \tilde{\phi}^{ik}) + L_M(\tilde{g}^{ik}, \phi_A), \qquad \text{(A)}$$

where $\tilde{\phi}^{ik} = \sqrt{-\gamma}\phi^{ik}$ is the density of the tensor of a field variable in the gravitational field ϕ^{ik}, $\tilde{g}^{ik} = \sqrt{-g}g^{ik}$ is the density of the metric tensor of the Riemannian space g^{ik}, $\tilde{\gamma}^{ik} = \sqrt{-\gamma}\gamma^{ik}$ is the density of the metric tensor of Minkowski space, ϕ_A are the fields of matter.

In this theory, the Lagrangian density of the gravitational field depends on the metric tensor γ^{ik} and the gravitational field ϕ^{ik}, that is why it crucially differs from the GR, where the Lagrangian density depends only on the metric tensor of the Riemannian space g^{ik}. Thus, in our theory, contrary to the GR, the geometrization of the Lagrangian density of the gravitational field is not complete.

1. GEOMETRIZATION PRINCIPLE AND GENERAL RELATIONS IN THE RELATIVISTIC THEORY OF A GRAVITATIONAL FIELD

Without any loss of generality, we shall assume the tensor density $\hat{g}^{ik}$ of the metric tensor of Riemannian space-time to be a local function depending on the tensor density $\tilde{\gamma}^{ik}$ of the metric tensor of Minkowski space and the tensor density $\tilde{\phi}^{ik}$ of the gravitational field.

Let the Lagrangian density L_M depend only on the fields ϕ_A, on their first-order covariant derivatives, and also on the tensor density $\tilde{g}^{ik}$ in virtue of the geometrization principle. The Lagrangian density of the gravitational field is taken to depend on the tensor density $\tilde{\gamma}^{ik}$, their first-order partial derivatives, as well as on the gravitational field density $\tilde{\phi}^{ik}$ and its first-order covariant derivatives with respect to the Minkowski metric. To obtain the conservation laws, we use the invariance of the action in the infinitesimal covariant shift. Indeed, since the action is a scalar, the variations of the actions of matter, δJ_M, and of the gravitational field, δJ_g, will be zeros under an arbitrary infinitesimal coordinate transformation. Calculate first the variation of the action of matter under the transformation

$$x'^i = x^i + \xi^i(x) \tag{1.1}$$

ξ^i being the infinitesimal shift four-vector

$$J_M = \int d^4x \left[\frac{\delta L_M}{\delta \tilde{g}^{\mu\nu}} \delta_L \tilde{g}^{\mu\nu} + \frac{\delta L_M}{\delta \phi_A} \delta_L \phi_A + \text{div}\right] = 0, \tag{1.2}$$

where div are divergence terms inessential for our consideration.

The Euler variation is defined as usual:

$$\frac{\delta L}{\delta \phi} = \frac{\partial L}{\partial \phi} - \partial_n \left(\frac{\partial L}{\partial(\partial_n \phi)}\right).$$

Under coordinate transformation (1.1) the variations $\delta_L \tilde{g}^{\mu\nu}$, $\delta_L \phi_A$, are easy to calculate, provided one uses their transformation law:

$$\delta_L \tilde{g}^{\mu\nu} = \tilde{g}^{\alpha\nu} D_\alpha \xi^\mu + \tilde{g}^{\alpha\mu} D_\alpha \xi^\nu - D_\alpha(\xi^\alpha \tilde{g}^{\mu\nu}); \tag{1.3}$$

$$\delta_L \phi_A = -\xi^\alpha D_\alpha \phi_A + F^{B;n}_{A;k} \phi_B D_n \xi^k. \tag{1.4}$$

Here and in what follows D_ν is the covariant derivative with respect to the Minkowski metric. Putting these expressions into (1.2) and integrating them by parts one obtains

$$\delta J_M = \int d^4x \Big\{ -\xi^\mu \Big[D_\alpha\Big(2\frac{\delta L_M}{\delta \tilde{g}^{\mu\nu}} \tilde{g}^{\alpha\nu}\Big) - D_\mu\Big(\frac{\delta L_M}{\delta \tilde{g}^{\alpha\beta}}\Big)\tilde{g}^{\alpha\beta} + D_\nu\Big(\frac{\delta L_M}{\delta \phi_A} F^{B;\nu}_{A;\mu} \phi_B\Big) + \frac{\delta L_M}{\delta \phi_A} D_\mu \phi_A \Big] + \mathrm{div} \Big\} = 0.$$

In virtue of arbitrariness of the vector ξ^μ, one finds from the condition $\delta J_M = 0$ a strong identity

$$D_\alpha\Big(2\frac{\delta L_M}{\delta \tilde{g}^{\mu\nu}} \tilde{g}^{\alpha\nu}\Big) - D_\mu\Big(\frac{\delta L_M}{\delta \tilde{g}^{\alpha\beta}}\Big)\tilde{g}^{\alpha\beta} = -D_\nu\Big(\frac{\delta L_M}{\delta \phi_A} F^{B;\nu}_{A;\mu} \phi_B\Big) - \frac{\delta L_M}{\delta \phi_A} D_\mu \phi_A. \tag{1.5}$$

Introduce the notations

$$\tilde{T}_{\mu\nu} = 2\frac{\delta L_M}{\delta \tilde{g}^{\mu\nu}}; \quad T_{\mu\nu} = 2\frac{\delta L_M}{\delta g^{\mu\nu}}. \tag{1.6}$$

Then the lhs of relation (1.5) may be represented as:

$$D_\alpha(\tilde{T}_{\mu\nu} \tilde{g}^{\alpha\nu}) - \frac{1}{2}\tilde{g}^{\alpha\beta} D_\mu \tilde{T}_{\alpha\beta} = \partial_\alpha(\tilde{T}_{\mu\nu}\tilde{g}^{\alpha\nu}) - \frac{1}{2}\tilde{g}^{\alpha\beta}\partial_\mu \tilde{T}_{\alpha\beta}$$

The rhs of this equality may easily be brought to the form

$$\partial_\alpha(\tilde{T}_{\mu\nu}\tilde{g}^{\alpha\nu}) - \frac{1}{2}\tilde{g}^{\alpha\beta}\partial_\mu \tilde{T}_{\alpha\beta} = \tilde{g}_{\mu\nu} \nabla_\alpha (\tilde{T}^{\alpha\nu} - \frac{1}{2}\tilde{g}^{\alpha\nu}\tilde{T}) \tag{1.7}$$

Proceeding from this, strong identity (1.5) may be written in the form

$$\tilde{g}_{\mu\nu} \nabla_\alpha(\tilde{T}^{\alpha\nu} - \frac{1}{2}\tilde{g}^{\alpha\nu}\tilde{T}) = -D_\nu\Big(\frac{\delta L_M}{\delta \phi_A} F^{B;\nu}_{A;\mu} \phi_B\Big) - \frac{\delta L_M}{\delta \phi_A} D_\mu \phi_A \tag{1.8}$$

Due to the least action principle, the equations for matter fields have the form

$$\frac{\delta L_M}{\delta \phi_A} = 0. \tag{1.9}$$

Taking into account the above equation, one may find from strong identity (1.8) the weak identity:

$$\nabla_\mu (\tilde{T}^{\mu\nu} - \frac{1}{2}\tilde{g}^{\mu\nu}\tilde{T}) = 0. \tag{1.10}$$

Note, that the density of the energy-momentum tensor of matter in Riemannian space, $T^{\mu\nu}$, is expressed via $\tilde{T}^{\mu\nu}$ as follows:

$$\sqrt{-g}\ T^{\mu\nu} = \tilde{T}^{\mu\nu} - \frac{1}{2}\tilde{g}^{\mu\nu}\tilde{T}. \tag{1.11}$$

Thus, expression (1.10) entails the covariant equation for matter conservation in Riemannian space:

$$\nabla_\mu T^{\mu\nu} = 0. \tag{1.12}$$

If the number of equations for matter is four, then instead of equations for matter (1.9) one can always use equivalent eqs. (1.12).

The variation of the action integral may be written in the equivalent form

$$\delta J_M = \int d^4x \left[\frac{\delta L_M}{\delta\tilde{\phi}^{\mu\nu}}\delta_L\tilde{\phi}^{\mu\nu} + \frac{\delta L_M}{\delta\tilde{\gamma}^{\mu\nu}}\delta_L\tilde{\gamma}^{\mu\nu} + \frac{\delta L_M}{\delta\phi_A}\delta_L\phi_A + \mathrm{div}\right] = 0. \tag{1.13}$$

Here the variations $\delta_L\tilde{\phi}^{\mu\nu}, \delta_L\tilde{\gamma}^{\mu\nu}$ under coordinate transformation (1.1) T are equal

$$\delta_L\tilde{\phi}^{\mu\nu} = \tilde{\phi}^{\alpha\nu}D_\alpha\xi^\mu + \tilde{\phi}^{\alpha\mu}D_\alpha\xi^\nu - D_\alpha(\xi^\alpha\tilde{\phi}^{\mu\nu}); \tag{1.14}$$

$$\delta_L\tilde{\gamma}^{\mu\nu} = \tilde{\gamma}^{\alpha\nu}D_\alpha\xi^\mu + \tilde{\gamma}^{\alpha\mu}D_\alpha\xi^\nu - \tilde{\gamma}^{\mu\nu}D_\alpha\xi^\alpha \tag{1.15}$$

Substituting expressions for the variations $\delta_L\tilde{\phi}^{\mu\nu}, \delta_L\tilde{\gamma}^{\mu\nu}, \delta_L\phi_A$ into (1.13) and integrating by parts in virtue of arbitrariness of ξ^μ one comes to the strong identity

$$D_\alpha\left(2\frac{\delta L_M}{\delta\tilde{\phi}^{\mu\nu}}\tilde{\phi}^{\alpha\nu}\right) - D_\mu\left(\frac{\delta L_M}{\delta\tilde{\phi}^{\alpha\beta}}\right)\tilde{\phi}^{\alpha\beta} + D_\alpha\left(2\frac{\delta L_M}{\delta\tilde{\gamma}^{\mu\nu}}\tilde{\gamma}^{\alpha\nu}\right) - D_\mu\left(\frac{\delta L_M}{\delta\tilde{\gamma}^{\alpha\beta}}\tilde{\gamma}^{\alpha\beta}\right) =$$

$$= -D_\nu\left(\frac{\delta L_M}{\delta\phi_A}F^{B;\nu}_{A;\mu}\phi_B\right) - \frac{\delta L_M}{\delta\phi_A}D_\mu\phi_A. \tag{1.16}$$

It should be noted that this identity is valid irrespective of the fulfilment of the equations of motion for matter and for gravitation field.

Let us introduce the notations

$$\tilde{t}_M^{\mu\nu} = -2\frac{\delta L_M}{\delta\tilde{\gamma}_{\mu\nu}} \tag{1.17}$$

Comparing identities (1.8) and (1.6), we find

$$\tilde{g}_{\mu\nu}\nabla_\alpha(\tilde{T}^{\alpha\nu} - \frac{1}{2}\tilde{g}^{\alpha\nu}\tilde{T}) = \hat{\gamma}_{\mu\nu}D_\alpha(\tilde{t}_M^{\alpha\nu} - \frac{1}{2}\tilde{\gamma}^{\alpha\nu}\tilde{t}_M) + D_\alpha(2\frac{\delta L_M}{\delta\tilde{\phi}^{\mu\nu}}\tilde{\phi}^{\alpha\nu}) - \\ - D_\mu(\frac{\delta L_M}{\delta\tilde{\phi}^{\alpha\beta}})\tilde{\phi}^{\alpha\beta}. \tag{1.18}$$

Analogously, it follows from the invariance of the action of the gravitation field under coordinate transformations (1.1) that:

$$\tilde{\gamma}_{\mu\nu}D_\alpha\left[\tilde{t}_g^{\alpha\nu} - \frac{1}{2}\tilde{\gamma}^{\alpha\nu}\tilde{t}_g\right] + D_\alpha(2\frac{\delta L_g}{\delta\tilde{\phi}^{\mu\nu}}\tilde{\phi}^{\alpha\nu}) - D_\mu(\frac{\delta L_g}{\delta\tilde{\phi}^{\alpha\beta}})\tilde{\phi}^{\alpha\beta} = 0. \tag{1.19}$$

Here

$$\tilde{t}_g^{\mu\nu} = -2\frac{\delta L_g}{\delta\tilde{\gamma}_{\mu\nu}}. \tag{1.20}$$

Adding expressions (1.18) and (1.19) we get

$$\tilde{g}_{\mu\nu}\nabla_\alpha(\tilde{T}^{\alpha\nu} - \frac{1}{2}\tilde{g}^{\alpha\nu}\tilde{T}) = \hat{\gamma}_{\mu\nu}D_\alpha(\tilde{t}^{\alpha\nu} - \frac{1}{2}\tilde{\gamma}^{\alpha\nu}\tilde{t}) + D_\alpha(2\frac{\delta L}{\delta\tilde{\phi}^{\mu\nu}}\tilde{\phi}^{\alpha\nu}) - \\ - D_\mu(\frac{\delta L}{\delta\tilde{\phi}^{\alpha\beta}})\tilde{\phi}^{\alpha\beta} \tag{1.21}$$

Here

$$\tilde{t}^{\mu\nu} = \tilde{t}_g^{\mu\nu} + \tilde{t}_M^{\mu\nu}. \tag{1.22}$$

In virtue of the least action principle, the equations of the gravitation field have the form

$$\frac{\delta L}{\delta\tilde{\phi}^{\mu\nu}} = \frac{\delta L_g}{\delta\tilde{\phi}^{\mu\nu}} + \frac{\delta L_M}{\delta\tilde{\phi}^{\mu\nu}} = 0. \tag{1.23}$$

If we take these equations into account we obtain from (1.21) an extremely important equality

$$\tilde{g}_{\mu\nu}\nabla_\alpha(\tilde{T}^{\alpha\nu} - \frac{1}{2}\tilde{g}^{\alpha\nu}\tilde{T}) = \hat{\gamma}_{\mu\nu}D_\alpha(\tilde{t}^{\alpha\nu} - \frac{1}{2}\hat{\gamma}^{\alpha\nu}\tilde{t}). \tag{1.24}$$

As may easily be verified, the density of the energy-momentum tensor of matter in Riemannian space is equal to

$$\sqrt{-g}\,T^{\mu\nu} = \tilde{T}^{\mu\nu} - \frac{1}{2}\tilde{g}^{\mu\nu}\tilde{T} \tag{1.25}$$

Similarly, the density of the total energy-momentum tensor in Minkowski space equals

$$\sqrt{-\gamma}\, t^{\mu\nu} = \tilde{t}^{\mu\nu} - \frac{1}{2}\tilde{\gamma}^{\mu\nu}\tilde{t}. \tag{1.26}$$

On applying these expressions, relation (1.24) takes the form

$$\nabla_\mu T^\mu_\nu = D_\mu t^\mu_\nu \tag{1.27}$$

This equality reflects the geometrization principle.

In a pseudo-Euclidean space, the covariant divergence of the sum of densities of the energy-momentum tensors of matter and gravitation field is exactly equal to that in an effective Riemannian space, but only of the density of the energy-momentum tensor of matter. Provided the equations of motion for matter are satisfied, one has:

$$D_\mu t^\mu_\nu = \nabla_\mu T^\mu_\nu = 0. \tag{1.28}$$

From the covariant equation for matter conservation in Riemannian space, it is not clear, what is namely conserved while from the conservation law for the total energy-momentum tensor t^μ_ν in Minkowski space it is clear that both the energy-momentum of matter and gravitational field are conserved. Thus, in this theory Riemannian space appears as a result of the action of the gravitation field on all forms of matter. That is why it is an effective Riemannian space of field origin. Minkowski space finds its exact physical reflection in the conservation laws for the energy-momentum tensor and the angular momentum of matter and gravitation field taken together.

Since there are ten Killing vectors in a flat space, there are, consequently, ten conserving integral quantities for a closed system of fields. If the number of equations of motion for matter is four, then instead of them we may use the equations expressing the total energy-momentum tensor conservation in Minkowski space:

$$D_\mu(t^\mu_{g\nu} + t^\mu_{M\nu}) = 0. \tag{1.29}$$

This equation, alongside with those for a gravitation field, defines all the unknown characteristics of matter and gravitation field. It is worth noting that both matter and gravitation field in our theory are characterized by energy momentum tensors. As a result, in our theory, contrary to the GR, no pseudotensors arise and, hence, there are no unphysical concepts of the nonlocalizability of gravitation energy.

If we, following Hilbert and Einstein, choose the Lagrangian density of a gravitation field in a completely geometrized form, i.e. depending on the metric tensor g^{ik} of Riemannian space and its derivatives only, e.g.

$$L_g = \sqrt{-g}R,$$

where R is the scalar curvature of Riemannian space, then in virtue of the field equations the energy-momentum tensor density of the free gravitation field in Minkowski space would always be equal to zero:

$$\frac{\delta L_g}{\delta \gamma^{\mu\nu}} = \frac{\delta L_g}{\delta g^{\sigma\tau}} \cdot \frac{\partial g^{\sigma\tau}}{\partial \gamma^{\mu\nu}} = 0. \tag{1.30}$$

Thus, it is impossible, in principle, to construct a completely geometrized Lagrangian of the tensor physical field, possessing energy and momentum. Therefore, the theory constructed on the basis of the completely geometrized Lagrangian cannot describe the physical gravitation field of Faraday-Maxwell type in Minkowski space. As was stated in literature (see e.g. ref. [8]), in Minkowski space one can in a unique way, find with the help of the spin-2 tensor field the GR gravitation field

Lagrangian which equals the scalar curvature R. However, these papers have no physical sense, as long as the energy-momentum tensor is zero, as is seen from (1.30). Therefore, all these papers are meaningless from the viewpoint of physics, and their results are mistaken.

2. RELATION BETWEEN CANONICAL ENERGY-MOMENTUM TENSOR AND HILBERT TENSOR

The Lagrangian density of the gravitational field depends on the metric tensor density $\tilde{\gamma}^{\mu\nu}$, the density $\tilde{\phi}^{\mu\nu}$ of the tensor gravitation field and on their first-order derivatives. Under coordinate transformation (1.1), the action variation δJ_g is zero and, hence:

$$\delta J_g = \int_\Omega d^4x \left[D_\lambda J^\lambda + \frac{\delta L_g}{\delta \tilde{\phi}^{\mu\nu}} \delta_L \tilde{\phi}^{\mu\nu} + \frac{\delta L_g}{\delta \tilde{\gamma}^{\mu\nu}} \delta_L \tilde{\gamma}^{\mu\nu} \right] = 0. \qquad (2.1)$$

Here

$$J^\lambda = -\xi^\alpha \tau^\lambda_\alpha + K^{\alpha\lambda}_\mu D_\alpha \xi^\mu \qquad (2.2)$$

where the density of the canonical tensor τ^λ_α is equal to

$$\tau^\lambda_\alpha = -\delta^\lambda_\alpha L_g + D_\alpha \tilde{g}^{\mu\nu} \cdot \frac{\partial L_g}{\partial(\partial_\lambda \tilde{\phi}^{\mu\nu})} = -\delta^\lambda_\alpha L_g + D_\alpha g^{\mu\nu} \cdot \frac{\partial L_g}{\partial(\partial_\lambda g^{\mu\nu})}; \qquad (2.3)$$

and the density of the tensor of rank three $K^{\alpha\lambda}_\mu$ is

$$K^{\alpha\lambda}_\mu = 2\frac{\partial L_g}{\partial(\partial_\lambda \tilde{\phi}^{\mu\nu})}\tilde{\phi}^{\alpha\nu} - \delta^\alpha_\mu \frac{\partial L_g}{\partial(\partial_\lambda \tilde{\phi}^{\sigma\tau})}\tilde{\phi}^{\sigma\tau} + 2\frac{\partial L_g}{\partial(\partial_\lambda \tilde{\gamma}^{\mu\nu})}\tilde{\gamma}^{\alpha\nu} -$$
$$- \delta^\alpha_\mu \frac{\partial L_g}{\partial(\partial_\lambda \tilde{\gamma}^{\sigma\tau})}\tilde{\gamma}^{\sigma\tau} \qquad (2.4)$$

Putting into (2.1) formulae (1.14) and (1.15) for the variations $\delta_L \tilde{\phi}^{\mu\nu}$, $\delta_L \tilde{\gamma}^{\mu\nu}$ we shall obtain in virtue of arbitrariness of the volume Ω the following strong identity:

$$0 = \xi^\alpha \left[D_\lambda \tau^\lambda_\alpha + \frac{\delta L_g}{\delta \tilde{\phi}^{\mu\nu}} \cdot D_\alpha \tilde{\phi}^{\mu\nu} \right] - K^{\alpha\lambda}_\mu D_\alpha D_\lambda \xi^\mu + D_\alpha \xi^\mu \left[\tau^\alpha_\mu - D_\lambda K^{\alpha\lambda}_\mu - \right.$$
$$\left. -2\frac{\delta L_g}{\delta \tilde{\phi}^{\mu\nu}}\tilde{\phi}^{\alpha\nu} + \delta^\alpha_\mu \frac{\delta L_g}{\delta \tilde{\phi}^{\sigma\tau}}\tilde{\phi}^{\sigma\tau} - 2\frac{\delta L_g}{\delta \tilde{\gamma}^{\mu\nu}}\tilde{\gamma}^{\alpha\nu} + \delta^\alpha_\mu \frac{\delta L_g}{\delta \tilde{\gamma}^{\sigma\tau}}\tilde{\gamma}^{\sigma\tau} \right] \qquad (2.5)$$

Since the shift vector ξ^α is arbitrary, the last expression leads to

$$D_\lambda \tau^\lambda_\alpha = -\frac{\delta L_g}{\delta \tilde{\phi}^{\mu\nu}} D_\alpha \tilde{\phi}^{\mu\nu} \tag{2.6}$$

$$\tau^\alpha_\mu - D_\lambda K^{\alpha\lambda}_\mu = 2\frac{\delta L_g}{\delta \tilde{\phi}^{\mu\nu}}\tilde{\phi}^{\alpha\nu} - \delta^\alpha_\mu \frac{\delta L_g}{\delta \tilde{\phi}^{\sigma\tau}}\tilde{\phi}^{\sigma\tau} + 2\frac{\delta L_g}{\delta \tilde{\gamma}^{\mu\nu}}\tilde{\gamma}^{\alpha\nu} - \delta^\alpha_\mu \frac{\delta L_g}{\delta \tilde{\gamma}^{\sigma\tau}}\tilde{\gamma}^{\sigma\tau} \tag{2.7}$$

$$K^{\alpha\lambda}_\mu = -K^{\lambda\alpha}_\mu . \tag{2.8}$$

Our theory is based on the linear relation between the metric tensor density $\tilde{g}^{\mu\nu}$ of the effective Riemannian space and the density of the tensor gravitation field $\tilde{\phi}^{\mu\nu}$

$$\tilde{g}^{\mu\nu} = \tilde{\gamma}^{\mu\nu} + \tilde{\phi}^{\mu\nu} \tag{2.9}$$

In this case we shall obtain the equalities

$$\frac{\delta L_g}{\delta \tilde{\phi}^{\mu\nu}} = \frac{\delta L_g}{\delta \tilde{g}^{\mu\nu}}; \quad \frac{\partial L_g}{\partial(\partial_\sigma \tilde{\phi}^{\mu\nu})} = \frac{\partial L_g}{\partial(D_\sigma \tilde{g}^{\mu\nu})}.$$

Using these equalities we find

$$\frac{\partial L_g}{\partial(\partial_\lambda \tilde{\gamma}^{\mu\nu})} = \frac{\partial L_g}{\partial(D_\lambda \tilde{g}^{\mu\nu})} - \tilde{g}^{\alpha\beta}\frac{\partial L_g}{\partial(D_\tau \tilde{g}^{\alpha\beta})}\frac{\partial \Gamma^\sigma_{\sigma\tau}}{\partial(\partial_\lambda \tilde{\gamma}^{\mu\nu})} + \tilde{g}^{\sigma\beta}\frac{\partial L_g}{\partial(D_\tau \tilde{g}^{\alpha\beta})}\frac{\partial \Gamma^\alpha_{\sigma\tau}}{\partial(\partial_\lambda \tilde{\gamma}^{\mu\nu})}.$$

Here $\Gamma^\sigma_{\mu\nu}$ are Christoffel symbols of Minkowski space

$$\Gamma^\sigma_{\mu\nu} = \frac{1}{2}\gamma^{\sigma\tau}[\partial_\mu \gamma_{\tau\nu} + \partial_\nu \gamma_{\tau\mu} - \partial_\tau \gamma_{\mu\nu}].$$

As a result of elementary calculations, one obtains for $K^{\alpha\lambda}_\mu$ the following expression

$$K^{\alpha\lambda}_\mu = \frac{\partial L_g}{\partial(D_\lambda \tilde{g}^{\mu\nu})}\tilde{g}^{\alpha\nu} - \frac{\partial L_g}{\partial(D_\alpha \tilde{g}^{\mu\nu})}\tilde{g}^{\lambda\nu} + \tilde{g}^{\sigma\nu}\tilde{\gamma}_{\sigma\mu}\Big[\frac{\partial L_g}{\partial(D_\alpha \tilde{g}^{\tau\nu})}\tilde{\gamma}^{\lambda\tau} - \frac{\partial L_g}{\partial(D_\lambda \tilde{g}^{\tau\nu})}\tilde{\gamma}^{\alpha\tau}\Big] + \frac{\partial L_g}{\partial(D_\tau \tilde{g}^{\sigma\nu})}\tilde{\gamma}_{\tau\mu}\big[\tilde{g}^{\alpha\nu}\tilde{\gamma}^{\lambda\sigma} - \tilde{g}^{\lambda\nu}\tilde{\gamma}^{\alpha\sigma}\big]. \tag{2.10}$$

Since the density of the energy-momentum tensor of the gravitation fields is

$$t_{g\mu}^{\alpha} = 2\frac{\delta L_g}{\delta \hat{\gamma}^{\mu\nu}} \tilde{\gamma}^{\alpha\nu} - \delta_{\mu}^{\alpha}\frac{\delta L_g}{\delta \hat{\gamma}^{\sigma\tau}} \hat{\gamma}^{\sigma\tau} \tag{2.11}$$

identity (2.7) may be written in the form

$$t_{g\mu}^{\alpha} = \tau_{\mu}^{\alpha} - D_{\lambda}K_{\mu}^{\alpha\lambda} - 2\frac{\delta L_g}{\delta \tilde{\phi}^{\mu\nu}} \tilde{\phi}^{\alpha\nu} + \delta_{\mu}^{\alpha}\frac{\delta L_g}{\delta \hat{\phi}^{\sigma\tau}} \tilde{\phi}^{\sigma\tau} \tag{2.12}$$

It is just the identity which establishes the relation between the Hilbert tensor density in Minkowski space and that of the canonical energy momentum tensor.

For further use it is convenient to introduce the following quantity as a characteristic of the gravitation field:

$$\overset{o}{t}{}_{g\mu}^{\alpha} = \tau_{\mu}^{\alpha} - D_{\lambda}K_{\mu}^{\alpha\lambda}. \tag{2.13}$$

In virtue of identity (2.12), this quantity coincides exactly with the density of the Hilbert energy-momentum tensor in the case of a free gravitation field.

3. THE BASIC IDENTITY

As was shown in ref. [9], the symmetric tensor ϕ_{ik} of rank two may be represented as the sum of the irreducible representations:one with spin 2, one with spin 1 and two with spin 0:

$$\phi_{ik} = [\mathbf{P}_2 + \mathbf{P}_1 + \mathbf{P}_o + \mathbf{P}_{o'}]_{ik}^{lm} \phi_{lm} \tag{3.1}$$

The quantities $\mathbf{P}_s$ are convenient to write in the momentum representation. Let us introduce the auxilary operators:

$$X_{ik} = \frac{1}{\sqrt{3}}\left[\gamma_{ik} - \frac{q_i q_k}{q^2}\right]; \quad Y_{ik} = \frac{q_i q_k}{q^2}, \tag{3.2}$$

with the help of which $\mathbf{P}_s$ may be represented in the form

$$\mathbf{P}_o = X_{ni}X^{ml}; \quad \mathbf{P}_{o'} = Y_{ni}Y^{ml}; \tag{3.3}$$

$$\mathbf{P}_1 = \frac{\sqrt{3}}{2}\left[X_i^{\ell}Y_n^m + X_n^m Y_i^{\ell} + X_i^m Y_n^{\ell} + X_n^{\ell} Y_i^m\right]; \tag{3.4}$$

$$\mathbf{P}_2 = \frac{3}{2}\left[X_i^{\ell} X_n^m + X_i^m X_n^{\ell}\right] - X_{ni} X^{m\ell}. \tag{3.5}$$

In the x-representation the projection operators $\mathbf{P}_s$ are nonlocal integrodifferential ones:

$$\mathbf{P}^{m\ell}_{s;ni}\,\phi_{m\ell} = \int d^4y\,\mathbf{P}^{m\ell}_{s;ni}(x,y)\,\phi_{m\ell}(y).$$

With the help of expressions (3.3)-(3.5) one may easily make sure that only operators $\mathbf{P}_2$ and $\mathbf{P}_o$ are conserved:

$$q_1\mathbf{P}^{m\ell}_{2;ni} = q_1\mathbf{P}^{m\ell}_{o;ni} = 0;\quad q_m\mathbf{P}^{m\ell}_{2;ni} = q_m\mathbf{P}^{m\ell}_{o;ni} = 0. \tag{3.6}$$

As may be easily verified, the tensor field has the only local operator of a lower order which is linear in field. It equals

$$f_{ik} = \Box\left[(\mathbf{P}_2 - 2\mathbf{P}_o)\phi\right]_{ik}, \tag{3.7}$$

and its divergence is identically equal zero.

$$\partial^i f_{ik} = 0. \tag{3.8}$$

The field f_{ik} describes only spin 2 and 0, i.e. in a more detailed form

$$f_{ik} = \Box\theta_{ik} - \partial_i\partial^m\theta_{mk} - \partial_k\partial^m\theta_{mi} + \gamma_{ik}\partial^m\partial^{\ell}\theta_{m\ell}; \tag{3.9}$$

$$\theta_{ik} = \phi_{ik} - \frac{1}{2}\gamma_{ik}\phi. \tag{3.10}$$

In terms of the covariant derivatives this operator becomes

$$J^{mn} = D_\mu D_\sigma(\tilde{g}^{\sigma n}\gamma^{\mu m} + \tilde{g}^{\sigma m}\gamma^{\mu n} - \tilde{g}^{mn}\gamma^{\mu\sigma} - \tilde{g}^{\mu\sigma}\gamma^{mn}). \tag{3.11}$$

Or, in another form

$$J^{mn} = -\gamma^{\mu\sigma}D_\mu D_\sigma\tilde{g}^{mn} + D^m D_\sigma\tilde{g}^{\sigma n} + D^n D_\sigma\tilde{g}^{\sigma m} - \gamma^{mn}D_\mu D_\sigma\tilde{g}^{\mu\sigma}.$$

One may easily get convinced that the following identity takes place

$$D_m J^{mn} \equiv 0. \tag{3.12}$$

We have called it basic, since it is of fundamental importance for the construction of the relativistic theory of gravity.

4. EQUATIONS OF THE RELATIVISTIC THEORY OF GRAVITATION

In the present section we construct relativistic equations for matter and for gravitational field in the framework of the special relativity and geometrization principle.

The relation between the effective metric of a field Riemannian space and gravitational field may be chosen in the simplest form:

$$\tilde{g}^{ik} = \sqrt{-g}g^{ik} = \sqrt{-\gamma}\gamma^{ik} + \sqrt{-\gamma}\phi^{ik} \qquad (4.1)$$

In our theory, the field variable of the gravitational field is the tensor ϕ^{ik}. Let us assume that in the general case the gravitational field has only spins 2 and 0 and that a free gravitational field has spin 2. Such physical requirements, as is seen in section 3, lead in Galilean coordinates to the following four equations for the gravitational field:

$$\partial_i \phi^{ik} = \partial_i \tilde{g}^{ik} = 0. \qquad (4.2)$$

Similar conditions were sometimes used earlier [7,10] in the GR as a specific class of harmonic coordinate conditions to solve "island-type" problems. It was Fock [7] who paid special attention to the importance of harmonic coordinate conditions in solving "island" problems. He wrote as follows: "The above remarks concerning the privileged character of the harmonic system of coordinates should not be understood, in any case, as some kind of prohibition of the use of other coordinate systems. Nothing is more alien to our point of view than such an interpretation". He went on: "Likewise, in the case of the Theory of Gravitation, the existence of harmonic coordinates, defined apart from a Lorentz transformation, though a fact of primary theoretical and practical importance, does not in any way preclude the use of other, non-harmonic, coordinate syetems". From the point of view of our theory, when solving "island" problems, Fock was unconsciously working with ordinary Galilean coordinates in an inertial reference frame, which are, as known from the special relativity, definitely distinguished. Therefore, in Fock's calculations for "island" systems the harmonic conditions were not the coordinate ones as he thought them to be but, as will be seen from our theory, they are field equations in Galilean coordinates of an inertial reference frame. It was due to this very fact that they played such an important role in his specific calculations, which neither Fock nor others even suspected.

Thus, Fock considered harmonic conditions no more than as privileged coordinate ones applicable for "island" type problems only. This is quite natural, since he and all his eminent predecessors were captured by the Riemannian geometry, which basically gave no possibility to make a deeper insight into the problem. In order to make a step ahead and impose these conditions as universal covariant field equations, it was necessary to give up the ideology of the GR, leave the maze of the Riemannian geometry, apply a special principle of relativity in defiance of the GR, as well as to introduce the concepts of a gravitational field as a Faraday-Maxwell physical field, possessing some energy and momentum. All of it was translated into reality in our theory, with the choice of coordinates being arbitrary and set only by the metric tensor γ^{ik} of Minkowski space, as is generally accepted in elementary particle theory. As for equations (4.2), in our theory they are comprehensive and universal because of being the gravitational field equations. They have nothing to do with the choice of coordinates. In Minkowski space these equations are written in the covariant form

$$\sqrt{-\gamma}\, D_i \phi^{ik} = D_i \tilde{g}^{ik} = 0. \quad (4.3)$$

Judging by section 3, we conclude that these field equations exclude automatically spins 1 and 0' from a gravitational tensor field. Thus, we have already constructed four covariant equations (4.3) for the fourteen unknown variables of a gravitation field and of matter. To construct other ten equations, we draw a simple, but far reaching analogy with an electromagnetic field. As is known, Maxwell electrodynamic equations may be written in the covariant form as follows:

$$\gamma^{ik} D_i D_k A^{\nu} - D^{\nu} D_k A^k = \frac{4\pi}{c} j^{\nu}. \quad (4.4)$$

Due to conservation of electromagnetic current one has

$$D_{\nu} j^{\nu} = 0. \quad (4.5)$$

The lhs of Maxwell equations (4.4) is constructed in such a way that its divergence is identically zero. It follows that spin 0 is excluded from the vector field with spins 1 and 0. By analogy with electrodynamics, we shall construct equations for the tensor gravitation field. The unique tensor of the second rank which conserves is the energy-momentum tensor of matter and a gravitation field in Minkowski space.

$$D_{\mu}(t_g^{\mu\nu} + t_M^{\mu\nu}) = 0. \quad (4.6)$$

Therefore, it would be natural to choose it as a complete source of the gravitation field. We have already obtained four equations (4.3) describing a gravitation field. Therefore, in order to achieve a complete description of the unknown ten field variables and of four variables of matter, it is necessary to write ten more equations. As proved in section 3, the only identically conserved tensor linear operator is $J^{\mu\nu}$. Hence, by analogy with electrodynamics, for the remaining ten equations one should necessarily take the following ones:

$$D_\sigma D_\tau\left[\tilde{g}^{\sigma\nu}\gamma^{\tau\mu}+\tilde{g}^{\tau\mu}\gamma^{\tau\nu}-\tilde{g}^{\mu\nu}\gamma^{\sigma\tau}-\tilde{g}^{\sigma\tau}\gamma^{\mu\nu}\right]=\lambda\left[t_g^{\mu\nu}+t_M^{\mu\nu}\right]. \quad (4.7)$$

This type of equations guarantees automatically that the conservation law for the energy-momentum tensor of matter and a gravitational field will be fulfilled in Minkowski space:

$$D_\mu(t_g^{\mu\nu}+t_M^{\mu\nu})=0. \quad (4.8)$$

Besides, as a consequence, the covariant equation, for matter conservation in Riemannian space is also satisfied:

$$\nabla_\mu T^{\mu\nu}=0. \quad (4.9)$$

Since the divergence of the lhs of eq. (4.7) is identically zero, four components corresponding to spins 1 and 0' are excluded from the tensor field ϕ^{ik} containing representations with spins 2,1,0,0'. Systems of equations (4.3) and (4.7) determine completely the unknown variables of matter and a gravitation field. The principal problem is to find out if there is a Lagrangian density for the gravitation field with spins 2 and 0 which would automatically lead to equations (4.7) due to the least action principle. The most general Lagrangian density of the gravitation field ϕ^{ik} describing spins 2 and 0 and quadratic in the first-order derivatives has the form

$$L_g = a\tilde{g}_{km}\tilde{g}_{nq}\tilde{g}^{lp}D_l\tilde{g}^{kq}D_p\tilde{g}^{mn} + b\tilde{g}_{kq}D_m\tilde{g}^{pq}D_p\tilde{g}^{km} + c\tilde{g}_{km}\tilde{g}_{nq}\tilde{g}^{lp}D_l\tilde{g}^{km}D_p\tilde{g}^{nq}. \quad (4.10)$$

The convolution of covariant derivatives taken with respect to Minkowski metric is realized with the help of the effective metric tensor $\tilde{g}^{ik}$ of Riemannian space. In this way we guarantee the action of the gravitation field on itself which is similar to the one of matter. The constants a, b and c in the Lagrangian are arbitrary so far.

Because of the geometrization principle, the Lagrangian density of matter is

$$L_M = L_M(\tilde{g}^{ik}, \phi_A). \qquad (4.11)$$

In virtue of the least action principle, the system of equations for a gravitation field is as follows:

$$\frac{\delta L_g}{\delta \tilde{\phi}^{ik}} + \frac{\delta L_M}{\delta \tilde{\phi}^{ik}} = \frac{\delta L_g}{\delta \tilde{g}^{ik}} + \frac{\delta L_M}{\delta \tilde{g}^{ik}} = 0. \qquad (4.12)$$

Here relation (4.1) was taken into account. For this system of equations to be presented in the form of (4.7), it is necessary to choose the constants a, b and c for the Lagrangian density in a definite and unique way.

This suggests that the Lagrangian of the gravitation field with spins 2 and 0 in Minkowski space is determined in a unique way. In order to make such a choice of the coefficients a, b and c let us calculate the density of the energy momentum tensor for matter and a gravitational field.

Let us introduce the notations

$$\tilde{t}_M^{mn} = -2\frac{\delta L_M}{\delta \tilde{\gamma}_{mn}}; \quad \tilde{t}_M = -2\tilde{\gamma}_{mn}\frac{\delta L_M}{\delta \tilde{\gamma}_{mn}}. \qquad (4.13)$$

Taking into account the definitions of $\tilde{\gamma}^{mn}$ we find

$$\frac{\partial \tilde{\gamma}^{mn}}{\partial \gamma^{pq}} = \sqrt{-\gamma}\,(\delta^{mn}_{pq} - \frac{1}{2}\gamma^{mn}\gamma_{pq}). \qquad (4.14)$$

Similarly,

$$\frac{\partial \tilde{g}^{mn}}{\partial g^{pq}} = \sqrt{-g}\,(\delta^{mn}_{pq} - \frac{1}{2}g^{mn}g_{pq}). \qquad (4.15)$$

Using equality (4.14) we obtain

$$t_M^{mn} = \frac{1}{\sqrt{-\gamma}}\left[\tilde{t}_M^{mn} - \frac{1}{2}\tilde{\gamma}^{mn}\tilde{t}_M\right]. \qquad (4.16)$$

Calculating the variation of the total Lagrangian in $\tilde{\gamma}_{mn}$ and with regarding the field equations

$$\frac{\delta L_g}{\delta \tilde{g}^{mn}} + \frac{\delta L_M}{\delta \tilde{g}^{mn}} = 0, \tag{4.17}$$

one finds

$$t^{mn} = D_\sigma \Big\{ (2a+b)\big[H^{\sigma n}_{\nu}\gamma^{\nu m} + H^{\sigma m}_{\nu}\gamma^{\nu n} - H^{mn}_{\nu}\gamma^{\nu\sigma}\big] - - 2(a+2c)\tilde{g}^{\nu\sigma}\tilde{g}_{\lambda k}\gamma^{mn}D_\nu \tilde{g}^{\lambda k}\Big\} + 2bJ^{mn} \tag{4.18}$$

where

$$H^{\sigma n}_{\nu} = (\tilde{g}^{\sigma l}D_l \tilde{g}^{kn} + \tilde{g}^{nl}D_l \tilde{g}^{k\sigma})\tilde{g}_{k\nu} \tag{4.19}$$

In order that no new equations on the field ϕ^{ik} would arise from the equality

$$D_m t^{mn} = 0,$$

which would otherwise lead to a redefinition of the system of equations, it is necessary and sufficient for the coefficients a, b and c to satisfy the following conditions:

$$a = -\frac{b}{2}; \quad c = +\frac{b}{4}. \tag{4.20}$$

Thus, with such a choice of the constants one comes to the identity

$$D_m t^{mn} \equiv 0,$$

which was enclosed in equations (4.7). With regard for the choice of the coefficients as in (4.20), expression (4.18) takes the form

$$D_\sigma D_\nu(\tilde{g}^{\sigma n}\gamma^{\nu m} + \tilde{g}^{\sigma m}\gamma^{\nu n} - \tilde{g}^{mn}\gamma^{\nu\sigma} - \tilde{g}^{\nu\sigma}\gamma^{mn}) = \frac{1}{2b}(t^{mn}_g + t^{mn}_M), \tag{4.21}$$

which coincides with equations (4.7) obtained earlier by analogy with electrodynamics, provided one puts

$$2b = 1/\lambda .$$

So, the only Lagrangian density of the form

$$L_g = -\frac{1}{32\pi}\Big[\tilde{g}_{kq}D_m\tilde{g}^{pq}D_p\tilde{g}^{km} - \frac{1}{2}\tilde{g}_{km}\tilde{g}_{nq}\tilde{g}^{\ell p}D_\ell\tilde{g}^{kq}D_p\tilde{g}^{mn} +$$

$$+ \frac{1}{4}\tilde{g}_{km}\tilde{g}_{nq}\tilde{g}^{\ell p}D_\ell\tilde{g}^{km}D_p\tilde{g}^{nq}\Big] \tag{4.22}$$

leads to the field equations in the form of (4.21). According to the correspondence principle, the constant λ is chosen to be equal

$$\lambda = -16\pi. \tag{4.23}$$

This Lagrangian density can be presented only in the form

$$L_g = \frac{1}{32\pi}\Big[\tilde{G}^{\ell}_{mn}D_\ell\tilde{g}^{mn} - \tilde{g}^{mn}\tilde{G}^{k}_{mk}\tilde{G}^{\ell}_{n\ell}\Big] \tag{4.24}$$

where the tensor of rank three, $\tilde{G}^{k}_{m\ell}$, is defined by the formula

$$\tilde{G}^{k}_{m\ell} = \frac{1}{2}\tilde{g}^{pk}(D_m\tilde{g}_{\ell p} + D_\ell\tilde{g}_{mp} - D_p\tilde{g}_{\ell m}). \tag{4.25}$$

It may also be presented in the form

$$L_g = -\frac{1}{16\pi}\sqrt{-g}g^{mn}\Big[G^{k}_{m\ell}G^{\ell}_{nk} - G^{\ell}_{mn}G^{k}_{\ell k}\Big]. \tag{4.26}$$

Such a Lagrangian was considered for the first time by Rosen in [2].

In (4.26) the tensor of the third rank, $G^{k}_{m\ell}$ is equal to

$$G^{k}_{m\ell} = \frac{1}{2}g^{kp}(D_m g_{p\ell} + D_\ell g_{pm} - D_p g_{\ell m}). \tag{4.27}$$

With consideration for eq. (4.3), the completele system of equations for matter and gravitational field will be [6]

$$\gamma^{\mu\sigma}D_\mu D_\sigma\tilde{g}^{mn} = 16\pi(t^{mn}_{g} + t^{mn}_{M}); \tag{4.28}$$

$$D_m\tilde{g}^{mn} = 0 \tag{4.29}$$

or in the Galilean coordinate system

$$\Box\tilde{g}^{mn} = 16\pi(t^{mn}_{g} + t^{mn}_{M}); \quad \partial_m\tilde{g}^{mn} = 0.$$

Should we confine ourselves only with the first system of equations, (4.28), then the separation of the Riemannian space metric into the Minkowski space metric and tensor gravitational field would be of a conditional character and would not have any physical meaning. The second system of four field equations (4.29), separates decisively all that relates to the inertia forces from all that is connected with gravitational field. The two systems of equations, (4.28) and (4.29), are generally covariant. Corresponding physical conditions are imposed, as usual, within a given, for example, Galilean coordinate system on the behaviour of gravitational field. In the framework of GR, one cannot formulate the conditions for the metric g^{ik} remaining in Riemannian space since the asymptotics of the metric always depends on the choice of the three-dimensional coordinate system. It should also be noted that the equations of matter motion are contained in the given system of equations. The density for the energy-momentum tensor of gravitational field in Minkowski space is equal to

$$t_g^{mn} = -\frac{1}{16\pi} J^{mn} - \frac{\sqrt{-\gamma}}{8\pi}\gamma^{mp}\gamma^{nq}\left[R_{pq} - \frac{1}{2}\sqrt{\frac{g}{\gamma}}\gamma_{pq}R\right] \quad (4.30).$$

for Lagrangian density (4.22). Here, as we can see, there automatically appears the second-rank curvature tensor R_{pq} in Riemannian space. Similarly, the tensor density of matter energy-momentum in Minkowski space is equal to

$$t_M^{mn} = \sqrt{\frac{\gamma}{g}}\gamma^{mp}\gamma^{np}\left[T_{pq} - \frac{1}{2}g_{pq}T\right] + \frac{1}{2}\gamma^{mn}T \quad (4.31)$$

for Lagrangian density (4.11).

In deriving expressions (4.30) and (4.31) we used the identity

$$\frac{\delta L_M}{\delta\gamma_{pq}} = \frac{\delta L_M}{\delta g^{mn}}\frac{\partial g^{mn}}{\partial\gamma_{pq}}, \quad (4.32)$$

and the equality

$$\frac{\partial g^{mn}}{\partial\gamma_{pq}} = -\sqrt{\frac{\gamma}{g}}\gamma^{pi}\gamma^{qk}\left[\delta^{mn}_{ik} - \frac{1}{2}g_{ik}g^{mn}\right] - \frac{1}{2}\gamma^{pq}g^{mn}, \quad (4.33)$$

which directly follows from the expression for coupling (4.1). Putting the expressions for the energy-momentum tensors of matter and gravitational field in field equations (4.7), we transform them to the form of Hilbert-Einstein equations

$$\sqrt{-g}\, R_{\mu\nu} = 8\pi\,(T_{\mu\nu} - \frac{1}{2}\, g_{\mu\nu}\, T). \qquad (4.34)$$

Thus, system of equation (4.7) is equivalent to the system of Hilbert-Einstein equations. The complete system of equations for matter and gravitational field, (4.28)-(4.29), is equivalent to the system of equations

$$\sqrt{-g}\, R_{\mu\nu} = 8\pi(T_{\mu\nu} - \frac{1}{2} g_{\mu\nu}\, T); \qquad (4.35)$$

$$D_{\mu} \tilde{g}^{\mu\nu} = 0. \qquad (4.36)$$

It is worth mentioning that equations (4.36) are general and universal since these are field equations describing a gravitational field with spins 2 and 0. The choice of a reference frame (or coordinate system) is determined by the metric tensor $\gamma^{\mu\nu}$ of Minkowski space. Hence, equations (4.36) do not impose any restrictions on the choice of a coordinate system. Consequently, the system of equations (4.36) excludes spins 1 and 0' in the density of tensor field ϕ^{ik}, leaving only spins 2 and 0. The required six components of gravitational field, corresponding to these spins, and four components of matter are defined from field equations (4.28) or from their equivalent Hilbert-Einstein equations (4.35). The system of equations for gravitational field, (4.28)-(4.29), may be expressed in a somewhat different form through the Hilbert energy-momentum tensor density in Riemannian space. However, for this purpose we will have to obtain some relations if use is made of a specific expression for the Lagrangian density of gravitational field obtained by us earlier, (4.24):

$$L_g = \frac{1}{32\pi}\left[\tilde{G}^{\ell}_{mn} D_{\ell} \tilde{g}^{mn} - \tilde{g}^{mn} \tilde{G}^{k}_{mk} \tilde{G}^{\ell}_{n\ell}\right],$$

where

$$\tilde{G}^{\lambda}_{\mu\nu} = \frac{1}{2} \tilde{g}^{\lambda\sigma} (D_{\mu} \tilde{g}_{\sigma\nu} + D_{\nu} \tilde{g}_{\sigma\mu} - D_{\sigma} \tilde{g}_{\mu\nu}).$$

Using these expressions, we calculate then the density of the tensor of the third rank, $K^{\alpha\lambda}_{\mu}$, with formula (2.10).

With account for the equality

$$\frac{\partial L_g}{\partial (D_{\lambda} \tilde{g}^{\mu\nu})} = \frac{1}{16\pi}\left[\tilde{G}^{\lambda}_{\mu\nu} + \frac{1}{2} \tilde{g}^{\lambda\sigma} \tilde{g}_{\mu\nu} \tilde{G}^{\tau}_{\tau\sigma}\right],$$

we shall have

$$16\pi K^{\alpha\lambda}_{\mu} = \left[\tilde{g}^{\alpha\nu} \tilde{G}^{\lambda}_{\mu\nu} - \tilde{g}^{\lambda\nu} \tilde{G}^{\alpha}_{\mu\nu}\right] + \tilde{g}^{\sigma\nu} \tilde{\gamma}_{\sigma\mu} \left[\tilde{\gamma}^{\lambda\tau} \tilde{G}^{\alpha}_{\tau\nu} - \tilde{\gamma}^{\alpha\tau} \tilde{G}^{\lambda}_{\tau\nu}\right] + \tilde{\gamma}_{\mu\tau} \tilde{G}^{\tau}_{\sigma\nu} \left[\tilde{g}^{\alpha\nu} \tilde{\gamma}^{\lambda\sigma} - \tilde{g}^{\lambda\nu} \tilde{\gamma}^{\alpha\sigma}\right].$$

Putting the value for $\tilde{G}^{\lambda}_{\mu\nu}$ into this equality we shall obtain

$$16\pi K^{\alpha\lambda}_{\mu} = -\gamma_{\mu\nu} D_{\sigma}\left[\tilde{g}^{\lambda\sigma} \gamma^{\alpha\nu} + \tilde{g}^{\alpha\nu} \gamma^{\lambda\sigma} - \tilde{g}^{\alpha\sigma} \gamma^{\lambda\nu} - \tilde{g}^{\lambda\nu} \gamma^{\alpha\sigma}\right] + \tilde{g}_{\mu\nu} D_{\sigma} (\tilde{g}^{\lambda\sigma} \tilde{g}^{\alpha\nu} - \tilde{g}^{\alpha\sigma} \tilde{g}^{\lambda\nu}). \quad (4.37)$$

Using this expression and definition (2.13) we shall have for $\overset{o}{t}{}^{\alpha}_{g\mu}$ the following expression

$$\overset{o}{t}{}^{\alpha}_{g\mu} = \tau^{\alpha}_{\mu} - \frac{1}{16\pi} D_{\lambda} \sigma^{\alpha\lambda}_{\mu} - \frac{1}{16\pi} \gamma_{\mu\nu} J^{\alpha\nu}, \quad (4.38)$$

where the density of the antisymmetric tensor $\sigma^{\alpha\lambda}_{\mu}$ is equal to

$$\sigma^{\alpha\lambda}_{\mu} = \tilde{g}_{\mu\nu} D_{\sigma} \left[\tilde{g}^{\lambda\sigma} \tilde{g}^{\alpha\nu} - \tilde{g}^{\alpha\sigma} \tilde{g}^{\lambda\nu}\right] \quad (4.39)$$

and the well-known expression (3.11) is denoted through $J^{\alpha\nu}$.

Expression (4.38) is right the one we will need in what follows. In preparing for further calculations we derive now an identity often used in literature. In Galilean coordinates, the Lagrangian density of gravitational field, (4.24), takes the form

$$L_g = \frac{1}{32\pi}\left[\tilde{G}^{\ell}_{mn} \partial_{\ell} \tilde{g}^{mn} - \tilde{g}^{mn} \tilde{G}^{k}_{mk} \tilde{G}^{\ell}_{n\ell}\right],$$

where in this case

$$\tilde{G}^{\lambda}_{\mu\nu} = \frac{1}{2}\, \tilde{g}^{\lambda\sigma} (\partial_\mu \tilde{g}_{\sigma\nu} + \partial_\nu \tilde{g}_{\sigma\mu} - \partial_\sigma \tilde{g}_{\mu\nu}).$$

The quantities $\tilde{G}^{\lambda}_{\mu\nu}$ are the tensors of the third rank with respect to the linear transformations of the coordinates, therefore L_g will be a scalar density with respect to the same transformations. From the invariance of action with respect to the linear transformations, we have

$$\delta J_g = \int_\Omega d^4x \left[\partial_\lambda J^{\lambda} + \frac{\delta L_g}{\delta \tilde{g}^{\mu\nu}} \delta_L \tilde{g}^{\mu\nu}\right] = 0. \qquad (4.40)$$

Here

$$J^{\lambda} = -\xi^{\alpha} \tau^{\lambda}_{\alpha} + \tilde{K}^{\alpha\lambda}_{\mu} \partial_\alpha \xi^{\mu}, \qquad (4.41)$$

where the density of the canonical tensor τ^{λ}_{α} is equal to

$$\tau^{\lambda}_{\alpha} = -\delta^{\lambda}_{\alpha} L_g + \partial_\alpha \tilde{g}^{\mu\nu} \frac{\partial L_g}{\partial(\partial_\lambda \tilde{g}^{\mu\nu})}, \qquad (4.42)$$

and the density of the tensor of the third rank, $\tilde{K}^{\alpha\lambda}_{\mu}$, is in this case

$$\tilde{K}^{\alpha\lambda}_{\mu} = 2\, \frac{\partial L_g}{\partial(\partial_\lambda \tilde{g}^{\mu\nu})} \tilde{g}^{\alpha\nu} - \delta^{\alpha}_{\mu} \frac{\partial L_g}{\partial(\partial_\lambda \tilde{g}^{\sigma\tau})} \tilde{g}^{\sigma\tau}. \qquad (4.43)$$

Putting into (4.40) the formula for the variation of $\delta_L \tilde{g}^{\mu\nu}$ with respect to linear transformations we shall, in virtue of the arbitrariness of the volume Ω, obtain the identity

$$\xi^{\alpha}\left[\partial_\lambda \tau^{\lambda}_{\alpha} + \frac{\delta L_g}{\delta \tilde{g}^{\mu\nu}} \partial_\alpha \tilde{g}^{\mu\nu}\right] + \partial_\alpha \xi^{\mu}\left[\tau^{\alpha}_{\mu} - \partial_\lambda \tilde{K}^{\alpha\lambda}_{\mu} - 2\, \frac{\delta L_g}{\delta \tilde{g}^{\mu\nu}} \tilde{g}^{\alpha\nu} + \delta^{\alpha}_{\mu} \frac{\delta L_g}{\delta \tilde{g}^{\sigma\tau}} \tilde{g}^{\sigma\tau}\right] = 0. \qquad (4.44)$$

Whereof there straightforwardly follow the identities

$$\partial_\lambda \tau^{\lambda}_{\alpha} = -\frac{\delta L_g}{\delta \tilde{g}^{\mu\nu}} \partial_\alpha \tilde{g}^{\mu\nu} \qquad (4.45)$$

$$\tau^{\alpha}_{\mu} - \partial_\lambda \tilde{K}^{\alpha\lambda}_{\mu} = 2\, \frac{\delta L_g}{\delta \tilde{g}^{\mu\nu}} \tilde{g}^{\alpha\nu} - \delta^{\alpha}_{\mu} \frac{\delta L_g}{\delta \tilde{g}^{\sigma\tau}} \tilde{g}^{\sigma\tau}. \qquad (4.46)$$

Since

$$\frac{\delta L_g}{\delta \tilde{g}^{\mu\nu}} = - \frac{1}{16\pi} R_{\mu\nu} \tag{4.47}$$

then from identity (4.46) we have

$$\tau_\mu^\alpha - \partial_\lambda \tilde{K}_\mu^{\alpha\lambda} = - \frac{\sqrt{-g}}{8\pi}\left[R_\mu^\alpha - \frac{1}{2}\delta_\mu^\alpha R\right]. \tag{4.48}$$

Considering the equality

$$\frac{\partial L_g}{\partial(\partial_\lambda \tilde{g}^{\mu\nu})} = \frac{1}{16\pi}\left[\tilde{G}_{\mu\nu}^\lambda + \frac{1}{2}\tilde{g}^{\lambda\tau}\tilde{g}_{\mu\nu}\tilde{G}_{\tau\sigma}^\sigma\right]$$

and the expressions for $\tilde{G}_{\mu\nu}^\lambda$ in Galilean coordinates we will come to

$$16\pi \tilde{K}_\mu^{\alpha\lambda} = \partial_\nu(\delta_\mu^\lambda \tilde{g}^{\alpha\nu} - \delta_\mu^\nu \tilde{g}^{\lambda\alpha}) + \sigma_\mu^{\alpha\lambda}, \tag{4.49}$$

where $\sigma_\mu^{\alpha\lambda}$ is the density of the antisymmetric tensor,

$$\sigma_\mu^{\alpha\lambda} = - \sigma_\mu^{\lambda\alpha} = \tilde{g}_{\mu\nu}\partial_\sigma(\tilde{g}^{\lambda\sigma}\tilde{g}^{\alpha\nu} - \tilde{g}^{\alpha\sigma}\tilde{g}^{\lambda\nu}). \tag{4.50}$$

Putting the expression for $\tilde{K}_\mu^{\alpha\lambda}$ into (4.48), we shall obtain the identity

$$\tau_\mu^\alpha - \frac{1}{16\pi}\partial_\lambda \sigma_\mu^{\alpha\lambda} = - \frac{\sqrt{-g}}{8\pi}\left[R_\mu^\alpha - \frac{1}{2}\delta_\mu^\alpha R\right]. \tag{4.51}$$

In the curvature tensor, one can always identically replace, leaving it unchanged, the conventional derivatives by the covariant ones in the Minkowski metric, therefore expression (4.51) may be presented in the covariant form:

$$\tau_\mu^\alpha - \frac{1}{16\pi}D_\lambda \sigma_\mu^{\alpha\lambda} = - \frac{\sqrt{-g}}{8\pi}\left[R_\mu^\alpha - \frac{1}{2}\delta_\mu^\alpha R\right]. \tag{4.52}$$

In this case the canonical tensor density in (4.52) will be equal to expression (2.3)

$$\tau_\mu^\alpha = - \delta_\mu^\alpha L_g + \frac{\partial L_g}{\partial(\partial_\lambda g^{\mu\nu})} D_\alpha g^{\mu\nu},$$

where the Lagrangian density L_g is already presented in terms of the derivatives covariant in the Minkowski metric (see (4.24)).

Using identity (4.52) we may present the expression for $\overset{0}{t}{}^{\alpha}_{g\mu}$ (4.38) in the form

$$\overset{0}{t}{}^{\alpha}_{g\mu} = \frac{-\sqrt{-g}}{8\pi}\left[R^{\alpha}_{\mu} - \frac{1}{2}\delta^{\alpha}_{\mu}R\right] - \frac{1}{16\pi}\gamma_{\mu\nu}J^{\alpha\nu}. \qquad (4.53)$$

As we have already established, the system of equations of matter and gravitational field, (4.28) and (4.29), is equivalent to the system of equations (4.35) and (4.36). With the help of expression (4.53) the system of equations of matter and gravitational field may also be rewritten in another equivalent form [6]:

$$\gamma_{\nu\sigma}\gamma^{\alpha\beta}D_{\alpha}D_{\beta}\,\tilde{g}^{\mu\sigma} = 16\pi(T^{\mu}_{\nu} + \overset{0}{t}{}^{\mu}_{g\nu}); \qquad (4.54)$$

$$D_{\mu}\tilde{g}^{\mu\nu} = 0. \qquad (4.55)$$

Here T^{μ}_{ν} is the Hilbert energy-momentum tensor density (1.6) for the matter in Riemannian space. It is quite obvious that in virtue of (4.54) and (4.55) the conservation law for the energy-momentum tensor of matter and gravitational field has the form

$$D_{\mu}(T^{\mu}_{\nu} + \overset{0}{t}{}^{\mu}_{g\nu}) = 0. \qquad (4.56)$$

The covariant matter conservation law in Riemannian space may identically be presented in the form

$$\nabla_{\mu}T^{\mu}_{\nu} = \partial_{\mu}T^{\mu}_{\nu} - \frac{1}{2}T^{\mu\sigma}\partial_{\nu}\,g_{\mu\sigma} \equiv D_{\mu}\,T^{\mu}_{\nu} - G^{\lambda}_{\mu\nu}\,T^{\mu}_{\lambda} = 0. \quad (4.57)$$

From comparison of (4.56) and (4.57), we have

$$G^{\lambda}_{\mu\nu}\,T^{\mu}_{\lambda} = -D_{\mu}\,\overset{0}{t}{}^{\mu}_{g\nu}. \qquad (4.58)$$

As seen from this expression, the matter acquires energy and momentum right from gravitational field, the total energy-momentum tensor of matter and gravitational field being always conserved rigorously. The construction of RTG on the basis of Minkowski space and geometrization principle allowed us to do deal only with covariant quantities at every stage of our reasonings. Here we give briefly some of our results following from this theory. The post-Newtonian parameters in it are equal to

$$\gamma = \beta = 1; \quad \alpha_1 = \alpha_2 = \alpha_3 = \xi_1 = \xi_2 = \xi_3 = \xi_4 = \xi_w = 0.$$

Therefore the theory describes the whole set of gravitation experiments available at present. It strictly satisfies the correspondence principle and predicts the existence of gravitational waves in the Faraday-Maxwell spirit, which possess both energy and momentum. In virtue of the geometrization principle the curvature of the effective field Riemannian space appears as a result of gravitational field action on matter. Gravitational waves in this theory propagate as electromagnetic ones. Since our theory is based on the spacial relativity principle, the inertial mass of the system is well defined

$$m = \int d^3x \left[t_g^{oo} + t_M^{oo} \right]$$

and is a scalar with respect to the three-dimensional transformations of coordinates. The quantity

$$P^{\nu} = \int d^3x \left[t_g^{o\nu} + t_M^{o\nu} \right]$$

is an energy-momentum four-vector with respect to any coordinate transformations; similarly, the angular momentum is also a tensor with respect to any coordinate transformations in four-dimensional Minkowski space. It may also be shown that for any "island" static system, the inertial mass is exactly equal to its active gravitating mass. The given theory provides a prediction of an extraordinary force - it leads to a strictly definite development of the Universe [6]. According to it, the Universe is not closed, it is "flat" in virtue of equations (4.29)

$$ds^2 = c^2 d\tau^2 - V(\tau)(dx^2 + dy^2 + dz^2).$$

The Universe expansion is defined by the function $V(\tau)$ which is easily calculated from the field equations. Eqs. (4.28)-(4.29) make us easily convinced that the total energy density of matter and gravitational field is always zero at any instant of time of the Universe development. The nowaday density of all the forms of matter, ρ_0 ,should be equal to its critical density ρ_c ,

$$\rho_0 = \rho_c = \frac{3H^2}{8\pi G},$$

where H is the Hubble constant.

The Universe expands infinitely, the deceleration parameter q being equal to

$$q = 1/2.$$

The Universe age T is defined by the formula

$$T = 2/3H.$$

The density of the observed usual matter is much smaller than the critical density ρ_c . Since the given theory is based on fundamental general physical principles and is constructed in a unique way, its predictions on the character of the Universe development are so general that they necessarily demand an obligatory existence of a "hidden" mass is some form of matter. Hence, in the Universe there must exist a "hidden" mass so that the total density of matter be equal to the critical value of ρ_c.

The authors express their deep gratitude to V.A.Ambartsumyan, N.N.Bogolubov, A.A.Vlasov, S.S.Gershtein, A.N.Tavkhelidze, for valuable discussions.

REFERENCES

1. A.A.Logunov et al., Theor. Math. Phys., v. 40 (1979); N. 3, p. 291;
V.I.Denisov, A.A.Logunov, Theor. Math. Phys., v. 50 (1982), N. 1, p. 3;
V.I.Denisov, A.A.Logunov, Elementary Particle and Atomic Nucleus Physics, v. 13, part 4, (1982), p. 757;
V.I.Denisov, A.A.Logunov, Modern Problems of Mathematics: M., VINITI of Acad. of Sciences, USSR, 1982, v. 21.
2. N.Rosen, Phys. Rev., v. 57, p. 147 (1940);
N.Rosen, Ann. of Phys., v. 22, p. 1 (1963);
A.Papapetrou, Proc.Roy Irish Acad.,v.A52,p. 11 (1948);
S.Gupta, Proc. Phys. Soc., v.A65, p. 608 (1952);
W.Thirring, Ann. of Phys.,v.16,p. 69 (1961).
3, H.Poincaré, Present and Future of Mathematical Physics, Bulletin des Sciences Mathematiques, December, v. 28, ser. 2, p. 302 (1904); The monist, January, v. XV, N. 1 (1905);
Principle of Relativity, ed. A.A.Tyapkin - M., Atomizdat, 1973.
4. A.A.Logunov. Lectures on the Theory of Relativity. - Moscow University, 1984 (In Russian).
5. A.Einstein. Collected works. - M., Nauka, 1965, v. 1;
W.Pauli. Theory of Relativity. - Pergamon Press,1965;
C.Møller. The Theory of Relativity, Oxford, Clarendon Press, 1972;
L.I.Mandelstam, Lectures on Optics, Relativity Theory and Quantum Mechanics, M., Nauka, 1972, p. 218-219.

6. A.A.Logunov, A.A.Vlasov, Minkowski Space as the Basis of the Physical Gravitation Theory. M., Moscow University, 1984; A.A.Logunov, A.A.Vlasov, TMF, v. 60, p. 3, (1984); A.A.Logunov, A.A.Vlasov. Spherically Symmetric Solution in Gravitation Theory Based on Minkowski Space, M., Moscow Univ., 1984; A.A.Logunov, A.A.Vlasov, TMF, v. 60, p. 163 (1984); A.A.Vlasov, A.A.Logunov, M.A.Mestvirishvili, IHEP Preprint 84-156, Serpukhov, 1984.

7. V.A.Fock. Theory of Space, Time and Gravitation. - London, Pergamon Press, 1959.

8. V.I.Ogievetsky, I.V.Polubarinov, Ann. of Phys., v. 35, p. 167 (1965); JINR preprint P-2106, 1965.

9. C.Fronsdal, Sup. Nuovo Cimento, v. 9, p. 416 (1958); K.J.Barnes, J. Math. Phys., v. 6, p. 788 (1965);

10. De-Donder. La Gravifique Einsteinienne. - Paris, 1921; Theorie des Champs Gravifiques, - Paris, 1926; V.A.Fock., Jour. of Phys., v. 1, p. 81 (1939); Rev. Modern. Phys., v. 29, p. 235 (1957).

QUANTIZED STRINGS AND CURVED SPACE-TIME

E.S.Fradkin and A.A.Tseytlin
Department of Theoretical Physics, P.N.Lebedev Physical Institute, Leninsky pr. 53, Moscow 117924 USSR

ABSTRACT

We develop field-theoretic description of closed Bose string theory giving a covariant definition of the effective action Γ for fields corresponding to different string "excitations". The problem of deriving the low-energy approximation for Γ is reduced to that of establishing the effective action for a generalized σ-model on a curved 2-dimensional background and that of subsequent integration over all 2-dimensional metrics. Our formalism makes possible a self-consistent study of ground state problem (including the problem of space-time compactification) in the quantum string theory. We suggest a solution to the old "tachyon problem" based on generation of non-trivial vacuum values for the lowest-mass "excitations" (scalar, metric and antisymmetric tensor). We also discuss generalizations to the Fermi string theory and theories of other extended objects (membranes, etc.).

1. INTRODUCTION

It presently appears very unlikely that the ultraviolet problem of quantum gravity can be solved within a perturbatively non-renormalizable field theory with a dimensional coupling constant and a finite number of fields, i.e. within a kind of "ordinary" supergravity theory. Thus one is either to start with a power counting renormalizable theory (like conformal supergravity) or to introduce an infinite number of fields so that to make the resulting theory finite. The latter approach may be realized (in a way preserving unitarity and causality) in a superstring theory [1/]. It was noted long ago that a theory of one-dimensional extended objects - strings may provide us a consistent fundamental theory of all interactions including the quantized gravitational one [2/]. A free (closed) string can be described in terms of the infinite number of its "oscillation modes" (scalar, symmetric 2-tensor, antisymmetric 2-tensor, ...). The basic observation was that the zero slope (zero "string size") limit ($\alpha' \to 0$) of scattering amplitudes for different string modes corresponds to on shell scattering amplitudes of some field theory for fields associated with the elementary string modes [2/]. The extrapolated covariant action of this theory was shown to contain the Einstein gravitational term for the symmetric 2-tensor (considered as a perturbation of the flat metric). In more "realistic" case of supersymmetric strings in D=10 this action contains an N=2, D=10 supergravity action [1/]. Given that the (closed) superstring theory is likely to be finite to all orders [1/], it can be considered as an interesting candidate for a fundamental theory.

There is, however, a number of conceptual as well as technical problems in a string theory considered from this fundamental point of view. The above mentioned connection between string and field theories was previously established in a non-covariant on shell way using expansions near a flat background. This made difficult to understand how a curved space-time metric can be built of "graviton" modes of the string and how "spontaneous compactification" from D=26 or 10 to four dimensions can take place. What was lacking is a covariant ("off shell") effective action Γ for infinite number of fields corresponding to string "excitations", which approximates the string dynamics. Had we such an effective action we could consistently formulate the ground state problem (including the problem of compactification of extra dimensions) in the string theory in complete analogy with the procedure known in ordinary field theories. If solution for a ground state metric would appear to be non-trivial this would be a manifestation of dynamical "condensation" of initial "graviton modes".

Our aim here is to present such a "background field method" formulation of quantum string theory starting with a general definition of the effective field theory action Γ in terms of a path integral over "internal" string variables (sect.2). We will show how one can compute Γ expanding in α' (sect.3). To determine the ground state values of fields one is to study the extrema of Γ. As a result of such a study we will find that the closed Bose string theory prefers compactifications to three dimensions with the ground state values of the metric and the antisymmetric tensor field being generally non-zero (sect.4). The final sect. 5 is devoted to a discussion of some generalizations of our approach.

2. DEFINITION OF EFFECTIVE ACTION

We follow the covariant approach to quantum string dynamics [3] (see also [4-7]) and consider mainly the case of closed Bose strings. The corresponding action is [8]

$$I_0 = \frac{1}{2\pi\alpha'}\int d^2x \, \frac{1}{2}\sqrt{g}\, g^{\mu\nu}\partial_\mu \varphi^i \partial_\nu \varphi^i \,, \quad [\alpha'] = cm^2 \qquad (1)$$

Here x^μ ($\mu = 1, 2$) are coordinates of some 2-dimensional compact space M^2, $g_{\mu\nu}$ is a metric on M^2, φ^i ($i = 1, \dots, D$) are coordinates of an "external" space-time M^D where strings propagate (external space metric G_{ij} is assumed to be flat in (1)). We start with recalling the expression for the tree amplitudes for the scattering of the ground state ("tachyon") scalar string modes [3] (cf. refs. [9])

$$G_N(\phi_1, \dots, \phi_N) = < \prod_{k=1}^{N} \int d^2x_k \sqrt{g(x_k)}\, \delta^{(D)}(\phi_k - \varphi(x_k)) > \quad (2)$$

,

$$< \dots > = \int [dg_{\mu\nu}] \int [d\varphi^i]\, e^{-\frac{1}{\hbar} I_0 [\varphi, g]} \dots \quad . \quad (3)$$

ϕ_n^i are coordinates of N points in M^D and M^2 is taken to be compact without boundary and simply connected. As was found in [3], [5], [10] G_N reproduce the Virasoro-Shapiro amplitudes (see refs. [11], [12] for reviews) in the case of $D = 26$. Let us now write down the expression for the generating functional, corresponding to G_N. Introducing a "source" field $\Phi(\varphi)$ we get

$$\Gamma^{(0)}[\Phi] = < \exp[- \int d^2x \sqrt{g}\, \Phi(\varphi(x))] > , \quad (4)$$

$$G_N(\phi_1, \dots, \phi_N) \sim \frac{\delta^N \Gamma^{(0)}[\Phi]}{\delta\Phi(\phi_1) \dots \delta\Phi(\phi_N)} \bigg|_{\Phi = 0} . \quad (5)$$

The interpretation we suggest for (4) is the following: $\Gamma^{(0)}[\Phi]$ is a "tree" effective action for the scalar field Φ corresponding to the ground state mode of the closed Bose string. Eq. (5) gives the amplitudes (more exactly, irreducible Green's functions) on a "naive" vacuum $\Phi = 0$; a true vacuum value of Φ is to be determined by minimizing the full effective action, thus hopefully solving the "tachyon problem" [11], [12] (see sect.4).

It is obvious how to generalize (4) to include fields for other closed string excitations: we are simply to add in the exponent of (4) all other possible "source" terms which preserve the reparametrizatinn invariance

$$I_{source} = \int d^2x \left[\sqrt{g}\,\Phi(\varphi(x)) + \frac{1}{2}\sqrt{g}\,g^{\mu\nu}\partial_\mu\varphi^i\partial_\nu\varphi^j H_{ij}(\varphi(x)) + \frac{1}{2}\varepsilon^{\mu\nu}\partial_\mu\varphi^i\partial_\nu\varphi^j A_{ij}(\varphi(x)) + \dots\right] \quad (6)$$

The symmetric tensor H_{ij} is a "source" for the "graviton" modes [13], the antisymmetric tensor A_{ij} is a "source" for the antisymmetric 2-tensor modes [14], while dots stand for higher tensor field terms (e.g. $\int d^2x\sqrt{g}\,g^{\mu\nu}g^{\rho\sigma}\times \partial_\mu\varphi^i\partial_\nu\varphi^j\partial_\rho\varphi^k\partial_\sigma\varphi^l B_{ijkl}(\varphi(x))$) corresponding to "spin $>$ 2" massive modes.

Eq. (4) is true in a "first quantized" string theory (a closed simply connected surface corresponds to a world sheet of a virtual string which appears at some point, propagates and then disappears at another point). To account for processes with a "cubic" interaction of virtual strings we are to sum over all closed oriented manifolds with an arbitrary number of handles h (h is thus a number of "loops" in the full second quantized string theory). This is natural given that h is the only topological characteristic of such 2-dimensional manifolds. Thus we a are led to the following expression for the effective field theory action corresponding to the "second quantized" string theory

$$\Gamma[\Phi, G_{ij}, A_{ij}, \dots] = \sum_{h=0}^{\infty} e^{\sigma\chi}\int_{M^2_\chi}[dg_{\mu\nu}]\int[d\varphi^i]\, e^{-\frac{1}{\hbar}I} \quad (7)$$

$$I = I_0 + I_{source} = \int d^2x \sqrt{g}\, \Phi(\varphi) \quad +$$
$$+ \frac{1}{2\pi\alpha'} \int d^2x \left[\frac{1}{2} \sqrt{g}\, g^{\mu\nu} \partial_\mu \varphi^i \partial_\nu \varphi^j G_{ij}(\varphi) + \right. \tag{8}$$
$$\left. + \frac{1}{2} \varepsilon^{\mu\nu} \partial_\mu \varphi^i \partial_\nu \varphi^j A_{ij}(\varphi) + \ldots \right]$$

Here $\chi = 2 - 2n$ is the Euler number of M^2_χ and $G_{ij} \equiv \delta_{ij} + 2\pi\alpha' H_{ij}$ is an arbitrary metric of the "external" space-time M^D. σ is a dimensionless coupling constant of the theory (as we will see in sect. 3 it is not subject to a renormalization, i.e. has a fixed value). The weight $\exp(\sigma\chi)$ with which different topologies are summed (this choice seems to be unique) is distinguished by the observation that $\sigma\chi$ can be rewritten as a local addition to the action (8) because

$$\chi = \frac{1}{4\pi} \int_{M^2_\chi} R\sqrt{g}\, d^2x \tag{9}$$

($R = R^\lambda_{\mu\lambda\nu} g^{\mu\nu}$, $R^\lambda_{\mu\nu\rho} = \partial_\nu \Gamma^\lambda_{\mu\rho} - \ldots$ is the curvature of $g_{\mu\nu}$).

It is important to stress that the structure of (6) or (8) is quite natural: we simply write down all possible local terms that respect covariance in two dimensions. The only "external" gauge invariances of I (8) (and hence of Γ) are general covariance in D dimensions (which follows from 2-dimensional covariance of I) and abelian gauge symmetry $\delta A_{ij} = \partial_i \lambda_j - \partial_j \lambda_i$, $\partial_i = \partial / \partial \varphi^i$. Hence the only fields which are massless in Γ are the metric and the antisymmetric tensor (in particular, "higher spin" fields $B_{ijkl\ldots}$ must be massive because of the lack of the corresponding gauge invariances of I necessary to provide their masslessness in Γ). In this way we deduce the free closed string spectrum without use of any a priori knowledge about it.

To determine Γ in (7) we are thus to compute the

partition function for quantized strings propagating in an arbitrarily curved space-time M^D and interacting (in addition to gravity G_{ij}) with the infinite number of local fields Φ , A_{ij} , ... The important property of Γ is that it can be written as an integral over the space-time M^D. The reason is that the free string theory is insensitive to a position of a string "center", i.e. the action (1) is invariant under $\varphi^i \to \varphi^i +$ + const. Hence the free string partition function contains this zero mode contribution (the volume of $\mathbb{R}^D$) as a factor. This translational invariance is broken in the presence of "external" fields so that D-dimensional "zero mode"

integral is no longer trivial one. It is useful to extract this integral over a "string centre" collective coordinate from the very beginning by splitting φ^i on a constant and non-constant parts

$$\varphi^i(x) = \phi^i + \eta^i(x) \ , \quad \phi^i = \text{const} \ ,$$

$$\int d\varphi\, F(\varphi) = \int d^D\phi \int [d\eta]\, F(\phi+\eta) \ , \tag{10}$$

$$[d\eta] = d\eta\ \delta^{(D)}\left(P^i[\phi,\eta]\right) Q[\phi,\eta] \ .$$

Here $P^i=0$ is a "gauge condition", breaking the invariance under $\eta^i \to \eta^i$ + const to avoid overcounting and $Q = \det\left(\partial P^i[\eta+a]/\partial a^j\right)_{a=0}$ is the "ghost determinant". Then

$$\Gamma = \int d^D\phi \sqrt{G(\phi)}\, \mathcal{L}\Big(\Phi(\phi), \mathcal{D}_i\Phi(\phi), \mathcal{D}_i\mathcal{D}_j\Phi(\phi), \ldots ; G_{ij}(\phi), R^k{}_{ijm}(\phi), \ldots ; F_{ijk}(\phi), \ldots ; \ldots\Big) \tag{11}$$

$$R^i{}_{jkm} = \partial_k \Gamma^i_{jm} - \ldots \ , \quad F_{ijk} = 3\, \partial_{[i} A_{jk]} \ ,$$

where $\mathcal{L}$ contains all powers of derivatives of the fields. Gauge invariances imply that the derivatives of G_{ij} and A_{ij} combine into the curvature tensors $R^i{}_{jkl}$ and F_{ijk} (Γ^i_{jk} is the Christoffel connection for G_{ij}). The factor $\sqrt{G(\phi)}$ comes from the covariant measure

$\prod d\varphi^i_{(x)} \sqrt{G(\varphi(x))}$ in (7) $(G = \det G_{ij})$. The representation (11) shows that it is Γ defined by (7) (and not, e.g., $\ln \Gamma$) that is the effective field action. The Lagrangian

$$\mathcal{L} = \sum_{\chi} e^{6\chi} \int [dg_{\mu\nu}] \int [d\eta] \times \qquad (12)$$
$$\times \exp\left\{-\frac{1}{\hbar} I[\phi + \eta ; \Phi(\phi+\eta), G(\phi+\eta), \dots]\right\}$$

is given by the path integrals over the "internal" string degrees of freedom. Thus $\mathcal{L}$ is effectively non-local in ϕ . This "smearing" should be responsible for a manifest finiteness of $\mathcal{L}$. The fields Φ , G_{ij}, A_{ij} , ... , can now be treated as some "bound state" excitations of string degrees of freedom. So we get a special sort of an "induced" (gravity, etc) theory where however the effective fields (the metric, etc) are not to be further quantized. The point is that loop effects of an approximate field theory (valid at energies $E << M = (2\pi\alpha')^{-1/2}$) are automatically accounted for by the string theory loop corrections (the infinite system of modes propagating in loops coincides with the system of the external fields $\Phi, G, A, \dots$).

It is instructive to compare (7), (8) with the analogous formulas in the quantum particle dynamics

$$\Gamma[\Phi, G_{ij}, A_{ij}, \dots] = \int [de(t)] \int [d\varphi^i(t)]\, e^{-\frac{1}{\hbar} I} , \qquad (13)$$

$$I = \int_0^1 dt\, e\, \Phi(\varphi(t)) + m \int_0^1 dt \left[e^{-1} \dot\varphi^i \dot\varphi^j G_{ij}(\varphi) \right. \qquad (14)$$
$$\left. + \dot\varphi^i A_i(\varphi) \right] + \dots , \qquad \dot\varphi = \frac{d\varphi}{dt} .$$

Here $g = e^2$ is a metric on a closed path M^1 $(\varphi(0) = \varphi(1))$ A_i is a vector potential (I is invariant under

$\delta A_i = \partial_i \lambda$) and Φ contains a constant part, $\Phi = m + \Phi'$. Going to a "proper time" gauge $e(t) = m\mathcal{T} =$ const, $\tau = \mathcal{T}t$, we observe that (13) is simply a "proper-time" representation for the one-loop effective action for the quantum (complex) scalar field interacting with the external fields,

$$\Gamma \sim \ln \det \left(-\nabla_i \nabla^i + m^2 + m\Phi' + \ldots \right) , \qquad (15)$$

$$\nabla_k = \partial_k + \Gamma^i_{kj} + iA_k .$$

The $1/m$ expansion gives $\Gamma \sim \int d^D\phi \, \{$ power series in derivatives of fields, $(\partial/m)^n \}$. Note that combining (7), (8) and (13) it is possible to generalize (7), (8) to the case of the theory, containing open as well as closed strings (here one is to integrate over compact 2-spaces with boundaries). The corresponding action contains boundary terms and we find a gauge vector A_i (known to be present as a massless mode in the open string spectrum) naturally interacting with the ends of an open string.

3. CALCULATION OF EFFECTIVE ACTION

Our aim here is to determine the leading terms in the "low energy" ($\alpha' \to 0$) expansion of Γ defined by (7), (8). As is clear from (12) we are first to expand fields near ϕ^i and to integrate over η^i. At this first stage the metric $g_{\mu\nu}$ (and the Euler number) of M^2_χ is taken to be arbitrary. This first-stage "effective action" is given by

$$\exp\left[-\tfrac{1}{\hbar} W\left([g], \Phi, G, \ldots\right)\right]$$
$$= \int [d\eta] \exp\left[-\tfrac{1}{\hbar} I\left[\phi+\eta, g, \Phi(\phi+\eta), G(\phi+\eta)\right]\right] \qquad (16)$$

Let us start with the formal case of all the "external" fields except the metric G_{ij} put equal to zero. The action for strings propagating in a curved space-time M^D

$$I_G = \frac{1}{2\pi\alpha'} \int d^2x \frac{1}{2} \sqrt{g}\, g^{\mu\nu} \partial_\mu \varphi^i \partial_\nu \varphi^j G_{ij}(\varphi) \tag{17}$$

coincides with the action for a generalized σ -model (with an "internal" space being M^D with the metric G_{ij}) defined on a curved 2-dimensional space M^2_x. (Note a reversal of the roles of M^D and M^2 on going from the string to the σ-model picture). Hence (16) is the quantum partition function for the σ -model, interacting with the external 2-metric. The important observation is that being interested in the $\alpha' \to 0$ expansion of Γ we are simply to compute W perturbatively in $\hbar$ ($\alpha'\hbar$ counts a loop order in (16)). To make perturbation theory manifestly covariant in D dimensions it is convenient to do a local change of the quantum variable : $\eta^i \to \xi^i(\eta, \phi)$ introducing geodesic normal coordinates at the point ϕ 15/-17/

$$\eta^i = \varphi^i - \phi^i = \xi^i - \frac{1}{2} \Gamma^i_{jk}(\phi) \xi^j \xi^k + \dots \tag{18}$$

Then (17) takes the form

$$\begin{aligned} I_G = \int d^2x \sqrt{g}\, g^{\mu\nu} \Big\{ & \frac{1}{2} \partial_\mu \xi^i \partial_\nu \xi^j G_{ij}(\phi) - \\ & - \frac{1}{6}(2\pi\alpha') R_{ikjl}(\phi)\, \xi^k \xi^l \partial_\mu \xi^i \partial_\nu \xi^j - \\ & - \frac{1}{12} (2\pi\alpha')^{3/2} \mathcal{D}_k R_{iljm}(\phi) \xi^k \xi^l \xi^m \partial_\mu \xi^i \partial_\nu \xi^j + O(\alpha'^2 R^2) \Big\} \end{aligned} \tag{19}$$

We made the rescaling $\xi \to (2\pi\alpha')^{1/2} \xi$ so that ξ in(19) is dimensionless. Higher order terms in (19) have the following structure: $\int d^2x \sqrt{g}\, g^{\mu\nu} [(\alpha' \xi^2)^{p+\frac{1}{2}q} \mathcal{D}^q R^p] \partial_\mu \xi \partial_\nu \xi$

("coupling constants" in brackets are dimensionless). By noting that ξ^i transforms as a vector under the point transformations of ϕ^i we can pass to the orthonormal basis, introducing $\xi^a(x) = e^a_i(\phi) \xi^i(x)$ $(G_{ij} = e^a_i e^a_j)$

so that the kinetic term in (19) takes the usual form $(\partial_\mu \xi^a)^2$. A convenient covariant choice of the "gauge" P^i in (10) is [16] $P^a = \int d^2x \sqrt{g}\, \xi^a(x)$. The "jacobians" like $\det(\partial \eta^i/\partial \xi^j)$ and Q contribute only in a local measure and can be ignored (in dimensional regularization).

The classical action (19) is invariant under the Weyl transformations $g_{\mu\nu} \to \lambda^2(x)\, g_{\mu\nu}$ as well as under the point transformations $x^\mu \to x'^\mu$. To preserve the latter invariance we employ dimensional regularization which breaks down the Weyl invariance. Considerations based on covariance and conformal (Weyl) anomaly in an interacting theory (cf. [18]) make possible to determine the general structure of $W([g], G)$ in (16)

$$W = \frac{1}{\varepsilon}\beta \int R\sqrt{g}\, d^2x + \gamma \int d^2x\, d^2x' (R\sqrt{g})_x\, \Box^{-1}_{xx'} (R\sqrt{g})_{x'} \tag{20}$$

$$\gamma = \frac{1}{4}\beta = a_1 + a_2 \hbar \alpha' \mathcal{R}(\phi) + a_3 \hbar^2 \alpha'^2 (\mathcal{R}_{\dots}(\phi))^2 + O(\alpha'^2) \tag{21}$$

Here $R = R^\mu{}_{\lambda\mu\nu}\, g^{\lambda\nu}$ and $\mathcal{R} = \mathcal{R}^i{}_{jil}\, G^{jl}$ are the scalar curvatures of $g_{\mu\nu}$ and G_{ij}, $\Box = \partial_\mu (g^{\mu\nu}\sqrt{g}\,\partial_\nu)$, $\Box_x \Box^{-1}_{xx'} = \delta^{(2)}(x-x')$ and $\varepsilon = d-2 \to 0^-$. The first term in (20) is the ultravalet infinity (infrared infinities are absent due to compactness of M^2_x). The second term in (20) corresponds to the Weyl anomaly. To the one loop order W is simply the effective action for D free scalar fields on a curved 2-dimensional background, so that $\gamma^{(1)} = D/96\pi$ (see e.g. [19, 3]). To determine the two-loop coefficient a_2 one can use the covariant two-loop methods of ref. [20]

or expand the metric near a flat background, $g_{\mu\nu} = \delta_{\mu\nu} + h_{\mu\nu}$ and compute the "self-energy" graphs for $h_{\mu\nu}$ (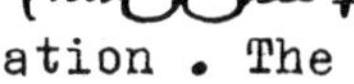 +  , etc) using momentum space representation . The result of this second approach reproduces the leading terms in the $h_{\mu\nu}$-expansion of (20) with $\gamma^{(2)} = -\frac{1}{128\pi^2}(2\pi\alpha' \mathcal{R})$. Thus in the two-loop approximation

$$\gamma = \frac{1}{4}\beta = \frac{D}{96\pi} - \frac{\hbar\alpha'}{64\pi}\mathcal{R}(\phi) + O(\hbar^2\alpha'^2\mathcal{R}^2) \qquad (22)$$

Next let us include in (20) the contribution of the antisymmetric tensor. The relevant part of the action (8)

$$I_{G,A} = \frac{1}{2\pi\alpha'}\int d^2x\,\Big\{\frac{1}{2}\sqrt{g}\,g^{\mu\nu}\partial_\mu\varphi^i\partial_\nu\varphi^j G_{ij}(\varphi) \qquad (23)$$
$$+ \frac{1}{2}\varepsilon^{\mu\nu}\partial_\mu\varphi^i\partial_\nu\varphi^j A_{ij}(\varphi)\Big\}$$

is exactly the action for the generalized σ -model with a generalized "Wess-Zumino term". In fact, the second Kalb-Ramond term [14] in (23) coincides with the Wess-Zumino term [21] for the special choices of M^D and $A_{ij}(\varphi)$ corresponding to the "standard" σ -models. Expanding φ^i near ϕ^i according to (18) we find that in addition to (19) (23) contains the terms depending on $F_{ijk} = 3\,\partial_{[i}A_{jk]}$, $\mathcal{R}^i{}_{jkl}$ and their covariant derivatives. The leading term is:

$\frac{1}{6}\int d^2x\,(2\pi\alpha')^{1/2} F_{ijk}(\phi)\,\xi^i\partial_\mu\xi^j\partial_\nu\xi^k\varepsilon^{\mu\nu}$. To establish the lowest order contribution of F_{ijk} to γ in (21) one is to compute the 2-loop $h_{\mu\nu}$ -self-energy diagrams with two $(\alpha')^{1/2}F$ - vertices ( , etc.). The result corresponds to the following substitution in (22)

$$\mathcal{R} \to \mathcal{R} - \frac{1}{12}F_{ijk}F^{ijk} \qquad (24)$$

(as will be clear, this is in agreement with the result found previously in the context of the old dual string model [22]; note also the connection with recent work [23] on the generalized σ -model (23)).

Finally, we are to account for the contribution in W of the scalar Φ in (8). [We shall ignore the contributions of higher tensor fields $B_{ijkl,...}$ because of their massiveness and apparent irrelevance for the ground state problem ($\langle B_{ijkl,...}\rangle \neq 0$ break the Lorentz invariance). From the σ -model point of view they correspond to dim $>$ 2 terms in (8) which are irrelevant in the "low temperature" ($\alpha' \to 0$) limit]. According to (18)

$$\Phi(\phi+\eta) = \Phi(\phi) + (\mathcal{D}_i\Phi)(\phi)\xi^i + \frac{1}{2!}(\mathcal{D}_i\mathcal{D}_j\Phi)(\phi)\xi^i\xi^j + \ldots \tag{25}$$

Integrating in (16) over ξ^i (regular non-constant functions on M_x^2 satisfying $\int d^2x\sqrt{g}\,\xi^i(x) = 0$) in the leading (one-loop) approximation we obtain (in addition to (20), (24))

$$W(\Phi) = \Phi(\phi)\int d^2x\sqrt{g} + \frac{1}{2}\ln\det\left[\delta_{ij}\Delta + 2\pi\alpha'(\mathcal{D}_i\mathcal{D}_j\Phi)(\phi)\right] , \tag{26}$$

$$\Delta = -\frac{1}{\sqrt{g}}\partial_\mu(\sqrt{g}\,g^{\mu\nu}\partial_\nu) \equiv -\frac{1}{\sqrt{g}}\Box .$$

Thus to the lowest order in α'

$$W(\Phi) = \Phi\int d^2x\sqrt{g} - \pi\alpha'\mathcal{D}^2\Phi\int d^2x\sqrt{g_x}\,\Box^{-1}_{xx} + \ldots \tag{27}$$

$$\int d^2x\sqrt{g_x}\,\Box^{-1}_{xx} = -\frac{1}{4\pi}\ln\Lambda^2\int d^2x\sqrt{g} + \frac{1}{4\pi}\int d^2x\,d^2x'\sqrt{g_x}\,\Box^{-1}_{xx'}(R\sqrt{g})_{x'} , \tag{28}$$

where $\Lambda \to \infty$ is a covariant ultra-violet cut-off $(1/\varepsilon = -\ln\Lambda)$. [Throughout this paper we ignore quadratic infinities that can be absorbed in redefinition of a constant part of Φ or cancelled by a proper choice of the path integral measure].

The apparent dependence of W on the ultraviolet cutoff of the two-dimensional theory could be a serious problem of the (Bose) string theory if it implied that Γ is also cutoff dependent and hence not unambiguously calculable. Happily, this does not happen. There exists a natural "built in" mechanism that leads to a well-defined Γ as a functional of fields rescaled by powers of the cutoff so that to make them dimensionless ($\Phi \to \Lambda^2 \Phi$, $G \to G$, $A \to A$, ...). The basic observation is that a rescaling of a covariant cutoff should be equivalent to a rescaling of a two-dimensional metric $g_{\mu\nu}$. But $g_{\mu\nu}$ itself is the integration variable in (7). Hence all the cutoff dependence left after the rescaling of the fields should be absorbable in a redefinition of $g_{\mu\nu}$ and hence should be absent in Γ . There of course may be additional infinities coming from the integral over the metric itself. For example, the integral over a "scale" of $g_{\mu\nu}$ may diverge at lower limit. However, for consistency, this integral must be cutoff at $1/\Lambda$ because a short distance cutoff Λ was already introduced in the theory defined on M^2_χ with a fixed metric. As a result, such an integral will be automatically convergent after the rescaling of the metric ($g_{\mu\nu} \to \Lambda^{-2} g_{\mu\nu}$) discussed above. In particular, no renormalization of σ in (7) will be needed (see also 7/). The work of this mechanism can be seen on the example of the "free partition function",

$$Z = \Gamma\left[\Phi = \mu_0^2 = \mathrm{const},\, G_{ij} = \delta_{ij},\, A_{ij,\dots} = 0\right] = \left\langle \exp\left(-\mu_0^2 \int d^2x \sqrt{g}\right)\right\rangle,$$

which is a well-defined function of $\tilde{\Phi} = \Phi \Lambda^{-2}$ (cf. 7/, 24/). We will see that it works also in the general case.

Let us now consider the integration of $e^{-\frac{1}{\hbar} W}$ over the metrics. We shall specialize to the "tree" approximation (the χ = 2 term in the sum in (7)), i.e. will integrate over metrics on closed simply connected manifolds M_2^2. We fix the coordinate gauge by the condition $g_{\mu\nu} = e^{2\rho}\hat{g}_{\mu\nu}$, where $\hat{g}_{\mu\nu}$ is a metric of the sphere S^2 and ρ is a regular function on S^2. Accounting for the contribution of the determinant of the ghost operator [3, 4, 7] [$\Delta W = -\frac{1}{2}\ln\det\Delta_{gh}$, $(\Delta_{gh})_{\mu\nu} = -\nabla^2_{\mu\nu} - \frac{1}{2}Rg_{\mu\nu}$; ΔW has the same structure as (20) with $\Delta\gamma = 26/96\pi$; after proper subtraction of the volume of the diffeomorphism group (containing the conformal subgroup generated by the six zero modes of Δ_{gh}, i.e. the conformal Killing vectors) we also get $\Delta\beta = 4\Delta\gamma$, see ref. [7] for details] and expanding $\exp(-W)$ in powers of α' we find from (7), (20), (22), (24), (27)

$$\Gamma[\Phi, G, A] \sim \mathcal{V}^{-1}\int d^D\phi\sqrt{G(\phi)}\int d\rho\, e^{-W_0/\hbar} \times$$
$$\times\Big\{1 - \frac{1}{4}\ln\Lambda^2\cdot\alpha'\mathcal{D}^2\Phi\int d^2x\sqrt{g} + \frac{1}{4}\alpha'\mathcal{D}^2\Phi\int\Box^{-1}R$$
$$+ \frac{1}{4}\alpha'\ln\Lambda^2\cdot\big(\mathcal{R} - \tfrac{1}{12}F^2\big) + \frac{\alpha'}{64\pi}\big(\mathcal{R} - \tfrac{1}{12}F^2\big)\int R\Box^{-1}R$$
$$+ O(\alpha'^2)\Big\}, \qquad (29)$$

where

$$W_0 = -\varkappa\ln\Lambda^2 + \frac{\varkappa}{16\pi}\int R\Box^{-1}R + \Phi\int\sqrt{g}\,d^2x, \qquad (30)$$

$$\varkappa = \tfrac{1}{6}(D-26), \quad F^2 = F_{ijk}F^{ijk},$$
$$\int R\Box^{-1}R = \int d^2x\,d^2x'(\sqrt{g}R)_x\,\Box^{-1}_{xx'}(\sqrt{g}R)_{x'}, \quad \int\Box^{-1}R = \int d^2x\,d^2x'\sqrt{g}_x\,\Box^{-1}_{xx'}(R\sqrt{g})_{x'}, \qquad (31)$$

and $\mathcal{V}$ is the (infinite) volume of the conformal SO(2,2) group (the invariance group of the integrand in (29) left after the gauge choice). The metric $g_{\mu\nu}$ in (29) is assumed to be equal to $e^{2\rho}\hat{g}_{\mu\nu}$, so that

$$\hat{g}_{\mu\nu} = \delta_{\mu\nu}\Big[1 + \frac{x^2}{4r^2}\Big]^{-2}, \quad \hat{A} = \int d^2x\sqrt{\hat{g}} = 4\pi r^2,$$
$$\hat{R} \equiv R(\hat{g}) = 8\pi/\hat{A}, \quad R\sqrt{g} = \hat{R}\sqrt{\hat{g}} - 2\hat{\Box}\rho, \quad \hat{\Box} = \Box(\hat{g}), \qquad (32)$$

$$\int R\,\Box^{-1}R = \int \hat{R}\,\hat{\Box}^{-1}\hat{R} + 4\int d^2x\,\rho\,\hat{\Box}\rho - 4\int \rho\hat{R}\sqrt{\hat{g}}\,d^2x \,,$$

$$\int \Box^{-1}R = \int \hat{\Box}^{-1}_{xx'}\,\hat{R}\,e^{2\rho(x)}\sqrt{\hat{g}_x}\sqrt{\hat{g}_{x'}}\,d^2x\,d^2x' - \\ - 2\int \rho\, e^{2\rho}\sqrt{\hat{g}}\,d^2x \,, \qquad (33)$$

$$\int \hat{R}\,\hat{\Box}^{-1}\hat{R} = 16\pi \,, \quad \int \hat{\Box}^{-1}\hat{R} = 2\hat{A}$$

The general structure of (29) remains the same for the case of an arbitrary M^2_{χ} , except that for $\chi \neq 2$ there are additional integrals over the parameters of "Teichmuller deformations" of the metric 7/ (dimension of Teichmuller space of metrics on M^2_{χ} is equal to zero only for $\chi = 2$; any traceless deformation of a metric on M^2_2 can be generated by a local diffeomorphism).

Note that if we did not expand W in α' then the total coefficient of the $\int R\,\Box^{-1}R$ -term in the exponent of (29) would be (cf. (22)): $\gamma = \frac{1}{96\pi}\left(D - 26 - \frac{3}{2}\alpha'\mathcal{R} + O(\alpha'^2\mathcal{R}^2)\right)$. Thus the effect of (positive) curvature is to increase the "critical dimension". [The problem of calculating γ for strings propagating on the sphere S^D was previously treated in ref. 25/. However, the result found there $\gamma \sim (D-27)$ appears to be incorrect being independent of $\alpha'\mathcal{R}$].

Using (33) it is easy to check that all the cutoff dependence in (29) can indeed be absorbed in the following redefinitions of the 2-dimensional metric and the scalar field: $g_{\mu\nu} = \Lambda^{-2}\tilde{g}_{\mu\nu}$, $\Phi = \Lambda^2\tilde{\Phi}$. However, there remains a question about the infinities which appear after the integration over ρ . To clarify this point and to develop a systematic procedure for computation of Γ in (29) it is useful to isolate first the integral over a constant scale of the metric (constant part of ρ) and then to do loop expansion for the remaining degrees of freedom. This can be done by inserting $1 = \int_0^\infty dA\, \delta\left(\int \sqrt{g}\, d^2x - A\right)$ under the integrals in (29), i.e. by extracting the integral over a surface area [26]. Starting with (29) and using (33) we get

$$\Gamma[\Phi, G, A] \sim Z^{-1} \int d^D\phi \sqrt{G} \int_0^\infty \frac{dA}{A}\, e^{-\Phi A} \int d\tilde{\rho}\, \times$$
$$\times\, \delta\left(\int d^2x \sqrt{\hat{g}}\left(e^{2\tilde{\rho}} - 1\right)\right) \exp\left[-\varkappa S(A, \tilde{\rho})\right] \times$$
$$\times \left\{ 1 - \frac{\alpha'}{2} \cdot \mathcal{D}^2 \Phi \cdot A \cdot \hat{A}^{-1} \int d^2x \sqrt{\hat{g}}\, e^{2\tilde{\rho}} \left[\tfrac{1}{2} \ln(\Lambda^2 A) + \tilde{\rho} - \tfrac{1}{2} \hat{\Box}^{-1} \hat{R}\right] + \right.$$
$$+ \frac{\alpha'}{4}\left(R - \frac{1}{12} F^2\right)\left[1 - \ln(\Lambda^2 A) + \frac{1}{4\pi} \int d^2x\, \tilde{\rho}\, \Box_0 \tilde{\rho} - \right. \tag{34}$$
$$\left.\left. - 2 \hat{A}^{-1} \int d^2x \sqrt{\hat{g}}\, \tilde{\rho} \right] + O(\alpha'^2) \right\} ,$$

where

$$S(A, \tilde{\rho}) = \frac{1}{4\pi}\left[\int d^2x\, \tilde{\rho}\, \Box_0 \tilde{\rho} + \right.$$
$$\left. + \frac{1}{2} R_0 \int d^2x \sqrt{g_0}\left(e^{2\tilde{\rho}} - 2\tilde{\rho} - 1\right)\right] + 1 - \ln(\Lambda^2 A) , \tag{35}$$

$$\varkappa = \frac{1}{6}(D - 26) .$$

Here $\Box_0 = \Box(g_0)$, $R_0 = R(g_0) = \mathrm{const} = 8\pi/A$ correspond to the metric $g_{0\mu\nu} = (A/\hat{A})\, \hat{g}_{\mu\nu}$ which is the stationary point of the action $\int R\, \Box^{-1} R$ (see (30)) under the constraint $\int d^2x \sqrt{g} = A$. The total metric is thus $g_{\mu\nu} = e^{2\tilde{\rho}} g_{0\mu\nu}$, where $\tilde{\rho} = \rho - \frac{1}{2}\ln(A/\hat{A})$ (a regular function on S^2) has the vanishing vacuum value. The Liouville-type action (35) is positive under the constraint $\int d^2x \sqrt{\hat{g}}\left(e^{2\tilde{\rho}} - 1\right) = 0$ [27]. All the A-dependence in (34), (35) is shown explicitly except that coming from the dependence on $g_{0\mu\nu} \sim A \hat{g}_{\mu\nu}$. In view of the Weyl invariance of $\Box_0$ and $R_0 \sqrt{g_0}$, the latter dependence can be only of the "anomalous" type, $f(\Lambda^2 A)$, where Λ is the cutoff regulating the $\hat{\rho}$ -theory. Hence, in agreement with the proposal made above, all the cutoff dependence in (34) can be eliminated by introducing the new variables $\widetilde{A}=\Lambda^2 A$ and $\widetilde{\Phi}=\Lambda^{-2}\Phi$ so that Γ expressed in terms of $\widetilde{\Phi}$, G_{ij} and A_{ij} is independent of the cutoff, i.e. is an unambiguously calculable functional. This conclusion is unchanged if the integral over A formally diverges at $A=0$ because as was already noted it must be cut off at $A_{min}\sim 1/\Lambda^2$.

We shall evaluate Γ in (34) using the simplest possible approximation, namely, including the one-loop contribution from the $\widetilde{\rho}$-action S' (35) but ignoring the dependence on $\widetilde{\rho}$ of the preexponential terms in (34). Then the integral over $\widetilde{\rho}$ is given by

$$\int d\widetilde{\rho}\, \delta\left(\int d^2x \sqrt{\hat{g}}\, \widetilde{\rho}\right) \exp\left[\frac{\varkappa}{4\pi}\int d^2x \sqrt{g_0}\, \widetilde{\rho}\, \widetilde{\Delta}_0 \widetilde{\rho}\right] \times \{1+\dots\},$$

$$\widetilde{\Delta}_0=\Delta_0-R_0 \ , \quad \Delta_0=-\frac{1}{\sqrt{g_0}}\partial_\mu\left(\sqrt{g_0}\, g_0^{\mu\nu}\partial_\nu\right) \qquad (36)$$

The operator $\widetilde{\Delta}_0$ defined on S^2 (with the metric $g_{0\mu\nu}=(A/\hat{A})\, g_{\mu\nu}$) has the spectrum $\lambda_n=4\pi A^{-1}[n(n+1)-2]$, i.e. has one negative mode ($\widetilde{\rho}=$ const) and three zero modes (reflecting the invariance of the integrand of (34) under the proper conformal transformations, which "shift" the $\widetilde{\rho}$). The negative mode is projected out by the δ-function in (36), while the infinite integral over the (normalized) zero modes is cancelled out by the $\mathfrak{Z}^{-1}$-factor in (34). The contribution of (36) is thus equal to

$$(\Lambda^2 A)^{\frac{1}{2}\cdot 4}\left(\det{}'\widetilde{\Delta}_0\right)^{-1/2} , \qquad (37)$$

$$\ln\det{}'\widetilde{\Delta}_0=-(B_2-4)\ln(\Lambda^2 A)+\text{const} , \qquad (38)$$

where the first factor in (37) comes from the normalization of the negative and zero modes. $B_2=\frac{1}{4\pi}\int d^2x\sqrt{g}\left(\frac{1}{6}R_0+R_0\right)=7/3$ is the corresponding De Witt-Seeley coefficient and the prime and (-4) in (38) indicate that the first four modes of $\widetilde{\Delta}_0$ are not included in the determinant. Introducing $\widetilde{A}=\Lambda^2 A$ and $\widetilde{\Phi}=\Lambda^{-2}\Phi$ and using (33) we can put the result of integration over $\tilde{\rho}$ in the form

$$\Gamma \sim \int d^D\phi \sqrt{G} \int d\tilde{A} \exp(-\tilde{\Phi}\tilde{A} + \nu \ln \tilde{A}) \times \{1 + \frac{\alpha'}{2} \mathcal{D}^2 \tilde{\Phi} \tilde{A} (1 - \frac{1}{2} \ln \tilde{A}) + \frac{\alpha'}{4} (R - \frac{1}{12} F^2)(1 - \ln \tilde{A}) + \ldots\} \quad (39)$$

Here $\nu = \frac{1}{6}(D-25)$ if $D < 26$ (for $D > 26$ $\tilde{\rho}$ is a ghost, see (36)). [$D = 26$ $(\varkappa = 0)$ is a special case, when $\tilde{\rho}$ is a (Weyl) gauge degree of freedom. The contribution of $\det'\tilde{\Delta}_0$ in (37) is thus absent and the integral over $\tilde{\rho}$'s expandable in higher ($n > 1$) eigenfunctions of $\tilde{\Delta}_0$ is to be regulated by inserting the Weyl gauge. Hence in this case $\nu = 1$]. Note that for $\Phi = \mu_0^2 = const$, $G_{ij} = \delta_{ij}$, $A_{ij} = 0$ eq. (39) is in agreement with the result of ref. [24].

The integral over $\tilde{A}$ in (39) is also to be evaluated in a "stationary point" approximation. The "stationary point" for the "action" in the exponent in (39) is $\tilde{A}_0 = \nu \tilde{\Phi}^{-1}$. Integrating over $\tilde{A} - \tilde{A}_0$ in the gaussian approximation (putting $\tilde{A} = \tilde{A}_0$ in the preexponential terms) we find

$$\Gamma \simeq -c M^D \int d^D\phi \sqrt{G} (\tilde{\Phi}/\nu)^{-\nu-1} \{1 - \frac{\alpha'}{2} \nu \frac{\mathcal{D}^2 \tilde{\Phi}}{\tilde{\Phi}} (1 - \frac{1}{2} \ln \frac{\tilde{\Phi}}{\nu}) + \frac{\alpha'}{4} (R - \frac{1}{12} F^2)(1 + \ln \tilde{\Phi}/\nu) + O(\alpha'^2)\} \quad (40)$$

This is our final result. Here c is an overall constant, M is the normalization mass $(\sim \hat{A}^{-1/2})$ that can be chosen equal to $(2\pi\alpha')^{-1/2}$ (note that Γ and $\tilde{\Phi}$ are dimensionless). Taken in the formal point $\tilde{\Phi}$ =const the action (40) contains the Einstein and the antisymmetric tensor terms with physical relative signs (and also contains a cosmological constant). Higher order terms in (40) are higher derivative ones, $\alpha'^2 R^2$, $\alpha'^2 F^4, \ldots$ The gravitational constant k is thus of order of $(\alpha')^{(D-2)/4}$, so that $(\alpha')^{-1/2}$ should be of order of the Planck mass [2]. The crucial question is of course about the ground state values of $\tilde{\Phi}$, G_{ij} and A_{ij}

4. GROUND STATE PROBLEM

As is clear from (4), (5) the scalar field Φ corresponds to a "ground state" mode of the free closedstring.

This mode is known to be a tachyon, what is the most serious difficulty of old dual string approach (this difficulty probably present also in the approach of ref. 3/ is absent in the case of superstrings in $D=10$ 1/). The presence of tachyon implies that the naive vacuum in which all string modes have vanishing expectation values, so that strings propagate in an empty flat space-time ($\Phi=0, G_{ij}=\delta_{ij}, A_{ij},\ldots=0$) is unstable. It was suggested previously that condensation of scalar modes may stabilize the vacuum (see e.g. 28/). It was, however, ignored that the flat metric $G_{ij}=\delta_{ij}$ and $A_{ij}=0$ may not correspond to a true vacuum of the theory. This was partly due to the attitude to the string theory as a theory of strong interactions (where the space-time metric is supposed to be flat) and partly due to the absence of an off-shell method for formulation and solution of the ground state problem (analogous to the effective action method in ordinary field theories). But considering a string theory as a fundamental theory of all interactions including gravity it is even necessary to admit a non-trivial ground state metric in order to have a possibility of compact extra dimensions (the known string theories do not distinguish $D=4$). The effective action (7) provides an adequate framework for a search for such non-trivial ground states. The equations that determine the ground state are

$$\frac{\delta\Gamma}{\delta\Phi}=0\ , \quad \frac{\delta\Gamma}{\delta G_{ij}}=0\ , \quad \frac{\delta\Gamma}{\delta A_{ij}}=0\ , \ldots \qquad (41)$$

We shall see that the general solution of (41) corresponds to $G_{ij} \neq \delta_{ij}$ (A_{ij} may also be non-vanishing in the vacuum with $F_{ijk} \sim \varepsilon_{ijk}$ thus a priori distinguishing compactification to 3 dimensions [29]). If the corresponding vacuum is a stable one, we are get a natural solution of the tachyon problem, based on a "condensation" of scalar, graviton and antisymmetric tensor modes. Then the conclusion is that the (closed) Bose string is non-pathological by itself, the error was in the prejudice that the vacuum metric can be flat in all D dimensions.

To analyze eqs. (41) for the action (40) let us rewrite (40) in a more transparent form, introducing instead of $\tilde{\Phi}$ the dimensionless scalar field $\Omega = (\tilde{\Phi}/\nu)^{-\frac{1}{2}-\nu/2}$

$$\Gamma \simeq -c M^D \int d^D\phi \sqrt{G} \left[\Omega^2 - \alpha' (\partial_i \Omega)^2 (a_1 + a_2 \ln \Omega) + \alpha' \left(R - \frac{1}{12} F_{ijk}^2 \right) \Omega^2 (b_1 + b_2 \ln \Omega) \right] . \tag{42}$$

Here
$a_1 = \nu(5+2\nu)(1+\nu)^{-2}$, $a_2 = -2\nu(1+\nu)^{-3}$,
$b_1 = 1/4$, $b_2 = -\frac{1}{2}(1+\nu)^{-1}$. Higher loop corrections (in the two-dimensional theory in (34)) may probably change the values of (some of) the numerical constants so it is useful not to specialize their values in (42) from the beginning. It is instructive first to analyze the special case of $a_2 = b_2 = 0$ ($\ln \Omega$-terms may be considered as coming from "radiative corrections" and hence "ignorable" at a first step). Assuming $a_1 > 0$, $b_1 > 0$ (and $c > 0$) we find that a naive flat space limit of (42) corresponds to a scalar field with a tachyonic mass ($m^2 = -(\alpha' a_1)^{-1}$), i.e. $\Omega = 0$ - "vacuum" is unstable. To find the true vacuum state we are to vary (42) with respect to Ω , G_{ij} and A_{ij}. Looking for maximally symmetric vacua, we assume that the vacuum va-

lue of Ω is $\Omega_0 =$ const $\neq 0$. Then eqs. (41) take the form

$$\mathcal{R} - \frac{1}{12} F^2 = -\mu^2/b_1 \, , \quad \mathcal{D}_i F^{ijk} = 0 \, , \tag{43}$$
$$\mathcal{R}_{ij} - \frac{1}{4} F_{imn} F_j{}^{mn} = 0 \, , \quad \mu^2 \equiv (\alpha')^{-1} .$$

Thus Ω_0 is arbitrary and no solutions exist with $F_{ijk} = 0$. No solutions exist also if G_{ij} has the euclidean signature)(i.e. if $F^2 > 0$). Hence M^D must have at least one "time" direction. Only one"time" is consistent with the absence of ghosts. Solutions with maximal symmetry are obtained in the case when $F_{ijk} \sim \varepsilon_{ijk}$ in some two space and one time dimensions. Then $M^D = \tilde{S}^3 \times M^{D-3}$ where $\tilde{S}^3$ is the 3-dimensional anti de Sitter space-time and M^{D-3} is an arbitrary Einstein space, i.e.

$$\mathcal{R}_{ab} = -\frac{1}{2} f_0^2 G_{ab} \, , \quad F_{abc} = f_0 \epsilon_{abc} \, , \tag{44}$$
$$\mathcal{R}_{\alpha\beta} = 0 \, , \qquad f_0^2 = \mu^2/b_1 \, , \quad \epsilon^{abc} \equiv |G_3|^{-1/2} \varepsilon^{abc}$$

where $a, b = 0, 1, 2$; $\alpha, \beta = 3, \ldots, D$ (G_{ij} is assumed to be block-diagonal). Note that M^{D-3} may be non-compact (e.g. $\mathbb{R}^{D-3}$). More general solutions correspond to $M^D = \tilde{S}^3 \times S^3 \times \ldots \times S^3 \times M^N$ where compactness of the S^3-factors is supported by $F_{\alpha\beta\gamma} \sim \varepsilon_{\alpha\beta\gamma}$,

$$\mathcal{R}_{\alpha_n\beta_n} = \frac{1}{2} f_n^2 G_{\alpha_n\beta_n} \, , \quad F_{\alpha_n\beta_n\gamma_n} = f_n \epsilon_{\alpha_n\beta_n\gamma_n} \, , \tag{45}$$
$$f_0^2 - \sum_n f_n^2 = \mu^2/b_1 \, .$$

Hence the model prefers the <u>three</u> effective space-time dimensions with the anti de Sitter metric (with the isometry group $SO(2,2)$ curiously coinciding with the conformal symmetry group of the "elementary" 2-dimensional theory in (34)). Direct analysis shows that there are no tachyonic modes in the spectrum of fluctuations of the fields (Ω, G, A) near their vacuum values, i.e. the true ground state corresponding to (43) is stable.

Now let us consider the general case of (42) with $a_2 \neq 0$, $b_2 \neq 0$. Assuming again that Ω_0 = const $\neq$ 0 (Ω = 0 is a singular point for (42)) we get the generalization of eqs. (43)

$$R - \frac{1}{12}F^2 = \mu^2 N_1 \quad , \quad \mathcal{D}_i F^{ijk} = 0,$$
$$R_{ij} - \frac{1}{4} F_{imn} F_j{}^{mn} = \mu^2 N_2 G_{ij} \quad , \quad \Delta \equiv b_1 + b_2 \ln \Omega_0, \qquad (46)$$
$$N_1 = -2(b_2 + 2\Delta)^{-1}, \quad N_2 = (b_2/2\Delta)(b_2 + 2\Delta)^{-1},$$

Let us first consider solutions with $F_{ijk} = 0$. They exist if $-4\Delta = D b_2$ so that $N_1 = D N_2 = -D(D-2)^{-1}\Delta^{-1}$. Thus the space-time metric is the de Sitter one for $\Delta < 0$ $(b_2 > 0$, i.e. $2 < D < 19$, see (42)) and the anti de Sitter one for $\Delta > 0$ ($b_2 < 0$, $19 < D \leq 26$). Note that there is <u>no</u> solutions corresponding to the flat D-dimensional space-time. Stability of these vacua deserves special study (depending on the values of a_1 and a_2 in (42)). There are also solutions with $F_{abc} \sim \varepsilon_{abc}$ in some three dimensions. In this case eqs. (46) reduce to

$$R_{ab} = \mu^2 N_3 G_{ab} \, , \; R_{\alpha\beta} = \mu^2 N_2 G_{\alpha\beta} \, , \; F^2 = 12(N_3 - N_2),$$
$$N_3 = [4\Delta - (D-2)b_2][4\Delta(b_2 + 2\Delta)]^{-1}, \quad a, b = 0, 1, 2 \quad i = (a, \alpha) \qquad (47)$$

Here Ω_0 is an arbitrary parameter. A particular solution of (47) corresponds to a space-time M^D being a product of the anti de Sitter space $\tilde{S}^3$ and the sphere S^{D-3}. For $4\Delta = (D-2)b_2$ the "effective" 3-dimensional space-time can be chosen to be flat (if $b_2 > 0$ the $(D-3)$-dimensional space has positive curvature but for $b_2 > 0$ $F^2 < 0$ so that the 3-space must have non-euclidean signature). There can also be a number of S^3-

factors.

Our basic conclusion is that it is necessary to account for the "condensates" of the D-dimensional metric and the antisymmetric tensor as well as of the "ground state scalar" to determine the true ground state of the closed Bose string theory. It is inconsistent to take D-dimensional space-time to be flat.

5. GENERALIZATIONS

One natural generalization of our approach is to the case of the Fermi string theory of refs. [8], [30], [31] with the classical action invariant under the 2-dimensional local supersymmetry. To couple the free Fermi string theory to the external metric G_{ij} amounts to construct a locally supersymmetric generalization of the N=1 supersymmetric generalized σ -model action (see e.g. [32]). The result reads (cf. [33] for N = 2 case)

$$I_F = \frac{1}{2\pi\alpha'} \int d^2x \sqrt{g} \Big\{ \frac{1}{2} g^{\mu\nu} \partial_\mu \varphi^i \partial_\nu \varphi^j G_{ij}(\varphi) + \frac{i}{2} \bar{\Psi}^i \gamma^\mu D_\mu \Psi^j G_{ij}(\varphi) + \frac{1}{12} R_{ijkl} \bar{\Psi}^i \Psi^k \bar{\Psi}^j \Psi^l + 2(\partial_\mu \varphi^i + \bar{\chi}_\mu \Psi^i) \bar{\Psi}^j \gamma^\rho \gamma^\mu \chi_\rho G_{ij}(\varphi) \Big\} , \quad D_\mu{}^i{}_j = \delta^i_j \partial_\mu + \Gamma^i_{jk} \partial_\mu \varphi^k \tag{48}$$

Here Ψ^i and χ_μ are the 2-dimensional Majorana spinors and the gravitino. It is straightforward also to include the coupling with A_{ij} starting with the rigidly supersymmetric extension of (23) constructed in [23]. What is not clear is how the Fermi string theory (apparently lacking D-dimensional supersymmetry) can be coupled to a sort of $D=10$ -dimensional supergravity anticipated as its $\alpha' \to 0$ limit, cf. [34] Such a coupling is perfectly possible for the covariant superstring action found in [35]. Here we only make a remark about the influence of the Fermi degrees of freedom in (48) on the effective action $\Gamma[G, A]$. Expanding φ^i near $\phi^i = const$ one can find the analog of (19) for the Fermi case (48) and thus to compute the leading contribution of fermions in β , γ in (20). The result is that fermions produce the vanishing two-loop contribution to (20), i.e. the analog of (22) for the Fermi string case is

$$\gamma = \frac{1}{4}\beta = \frac{1}{36\pi}\cdot\frac{3}{2}D - \frac{\hbar\alpha'}{64\pi}\mathcal{R} + O(\alpha'^2\hbar^2\mathcal{R}^2) \tag{49}$$

All the following steps are the same as in the Bose case, except that $\varkappa$ in (30), (35) is now [31], [36] $\varkappa_F = \frac{1}{4}(D-10)$ (there is the additional contribution of the gravitino gauge ghost determinant). A_{ij} is again included by the "shift" (24). We would like to remark that the known finiteness properties of the supersymmetric σ-models (see e.g. [17]) may be important for the Fermi string effective field theory (in particular, may open a possibility of an exact computation of Γ).

Next, we would like to suggest a more general interpretation of the effective action Γ in (7), (8). Let us forget about the string theory and consider (6) as a basic definition of an effective action in a fundamental theory of an infinite number of fields with the massless ones being the metric and the antisymmetric tensor (a hope is that the number of massless fields in the supersymmetrical case will be sufficient for a low-energy correspondence with the "standard model"). The "internal" variables which are quantized in this theory are the coordinates φ^i (and also the internal metric). The "classical" space-time coordinates ϕ^i appear as mean values of $\varphi^i(x)$ (see (10), (11)). All "ordinary" fields are not explicitly quantized but their quantum dynamics is accounted for through their dependence on the fluctuating space-time coordinates.

A reasonable point of view is that all we need to know from a fundamental "quantum gravity" theory is an effective action $\Gamma[G_{ij}, \psi]$ (G_{ij} is a background metric and ψ stands for the background values of all other fields) that should satisfy several conditions: (i) it should be well-defined, i.e. finite, calculable, etc. (ii) it should possess a consistent unitary low-energy limit, i.e. should contain the Einstein term and kinetic terms for ψ with correct physical signs (no ghosts and tachyons, positive energy, etc.) (iii) it should be suited for a study of fundamental problems of a Planck scale physics (including quantum cosmology and compactification of extra dimensions).

The Γ in eq. (6) may be viewed as a candidate for such an effective action. It is free of the usual problems of effective actions for ordinary field theories: Γ is manifestly covariant, "gauge independent" (no gauge is needed to be fixed for the D-dimensional metric) and ultraviolet finite. There is no also other problems (like indefiniteness of the classical euclidean action) arizing in a naive quantization of the Einstein theory itself.

Adopting this general point of view on the action in (7) one can raize the following question: why the action (8) standing in the exponent in (7) is given by a two-dimensional integral, i.e. corresponds to one- dimensional extended objects - string? Historically the string theory was first developed in the context of strong interaction theory and only then it was understood that the presence of "graviton" in the string excitation spectrum suggests a more fundamental role of strings. Nowdays it is clear that a quark-antiquark pair connected by an open string is probably an adequate description of mesons. But there seems to be no general arguments distinguishing strings as elementary objects for construction of a fundamental theory based on (7), (8). In fact, it is straightforward to write down a generalization of

(7), (8) to the case of closed extended objects of dimension d-1 = $0, 1, 2, \ldots$ (particle, string, membrane, ...) (cf. (7), (13))

$$\Gamma[\Phi, G_{ij}, A_{i_1 \ldots i_d}, \ldots] = \sum_{\text{topologies}} \int [dg_{\mu\nu}] \int [d\varphi^i] \, e^{-\frac{1}{\hbar} I} ,$$

$$I = \int d^d x \sqrt{g} \, \Phi(\varphi) + M^d \int d^d x \left[\tfrac{1}{2} \sqrt{g} \, g^{\mu\nu} \partial_\mu \varphi^i \partial_\nu \varphi^j \, G_{ij}(\varphi) + \right. \quad (50)$$
$$\left. + \frac{1}{d!} \varepsilon^{\mu_1 \ldots \mu_d} \partial_{\mu_1} \varphi^{i_1} \ldots \partial_{\mu_d} \varphi^{i_d} A_{i_1 \ldots i_d}(\varphi) \right] + \ldots .$$

Here $\mu, \nu = 1, \ldots, d$, $i, j = 1, \ldots, D$, $M = (2\pi\alpha')^{-1/2}$ is a "fundamental mass" and dots stand for "higher spin" tensor field terms ($\int \partial\varphi^{i_1} \ldots \partial\varphi^{i_N} B_{i_1 \ldots i_N}$). The action I (and hence $\Gamma = \int d^D\phi \, \mathcal{L}(\Phi(\phi), \ldots)$, cf. (11)) possesses only two kinds of "external" gauge invariances: the D-dimensional covariance (which is simply a consequence of the d-dimensional covariance of I) and the abelian gauge invariance $\delta A_{i_1 \ldots i_d} = \partial_{[i_1} \lambda_{i_2 \ldots i_d]}$. As a result, we deduce that the spectrum of excitations in a theory of closed $d-1 > 0$ -dimensional objects should contain only two massless (gauge) fields (represented by the metric and the antisymmetric rank d-tensor) in addition to the infinite number of massive fields. Thus any theory based on extended objects of an arbitrary dimension should contain (when interpreted in four dimensions) <u>massless</u> fields <u>only</u> of spins $S \leq 2$. The point is that "higher spin" fields in (50) lack the corresponding gauge groups needed to ensure their masslessness in Γ. The distinguished role played by G_{ij} and $A_{i_1 \ldots i_d}$ is due to the existence of only <u>two</u> covariant objects ($g_{\mu\nu}$ and $\varepsilon^{\mu_1 \ldots \mu_d}$) on arbitrary finite dimensional riemann manifold. To get massless higher spin fields we probably need a new (infinite dimensional?) geometry with higher symmetric invariants.

There of course remains a number of technical problems connected with the effective action (50) for $d > 2$ (some of them are: how to sum over topologies?; how to compute a "first stage" effective action W (16) in the absence of the Weyl invariance of I for $d>2$?; is there a "duality" between states propagating in "loops" and the external fields in (50)? Leaving aside these complications one can try to guess a low energy ($M\to\infty$) structure of Γ (50). In analogy with the string case it is reasonable to suppose that the $1/M$ expansion of Γ looks like (cf.(40))

$$\Gamma[\Phi, G, A, \dots] \sim M^D \int d^D\phi \sqrt{G}\, \Big\{ V_0(\Phi) + \frac{1}{M^2} V_1(\Phi)\partial_i\Phi\,\partial^i\Phi + \frac{1}{M^2} V_2(\Phi)\mathcal{R} + \frac{1}{M^2} V_3(\Phi) F^2_{i_1\dots i_{d+1}} + O(1/M^4)\Big\} \tag{51}$$

Then the ground state values of the fields are again likely to be: Φ = const, G_{ij} = { block-diagonal metric, corresponding to a D-dimensional space-time being a product space containing one or several $(d+1)$-dimensional sphere or (anti) de Sitter factors } and $F_{i_1\dots i_{d+1}} \sim \varepsilon_{i_1\dots i_{d+1}}$ (for indices from $(d+1)$ -subspaces). Thus compactification to four dimensions is prefered in the case of membranes ($d=3$). We would like to note that "four-dimensional" compactification may be also prefered already in the superstring case, where several (second and third rank) antisymmetric tensors will appear in (50), (51). Thus a generalization of our approach to the superstring case remains the important problem for future.

ACKNOWLEDGEMENTS

One of the authors (A.A.Ts.) acknowledges discussion with S.M.Apenko, A.M.Semikhatov, I.V.Tyutin, M.A. Vasiliev and B.L.Voronov.

Note Added:

After this paper was written we have learned about the preprint [37] where a "phenomenological" approach to a study of a space compactification in the Bose string theory is discussed. It seems that the "first principles" approach to a field-theoretic description of a string theory developed in our paper provides a consistent framework for solution of a ground state problem (including space compactification) in this theory and gives answers to the questions raised in ref. 37/.

REFERENCES

1. Schwarz, J.H., Phys. Repts 89, 223 (1982). Green, M.B., Surveys in H.E.Physics 3, 127 (1983).
2. Scherk, J. and Schwarz, J.H., Nucl. Phys. B81, 118 (1974). Yoneya, T., Progr. Theor. Phys. 51, 1907(1974).
3. Polyakov, A.M., Phys. Lett. 103B,207 (1981).
4. Friedan, D., Lectures at Copenhagen Workshop (Oct. 1981); preprint EFI-82-50 (1982).
5. Fradkin, E.S. and Tseytlin, A.A., Annals of Phys.143, 413 (1982).
6. Durhuus, B., Olesen, P. and Petersen, J.L., Nucl. Phys. B198,157 (1982).
7. Alvarez, O., Nucl. Phys. B216,125 (1983).
8. Brink, L., Di Vecchia, P. and Howe, P.S., Phys. Lett. 65B,471 (1976).
9. Gervais, J.L. and Sakita, B., Phys. Rev. Lett. 30, 706 (1973). Mandelstam, S., Nucl. Phys. B64,205(1973).
10. Nepomechie, R.I., Phys. Rev. D25,2706 (1982).
11. Mandelstam, S., Phys. Repts. 13,259 (1974).
12. Rebbi, C., Phys. Repts. 12,1 (1974). Scherk, J., Revs. Mod. Phys. 47, 123 (1975).
13. Ademollo, M., D'Adda, A., D'Auria, R., Napolitano, E., Sciuto, S., Di Vecchia, P., Gliozzi, F., Musto, R. and Nicodemi, F., Nuovo Cim. 21A, 77 (1974).
14. Kalb, M. and Ramond, P., Phys. Rev. D9,2273 (1974).
15. Honerkamp, J., Nucl. Phys. B36,130 (1972).
16. Friedan, D., Nonlinear models in $2+\epsilon$ dimensions, Ph. D. Thesis, LBL-11517 (1980); Phys. Rev. Lett. 45, 1057 (1980).
17. Alvarez-Gaume, L., Freedman, D.Z. and Mukhi, S., Annals of Phys. 134, 85 (1981).
18. Brown, L.S. and Collins, J.C., Annals of Phys. 130, 215 (1980). Hathrell, S.J., Annals of Phys. 142, 34 (1982).
19. Duff, M.J., Nucl. Phys. B125, 334 (1977).
20. Jack, I. and Osborn, H., Nucl. Phys. B234,331 (1984).
21. Polyakov, A.M. and Wiegman, P.B., Phys. Lett. 131B, 121 (1983). Witten, E., Commun. Math. Phys. 92, 455 (1984).
22. Scherk, J. and Schwarz, J.H., Phys. Lett. 52B,347 (1974).
23. Curtright, T.L. and Zachos, C.K., Phys. Rev. Lett. 53, 1799 (1984).
24. Zamolodchikov, A.B., Phys. Lett. 117B,87 (1982).
25. Lovelace, C., Phys. Lett. 135B,75 (1984).
26. Onofri, E. and Virasoro, M.A., Nucl. Phys.B201,159 (1982).

27. Onofri, E., Commun. Math. Phys. 86,321 (1982).
28. Bardakci, K. and Halpern, M.B., Nucl. Phys. B96, 285 (1975). Bardakci, K., Nucl. Phys. B133,297 (1978),
29. Freund, P.G.O. and Rubin, M.A., Phys. Lett. 97B, 223 (1980).
30. Deser, S. and Zumino, B., Phys. Lett. 65B, 369 (1976).
31. Polyakov, A.M., Phys. Lett. 103B,211 (1981).
32. Freedman, D.Z. and Townsend, P.K., Nucl. Phys. B177, 282 (1981).
33. Yamagishi, K., Annals of Phys. 150, 439 (1983).
34. Gliozzi, F., Scherk, J. and Olive, D.I., Nucl. Phys. B122, 253 (1977).
35. Green, M.B. and Schwarz, J.H., Phys. Lett. 136B, 367 (1984); Nucl. Phys. B243, 285 (1984).
36. Fradkin, E.S. and Tseytlin, A.A., Phys. Lett. 106B, 63 (1981).
37. Freund, P.G.O., Oh, P. and Wheeler, J.T., String induced space compactification, Fermi Institute preprint, EFI 84/14 (1984).

Additional note

In eq.(8) we have missed the term corresponding to the "dilaton" (the massless scalar mode of the closed string spectrum), $\Delta I = \frac{1}{4\pi}\int d^2x \sqrt{g} R\, \phi(\varphi(x))$. Thus the dilaton ϕ couples to the scalar curvature of g_{ab} . As a result, a constant shift of $\phi \to \phi + a$ produces a redefinition of the dimensionless closed string coupling constant $g_c = \exp(-6)$, $g_c \to g_c \exp(a)$ (see (7), (9)). From this follows an observation made by Witten that if the dilaton's vacuum value is fixed after an account of loop corrections, g_c is not an independent coupling constant. It is straightforward to establish the contribution of ϕ in the low-energy effective action (following the steps analogous to (25)-(29)). In the tree approximation (χ = 2) and ignoring a non-trivial dependence on ρ (e.g. taking D = 26) we get (setting Φ =0)

$$\Gamma \sim g^{-2} M^D \int d^D\phi \sqrt{G}\, e^{-2\phi}\left\{1 + \frac{\alpha'}{4}\left[R + 4(\partial_i\phi)^2 - \frac{1}{12}F_{ijk}^2\right] + \ldots\right\} \tag{A1}$$

Thus ϕ indeed couples as a dilaton. After the Weyl rescaling $G_{ij} \to G_{ij}\exp[4\phi/D-2]$ we obtain (this rescaling is also necessary in (8) in order to obtain the standard dilaton vertex)

$$\Gamma = \frac{1}{K^2}\int d^D\phi \sqrt{G}\left\{\frac{4}{\alpha'} e^{4\phi/D-2} + \left(-R + \frac{4}{D-2}(\partial_i\phi)^2 + \frac{1}{12}F_{ijk}^2 e^{-8\phi/D-2} + \ldots\right)\right\}, \tag{A2}$$

$$K \sim g_c (\alpha')^{(D-2)/4}$$

Up to the potential term this result precisely agrees with the $\alpha' \to 0$ action deduced from the 3-point string amplitudes in [2, 22] . Note that our covariant off-shell method makes possible to establish the full non-polynomial form of the action, while in [2,22] it was found only to the third order in the fields. Eq. (A2) can be generalized to include the contribution of the vector fields (in the case of the theory containing both open and closed strings the effective action is defined by summing first over all surfaces with a given boundary and then over all boundaries; the vector field couples to the boundary "particle" so that $\Gamma[A_i,...]$ is given by the average of the Wilson factor $tr\, P\exp \int d\varphi^i A_i(\varphi)$). In this case the curly bracket in (A1) receives the additional term $\sim g_0^{-2}\, tr\, F_{ij}^2$ ($g_0 = \exp(-\frac{1}{2}\sigma)$ is the open string coupling constant) and hence (A2) generalize to

$$\Gamma = \frac{1}{k^2}\int d^D\phi \sqrt{G}\Big\{-\mathcal{R} + \frac{4}{D-2}(\partial_i \phi)^2 + \frac{1}{12}F_{ijk}^2 e^{-8\phi/D-2} - \frac{k^2}{8g_{YM}^2}\, tr\, F_{ij}^2 e^{-4\phi/D-2} + \ldots\Big\}, \quad g_{YM} \sim g_0(\alpha')^{\frac{D-4}{4}} \tag{A3}$$

For D = 10 these couplings of ϕ are exactly the same as present in N=1 D=10 supergravity - super Yang-Mills system known to correspond to the $\alpha' \to 0$ limit of type I superstrings. This coincidence is partly accidental because for the non-oriented type I string there is no $\partial x^i \partial x^j A_{ij}$ - coupling (the antisymmetric tensor appears from the fermionic sector, $\bar\theta\gamma^{ij}..\theta A_{ij}$). F_{ijk} - ϕ - coupling in (A2) is also the same present in N=2, D=10 supergravity.

Having included the dilaton we may reconsider the ground state problem of sect.4 setting $\Phi = 0$ and ϕ =const in the vacuum. Then due to the potential term there are non-trivial vacuum solutions, equivalent to those following from (43) (now with $b_1 = 4$), e.g. $M^D = (adS)_3 \times R^{D-3}$. It is remarkable that though higher order terms in (A1) ($\sim \alpha'^k \mathcal{R}^k$) all are of the same order on such a solution (in terms of unrescaled metric $\alpha' R \sim 1$) the string action (8) (with fields equal to their vacuum values) corresponds to the sigma-model with the Wess-Zumino term with the coefficients just the needed to have zero β -function (Witten [21], see also [23]). Hence this background may satisfy the full (unexpanded in α') equations (41). Yet a more promising approach to compactification (in a superstring context) seems to look for classical solutions with unfixed compactification scale r and try to fix r by account of loop corrections (non-analytic in α') in order to obtain a hierarchy $r \gg \sqrt{\alpha'}$ of a kind of $r \sim \sqrt{\alpha'}\exp 1/g_0^2$

DYNAMICAL SUPPRESSION OF TOPOLOGICAL CHANGE

Bryce DeWitt

The University of Texas at Austin

INTRODUCTION

This is a brief report of some work that I have carried out with my student, Arlen Anderson, concerning spontaneous changes of 3-space topology.

The idea that the topology of 3-space might undergo quantum fluctuations was first put forward in 1957 by J.A. Wheeler[1], who reasoned as follows: If one attempts to measure the average electric field in a spacetime volume L^4 in a vacuum at $0^\circ K$ (e.g., by observing during a time L the motion of a charged test body of volume L^3) one will not generally obtain a vanishing result. In a sequence of such measurements one will obtain a statistical distribution of values with magnitudes peaked around $1/L^2$ ($\hbar = c = 1$).

Similarly, quantization of the linearized theory of gravity suggests that if one attempts to measure the Riemann tensor averaged over a spacetime volume L^4 in a vacuum at $0^\circ K$ (e.g., by observing during a time L the change of strain of a gravitational attenna[2] of

[1] J.A. Wheeler in Conference on the Role of Gravitation in Physics, Chapel Hill, North Carolina, January 18-23, 1957 (WADC technical report 57-216, ASTIA document No. AD 118180).

[2] For the design of such measurement see B.S. DeWitt in Gravitation: An Introduction to Current Research, L. Witten, ed. (Wiley, 1962).

volume L^3) one will probably not obtain zero but rather a value of magnitude in the neighborhood of λ_P^2/L^4 (λ_P = Planck length) corresponding to a spacetime radius of curvature of order L^2/λ_P. Wheeler noted that this radius of curvature becomes comparable to L itself when $L \approx \lambda_P$. Although this is precisely the realm in which the linearized theory fails, the result suggests that the curvature at the Planck scale fluctuates so violently as practically to cause portions of space to "pinch off" or to become multiply connected, developing structures such as "wormholes". Wheeler proposed that this is indeed what happens.

DIFFICULTIES

Wheeler's proposal faces two immediate difficulties. First of all, if wormholes can develop, then wormholes themselves can acquire wormholes ad infinitum, and space can take on the character of a "foam". But, on a coarse scale ($\gg\lambda_P$), a foam can appear to have a dimensionality arbitrarily greater than that of the underlying manifold out of which it is built. The observed stability of the dimensionality of space is thus called into question, and any theory that predicts quantum fluctuations in the topology of space must also explain why nature seems so rigidly to prefer three dimensions.

Secondly, if a wormhole forms, the two entrances to it can move arbitrarily far apart before the wormhole dissolves. Indeed, classically, if time is reckoned by means of a foliation of spacetime into maximal spacelike hypersurfaces, then, once the two entrances get far enough apart the tunnel that connects them will "never" dissolve, and the distance between the entrances can become vastly greater on the "outside" than through the tunnel itself. Of course, one must expect quantum fluctuations to violate classical laws. But if a wormhole can fluctuate out of existence when its entrances are far apart ----- a transition that seemingly ought to take place readily enough ----- then, by the principle of microscopic reversibility, the fluctuation into existence of a wormhole having widely separated entrances ought to occur equally readily. This means that every region of space must, through the quantum principle, be potentially "close" to every other region, something that is certainly not obvious from the operator field equations which, like their classical counterparts, are strictly local.

It is difficult to imagine any way in which widely separated regions of space can be "potentially close" to each other unless spacetime itself is embedded in a convoluted way in a higher dimensional manifold. Additionally, a dynamical agency in that higher dimensional manifold must exist which can transmit a sense of that closeness. A model of this kind could well be called a <u>string theory</u> of spacetime. My chief objection to a string theory of spacetime is that it is anti-Riemann, anti-Clifford, and anti-Einstein. It was Riemann who first insisted that a curved manifold needs no higher dimensional embedding space for its definition. Clifford and Einstein went even further, insisting that spacetime itself is <u>all</u>. In a string theory of spacetime the Universe is not everything; there is something physical <u>outside</u>.

Even if one embraces the string idea one is by no means free of difficulties. The idea by itself gives no clue to the nature of the "closeness transmitting" agency in embedding space. The trouble with changing the rules of a game (in this case the standard rules of quantum gravity) is that there are no rules about changing the rules. The same difficulty arises with other suggestions that have been put forward from time to time, such as the proposal that one should cease to regard spacetime as a differentiable manifold at scales below the Planck length. If not a differentiable manifold then <u>what</u>? Since speculation is generally profitless I believe that the only useful procedure is to challenge the theory on its own ground, to see where its real weaknesses and strengths lie. One might then learn something.

CONTINUITY AND THE EUCLIDEAN OPTION.

It is a general rule in the quantum theory that the classically forbidden regions into which a system can fluctuate can always be connected by continuous paths (in configuration space) to classically allowed regions. There is no reason to suppose that quantum gravity is an exception to this rule. But this means that if one wishes to calculate, by means of a Feynman functional integral, the amplitude for a transition between states having different topologies, then one must have available a set of continuous field histories that connect those topologies. Now it is well known that the metric of any Lorentzian manifold that has no closed timelike curves, or other causal anomalies, necessarily becomes singular at points where it spatial sections undergo topological change. How does one handle these singularities?

One answer is to regard Lorentzian spacetime as an analytic continuation of a Euclidean manifold. In a Euclidean manifold the metric need not become singular at points where cross sections undergo topological change. The search for interpolating histories becomes a standard <u>cobordism</u> problem. This "Euclidean option" thus converts the problem of topological change into what looks like a standard quantum tunneling problem. However, it has difficulties of its own. The first is the well known failure of the Euclidean action to be positive definite: Conformal changes in the metric contribute negatively to it. The second is the seldom emphasized and much less well known fact that in applications of the Euclidean option one generally avoids asking for the amplitude for a topological change to be <u>observed</u>. It seems to be very difficult to ask for this, and hence, because physicists like to solve easy problems, a real test of Wheeler's idea has never been achieved.

In my view it is scandalous that, after more than a quarter of a century, Wheeler's idea remains untested. The aim of the present work is to take a tentative step towards improving the situation. Euclidean techniques will be avoided. Historically, most problems in quantum mechanics that can be illumined by analytic continuation were first solved by other more straightforward methods. The advantage of proceeding in a direct manner is that the physics remains in the foreground and our understanding is enhanced. The results described in the following paragraphs turn out to suggest that Wheeler's topological transitions may be both energetically and dynamically forbidden.

A (1+1)-DIMENSIONAL MODEL. THE TROUSERS TOPOLOGY.

The full range of topological varieties that can be connected to a given 3-manifold by interpolating 4-manifolds is daunting. Where shall we begin? It seems obvious that we should begin by making drastic simplifications, for example, by reducing the dimensionality of spacetime from 4 to 2. The trouble with this is that the gravitational field in a 2-dimensional spacetime has -1 degrees of freedom! A direct analog of Einstein's theory simply does not exist in two dimensions.

We shall avoid this difficulty for the moment by replacing the gravitational field by a linear massless scalar field. This field will not be regarded as describing a dynamics of the underlying spacetime; the metric of spacetime will be taken as given a priori. The topology we shall choose is shown in Figure 1. This may be called the trousers topology: Space, which is initially diffeomorphic to the circle, splits into two circles: $S^1 \to S^1 + S^1$. Our initial task will not be to ask how this topological transition can come about but rather to determine reasonable laws of propagation for the scalar field on such a background.

The only serious problem confronting us is how to deal with the metric singularity that inevitably occurs in the crotch region. A possible solution is to allow the metric to become complex (in a suitable patch) in this region. The metric then need not be singular and one may study the propagation problem as an ordinary problem in the theory of partial differential equations. A difficulty with this however, is that the scalar field, if initially real in the trunk region, will not generally remain real in the legs. Moreover, if the field is quantized, the coefficients connecting "in" mode functions (in the trunk) to "out" mode functions (in the legs) will not satisfy the standard Bogoliubov relations, and unitarity will be violated.

Unitarity could, of course, be regained by allowing the metric in the end to become real and the singularity to reappear. But this would be equivalent to simply excising the singular point from spacetime and specifying how waves are to propagate around it (different laws corresponding to different ways in which the imaginary part of the metric is allowed to tend to zero). This is in fact the procedure that will be adopted. In two dimensions there is also no need to introduce nonzero curvature. The purely topological aspect of the effects to be studied can therefore be cleanly isolated.

Figure 2 shows a convenient way to display the flat trousers manifold. The metric is everywhere Minkowskian and static except at the singular point, which is indicated by crosses. Below the crosses in the trunk region the two sides of the figure are to be identified. Similar identifications occur across each leg above the crosses. The circumferences of the legs need not be equal, but their sum must equal the circumference of the trunk.

Figures 3a and 3b show one possible law of propagation past the singular point, which may be called the <u>shadow law</u>. In figure 3a an "in" mode function propagating to the right splits into components propagating to the right in each leg. Although continuous in the trunk region such mode functions generally have discontinuities (located along the dotted lines) in the legs. Figure 3b shows the corresponding discontinuities possessed by right-propagating "out" mode functions associated with either leg. Every "out" mode function is continuous in each leg (vanishing in one of them) but has discontinuities in the trunk region. Mirror-image figures can be drawn for left-propagating mode functions.

THE "IN" AND "OUT" STATE VECTOR SPACES.

The scalar field can be expressed in terms of either "in" or "out" mode functions as follows:

$$\begin{aligned}\underset{\sim}{\phi}(x) &= \sum_{\underset{\sim}{p}} [u_{in}(x,\underset{\sim}{p})\underset{\sim}{a}_{in}(\underset{\sim}{p}) + u_{in}^*(x,\underset{\sim}{p})\underset{\sim}{a}_{in}^*(\underset{\sim}{p})] \\ &\quad + u_{0in}(x)\underset{\sim}{Q}_{in} + v_{0in}(x)\underset{\sim}{P}_{in} \qquad (1a)\\ &= \sum_{\underset{\sim}{p}'} [u_{out}(x,\underset{\sim}{p}',R)\underset{\sim}{a}_{out}(\underset{\sim}{p}',R) + u_{out}^*(x,\underset{\sim}{p}',R)\underset{\sim}{a}_{out}^*(\underset{\sim}{p}',R)] \\ &\quad + u_{0out}(x,R)\underset{\sim}{Q}_{out}(R) + v_{0out}(x,R)\underset{\sim}{P}_{out}(R) \\ &\quad + \sum_{\underset{\sim}{p}''} [u_{out}(x,\underset{\sim}{p}'',L)\underset{\sim}{a}_{out}(\underset{\sim}{p}'',L) + u_{out}^*(x,\underset{\sim}{p}'',L)\underset{\sim}{a}_{out}^*(\underset{\sim}{p}'',L)] \\ &\quad + u_{0out}(x,L)\underset{\sim}{Q}_{out}(L) + v_{0out}(x,L)\underset{\sim}{P}_{out}(L) \qquad (1b)\end{aligned}$$

In addition to standard Fock-space modes, massless fields in compact universes always possess zero-frequency modes. Here the $u_{in}(x,\underset{\sim}{p})$ are the Fock-space mode functions associated with the "in" region (the trunk) and $u_{0in}(x)$ and $v_{0in}(x)$ are zero-frequency mode functions. (u_{0in} is actually constant in space and time while v_{0in} is constant in space and varies linearly with time.) The $\underset{\sim}{a}_{in}(\underset{\sim}{p})$ and $\underset{\sim}{a}_{in}^*(\underset{\sim}{p})$ satisfy the standard commutation relations for annihilation and creation operators for particles of momentum $\underset{\sim}{p}$. They commute with the operators $\underset{\sim}{Q}_{in}$ and $\underset{\sim}{P}_{in}$, which are Hermitian and canonically conjugate. The momentum $\underset{\sim}{p}$ is quantized by the periodic boundary conditions in the trunk region, except that the value $\underset{\sim}{p} = 0$ is excluded.

The ground state vector in the trunk region is defined by

$$\underset{\sim}{a}_{in}(\underset{\sim}{p})|in, vac\rangle = 0 \quad , \qquad \underset{\sim}{P}_{in}|in, vac\rangle = 0 \quad . \qquad (2)$$

Although the symbol "vac" is used here, $|\text{in, vac}\rangle$ is not a true Fock-space vacuum state vector, for the second of equations (2) prevents it from being normalizable. In fact, the "in" state vector space is not a true Fock-space but rather the direct product of a Fock-space and an ordinary Hilbert space of complex functions of a single real variable Q_{in}.

The operators $\mathbf{a}_{out}(\mathbf{P}',R)$, $\mathbf{a}_{out}^{*}(\mathbf{p}',R)$, $\mathbf{Q}_{out}(R)$, $\mathbf{P}_{out}(R)$, $\mathbf{a}_{out}(\mathbf{p}'',L)$, $\mathbf{a}_{out}^{*}(\mathbf{p}'',L)$, $\mathbf{Q}_{out}(L)$, $\mathbf{P}_{out}(L)$ have analogous properties and interpretations in the leg regions. The momenta $\mathbf{p}'$ and $\mathbf{p}''$ are quantized by the periodic boundary conditions in the right and left legs respectively. Again the values $\mathbf{p}' = 0$ and $\mathbf{p}'' = 0$ are excluded. The ground state vector in the leg regions is defined by

$$\mathbf{a}_{out}(\mathbf{p}',R)|\text{out, vac}\rangle = 0 \quad , \qquad \mathbf{P}_{out}(R)|\text{out, vac}\rangle = 0 \quad ,$$

$$\mathbf{a}_{out}(\mathbf{p}'',L)|\text{out, vac}\rangle = 0 \quad , \qquad \mathbf{P}_{out}(L)|\text{out, vac}\rangle = 0 \quad . \tag{3}$$

Alternatively, one may introduce ground state vectors associated with each leg and regard $|\text{out, vac}\rangle$ as their Cartesian product. The "out" state vector space is the direct product of the Fock spaces and Hilbert spaces for the two legs.

GROUND STATE ENERGY AND UNITARY INEQUIVALENCE

The renormalized energy operator (i.e., with the vacuum energy of infinite Minkowski space subtracted off) for the scalar field in the trunk region is

$$\mathbf{E}_{in} = \sum_{\mathbf{p}} |\mathbf{p}|\mathbf{q}_{in}^{*}(\mathbf{p})\mathbf{a}_{in}(\mathbf{p}) + \frac{1}{2}\mathbf{P}_{in}^{2} + C \quad , \tag{4}$$

where C is the Casimir energy in the trunk. The corresponding operator for the legs is

$$\underset{\sim}{E}_{out} = \sum_{\underset{\sim}{p}'} |\underset{\sim}{p}'|\underset{\sim}{a}_{out}^{*}(\underset{\sim}{p}',R)\underset{\sim}{a}_{out}(\underset{\sim}{p}',R) + \frac{1}{2}\underset{\sim}{P}_{out}(R)^{2} + C_{R}$$

$$+ \sum_{\underset{\sim}{p}''} |\underset{\sim}{p}''|\underset{\sim}{a}_{out}^{*}(\underset{\sim}{p}'',L)\underset{\sim}{a}_{out}(\underset{\sim}{p}'',L) + \frac{1}{2}\underset{\sim}{P}_{out}(L)^{2} + C_{L} \quad ,(5)$$

where C_R and C_L are the Casimir energies in the right and left legs respectively. Evidently

$$\underset{\sim}{E}_{in}|in, vac\rangle = C|in, vac\rangle \quad ,$$

$$\underset{\sim}{E}_{out}|out, vac\rangle = (C_R + C_L)|out, vac\rangle \qquad (6)$$

C, C_R and C_L are generally all negative (if the scalar field is "untwisted") and both C_R and C_L are more negative than C. Therefore the Casimir energy is not conserved.

The situation is actually more dramatic than this. Because of the topological change the trousers manifold is not stationary, and there is no reason for $\underset{\sim}{E}_{in}$ and $\underset{\sim}{E}_{out}$ to be equal. Their expectation values, in fact, turn out to differ from one another, in all states, by infinite amounts. One way (the hard way) to show this is to compute the Bogoliubov coefficients that connect the "in" and "out" mode functions (or alternatively the operators $\underset{\sim}{a}_{in}$, $\underset{\sim}{a}_{in}^{*}$, $\underset{\sim}{Q}_{in}$, $\underset{\sim}{P}_{in}$ with the operators $\underset{\sim}{a}_{out}$, $\underset{\sim}{a}_{out}^{*}$, $\underset{\sim}{Q}_{out}$, $\underset{\sim}{P}_{out}$) and to reexpress $\underset{\sim}{E}_{out}$ in terms of the "in" operators, or vice versa. This has been done, although I shall refrain from displaying the not particularly illuminating details.

Another result that can be obtained directly from the Bogoliubov coefficients is

$$\langle out,vac|in,vac\rangle = 0 \quad . \tag{7}$$

It is a corollary of this result that the "in" and "out" state-vector spaces are unitarily inequivalent: A state having a finite number of "in" particles has an infinite number of "out" particles with an infinite amount of "out" energy, and vice versa.

The easy way to demonstrate the above results is to look at ground state expectation values of the renormalized stress-energy tensor. These may be obtained from the Wightman functions of the theory by suitable limiting procedures. Actually, Wightman functions do not exist for massless fields in 1+1 dimensions. However, expectation values of the stress-energy tensor involve spacetime derivatives of Wightman functions, and these do exist.

Figure 4 shows the distribution of energy in the "in" ground state, i.e., the structure of $\langle in,vac|\underset{\sim}{T}^{00}_{ren}|in,vac\rangle$. In the trunk region the energy density is simply the Casimir energy density C/L, where L is the circumference of the trunk. In the legs the energy density is also equal to C/L everywhere, except along the 45° lines emanating from the crotch singularity. Along these lines the energy density behaves like the square of the delta function and hence is not only infinite there but also corresponds to infinite <u>total</u> energy. That is to say, the crotch singularity produces an infinitely bright flash that propagates forward in time into the legs.

The reason for the infinitely bright flash is easily seen if one expresses $\langle in,vac|\underset{\sim}{T}^{00}_{ren}|in,vac\rangle$ as a sum over "in" mode functions. In the legs the "in" mode functions have discontinuities. When these functions are differentiated the discontinuities give rise to delta functions. Since the terms of the mode sum for $\langle in,vac|\underset{\sim}{T}^{00}_{ren}|in,vac\rangle$ are bilinear in differentiated mode functions, the square of the delta function automatically appears.

LINEARIZED GRAVITY IN A 3-TORUS.

One can make the above results bear directly on the real problem of dynamically induced topological change by increasing the dimensionality of spacetime from two to four and replacing the scalar field by the gravitational field. We shall work in a 3-torus universe and write the metric operator in the form

$$\underset{\sim}{g}_{\mu\nu} = \eta_{\mu\nu} + \underset{\sim}{\phi}_{\mu\nu} \quad , \tag{8}$$

where $\eta_{\mu\nu}$ is the Minkowski metric. In linear approximation one expands $\underset{\sim}{\phi}_{\mu\nu}$ in the form

$$\begin{aligned}\underset{\sim}{\phi}_{\mu\nu} = &\sum_{\underset{\sim}{p},s} [u_{\mu\nu}(x,\underset{\sim}{p},s)\underset{\sim}{a}(\underset{\sim}{p},s) + u_{\mu\nu}{}^{*}(x,\underset{\sim}{p},s)a^{*}(\underset{\sim}{p},s)] \\ &+ \sum_{A} [u_{0\mu\nu}(x,A)\underset{\sim}{Q}_{A} + v_{0\mu\nu}(x,A)\underset{\sim}{P}_{A}] \\ &+ \text{gauge terms}.\end{aligned} \tag{9}$$

Here s is a helicity label, which can assume the values +2 and -2, and $\underset{\sim}{p}$ is a 3-momentum quantized by the boundary conditions of the 3-torus, the value $\underset{\sim}{p} = 0$ being excluded. The $u_{\mu\nu}(x,\underset{\sim}{p},s)$ are Fock-space mode functions having the form

$$u_{\mu\nu}(x,\underset{\sim}{p},\pm 2) = \mu_{P}{}^{-1}V^{-1/2}(2\omega)^{-1/2}e_{\mu}(\underset{\sim}{p},\pm 1)e_{\nu}(\underset{\sim}{p},\pm 1)e^{ip\cdot x} \tag{10}$$

where μ_P is the Planck mass, V is the volume of the 3-torus, $\omega = p^0 = |\underset{\sim}{p}|$, and the e_μ are complex vectors satisfying

$$\begin{aligned}&e_0(\underset{\sim}{p},\pm 1) = 0 \;, \quad \underset{\sim}{e}(\underset{\sim}{p},\pm 1) = \underset{\sim}{e}(\underset{\sim}{p},\mp 1)^{*} \;, \quad \underset{\sim}{p}\cdot\underset{\sim}{e}(\underset{\sim}{p},\pm 1) = 0 \;, \\ &\underset{\sim}{e}(\underset{\sim}{p},\pm 1)\cdot\underset{\sim}{e}(\underset{\sim}{p},\pm 1)^{*} = 1 \;, \quad \underset{\sim}{e}(\underset{\sim}{p}1)\cdot\underset{\sim}{e}(\underset{\sim}{p},\pm 1) = 0 \;.\end{aligned} \tag{11}$$

In addition to the Fock-space mode functions there are six independent zero-frequency mode functions given by

$$u_{0\mu\nu}(x,A) = \mu_P^{-1}V^{-1/2}\chi_{\mu\nu}(A) \;,$$

$$A = 0,1,2,3,4,5, \qquad (12)$$

$$v_{0\mu\nu}(x,A) = \mu_P^{-1}V^{-1/2}\chi_{\mu\nu}(A)x^0 \;,$$

where

$$\chi_{\mu\nu}(0) = \frac{1}{\sqrt{6}}\,(\eta_{\mu\nu} + \eta_{\mu 0}\eta_{\nu 0}) \qquad (13)$$

and the $\chi_{\mu\nu}(A)$ with A = 1,...5 are any five other constant real symmetric tensors satisfying

$$\chi_{\mu 0}(A) = 0 \;, \quad \chi_{\mu}{}^{\mu}(A) = 0 \;,$$

$$A,A'' = 1,...5. \qquad (14)$$

$$\chi_{\mu\nu}(0)\chi^{\mu\nu}(A) = 0 \;, \quad \chi_{\mu\nu}(A)\chi^{\mu\nu}(A') = \delta_{AA'}$$

$\chi_{\mu\nu}(0)$ describes pure dilation of the 3-metric, whereas the $\chi_{\mu\nu}(A)$ with A = 1,...5 describe pure shear.

The $\underset{\sim}{a}$'s and $\underset{\sim}{a}^*$'s are standard annihilation and creation operators and the $\underset{\sim}{Q}$'s and $\underset{\sim}{P}$'s form canonically conjugate Hermitian pairs. Since the static 3-torus possesses a timelike Killing vector field there is a gauge invariant conserved total energy operator. Its expression in terms of the above operators may be shown to be

$$\underset{\sim}{E} = \sum_{\underset{\sim}{p},s} \omega\, a^*(\underset{\sim}{p},s)\,a(\underset{\sim}{p},s)$$

$$+ \frac{1}{2}\left(-\underset{\sim}{P}_0^{\,2} + \underset{\sim}{P}_1^{\,2} + \underset{\sim}{P}_2^{\,2} + \underset{\sim}{P}_3^{\,2} + \underset{\sim}{P}_4^{\,2} + \underset{\sim}{P}_5^{\,2}\right)$$

$$+ \text{Casimir energy}\,.^3 \qquad (15)$$

The noteworthy feature of this expression is that the dilation mode contributes negatively to the total energy whereas all other modes contribute positively.

LINEARIZATION CONSTRAINT.

If one goes beyond the linearized theory one finds that the total energy, through the operator field equations, can be expressed as an integral over a 2-dimensional surface at spatial infinity.[3] Since in a compact universe such as the 3-torus there is no spatial infinity it follows that the total energy must vanish. In the classical theory this fact may be restated as follows: Let Φ be the space of solutions of the classical field equations and let η be the point in Φ corresponding to the flat 3-torus. Let $\eta+\phi$ be a point in Φ close to η. Then ϕ is a solution of the linearized field equations and may be regarded as the tangent vector at η to a curve in Φ, i.e., to a one-parameter family of solutions of the nonlinear field equations. Not all solutions of the linearized equations are tangent to curves in Φ. The only solutions of the linearized equations that are tangent to curves in Φ are those for which the total energy vanishes. Because of the spatial homogeneity of the present model one may show that the total 3-momentum too must vanish. This means

[3]B.S. DeWitt in Relativity, Groups and Topology II, DeWitt and Stora, eds. (North Holland, 1984). See especially pp. 454-457 and problem 95 on pp. 672-674.

that the coefficients in the decomposition (9) cannot all be chosen independently but must be subjected to quadratic constraints, one for each independent Killing vector field that the 3-torus possesses.

The existence of the energy constraint implies that if any Fock modes or shearing modes are excited then, to the extent that their energy content together with the Casimir energy is positive, the dilation mode must be excited too. This phenomenon is well known in relativistic cosmology. The universe as a whole reacts to the energy that is in it by expanding or contracting. It cannot stand still.

Quantum mechanically the only way to impose the energy constraint is as a condition on the state vector Ψ:

$$\underset{\sim}{E}\Psi = 0 . \tag{16}$$

There are also three other similar constraints corresponding to the condition of vanishing total 3-momentum.

THE TRANSITION $T^3 \to T^3 + T^3$.

Suppose now that the 3-torus undergoes a topological transition in which it splits into two 3-tori. One may imagine this split occurring along one of the principal axes of the 3-torus so that the transition may be expressed more precisely as $T^2 \times S^1 \to T^2 \times (S^1 + S^1)$. In this case there will be a submanifold of singular points, having the topology T^2. Again one must propose laws by which the (linearized) gravitational field propagates around these points.

Here a naive shadow law is no longer adequate since not all plane waves propagate parallel to the splitting axis. However, there is a simple alternative that is applicable also to the (1+1)-dimensional trousers model previously considered. Consider an "in" mode function along a line t = constant just above the singularity point in figure 3a. The Cauchy data are ϕ and $\partial\phi/\partial t$ along this line. The derivative $\partial\phi/\partial t$ contains a delta function in each leg, introduced by the discontinuities along the dotted lines. The alternative proposal is simply to eliminate these delta functions at the entrances to each leg. Initially ϕ and $\partial\phi/\partial t$ then possess only discontinuities but no delta functions, and this rule may be applied to the transition $T^3 \rightarrow T^3 + T^3$ also.

In the case of the trousers topology although $\partial\phi/\partial t$ possesses no delta function initially (according to the new rule) it immediately thereafter acquires one ----- four, in fact; two in each leg. Even though the "in" mode function was propagating to the right in the trunk region it now acquires, in the legs, discontinuities that propagate to both the left and the right from the singular point.

DYNAMICAL SUPPRESSION OF TOPOLOGICAL CHANGE.

Examination of yet more complicated topological transitions suggests that discontinuities in the Caucy data will always emanate from the singular points under almost any conceivable law of propagation. These discontinuities will immediately give rise to delta functions in the derivatives of the fields and to squares of delta functions in the stress-energy tensor (or pseudo-tensor in the case of gravity). From this it follows that the singular points will always be the source of infinitely bright flashes and that a change of topology will always be accompanied by infinite particle and energy production.

In the case of the gravitational field the energy constraint must hold both before and after the topological change. The production of a infinite amount of energy will cause the dilation mode to come instantly into play. Indeed the entire burden of compensating for the infinite energy falls on the dilation mode alone. This means that the universe must either expand or contract infinitely rapidly immediately after the change. It cannot expand infinitely rapidly because of causality requirements. It therefore collapses, aborting the topological change.

The collapse, in fact, will occur locally. In the full nonlinear theory the condition of vanishing total energy and momentum is replaced by an infinity of local constraints: one Hamiltonian constraint and three momentum constraints at each point of space. These constraints control not merely the total energy and momentum but the distribution of energy and momentum. The infinite energy density associated with field discontinuities is thus locally suppressed, which is to say that the Hamiltonian and momentum constraints are not merely manifestations of the diffeomorphism invariance of the theory but also serve locally to maintain everywhere the integrity of the topology of 3-space. Because these constraints are local it is obvious that the topology of 3-space is dynamically preserved even if 3-space is noncompact and its volume is infinite.

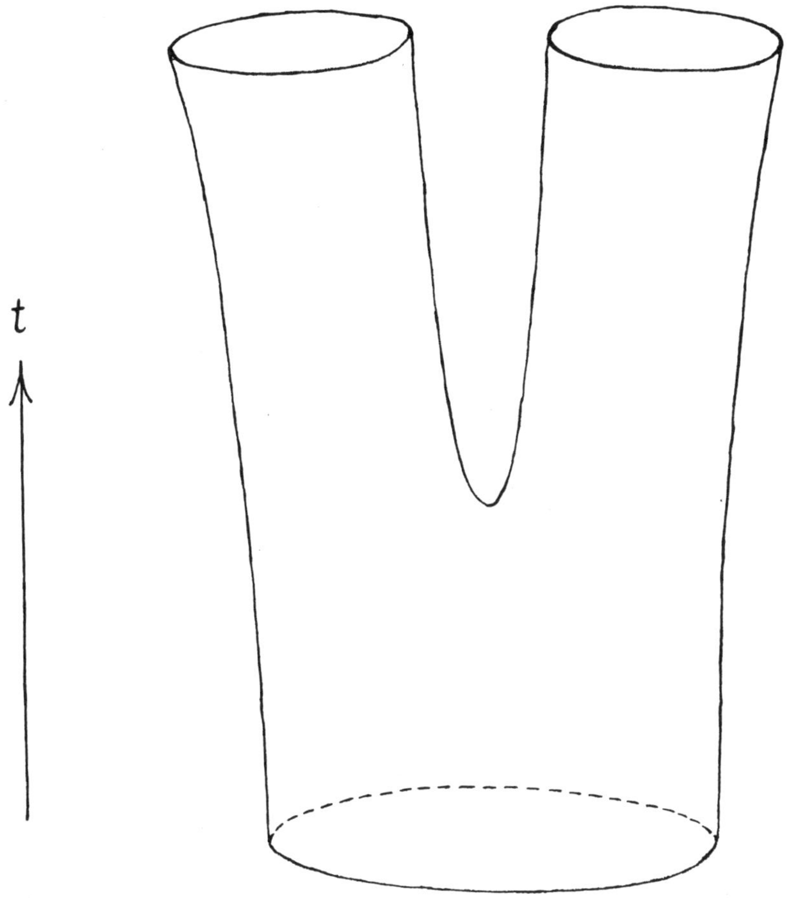

Fig. 1. The trousers topology, corresponding to the topological transition $S^2 \rightarrow S^1 + S^1$.

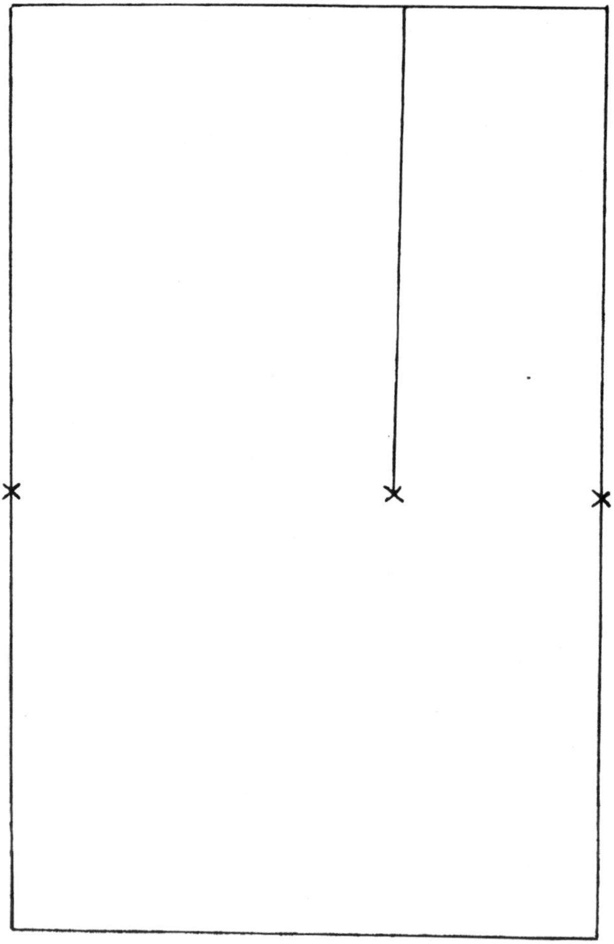

Fig. 2. Representation of the flat trousers topology. The crosses indicate the singular point. Below the crosses, opposite sides of the figure are to be identified. Above the crosses, identifications occur across each "leg".

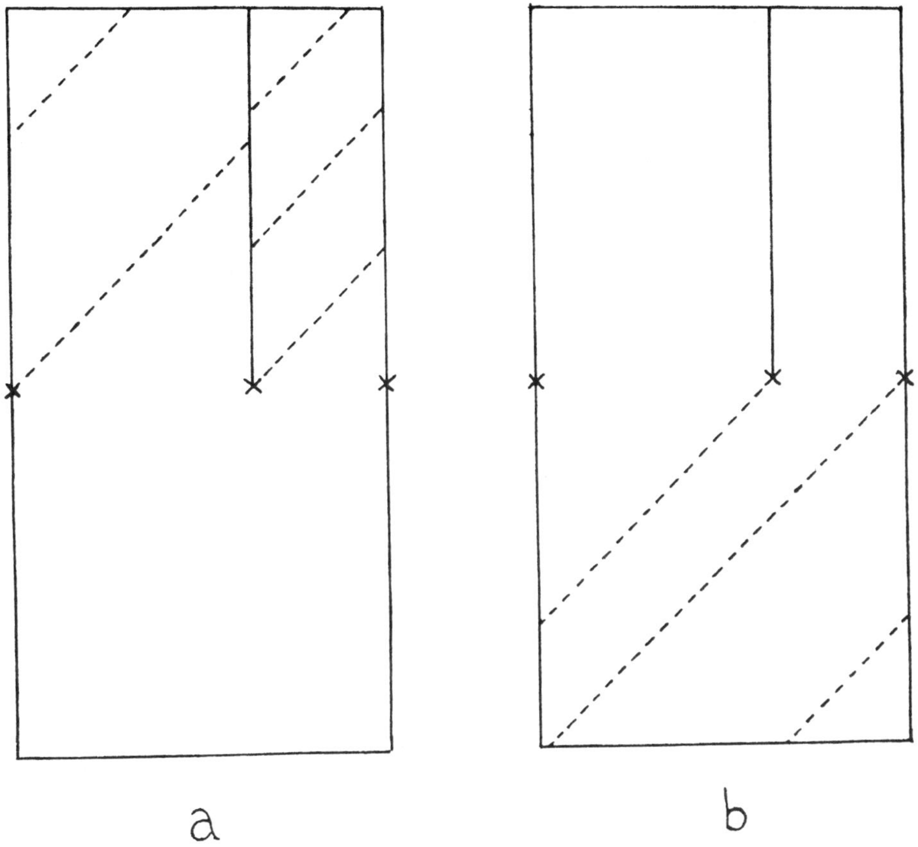

Fig. 3. Right-propagating mode functions: (a) "in" mode functions; (b) "out" mode functions. The dotted lines show the location of discontinuities.

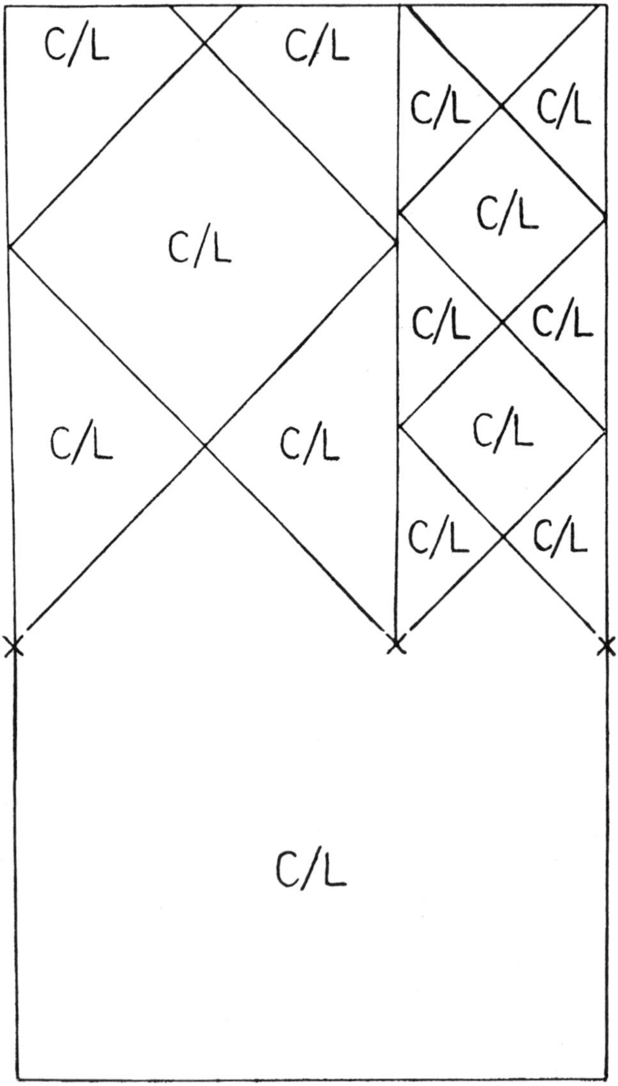

Fig. 4. Distribution of energy in the "in" ground state. The lines at 45° represent the infinitely bright flash emitted by the crotch singularity.

SIMPLICIAL QUANTUM GRAVITY

James B. Hartle
Department of Physics, University of California
Santa Barbara, California 93106, U.S.A.

ABSTRACT

Simplicial approximation and the ideas associated with the Regge calculus provide a concrete way of implementing a sum over histories formulation of quantum gravity. A simplicial geometry is made up of flat simplices joined together in a prescribed way together with an assignment of lengths to their edges. A sum over simplicial geometries is a sum over the different ways the simplices can be joined together with an integral over their edge lengths. The construction of the simplicial Euclidean action for this approach to quantum general relativity is illustrated. The recovery of the diffeomorphism group in the continuum limit is discussed. Some possible classes of simplicial complexes with which to define a sum over topologies are described. In two dimensional quantum gravity it is argued that a reasonable class is the class of pseudomanifolds.

The sum over histories formulation of quantum mechanics provides a direct and general framework for the construction of a quantum theory of gravity. Quantum amplitudes are specified by sums over geometries in a class appropriate to the particular amplitude of interest. For example, the amplitude for a given three geometry $^{(3)}\mathcal{G}$ to occur in the state of minimum excitation of a closed cosmology is[1)]

$$\Psi[^{(3)}\mathcal{G}] = \sum_{\mathcal{G}} \exp(-I[\mathcal{G}]) \quad . \tag{1}$$

Here, I is the Euclidean gravitational action and the sum is over all connected, compact Euclidean four geometries $\mathcal{G}$ which have the given three geometry as a boundary. It has been conjectured that this is the wave function of our universe.[2)]

A four geometry is a four dimensional manifold with a metric. A sum over geometries therefore means a sum over four manifolds and a functional integral over physically distinct four metrics. To understand what such sums and integrals mean, one should have a practical method of implementing them. Simplicial approximation and the ideas associated with the Regge calculus[3)] provide such a method. In this talk I would like to illustrate their utility. I shall emphasize the use of simplicial methods as tools for definition and approximate calculation in a continuum theory of gravity. It may be, however, that the discrete version is more fundamental and the continuum only an approximation as, for example, in a theory with a fundamental length.[4)]

A simplicial geometry is made up of flat simplices joined together. A two dimensional surface can be made out of flat triangles. A three dimensional manifold can be built out of tetrahedra; in four dimensions one uses 4-simplices and so on. The information about topology is contained in the rules by which the simplices are joined together. A metric is provided by an assignment of edge lengths to the simplices and a flat metric to their interiors. With this information we can, for example, calculate the distance along any curve threading the simplices.

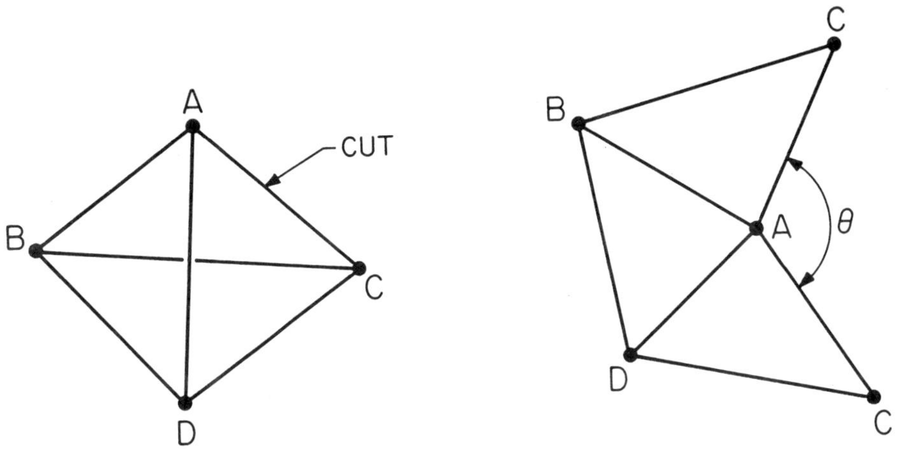

Figure 1. The surface of a tetrahedron is a two dimensional surface whose curvature is concentrated at its vertices. To flatten the three triangles meeting at vertex A one could cut the tetrahedron along edge AC. The angle θ by which the edges AC fail to meet when flattened is a measure of the curvature at A called the deficit angle.

A two dimensional surface made up of triangles is in general curved as, for example, the surface of the tetrahedron in Figure 1. The curvature is not in the interior of the triangles; they are flat. It is not on the edges; two triangles meeting in a common edge can be flattened without distorting them. Rather, the curvature of a two dimensional simplicial geometry is concentrated at its vertices, because one cannot flatten the triangles meeting in a vertex without cutting one of the edges. If one does cut one of the edges and flatten then the angle by which the separated edges fail to meet is a measure of the curvature called the deficit angle. (See Figure 1.) It is the angle by which a vector would be rotated if parallel transported around the vertex. Concretely the deficit angle is 2π minus the sum of the interior angles of the triangles meeting at the vertex. It can thus be expressed as a function of their edge lengths.

In four dimensions the situation is similar with all dimensions increased by 2. The geometry is built from flat 4-simplices. Curvature is concentrated on the two dimensional triangles in which they intersect. There is a deficit angle associated with each triangle which is 2π minus the sum of the interior angles between the bounding tetrahedra of the 4-simplices which intersect the triangle.

The gravitational action may be expressed as a function of the deficit angles and the volumes of the simplices. For example, the Euclidean Einstein action with cosmological constant for a connected closed manifold in n-dimensions is,

$$g_n \ell_p^{n-2} I_n = -\int d^n x (g)^{1/2} (R - 2\Lambda) \quad . \tag{2}$$

Here, $\ell_p = (16\pi G)^{1/2}$ is the Planck length and g_n is a dimensionless coupling. On a simplicial geometry (2) becomes exactly[3)]

$$g_n \ell_p^{n-2} I_n = -2 \sum_{\sigma \in \Sigma_{n-2}} V_{n-2} \theta_{n-2} + 2\Lambda \sum_{\tau \in \Sigma_n} V_n \quad . \tag{3}$$

Here, Σ_k is the collection of k-simplices and V_k is the volume of a k-simplex. The deficit angle θ_k is defined by

$$\theta_k(\sigma) = 2\pi - \sum_{\tau \supset \sigma} \theta_k(\sigma,\tau) \quad , \tag{4}$$

where the sum is over all the (k+2)-simplices τ which meet σ and the $\theta_k(\sigma,\tau)$ are their interior angles at σ. Both V_k and $\theta_k(\sigma,\tau)$ are simply expressible in terms of the edge lengths through standard flat space formulae. By using these expressions in (3) the action becomes a function of the edge lengths. Other gravitational actions, such as curvature squared Lagrangians, may be similarly expressed -

not exactly as here, but in an approximate form which becomes exact in the continuum limit.[5)]

Sums over geometries may be given concrete meaning by taking limits of sums of simplicial approximations to them. This is analogous to defining the Riemann integral of a function as the limit of sums of the area under piecewise linear approximations to it. Consider, by way of example, the sum over four geometries which gives the expectation value of physical quantity $A[\mathcal{G}]$ in the state of minimum excitation for closed cosmologies (1),

$$\langle A \rangle = \frac{\Sigma_{\mathcal{G}} A[\mathcal{G}]\exp(-I[\mathcal{G}])}{\Sigma_{\mathcal{G}} \exp(-I[\mathcal{G}])} \quad , \tag{5}$$

where the sum is over compact, closed Euclidean four geometries. We are accustomed to think of a geometry as a manifold with a metric, and one might therefore want to think of the sum in (5) as a sum over closed manifolds and a sum over physically distinct metrics on those manifolds. Simplicial approximation could be used to give a concrete meaning to such a sum as follows: (1) Fix a number of vertices n_0. (2) Approximate the sum over manifolds as the sum over the number of ways of putting together 4-simplices so as to make a simplicial manifold with n_0 vertices. (3) Approximate the sum over physically distinct metrics by a multiple integral over the squared edge lengths s_i. (4) Take the limit of these sums as n_0 goes to infinity. In short, express $\langle A \rangle$ as

$$\langle A \rangle = \lim_{n_0 \to \infty} \frac{\sum\limits_{M(n_0)} \int_C d\Sigma_1 A(s_i, M) \exp[-I(s_i, M)]}{\sum\limits_{M(n_0)} \int_C d\Sigma_1 \exp[-I(s_i, M)]} \tag{6}$$

There remains the specification of the measure $d\Sigma_1$ and the contour C for the integral over edge lengths. Of course, today we understand little about the convergence of such a process but it is at least definite enough to be discussed.

The central ingredient in weighting the sum over geometries is the action. A variety of gravitational actions could be considered which correspond in the continuum limit to Einstein's action, curvature squared Lagrangians and more complicated actions. The extrema of the action are the solutions of a finite set of algebraic equations

$$\frac{\partial I}{\partial s_i} = 0 \quad . \tag{7}$$

For the Regge action (3) in four dimensions these are the discrete version of Einstein's equation

$$\sum_{\sigma \in \Sigma_4} \theta(\sigma) \frac{\partial V_2}{\partial s_i} = \Lambda \sum_{\tau \in \Sigma_4} \frac{\partial V_4}{\partial s_i} \quad . \tag{8}$$

These extrema can be used to construct the semiclassical approximation to the quantum theory.

Figures 2 and 3 show a few simple numerical calculations of Regge's action on the four sphere. The simplest triangulations of S^4 are the four dimensional surface of a 5-simplex (α_5) and the four dimensional surface of the 5-cross polytope (β_5) - the 5 dimensional generalization of the octohedron. These are the only regular solids in five dimensions. The 5 simplex has 6 vertices, 15 edges, 20 triangles, 15 tetrahedra and 6 4-simplices. The cross polytope has 10 vertices, 40 edges, 80 triangles, 80 tetrahedra and 32 4-simplices. Figure 2 shows the action for these triangulations as a function of four volume when all their edges are equal and the cosmological constant is unity in Planck units.[5)] The action is always lower than the "continuum" value corresponding to the round four sphere but becomes closer to it as we move from the coarsest

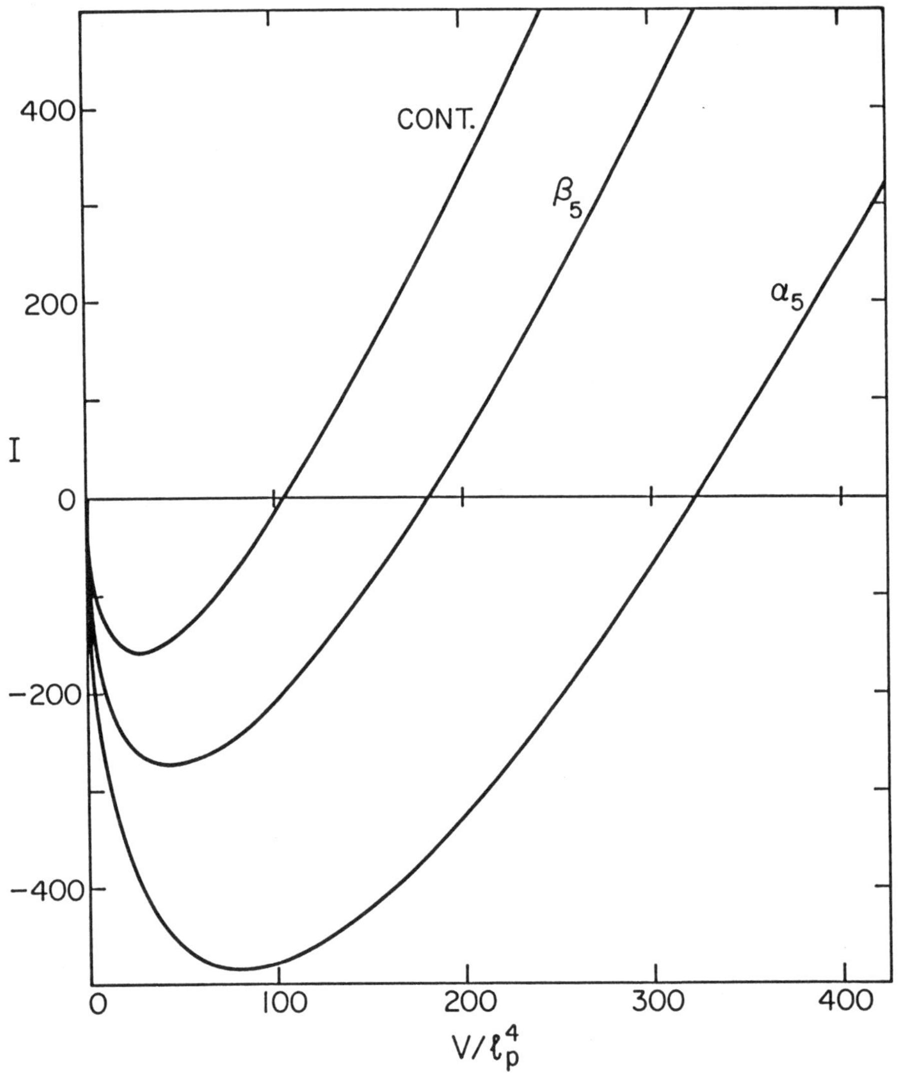

Figure 2. The action for some homogeneous isotropic four geometries as a function of volume. The figure shows the action for the 4-geometries which are the boundary of a 5-simplex (α_5) and the 5-dimensional cross polytope (β_5) (the 5-dimensional generalization of the octohedron) when all of their edges are equal. Also plotted is the "continuum" action for the 4-sphere.

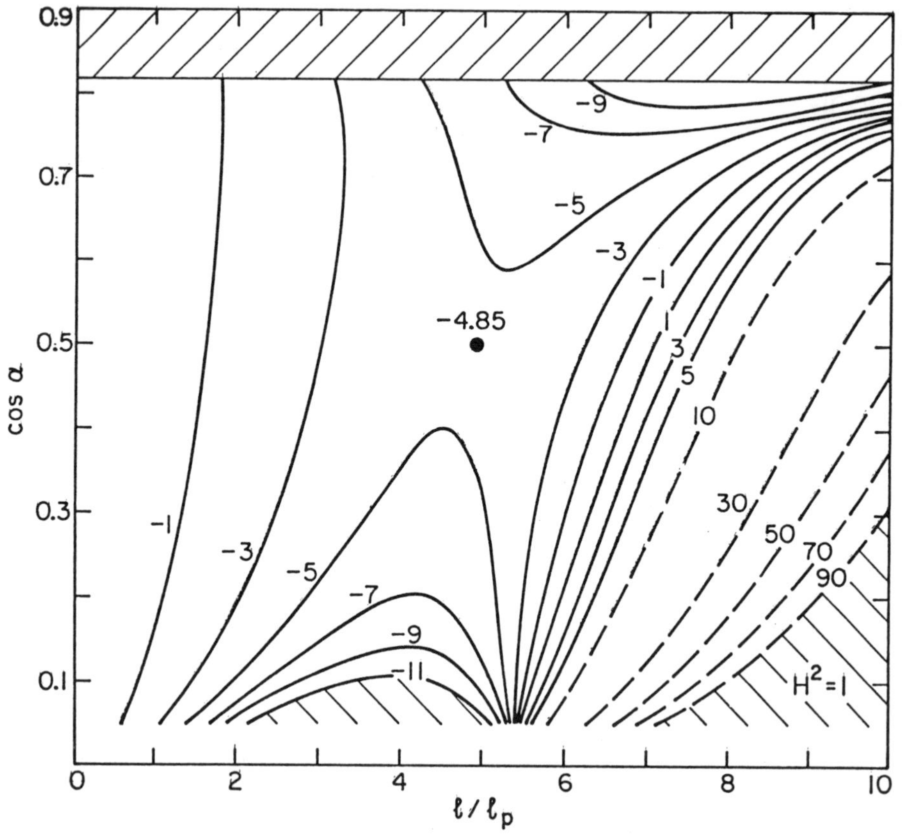

Figure 3. The action for distorted 5-simplices. The figure shows the action (divided by 100) for a two parameter family of 5-simplices in which all the edge lengths are ℓ except for the edges emerging from one vertex which are $\ell/(2\cos\alpha)$. α near $\pi/2$ corresponds to long thin 5-simplices. α near 0 corresponds to nearly flat 5-simplices. There are no 5-simplices with $\cos\alpha$ greater than .81 because the 4-simplex inequalities would be violated. There is a saddle point corresponding to equal edges of value about 4.9. This is a solution of the Regge equations. The figure displays the negative gravitational action arising from conformal distortions.

triangulation α_5 to the finer β_5.

Figure 3 shows a family of distorted 5-simplices. All the edges have the value ℓ except those leading to a particular vertex which have the value $\ell/(2\cos\alpha)$. $\cos\alpha$ near 0 thus corresponds to "long and thin" 5-simplices while large $\cos\alpha$ 5-simplices are "short and squat." $\cos\alpha$ cannot be too large because the analog of the triangle inequality for 4-simplices would not be satisfied. The two parameter family shows the characteristic saddle behavior of Einstein's action. There is an extremum when all the edges are equal to about $4.84\ell_p$. This is a solution of the discrete field equations corresponding to Euclidean de Sitter space. At this solution the action is neither a maximum nor a minimum but a saddle point.

One of the central features of geometric theories of gravity is their invariance under the diffeomorphism group. In a simplicial approximation the diffeomorphism group is broken in the sense that each different assignment of edge lengths will, in general, correspond to a physically distinct geometry with distinguishable curvatures. The diffeomorphism group reemerges in the limit of large n_0 because in this limit there are many simplicial geometries which <u>approximately</u> correspond to a given continuum geometry and whose actions are <u>approximately</u> equal. The integrals in the numerator and denominator of (6) thus approximately overcount continuum geometries in same way that a sum over different continuum metrics would overcount physically distinct geometries.

Let us see in more detail how this comes about. While in general one expects different assignments of edge lengths to be different geometries, there is one special case where this is certainly not true. This is flat space. Imagine distributing vertices about a region of n-dimensional flat space, connecting them so they form a simplicial manifold and assigning the appropriate flat space distances between the vertices as edge lengths. If the vertices are now moved about in flat space there will result a different assignment of edge lengths, but this new assignment results in the same flat geometry.[6] If there are n_0 vertices in this part of the manifold there will be an nn_0 parameter family of transformations of the edge lengths which leave the geometry unchanged.

Consider a curved simplicial geometry with many vertices such that the typical edge length is much smaller than the characteristic curvature scale L. For example, in the process of solving the Regge equations on an increasingly subdivided simplicial manifold to approach a continuum solution, one would expect to reach such a geometry. (The Regge equations can, however, exhibit solutions which do not correspond to a continuum one.[7]) In this situation, regions small compared to the curvature scale will contain many vertices and be <u>approximately</u> flat. There will therefore be nn_0 directions in the space of edge lengths in which the action is <u>approximately</u> constant for changes in the edge lengths smaller than the curvature scale. These are the "approximate diffeomorphisms" of the simplicial geometry. Their number is correct - n directions for each spacetime point. In the numerator and denominator of an expression like (6) we expect each sum over edge lengths to diverge like $(L/\ell)^{nn_0}$ as n_0 becomes large. For physical quantities, however, the ratio should remain finite.

Summing over metrics is only one of two parts of a sum over geometries even as the metric is only one of two parts in the specification of a geometry. The other part might be loosely called the "topology" and it is therefore of interest to investigate sums over topologies. Simplicial approximation is a natural framework in which to do this, because the topological and metrical aspects of a simplicial geometry are very clearly separated. The topological information is contained in the rules by which the simplices are joined together. The metrical information is contained in the assignment of edge lengths. It is, in particular, possible to consider geometries with complicated topologies but with relatively few edges.

To sum over the topologies of simplicial geometries with n_0 vertices is to sum over some collection of simplices with a total of n_0 vertices. The widest reasonable framework in which to discuss such collections is provided by the connected simplicial complexes. A connected simplicial complex is a collection of simplices such that if a simplex is in the collection then so are all its faces, and such that any two vertices can be connected by a sequence of edges. What connected complexes should be allowed? A natural restriction is to sum only over complexes which are manifolds - that is, such that each point has a neighborhood which is topologically equivalent (homeomorphic) to an open ball in $\mathbb{R}^n$. In classical general relativity, geometries on manifolds are the mathematical implementation of the principle of equivalence. That principle tells us that locally spacetime is indistinguishable from flat space, and this is the defining characteristic of a manifold. It would, therefore, seem reasonable to consider geometries on manifolds in the quantum regime although it is less clear that on the scale of the Planck length the principle of equivalence should be enforced in this strong way.

It is not straightforward to define a sum over manifolds. To do so there must at least be an effective procedure for listing those manifolds which contribute to the sum. We cannot do this by classification i.e. by by taking "one of type A", "two of type B", etc. because in four (and higher) dimensions the classification problem for manifolds without additional structure is unsolvable. That is, their does not exist an algorithm for deciding when two manifolds are topologically equivalent.[7)] This does not mean that one could not list manifolds with more structure. Neither does it mean that one could not construct a list on which every manifold of a given dimension would be guaranteed to occur at least once. One would simply not be able to tell when two manifolds on the list were the same. Both of these outs have been suggested as possibilities for constructing the sum over topologies. There is, however, another possibility: that we should sum over a more general class of objects than manifolds.

In the sum over histories formulation of quantum mechanics we are familiar with the idea of "unruly histories." These are histories which contribute significantly to the sums for quantum amplitudes but which are less regular than the classical histories. For example, in particle quantum mechanics the dominant paths are non-differentiable while the classical path is always differentiable. One would perhaps be comfortable with admitting to a sum over topologies a larger class of geometries than those defined on manifolds if one recovered manifolds in the classical limit. The question is then: Is there a class of simplicial complexes such that:

(1) the action for general relativity can be defined,

(2) there is an algorithm for listing the members of the class,

(3) manifolds are the dominant contribution to the sum over histories in the classical limit?

I cannot yet answer this question in general. In two dimensions, however, it is easily addressed. This is because two dimensional Einstein gravity has no metric degrees of freedom. It is not, however, topologically trivial.

The Regge action extends naturally to any simplicial complex in two dimensions. Recall that

$$g_2 I_2 = -2 \sum_{\sigma \in \Sigma_0} \theta(\sigma) + 2\Lambda \sum_{\tau \in \Sigma_2} V_2(\tau) \quad , \tag{9}$$

where the first sum is over the vertices and the second is over the triangles. Insert the definition (4) in this expression, interchange orders in the resulting double sum over vertices and triangles and note that the sum of the interior angles of a triangle is π. One finds

$$g_2 I_2 = -4\pi(n_0 - n_2/2) + 2\Lambda A \quad , \tag{10}$$

where n_0 is the number of vertices, n_2 the number of triangles and A is the total area. The curvature part of the action is independent of the edge lengths and is therefore metrically trivial. The action, however, does depend on how the simplices are joined together, that is, on the topology. This clean separation of metric and topology makes two dimensional Einstein gravity less interesting than the higher dimensional cases but it also makes topological questions easier to analyze.

Let us start with simplicial complexes which are two manifolds and enlarge the class by giving up as little as possible until a larger class is found which satisfies our criteria (1), (2) and (3). If a complex is going to fail to be a manifold it must fail on some collection of points. We give up least if we allow failure only at some discrete number of vertices of the complex and do not permit failure

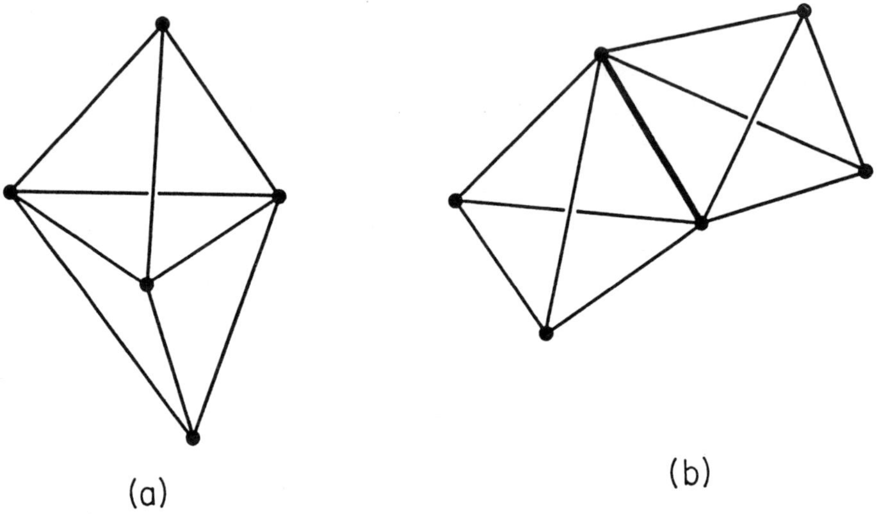

Figure 4. Branching and non-branching complexes. Non-branching two dimensional complexes like that in (a) have exactly two triangles intersecting at any one edge. The complex in (b) has four triangles intersecting along the more heavily drawn edge and is therefore a branching complex. Branching complexes fail to be manifolds at the edges on which they branch.

along the edges. This means we require every edge to be the face of exactly two triangles as in the complex in Figure 4a. We thus exclude complexes like Figure 4b which branch on an edge but permit those like Figure 5 which fail at vertices. For non-branching complexes, $3n_2 = 2n_1$ and the action is

$$g_2 I_2 = -4\pi\chi + 2\Lambda A \tag{11}$$

where $\chi = n_0 - n_1 + n_2$ is the Euler number, a topological invariant.

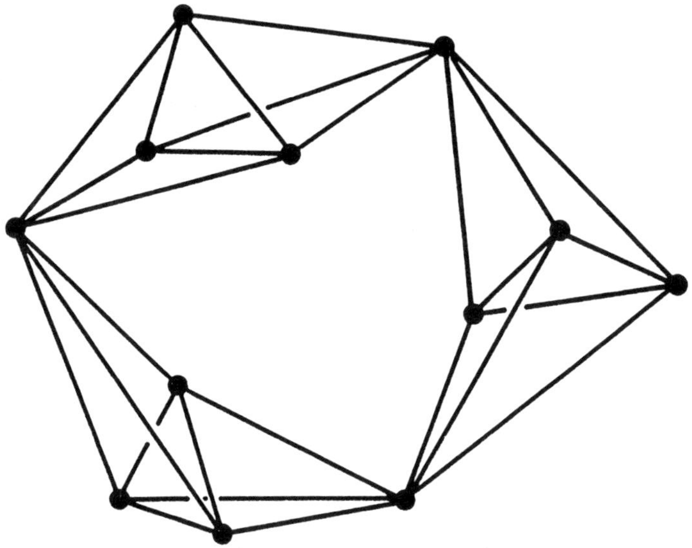

Figure 5. A two dimensional non-branching complex which fails to be a manifold at three vertices. This complex is not strongly connected and is thus not a pseudomanifold. It has Euler number $\chi = 3$.

If we were to stop here we could easily violate our criterion that a manifold have the smallest action. Compare the sphere in Figure 4a which has $\chi = 2$, with the complex in Figure 5. It has $\chi = 3$ and so a smaller action. This is because it consists of almost disconnected pieces. To prevent this we require that the complexes be strongly connected in the sense that any pair of triangles can be joined by a sequence of triangles connected along edges. The resulting complexes are called pseudomanifolds.[9)] The complex in Figure 6 is a pseudomanifold whereas the one in Figure 5 is not. In two dimensions, pseudomanifolds have $\chi \leq 2$, and the pseudomanifold with $\chi = 2$ is the sphere. Thus the pseudomanifold with the smallest action is a manifold and we recover manifolds in the classical limit.

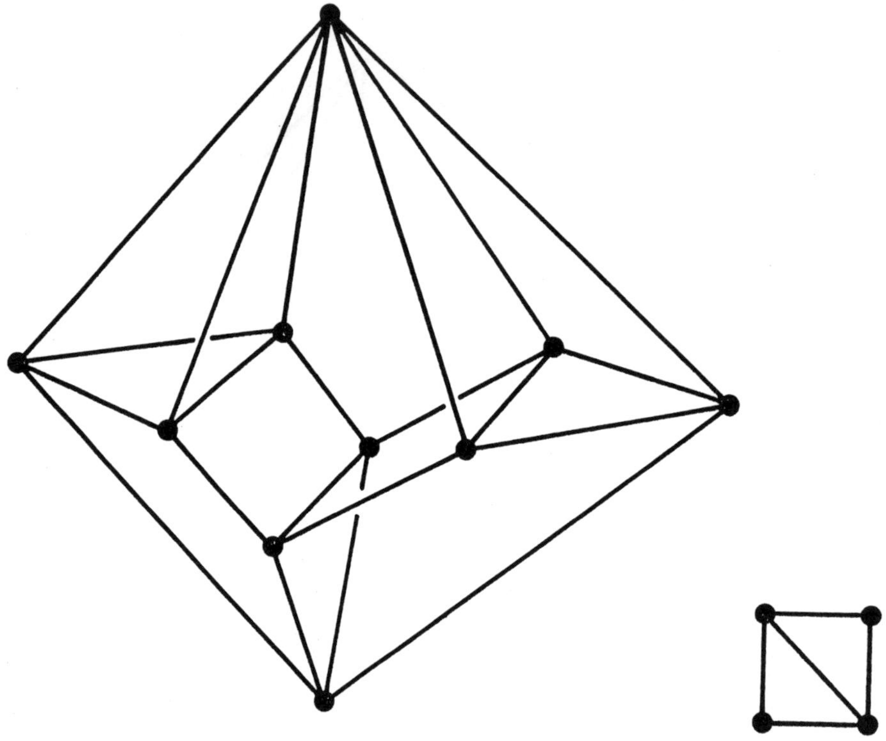

Figure 6. A pseudomanifold which fails to be a manifold at one vertex. The complex is two dimensional, non-branching and strongly connected. It is thus a pseudo-manifold. It may be thought of as a sphere with two points identified. The complex has Euler number $\chi = 1$ so that its action is larger than a sphere of equal area. For pictorial clarity some of the edges triangulating quadralaterals have been omitted but they should be imagined as in the example at lower right.

Most importantly for us, however, pseudomanifolds are easily enumerable. Their defining properties in n dimensions are

(1) Pure dimension - a simplex of dimension $k < n$ is contained in some n-simplex.

(2) Nonbranching - an (n-1)-simplex is the face of exactly two n-simplices.

(3) Strongly connected - any two n-simplices can be connected by a sequence of n-simplices connected along (n-1)-simplices.

These defining properties are essentially combinatorial. Given n_0 vertices one can imagine listing all the possible collections of n-simplices and checking to see which are pseudomanifolds and which are not in a finite number of steps.

In two dimensions pseudomanifolds satisfy all three criteria for a class of complexes with which to define a sum over topologies. The Regge action is defined for them, there is an algorithm for enumerating them, and the pseudomanifold of least action is a manifold. In higher dimensions, finding a class which meets these criteria is a deeper question. Pseudomanifolds can be defined in higher dimensions as described above. The action can be extended to them and they are enumerable. Finding the configurations of least action, however, is now not only a question of topology but also of metric. The possibilities for pseudomanifolds are so varied in higher dimensions that it seems likely that one must restrict the class of complexes further in order to have manifolds dominate in the classical limit. If this can be done then by relaxing the principle of equivalence at the quantum level we will have an attractive class of geometries with which to define a sum over topologies in quantum gravity

Preparation of this report was supported in part by the National Science Foundation under grant PHY 81-07384.

References

1. Hartle, J.B. and Hawking, S.W., Phys. Rev. D28, 2960 (1983).
2. See, e.g. Hawking, S.W., Nucl. Phys. B239, 257 (1984).
3. Regge, T., Nuovo Cimento 19, 558 (1961).
4. Friedberg, R. and Lee, T.-D., Nucl. Phys. B242, 145 (1984) and Feinberg, G., Friedberg, R., Lee, T.-D. and Ren, H.C., (to be published).
5. Hamber, H. and Williams, R. (to be published).
6. Rocek, M. and Williams, R., Phys. Lett. 104B, 21 (1981), Z. Phys. C21, 371 (1984).
7. Piran, T. and Strominger, A. (to be published).
8. For a review of this result of A.A. Markov see Haken, W. in Word Problems, ed. by Boone, C.M., Cannonito, F.B. and Lyndon, R.C. (North Holland, Amsterdam, 1973).
9. See, e.g. Seifert, H. and Threlfall, W., Textbook of Topology, (Academic Press, New York, 1980).

THE GENERALIZED SCHWINGER-DEWITT TECHNIQUE AND THE UNIQUE EFFECTIVE ACTION IN QUANTUM GRAVITY

A.O.Barvinsky
Moscow State Pedagogical Institute, Moscow, USSR
and
G.A.Vilkovisky
State Committee of Standards, Moscow, USSR

The purpose of this talk is to discuss how one can define and compute the effective action. We shall consider the following subjects:

1. The parameterization-independent and gauge-independent definition of the effective action in QFT.

2. Extension of the Schwinger-DeWitt technique to nonminimal operators, higher-order operators and higher loop orders. The method of generating functional traces.

3. Summation of the proper-time series. Nonlocal terms in the effective action.

4. Applications.

The complete algorithm for divergences of a fourth-order operator.

The calculational technique for the unique effective action in gauge theories.

The unique counterterms for Einstein's theory.

Covariant calculation of divergences in two-loop graphs and the absence of leading divergences in two-loop quantum gravity.

1. THE UNIQUE EFFECTIVE ACTION IN QFT.

The effective action contains all the information about the quantized field. It gives the S-matrix and generates the effective equations for the mean field. However the effective equations are not unique. Many people think that when the external sources are switched off the ambiguities in effective equations disappear, but this is not the case.

An example of nonuniquiness of the effective equations is their dependence on the choice of conditions fixing the gauge of quantized fields in gauge theories. Thanks to DeWitt [1,2] we have a covariantly defined effective action in gauge theories, but the problem is that this action still depends on the choice of the mean-field gauge conditions. The gauge dependence of the effective action is however the part of a more general problem: even in non-gauge theories the effective equations depend on the choice of parameterizations for the quantized fields.

The solution of the problem of constructing the unique effective action was recently proposed in these references [3,4].

If the mean field is naively defined as the functional average $g^i = \langle \varphi^i \rangle$, then the corresponding effective action $W[g]$ satisfies the equation

$$\exp iW[g] = \int d\varphi \, M[\varphi] \exp i\Big\{ S[\varphi] + (g^i - \varphi^i)\frac{\delta W[g]}{\delta g^i}\Big\}, \tag{1}$$

where $S[\varphi]$ is the classical action and $M[\varphi]$ is the local measure. Its parameterization dependence is seen already in the one-loop approximation

$$iW_{\text{one-loop}}[g] = -\frac{1}{2}\mathrm{Tr}\ln\frac{\delta^2 S[g]}{\delta g^m \delta g^n} + \ln M[g]. \tag{2}$$

Under local reparametrizations of the quantized field $\varphi \to f(\varphi)$ additional terms proportional to $\delta S/\delta g^i$ arise in Eq. (2). Thus only the S-matrix is unique while the effective equations are parameterization-dependent.

In order to define the parameterization-independent effective equations we shall consider S as a configuration-space scalar and require that W be also a scalar.

This may be attained by introducing a connection Γ^i_{mn} in configuration space and replacing the ordinary functional derivatives in (2) by covariant derivatives.

$$\frac{\delta^2 S}{\delta g^m \delta g^n} \to \nabla_m \nabla_n S \equiv \frac{\delta^2 S}{\delta g^m \delta g^n} - \Gamma^i_{mn} \frac{\delta S}{\delta g^i} . \quad (3)$$

This modification will not spoil the S-matrix if the connection will be ultralocal, that is proportional to the undifferentiated δ -functions and containing only undifferentiated fields. In this case the connection - dependent term vanishes on the mass shell.

The way to introduce this modification is to replace the naive source term, which is nonsense from the geometrical viewpoint, since φ is just a coordinate, by $\sigma^i[g,\varphi]\,\delta W[g]/\delta g^i$, where $\sigma^i[g,\varphi]$ is the geometrical two-point quantity considered by DeWitt [5] in a different context,

$$\exp iW[g] = \int d\varphi\, \mathcal{M}[\varphi] \exp i\Big\{ S[\varphi] + + \sigma^i[g,\varphi] \frac{\delta W[g]}{\delta g^i} \Big\} . \quad (4)$$

The quantity $\sigma^i[g,\varphi]$ is a scalar with respect to the quantum field φ and a vector with respect to the mean field g. This vector is tangent to the geodesic connecting the two points in configuration space:

$$\sigma^k \nabla_k \sigma^i = \sigma^i, \; \sigma^i|_{\varphi=g} = 0, \; \det \nabla_k \sigma^i|_{\varphi=g} \neq 0. \quad (5)$$

Thus we arrive at the parameterization-invariant effective action. This corresponds to the following new definition of the mean field: $\langle \sigma^i[g,\varphi] \rangle = 0$.

How to define the configuration-space connection? Obviously, it should be defined by the classical action itself and satisfy the following conditions. Firstly, there should be no radiative corrections in free-field theories. That is for free-field theories the connection should be zero in a parameterization in which the action is quadratic. Secondly, the modification should not spoil the S-matrix. That is the connection should be ultralocal. For the class of theories defined as ($\varphi^i = \varphi^a(t)$)

$$S[\varphi] = \int dt \{ A_{ab}(\varphi)\dot{\varphi}^a\dot{\varphi}^b + B_a(\varphi)\dot{\varphi}^a + C(\varphi) \},$$

$$\det A_{ab}(\varphi) \neq 0 \qquad (6)$$

the connection is the Christoffel symbol of the configuration - space metric $G_{mn} = A_{ab}(\varphi)\delta(t_a - t_b)$.

Let us turn to gauge theories. The modified equation for the effective action is now of the form

$$\exp iW[g] = \int d\varphi\, \mathcal{M}[\varphi] \exp i\{ S[\varphi] - -\frac{1}{2}(\chi^\mu)^2 + \frac{1}{i}\mathrm{Tr}\ln\left(R^i_\nu \frac{\delta\chi^\mu}{\delta g^i}\right) + \sigma^i[g,\varphi]\frac{\delta W[g]}{\delta g^i} \},$$

$$(7)$$

where R^i_ν are the generators of gauge transformations, and χ^μ are gauge conditions. In gauge theories the connection may be nonlocal because nonlocal field reparameterizations are admissible in the directions of vectors R^i_ν. For the unitarity of the physical S-matrix it is sufficient that the metric be local only in the space of group orbits. This is in addition to the requirement that the configuration-space metric must agree with the gauge group.

For theories with closed gauge algebras this leads to the following expression for the connection:

$$\Gamma^i_{mn} = \tau^i_{mn} - 2\gamma_{(mk}R^k_\nu N^{\nu\mu}\mathcal{D}_{n)}R^i_\mu + + (R^k_\alpha \mathcal{D}_k R^i_\beta) N^{\alpha\mu} R^l_\mu \gamma_{l(m} N^{\beta\nu} R^j_\nu \gamma_{jn)}, \qquad (8)$$

where γ_{mn} is some auxiliary metric, τ^i_{mn} is the Christoffel symbol of γ_{mn}, $\mathcal{D}_m$ is the covariant derivative with respect to γ_{mn}, and $N_{\mu\nu} = R^m_\mu \gamma_{mn} R^n_\nu$, $N^{\mu\alpha}N_{\alpha\nu} = \delta^\mu_\nu$. The γ_{mn} defines the metric in the space of orbits and satisfies the following conditions: i) γ_{mn} should be ultralocal, ii) local generators R^i_α should be the (generalized) Killings of γ_{mn}, iii) $\det N_{\mu\nu} \neq 0$.

This has the following consequences, crucial for the uniqueness of the effective action:

$$\nabla_m R^i_\mu \propto R^i_\nu ,$$

$$R^k_\mu(g)\, \nabla_k \sigma^i[g, \varphi] \propto R^i_\nu(g),$$

$$R^k_\mu(\varphi)\, \frac{\delta}{\delta \varphi^k} \sigma^i[g, \varphi] \propto R^i_\nu(g).$$

The latter relation means that the gauge transformation of the quantized field reduces to the gauge transformation of the mean field. As a result the modified effective action is gauge-independent, parameterization-independent and gauge-invariant.

The question of the metric in the space of group orbits was considered by DeWitt many years ago [6]. For the Yang-Mills field the above conditions define this metric uniquely

$$\gamma_{ik}\, d\varphi^i d\varphi^k = \int d^4x\, g^{1/2} g^{\mu\nu} \gamma_{ab}\, dA^a_\mu\, dA^b_\nu ,$$

where $\gamma_{ab} = -C^f_{ag} C^g_{bf}$ is the Cartan-Killing metric. It is pleasant that this is the metric which contracta the highest derivatives in the Yang-Mills action. For the gravitational field there is only a single one-parameter family:

$$\gamma_{ik}\, d\varphi^i d\varphi^k = \int d^4x\, g^{1/2} \left(2 g^{\mu(\alpha} g^{\nu\beta)} + a\, g^{\mu\nu} g^{\alpha\beta} \right) dg_{\mu\nu}\, dg_{\alpha\beta} .$$

If we want to connect the metric with the highest-derivative term of the classical action, then uniquely $a = -1$.

However, the locality of the metric in the space of orbits is sufficient but not necessary for the unitarity of the physical S-matrix. It is possible that for some exceptional theories very specific nonlocal γ may exist, which does not spoil the physical S-matrix. The Einstein quantum gravity seems to be such a theory. The correct choice of γ may help to solve such problems of this theory as the problem of the conformal mode and the tachyon ghost in the effective equations.

2. THE GENERALIZED SCHWINGER-DEWITT TECHNIQUE.

Let us turn to the calculational methods for the effective action. Unfortunately we continue to work in the framework of loop expansion. How the effective action can be computed in a given loop order? The simplest method is the expansion in powers of the background dimensionality. This means the following. When the theory has a mass parameter, the effective action can be expanded in inverse powers of this parameter. The coefficients of this expansion are local functions of background fields with growing dimensionality in units of inverse length $1/l$.

The basis of this expansion is the Schwinger-DeWitt technique [5, 7] . However this technique is directly applicable only to theories in which the inverse propagator is a minimal second-order operator

$$\hat{F}(\nabla)\varphi = (\hat{1}\Box + \ldots)\varphi, \quad \hat{1} = \delta^A_B, \quad \varphi = \varphi^B$$

(minimal means the box form of second derivatives) $\Box = g^{\mu\nu}\nabla_\mu\nabla_\nu$). We generalize this technique to nonminimal operators of any order [8, 9] and multiloop diagramms.

In the case of a general operator

$$\hat{F}(\nabla)\varphi = (\hat{D}^{\mu_1\ldots\mu_{2k}}\nabla_{\mu_1}\ldots\nabla_{\mu_{2k}} + \ldots)\varphi \quad (9)$$

it turns out that the fundamental assumption underlying the Schwinger-DeWitt technique is the causality condition for the matrix of highest derivatives

$$\det\hat{D}(n) = C(g^{\mu\nu}n_\mu n_\nu)^{k\delta^A_A}, \quad C \neq 0,$$
$$\hat{D}(n) \equiv \hat{D}^{\mu_1\ldots\mu_{2k}} n_{\mu_1}\ldots n_{\mu_{2k}}, \quad (10)$$

where n_μ is an arbitrary 4-vector and C is a constant independent of n_μ. This condition allows to invert the finite-dimensional matrix $\hat{D}(n)$

$$\hat{D}^{-1}(n) = \frac{\hat{K}(n)}{(n^2)^m}\ , \quad n^2 \equiv g^{\mu\nu} n_\mu n_\nu \qquad (11)$$

($\hat{K}(n)$ is a polynomial in n_μ) and in this way find the zero'th - order approximation for Green's function

$$\hat{G}(\nabla) = \hat{K}(\nabla)\frac{\hat{1}}{\Box^m} + O\left[\frac{1}{\ell}\right]. \qquad (12)$$

The further terms of the expansion in the background dimensionality are then easily obtained by iterations. The coefficients of this expansion are dimensional background quantities such as the commutator of covariant derivatives

$$(\nabla_\mu \nabla_\nu - \nabla_\nu \nabla_\mu)\varphi = \hat{R}_{\mu\nu}\varphi\ , \quad \hat{R}_{\mu\nu} \sim \frac{1}{\ell^2}\ . \qquad (13)$$

All calculations are carried out in a universal manner: the covariant derivatives with an arbitrary connection act on any set of fields φ of arbitrary spintensor structure.

In the one-loop order we obtain the following diagrammatic structure

$$iW_{\text{one-loop}} = -\frac{K}{2}\mathrm{Tr}\ln\Box + \sum_{r,s}\mathrm{Tr}\left[X^{\mu_1\ldots\mu_r}\nabla_{\mu_1}\ldots\nabla_{\mu_r}\frac{1}{\Box^s}\right]$$

$= \frac{K}{2}$ 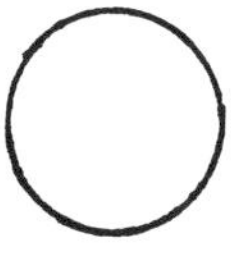 $+ \sum_{r,s}$ 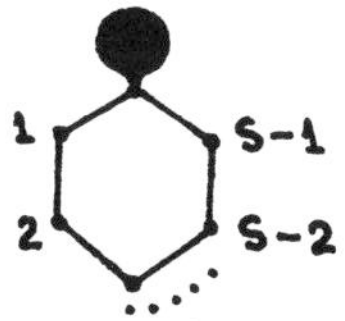

. (14)

This structure is given by the minimal contribution $-\frac{K}{2}\mathrm{Tr}\ln\Box$ and contributions of loops with many minimal propagators and one external blot. This blot corresponds to the dimensional background quantity $X^{\mu_1\ldots\mu_r}$. The background dimensionality of the blot increases with the growth of the number of propagators in the loop. Therefore the expansion is efficient.

All calculations reduce to the computation of the following universal functional traces

$$\begin{cases} \mathrm{Tr}\ln\Box\,, \\ \nabla_{\mu_1}\ldots\nabla_{\mu_r}\dfrac{\hat{1}}{\Box^s}\,\delta(x,y)\Big|_{y=x}\,. \end{cases} \tag{15}$$

The computation of these quantities of which the first one is well known, can be carried out in the standard manner using the Schwinger-DeWitt technique. This technique is merely a means to expand these quantities in powers of the background dimensionality. What one must do is to compile a table of the universal functional traces computed with a given accuracy. Using this table one may compute any one-loop graph in any theory.

If we confine ourselves to the lowest -order approximation, then we obtain the following table, where we keep only the divergent contributions

$$\mathrm{Tr}\ln\Box\Big|^{div} = -\frac{i}{16\pi^2(2-\omega)}\int d^4x\, g^{1/2}\,\mathrm{tr}\Big\{\hat{1}\times$$

$$\times\Big(\frac{1}{180}R^2_{\mu\nu\alpha\beta} - \frac{1}{180}R^2_{\mu\nu} + \frac{1}{72}R^2 + \frac{1}{30}\Box R\Big) +$$

$$+ \frac{1}{12}\hat{\mathcal{R}}^2_{\mu\nu}\Big\},$$

$$\nabla_\mu\frac{\hat{1}}{\Box}\,\delta(x,y)\Big|^{div}_{y=x} = -\frac{i}{16\pi^2(2-\omega)}\,g^{1/2}\times$$

$$\times\Big\{\frac{1}{12}\nabla_\mu R\,\hat{1} - \frac{1}{6}\nabla^\nu\hat{\mathcal{R}}_{\nu\mu}\Big\},$$

$$\frac{\hat{1}}{\Box}\,\delta(x,y)\Big|^{div}_{y=x} = -\frac{i}{16\pi^2(2-\omega)}\,g^{1/2}\Big(\frac{1}{6}R\hat{1}\Big),$$

$$\nabla_\mu \nabla_\nu \frac{\hat{1}}{\Box^2} \delta(x,y)\Big|^{div}_{y=x} = \frac{i g^{1/2}}{16\pi^2 (2-\omega)} \Big\{ \frac{1}{6}(R_{\mu\nu} - \frac{1}{2} g_{\mu\nu} R)\hat{1} + \frac{1}{2} \hat{R}_{\mu\nu} \Big\}, \qquad (16)$$

. . . .

(see [8, 9] for a more complete table). This table is universal in a sense that it is applicable to any theory: one must only substitute on the right the corresponding quantities of a given theory.

The extension of the Schwinger-DeWitt technique to higher-loop order is quite straight-forward. One expands everything in the covariant Taylor series and computes intermediate Gaussian integrals reserving the proper-time integrations to the last. However all this can be considerably improved. In fact, one can reduce the multi-loop calculations to the above table of one-loop functional traces. Consider for example the two-loop approximation. Here one has generally two topologically different diagrams which we shall call "eight" and "fish":

⚭ ← "eight" ,   ⊕ ← "fish" .

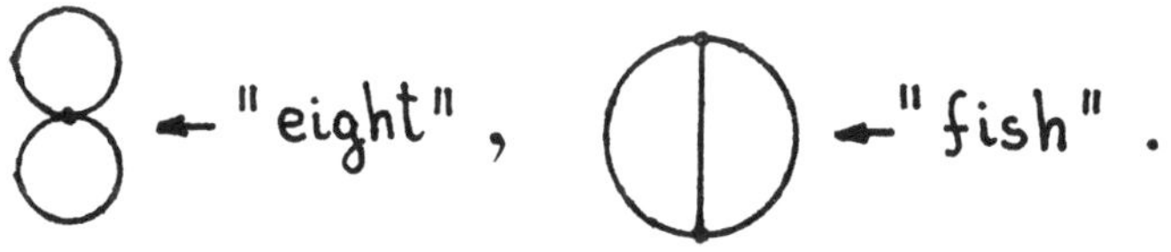

They are constructed from one-loop tadpoles having one external point and one-loop polarization operators having two external points:

$$\infty = \int dx \, [\underset{x}{\bigcirc}] \times [\overset{x}{\bigcirc}], \qquad (17)$$

$$\oplus = \int dx\, dy \, G(x,y) \times [\underset{y}{\overset{x}{\bigcirc}}]. \qquad (18)$$

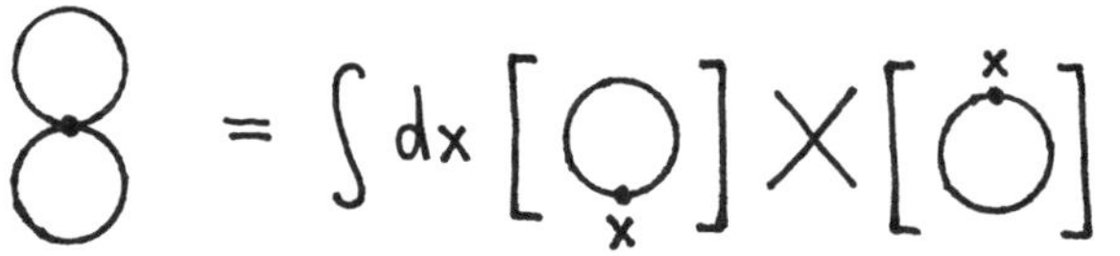

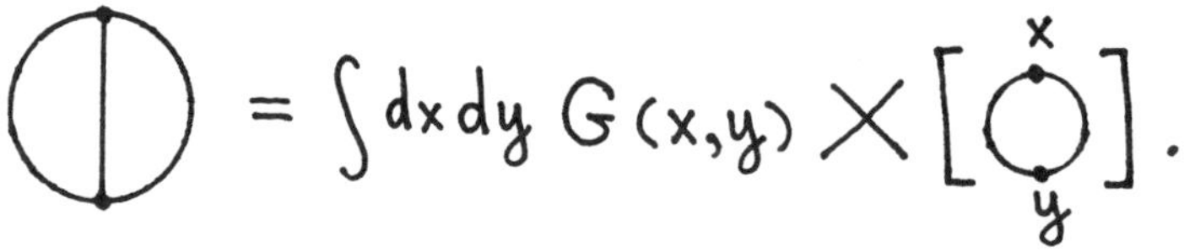

Tadpoles are simply the universal functional traces themselves:

$$= \nabla\ldots\nabla \frac{\hat{1}}{\Box}\,\delta(x,y)\Big|_{y=x}. \tag{19}$$

Thus any "eight" is just a product of two universal traces. For "fishes" one needs polarization operators. But polarization operators can be obtained by functionally differentiating the universal functional traces with respect to the coefficients of the inverse propagator. For example:

$$= \Pi(x,y) = \nabla_{\mu_1}\ldots\nabla_{\mu_n}\frac{\delta^A_C}{\Box}\delta(x,y)\frac{\delta^D_B}{\Box}\delta(y,x) = $$
$$= -\frac{\delta}{\delta P^C_D(y)}\nabla_{\mu_1}\ldots\nabla_{\mu_n}\frac{\delta^A_B}{\Box+P}\delta(x,z)\Big|_{\substack{z=x\\P=0}}, \tag{20}$$

$$= \Pi(x,y) = \nabla_{\mu_1}\ldots\nabla_{\mu_n}\frac{\delta^A_C}{\Box}\delta(x,y)\nabla_\alpha\frac{\delta^D_B}{\Box}\delta(y,x) = $$
$$= -\frac{\delta}{\delta \Gamma^{\alpha C}{}_D(y)}\nabla_{\mu_1}\ldots\nabla_{\mu_n}\frac{\delta^A_B}{\Box+\Gamma^\nu\nabla_\nu}\delta(x,z)\Big|_{\substack{z=x\\ \Gamma=0}}. \tag{21}$$

Sometimes one can do even better and obtain the polarization operator as the second functional derivative of a one-loop diagram without external points at all:

$$-\frac{\delta^2 \mathrm{Tr}\ln(\Box+P)}{\delta P^A_B(x)\,\delta P^C_D(y)}\bigg|_{P=0} = \frac{\delta^B_C}{\Box}\delta(x,y)\frac{\delta^D_A}{\Box}\delta(y,x). \tag{22}$$

There are very general formulas of this sort. If one takes the one-loop effective action and differentiates it twice, one gets the sum of "eight" and "fish"

$$2\int dx\,dy\,G(x,y)\frac{\delta^2 i W_{\text{one-loop}}}{\delta g(x)\,\delta g(y)} =$$  $+$  . (23)

So if one has "eight", one does not need to compute "fish". Compute instead the left-hand side of (23). However sometimes one does not want to compute "eight" too. Then one needs "eight" and "fish" separately, because they enter the two-loop action with different coefficients

$$-W_{\text{two-loop}} = \frac{1}{8}$$  $+\frac{1}{12}$  . (24)

The most general generating expression can be obtained as follows. Consider a minimal operator $\hat{F}(g|\nabla)$ and its Green's function $\hat{G}(g|\nabla)$. Their product at coincident field arguments gives one: $\hat{F}(g|\nabla)\,\hat{G}(g|\nabla) = -\hat{1}$. Consider now the trace of their product at different field arguments

$$E[g',g] = \mathrm{Tr}\left[\hat{F}(g'|\nabla)\,\hat{G}(g|\nabla)\right]. \tag{25}$$

The differentiations with respect to g and g' give "eight" and "fish" separately

$$\int dx\,dy\,G(x,y)\frac{\delta^2 E[g',g]}{\delta g'(x)\,\delta g'(y)}\bigg|_{g'=g} =$$  ,

$$\int dx\,dy\, G(x,y) \frac{\delta^2 E[g',g]}{\delta g'(x)\,\delta g(y)}\bigg|_{g'=g} =$$  . (26)

On the other hand (25) is just the tabular functional trace. In this way one may obtain a table of polarization operators. Here are several examples:

$$\frac{\delta^A_C}{\Box}\delta(x,y)\frac{\delta^D_B}{\Box}\delta(y,x) = \frac{i}{16\pi^2(2-\omega)}\delta^A_C\delta^D_B\,\delta(x,y),$$

$$\nabla_\alpha\frac{\delta^A_C}{\Box}\delta(x,y)\frac{\delta^D_B}{\Box}\delta(y,x) = \frac{i}{16\pi^2(2-\omega)}\delta^A_C\delta^D_B\nabla_\alpha\delta(x,y),$$

$$\nabla_\alpha\nabla_\beta\frac{\delta^A_C}{\Box}\delta(x,y)\frac{\delta^D_B}{\Box}\delta(y,x) = -\frac{i}{32\pi^2(2-\omega)}\times$$

$$\times\Big\{\Big[\frac{1}{6}g_{\alpha\beta}\Box - \frac{2}{3}\nabla_{(\alpha}\nabla_{\beta)} - \frac{1}{3}\Big(R_{\alpha\beta} -$$

$$-\frac{1}{2}g_{\alpha\beta}R\Big)\Big]\delta^A_C\delta^D_B - \mathcal{R}^A{}_{C\alpha\beta}\,\delta^D_B\Big\}\,\delta(x,y),$$

$$\cdot\quad\cdot\quad\cdot\quad\cdot\quad\cdot\quad\cdot\quad\cdot\quad\cdot\quad\cdot\quad\cdot\quad\cdot\qquad(27)$$

Here we again kept only divergent contributions.

If we substitute these polarization operators into (18) we arrive again at the tabular one-loop functional trace. So one simply uses the one-loop table twice.

Above all the use of the generating expressions helps to avoid the algebraic problems in theories like gravity theory, where the vertices are very long. For example in a "fish" one has to multiply by each other two such vertices. This amounts to thousands of terms, while the generating expressions are algebraically simple, because they do not have external indices. Thus our strategy is this: compute loops without external legs, external points and external indices and obtain everything else by functional differentiation. This saves one from a tedious algebra.

The extension of the Schwinger-DeWitt technique to superspace and supergraphs can be found in [10] .

3. SUMMATION OF THE PROPER-TIME SERIES.

This is what concerns the expansion in the background dimensionality. But this expansion is valid only if the dimensional background quantities are smaller than the mass parameter, and this expansion blows up in massless theories. The reason is that in massless theories one cannot integrate the proper-time series term by term.

Another approximation technique is an expansion in powers of the curvature. If we begin with the proper-time expansion then this new technique corresponds to the summation of all terms having a given power of the curvature and any number of derivatives. In other words this is an expansion in n-point Green's functions on flat background, but this expansion is carried out in a completely covariant fashion. Let us see how it goes on.

In each DeWitt's coefficient one may single out terms of a given power in the curvature:

$$s^n a_n(x,x) = [s^n \underbrace{\nabla\ldots\nabla}_{2n-2} R] + [s^n \underbrace{\nabla\ldots\nabla}_{} R \underbrace{\nabla\ldots\nabla}_{2n-4} R] + $$
$$+ [s^n \underbrace{\nabla\ldots\nabla}_{} R \underbrace{\nabla\ldots\nabla}_{2n-6} R \underbrace{\nabla\ldots\nabla}_{} R] + \ldots .$$

All the derivatives acting on R 's can be shown to reduce to the powers of D'Alambertian

$$\int dx \sum_n s^n a_n(x,x) = \int dx \sum_n \{ \alpha_n (s\Box)^{n-1} R s$$

$$+\left[\beta_n R_{\mu\nu}(s\Box)^{n-2}R^{\mu\nu} + \text{other structures}\right]s^2$$

$$+\left[\gamma_n R_{\mu\nu}(s\Box)^{n-3}R^{\mu\nu}R + \text{other structures}\right]s^3$$

$$+ \ldots \Big\}.$$

Then the integration over the proper time yields the nonlocal effective action

$$W_{\text{one-loop}} = \int dx \Big\{ \Big[\frac{\beta_2}{2-\omega} R_{\mu\nu}^2 - \beta_2 R_{\mu\nu} \ln \Box R^{\mu\nu} +$$

$$+ \text{other } R^2\text{-structures}\Big] + \Big[\gamma R_{\mu\nu} \frac{1}{\Box} R^{\mu\nu} R +$$

$$+ \text{other } R^3\text{- structures}\Big] + \ldots \Big\}. \qquad (28)$$

Frolov and Vilkovisky [11] show that these nonlocal terms are responsible for the effects of particle production such as the Hawking effect.

However the above nonlocal expansion in powers of the curvature is also of a limited applicability. If the condition

$$\nabla^2 R \gg R^2 \qquad (29)$$

does not hold, a further summation is needed, which is likely to result in an expression like

$$W_{\text{one-loop}} = \int dx\, R^{\cdot\cdot}_{\cdot\cdot} \ln(\Box + X_1) R^{\cdot\cdot}_{\cdot\cdot}\,, \qquad (30)$$

$$X_1 = R^{\cdot\cdot}_{\cdot\cdot} + R^{\cdot\cdot}_{\cdot\cdot} \frac{1}{\Box + X_2} R^{\cdot\cdot}_{\cdot\cdot}\,,$$

$$\cdot \quad \cdot \quad \cdot \quad \cdot \quad \cdot \quad \cdot \quad \cdot \quad \cdot \quad \cdot \quad \cdot$$

$$X_n = R^{\cdot\cdot}_{\cdot\cdot} + R^{\cdot\cdot}_{\cdot\cdot} \frac{1}{\Box + X_{n+1}} R^{\cdot\cdot}_{\cdot\cdot}\,.$$

If condition (29) holds one may expand (30) in X's and arrive at the nonlocal expansion discussed above. If the opposite condition holds one may expand (30) in boxes and obtain a local but nonanalitic expansion valid for large but slowly varying curvatures.

4. SEVERAL APPLICATIONS.

4.1. THE COMPLETE ALGORITHM FOR DIVERGENCES OF A FOURTH-ORDER OPERATOR.

One place where the generalized Schwinger-DeWitt technique works is the one-loop effective action for higher-order operators. For example, the general minimal fourth-order operator is of the form

$$\hat{F}(\nabla)\varphi = \{\Box^2 + \hat{\Omega}^{\mu\nu\alpha}\nabla_\mu\nabla_\nu\nabla_\alpha + \hat{D}^{\mu\nu}\nabla_\mu\nabla_\nu + \hat{H}^\mu\nabla_\mu + \hat{P}\}\varphi.$$

Since the coefficients here are dimensional, one may expand in the lower-order terms. Using the table of functional traces we thus obtain the general algorithm for divergences for this operator [9]. In the particular case when the third-order term is absent this result was previously obtained by Fradkin and Tseytlin [12].

4.2. THE CALCULATIONAL TECHNIQUE FOR THE UNIQUE EFFECTIVE ACTION.

In the <u>unique</u> effective action for gauge theories one encounters the determinant of a nonlocal operator because the configuration-space connection Γ^i_{mn} is nonlocal

$$iW_{one-loop} = -\frac{1}{2}\text{Tr}\ln\left[\frac{\delta^2 S}{\delta g^m \delta g^n} - \Gamma^i_{mn}\mathcal{E}_i - \frac{\delta x^\mu}{\delta g^m}\frac{\delta x^\mu}{\delta g^n}\right] +$$
$$+ \text{Tr}\ln(R^i_\nu \delta x^\mu/\delta g^i). \quad (31)$$

The nonlocal term is however proportional to the "extremal" (the left-hand side of field equations) which is a dimensional quantity $\mathcal{E}_i \equiv \delta S/\delta g^i$. Expanding in powers of the "extremal" one arrives at a diagrammatic technique with minimal propagators only

$$iW_{one-loop} = -\frac{1}{2}\text{Tr}\ln\mathcal{G}^{-1}_{mn} + \text{Tr}\ln N_{\mu\nu} +$$
$$+ \frac{1}{2}\bigcirc - \frac{1}{2}\triangle + \frac{1}{4}\square + O[\mathcal{E}^3], \quad (32)$$

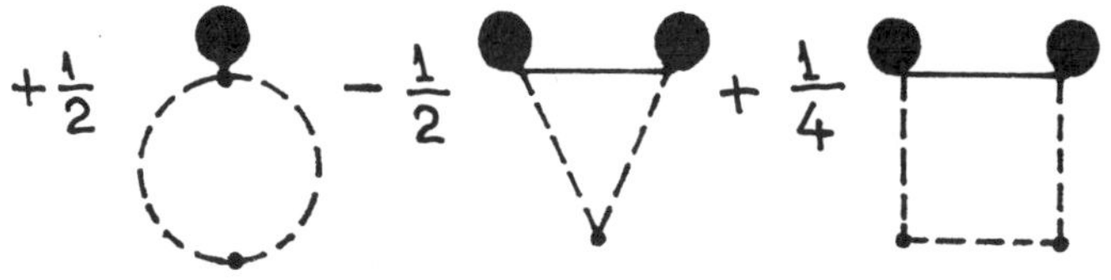

— $g^{mn} = -(\mathcal{D}_m \mathcal{D}_n S - \gamma_{mi} R^i_\nu \delta^{\nu\mu} R^\kappa_\mu \gamma_{\kappa n})^{-1}$,

--- $N^{\mu\nu}$, ∧ $\delta^{\nu\mu}$,

⊗ $\mathcal{E}_i \mathcal{D}_m R^i_\nu$, ⊗ $\mathcal{E}_i R^\kappa_{(\mu} \mathcal{D}_\kappa R^i_{\nu)}$.

All diagrams here can be easily reduced to the tabular functional traces.

In this way we obtain for example the unique counter-terms for Einstein's gravity theory

$$W^{div}_{one-loop} = \frac{1}{96\pi^2(2-\omega)} \int d^4x\, g^{1/2} \left\{ \frac{53}{15}(R^2_{\mu\nu\alpha\beta} - 4R^2_{\mu\nu} + R^2) + \frac{121}{10}(R^2_{\mu\nu} - \frac{1}{3}R^2) + \frac{31}{2}R^2 \right\}. \quad (33)$$

The coefficients here are parameterization-independent and gauge-independent. These coefficients can be now used to discuss the problem of asymptotic freedom.

4.3. Covariant Calculation of Divergences in Two-Loop Graphs and the Absence of Leading Divergences in Two-Loop Quantum Gravity.

We have discussed already how to compute the divergences of tadpoles and polarization operators. For non-local divergences of two-loop graphs we obtain the following algorithms:

$$\underset{2}{\overset{1}{\infty}} = \underset{2}{\overset{1}{\otimes}} + \underset{2}{\overset{1}{\otimes}} + \text{local divergences} + \text{finite part}, \quad (34)$$

$$\underset{1}{\bigcirc}\underset{2}{\bigcirc}3 = \left[\underset{1}{\bigcirc}\underset{2}{\otimes}3 + \text{cycl}(1,2,3) \right] + \quad (35)$$

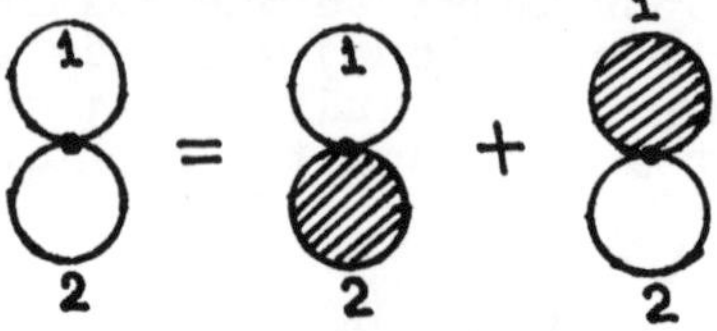

+ local divergences + finite part.

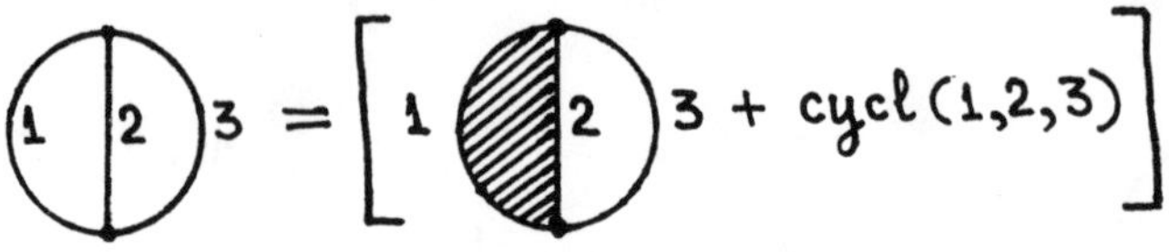

A shaded loop here means the divergent part of a loop. These identities of course underlie the Bogoljubov-Parasjuk-Hepp theorem. In our case, when we have the background field instead of external lines and vertices contain derivatives, there are subtleties about these relations, but still they hold.

For the leading local divergences we prove similar algorithms with one peculiarity however. This peculiarity is an extra $1/2$ coefficient

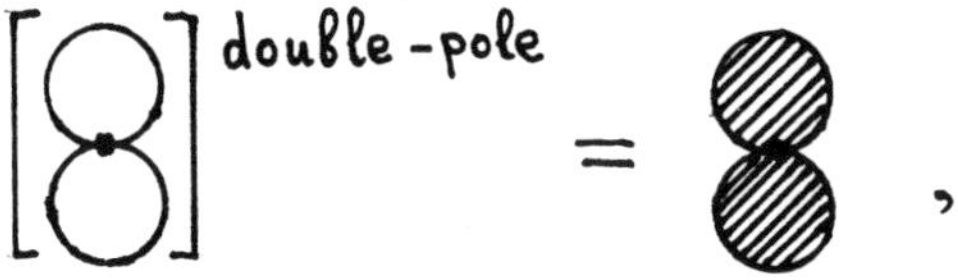

$$\left[\vcenter{\hbox{⧖}}\right]^{\text{double-pole}} = \vcenter{\hbox{⧖}} \quad , \tag{36}$$

$$\left[\vcenter{\hbox{(1,2,3)}}\right]^{\text{double-pole}} = \frac{1}{2}\left[1\,\vcenter{\hbox{▨}}\,2\,\,3 + \text{cycl}(1,2,3) \right]^{\text{double-pole}}. \tag{37}$$

The virtue of these algorithms is that they reduce the calculation of two-loop divergences to one-loop calculations:

$$1\,\vcenter{\hbox{▨}}\,2\,\,3 = \text{Tr}\left[\Pi_{12}^{\text{div}} G_3 \right].$$

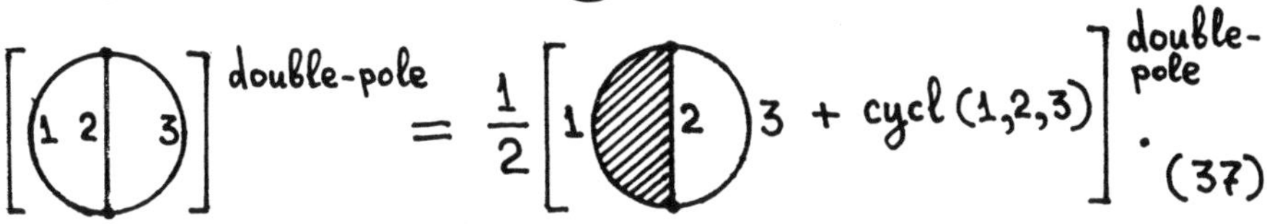

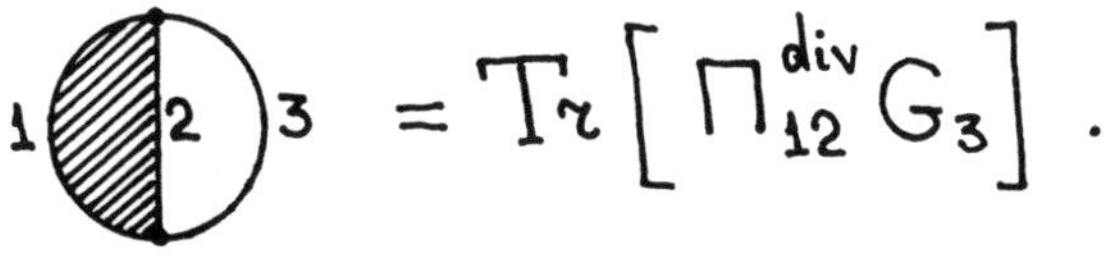

The right-hand side here is again a universal functional trace.

However when the vertices are very long one has also algebraic problems because there are many such contributions. Therefore one needs some generating expressions. We prove that for any theory one can combine all contributions as follows:

$$W_{\text{two-loop}}^{\text{nonlocal}} = -\frac{1}{2}\left[G^{mn}\frac{\delta^2}{\delta g^m \delta g^n}(iW_{\text{one-loop}}^{\text{div}}) \right]^{\text{nonlocal}}, \tag{38}$$

$$W_{\text{two-loop}}^{\text{double-pole}} = -\frac{1}{4}\left[G^{mn}\frac{\delta^2}{\delta g^m \delta g^n}(iW_{\text{one-loop}}^{\text{div}}) \right]^{\text{double-pole}}. \tag{39}$$

For non-gauge theories these relations are immediate consequences of algorithms (34)-(37). For gauge theories things are not so simple, but relations (38) and (39) still hold <u>on shell</u> provided that covariant gauge conditions are used.

To illustrate the above methods consider Einstein's gravity theory. It is easy to work out the generating trace (25) with different field arguments, because it is just the tabular trace. This generating trace produces "eight" and "fish" separately. For example, the double-pole contribution of the graviton "eight" on shell is

$$\text{double-pole} = -\frac{24}{[32\pi^2(2-\omega)]^2}\int d^4x\, g^{1/2}\left(R^{\mu\nu}_{\cdot\cdot\alpha\beta}\right)^3,$$

$$R^{\mu\nu}_{\cdot\cdot\alpha\beta} = g^{\nu\lambda}\partial_\alpha\Gamma^{\mu}_{\lambda\beta} - \dots \qquad (40)$$

The use of the generating expression saves us from the awful algebra of the graviton vertices. If we sum up all contributions, the result for the leading divergences turns out to be zero.

This result can in fact be obtained without any explicite calculation. Since the divergent part of the one-loop effective action is quadratic in the extremal

$$W^{div}_{one\text{-}loop} = \frac{1}{2-\omega}\int d^4x\, g^{1/2}(aR^2_{\mu\nu} + bR^2) =$$

$$= \frac{1}{2-\omega}\,\mathcal{E}_i c^{ik}\mathcal{E}_k,$$

$$\mathcal{E}_i \sim R_{\mu\nu}, \quad c^{ik} \sim \text{local},$$

one immediately finds that the generating expression is $\delta(0)$ on shell

$$G^{mn}\frac{\delta^2}{\delta g^m \delta g^n} W^{div}_{one\text{-}loop} = \frac{1}{2-\omega}\,\frac{\delta^2 S}{\delta g^m \delta g^n}\,c^{nm} \sim \delta(0)$$

and vanishes in dimensional regularization. This means that both nonlocal and double-pole divergences vanish on shell by (38) and (39).

This fact is very natural and was first revealed

by Chase [13] using a different argument. Since the one-loop divergences vanish on shell, no counterterm at the one-loop level is needed. Then it is easy to believe that the work of this counterterm at the two-loop level is neither needed. This means that nonlocal divergences at the two-loop order must vanish on shell by themselves. And this is indeed the case. There remains to be noted that the double-pole divergences are intimately connected with the nonlocal divergences by dimensional arguments. Indeed, the two-loop nonlocal divergences contain the $\ln\Box$ term, which we discussed previously

$$W_{\text{two-loop}} \sim A\int dx\left\{R_{\cdot\cdot}^{\cdot\cdot}\left(\frac{1}{2-\omega}-\ln\Box\right)R_{\cdot\cdot}^{\cdot\cdot}\times\left(\frac{1}{2-\omega}-\ln\Box\right)R_{\cdot\cdot}^{\cdot\cdot}\right\}+\ldots=$$

$$=\int dx\left\{\frac{A}{(2-\omega)^2}(R_{\cdot\cdot}^{\cdot\cdot})^3-\frac{2A}{2-\omega}(R_{\cdot\cdot}^{\cdot\cdot})^2\ln\Box\, R_{\cdot\cdot}^{\cdot\cdot}+\ldots\right\}.$$

The coefficient of the divergent $\ln\Box$ term should be exactly twice greater than the coefficient of the corresponding double-pole term. Therefore the double-pole terms vanish if the nonlocal divergences vanish. These two-loop cancellations are the direct consequences of the one-loop finiteness.

REFERENCES.

1. DeWitt B.S., Phys. Rev. 162, 1195 (1967).

2. DeWitt B.S., in: Quantum Gravity 2, ed. C.Isham, R. Penrose and D.Sciama (Oxford: Oxford University Press, 1981).

3. Vilkovisky G.A., Nucl. Phys. B234, 125 (1984).

4. Vilkovisky G.A., chapter 4 of "The Gospel According to DeWitt", in: Quantum Theory of Gravity, ed. S.M. Christensen (Adam Hilger Ltd., Bristol, 1984).

5. DeWitt B.S., Dinamical Theory of Groups and Fields (New York: Gordon and Breach, 1965).

6. DeWitt B.S., Phys. Rev. 160, 1113 (1967).

7. Schwinger J.S., Phys. Rev. 82, 664 (1951).

8. Barvinsky A.O. and Vilkovisky G.A., Phys. Lett. 131B, 313 (1983).

9. Barvinsky A.O. and Vilkovisky G.A., Phys. Repts.C

(in press).

10. Buchbinder J.L., Journal of Nuclear Physics (USSR) 36,509 (1982).

11. Frolov V.P. and Vilkovisky G.A., "Spherically Symmetric Collapse in Quantum Gravity", preprint of the University of Texas at Austin (1982) ; in Quantum Gravity, ed. M.A.Markov (Plenum Press, 1983).

12. Fradkin E.S. and Tseytlin A.A., Phys. Lett. 104B, 377 (1981).

13. Chase M.K., Nucl. Phys. B203 , 434 (1982).

CONFORMAL ANOMALY FOR GRAVITONS AND THE COUPLED GRAVITY-MATTER EXCITATIONS

L.P.Grishchuk, A.D.Popova
Sternberg Astronomical Institute, 119899 Moscow, V-234, U.S.S.R.

Conformal anomaly for gravity and the coupled gravity-matter excitations is computed in general form for an arbitrary background space-time.

Quantized fields given at a classical background present an important problem for theoretical investigations and astrophysical applications. The adequate and powerful technique is the background field method [1]. Here we consider the coupled quantized excitations for matter and gravitational fields. Only under special conditions the excitations can be decoupled and considered separately. The background matter field Φ^a (where a is some general index) and the background metric tensor $\gamma^{\mu\nu}$ are assumed to satisfy the classical field equations.

The previous calculations of conformal anomaly for gravitons [2,3] can only be applied to the Ricci-flat, $R_{\alpha\beta} = 0$, or the Einstein, $R_{\alpha\beta} = \varkappa \gamma_{\alpha\beta}$, background space-times. These restrictions are caused by the well known inconsistency problem for the higher spin field equations in a curved space-time. Since we are interested in cosmological backgrounds which, obviously, do not satisfy the aformentioned restrictions, we have to take into account the background matter fields.

In this work we apply the results of the previous paper [4] where it has been shown how the general relativity can be treated as an exact nonlinear theory for dynamical gravitational $h^{\mu\nu}$ and matter φ^a variables. Here we will need only the quadratic approximation to this theory since, in the quantum domain, our calculations correspond to the one loop approximation.

The dynamical (quadratic) Lagrangian has the following general form

$$L = \frac{1}{2}\sqrt{-\gamma}\Big\{h^{\mu\nu}\Big[\Big(-I_{\mu\nu}{}^{\lambda\sigma} + \frac{1}{2}\gamma_{\mu\nu}\gamma^{\lambda\sigma}\Big)\Box h_{\lambda\sigma} + X_{\mu\nu}{}^{\lambda\sigma} h_{\lambda\sigma} + Y_{\mu\nu}{}^{\lambda\sigma\rho}\nabla_\rho h_{\lambda\sigma} + X_{\mu\nu}{}^{a}\varphi_a + Y_{\mu\nu}{}^{a\rho}\nabla_\rho\varphi_a\Big] + \varphi^a\Big[-I_a^b\Box\varphi_b + X_a{}^b\varphi_b + Y_a{}^{b\rho}\nabla_\rho\varphi_b + X_a{}^{\mu\nu}h_{\mu\nu} + Y_a{}^{\mu\nu\rho}\nabla_\rho h_{\mu\nu}\Big]\Big\}, \quad (1)$$

where all the coefficients X, Y depend on the background quantities $\gamma^{\mu\nu}$, Φ^a only; X and Y can be expressed in terms of derivatives from the background Lagrangian. Here $I_{\mu\nu}{}^{\lambda\sigma} = \frac{1}{2}\left(\delta^\lambda_\mu\delta^\sigma_\nu + \delta^\lambda_\nu\delta^\sigma_\mu\right)$ and I^b_a are the unity matrices, ∇_ρ denoted the usual covariant differentiation and $\Box \equiv \nabla_\rho\nabla^\rho$. The Lagrangian L includes the gauge fixing term

$$L^{g\text{-}f} = \sqrt{-\gamma}\left(\nabla^\mu h_{\mu\nu} - \frac{1}{2}\nabla_\nu h\right)\left(\nabla_\alpha h^{\nu\alpha} - \frac{1}{2}\nabla^\nu h\right)$$

and it should be supplied with the ghost term

$$L^{gh} = -\sqrt{-\gamma}\left(\nabla_\mu\psi_\nu^*\nabla^\mu\psi^\nu - \psi^*_\mu\psi_\nu R^{\mu\nu}\right)$$

The coupled equations for $h^{\mu\nu}$ and φ^a do easily follow from (1) . It is convenient to treat the fields

$h^{\mu\nu}$ and φ^a as a single multicomponent field H^A where the index A runs out the values $\mu\nu$ and a. The equations for H^A acquire the form

$$F_A{}^B H_B = 0 . \tag{2}$$

By some additional rearrengements the differential operator $F_A{}^B$ can be reduced to the form which allows the application of the standard technique [5), 1)] for finding the first coefficients E_{2n} ($n = 0,1,2$) in the asymptotic Shwinger- DeWitt expansion of the Feynman Green's function for H_A field. These coefficients are determined by X and Y and can be expressed entirely in terms of the background quantities. We will not write them down explicitly here, but just call attention to their general form. The coefficients E_2 and E_4 consist of three contributions - those defined by the fields $h^{\mu\nu}$ and φ^a separately, as if they were free, and by their interaction terms. The coefficients E_2^{gh}, E_4^{gh} for the ghost field are standard [2)]. It is known [1),6)] that the trace of E_0, E_2, E_4 including the ghost contribution, describe the divergent part of the effective action.

The renormalized trace of the total energy momentum tensor T^{ren} can be derived from the renormalized effective action, W^{ren}. By applying a conformal transformation to the background metric $\gamma_{\mu\nu} \to a(x)\gamma_{\mu\nu}$ one can transform W^{ren}. The subsequent functional differentation, $\delta W^{ren}/\delta a$ at $a = 1$, gives T^{ren} [7)]. For conformally invariant fields T^{ren} reduces to $E_4 - 2E_4^{gh}$ which is the conformal anomaly. In our case this combination is only a local part of T^{ren}, which is still reasonable to call conformal anomaly.

We have explicitly calculated E_4 in general form for the Lagrangian (1) what allowed us to find conformal anomaly for different specific examples. Among them are coupled excitations for gravity and self-interacting scalar field and free gravitational waves (gravitons). In the later case one has, first, to find conditions under which there exist a consistent solution to equations (2) with $h^{\mu\nu} \neq 0$ and $\varphi^a = 0$. It is known [8] that such solutions do exist for some background space-times to which belongs the most interesting case of the Friedman-Robertson-Walker metrics. The conformal anomaly, for gravitons, alongside with the usual terms such as the topological invariant G and $\Box R$ includes also the term R^2 and different combinations of the background matter quantities.

The full paper will be published elsewhere.

REFERENCES

1) DeWitt B.S. Dynamical theory of groups and fields. Gordon and Breach, N.Y. (1965); In: General Relativity, eds S.W.Hawking, W.Israel, CUP (1979).

2) Critchley R. Phys. Rev. D18, 1849 (1978).

3) Christensen S.M. and Duff M.Y. Nuclear Phys. B154, 301 (1979).

4) Grishchuk L.P., Petrov A.N., Popova A.D. Comm. Math. Phys. 94, 378 (1984).

5) Gilkey P.B. Proc. Symp. Pure Math. 27, 265 (1975); J. Diff. Geom. 10, 601 (1975).

6). Birrell N.D., Davies P.C.W. Quantum fields in curved space . CUP (1982).

7) Parker L. In: Recent developments in gravitation, eds. S.Deser, M.Levy, N.Y.Plenum (1979).

8) Grishchuk L.P., Popova A.D. Sov. Phys. JETP 53, 1 (1981).; J.Phys. A, 15, 3525 (1982).

ANOMALIES, COHOMOLOGY AND GENERALIZED SECONDARY CLASSES*

Han-ying GUO, Ke WU
Institute of Theoretical Physics, Academia Sinica
P.O. Box 2735, Beijing
Bo-Yu HOU
Institute of Morden Physics, Northwest University
Xian, Shaanxi
Shi-kun WANG
Institute of Applied Mathematics, Academia Sinica
P.O. Box 300, Beijing
CHINA

ABSTRACT

We introduce, in this paper, our recent works on the generalized secondary characteristic classes, the cohomologies of gauge groups realized upon these classes and the degenerate forms of these classes, as well as their applications to the anolyses on both gauge and gravitational anomalies in spacetimes of different dimensions. We also show the relations between our works and Faddeev's, Song's and Zumino's approaches.

1. INTRODUCTION

We will introduce, in this paper, our recent works[1-4] and their relations to these by Faddeev[5], Song[6], and Zumino[7]. These consti-

*Invited talk presented by the first author at the 3rd Seminar on Quantum Gravity, Moscow, October 23-25, 1984.

tute in certain sense some deepgoing of the previous works on the global aspects of the anomalies in spacetimes of different dimensions by means of differential geometric methods[8-11].

On the other hand, the topological origin of the non-abelian anomalies and its relation to the abelian or singlet anomalies in the spacetime of two more dimensions had been found in the previous works where the well-known Chern-Simons secondary classes[12-15] have played an important role. On the other hand, however, there have still been some problems to be solved. For instance, if the spacetime has boundary such as the hybrid bag model[16], some gravitational fields[14] and so on, how to write down the Wess-Zumino-Witten anomalous effective action[17,18], the generating functionals of the non-abelian anomalies and so on? Do the non-abelian anomalies have certain topological meaning in addition to their topological origin? What are the physical implications for those polynomials with respect to higher order of ghosts in the series expansions of certain secondary characteristic classes[9,10]?

Up to the present time, we would address that these problems can definitely be solved based upon the recent works mentioned above, although some concrete analyses have not been worked out yet. As a matter of fact, the conception of the secondary characteristic classes associated to certain characteristic polynomials in the curvature[13,14] can be generalized to introduce a sequence of new characteristic classes[1], the generalized secondary classes. And these new classes in addition to the classes in the curvature and the secondary classes have definitely cohomological meaning so that the cohomology of gauge groups can be realized upon either these classes or their degenerate forms[1-3]. The cohomology of gauge groups proposed by Faddeev[5] is the one upon the degenerate forms of these classes. By means of these generalized secondary classes and relevant cohomologies of gauge groups, one can show that for instance the Wess-Zumino-Witten term, generating functionals and the finite forms of the non-abelian anomalies are 1-degenerate cocycles[5,2,3,6,7,] and the boundary effects can be calculated as well[3,4].

In this paper, we will systematically clarify these issues. In section 2, we introduce the generalized secondary characteristic classes. In section 3, we underline some concepts on cohomology and show the cohomologies of gauge groups realized upon either these generalized classes or their degenerate forms. The cohomological analyses on both gauge and gravitational anomalies in spacetimes of different dimensions are given in section 4.

2. GENERALIZED SECONDARY CHARACTERISTIC CLASSES

We consider a gauge theory with gauge group G on a manifold $\mathcal{M}$ which can be spacetime itself or contains the spacetime as a submanifold. The gauge potential specifies a connection 1-form

$$A = A_\mu(x)dx^\mu , \qquad A_\mu = A_\mu^a\lambda_a , \qquad \forall x \in \mathcal{M} , \tag{2.1}$$

where A_μ is a Lie algebra-valued potential, λ^a the generators of the Lie algebra of G. The curvature F, defined by

$$F = dA + A^2 , \tag{2.2}$$

is a Lie algebra-valued 2-form (the wedge product symbol is suppressed). The Bianchi identity

$$DF = dF + [A,F] = 0 \tag{2.3}$$

is a simple consequence of the definitions above. Under a gauge transformation, the changes in A and F are given by

$$\begin{aligned} A &\longrightarrow {}^gA = g^{-1}Ag + g^{-1}dg. \\ F &\longrightarrow {}^gG = g^{-1}Fg. \end{aligned} \tag{2.4}$$

Let $P(F^n)$ be a characteristic polynomial of degree n in the curvature satisfying the invariance property under the gauge transformation

$$P(g^{-1}Fg) = P(F). \tag{2.5}$$

Such a polynomial is closed with respect to the exterior differential and has a topologically invariant integral over a boundaryless manifold $M^{2n} \subseteq \mathcal{M}$[13,14].

Let $A^o,\ldots,A^k$ be k+1 connection 1-forms on a submanifold $M^{2n-k+1} \subset \mathcal{M}$; $t^1,\ldots,t^k$ be k parameters with constraints $0 \le t^1,\ldots,$

$t^k \le 1$. Introduce an interpolation 1-form on M^{2n-k+1}

$$A_{o,t^1\ldots t^k} = A^o + \sum_{j=1}^{k} t^i \eta^{j,o} \ , \quad \eta^{j,i} = A^j - A^i \ , \tag{2.6}$$

and a corresponding 2-form

$$F_{o,t^1\ldots t^k} = dA_{o,t^1\ldots t^k} + A^2_{o,t^1\ldots t^k} \ , \tag{2.7}$$

Extend (2.6) as an interpolation 1-form, $\mathcal{A}_{o,t^1\ldots t^k}$, on $M^{2n-k+1} \times \Delta^k$ with t-components of $\mathcal{A}$ taken to be zeros, where Δ^k is a simplex of dimension k in $R^k(t)$ defined by

$$\Delta^k = \{(t^1,\ldots,t^k) | \sum_{j=1}^{k} t^i = 1\} \ , \tag{2.8}$$

and correspondingly

$$\begin{aligned} \mathcal{F}_{o,t^1\ldots t^k} &= d\mathcal{A}_{o,t^1\ldots t^k} + \mathcal{A}^2_{o,t^1\ldots t^k} \\ &= F_{o,t^1\ldots t^k} + H_o \ , \end{aligned} \tag{2.9}$$

where $H_o = \sum_{j=1}^{k} dt^j \eta^{j,o}$, $d = d_{(x)} + d_{(t)}$; d, $d_{(x)}$, and $d_{(t)}$ is the exterior differential operator defined on $M^{2n-k+1} \times R^k(t)$, M^{2n-k+1}, and $R^k(t)$ respectively.

We now introduce the density of the k-th generalized secondary characteristic class, $q^{(k)}_n$, associated to the invariant polynomial $P(\mathcal{F}^n)$, a 2n-form on $M^{2n-k+1} \times R^k(t)$, as follows

$$q^{(k)}_n (A^o,\ldots,A^k) = \frac{n!}{k!(n-k)!} P(H^k_o, \mathcal{F}^{n-k}_{o,t^1\ldots t^k}),$$

$$q^{(o)}_n = P_n \ . \tag{2.10}$$

And define the k-th generalized secondary class as the integral of the class density $q^{(k)}_n$ over the k-simplex Δ^k

$$Q^{(k)}_n (A^o,\ldots,A^k;\Delta^k) = \int_{\Delta^k} q^{(k)}_n (A^o,\ldots,A^k) \ , \quad Q^{(o)}_n = P_n \ , \tag{2.11}$$

whose integral over $M^{2n-k} \subset \mathcal{M}$ is defined as the k-th characteristic invariant

$$\eta_n^{(k)}(A^o,\ldots,A^k;\Delta^k,M^{2n-k}) = \int_{M^{2n-K}} Q_n^{(k)}(A^o,\ldots,A^k;\Delta^k). \qquad (2.12)$$

It is easy to show that $q_n^{(k)}$ can be reexpressed as

$$q_n^{(k)}(A^o,\ldots,A^k) = \frac{n!}{k!(n-k)!} P(H_o^k,F_{o,t^1}^{n-k}\ldots t^k), \qquad (2.13)$$

and all $q_n^{(k)}$, $Q_n^{(k)}$, and $\eta_n^{(k)}$ satisty

$$\mathcal{Q}_n^{(k)}(A^o,\ldots,A^k) = \epsilon^{o\ldots k}_{i_o\ldots i_k}\mathcal{Q}_n^{(k)}(A^{i_o},\ldots,A^{i_k}),$$

$$\mathcal{Q}_n^{(k)} = (q_n^{(k)},Q_n^{(k)},\eta_n^{(k)}). \qquad (2.14)$$

where $(i_o\ldots i_k)$ is a permutation of $(o\ldots k)$.

The most important properties of these polynomials and quantities are discribed by the following theorems:

Theorem 1. The density polynomials q's satisfy a hierarchy of equations

$$d_{(x)}q_n^{(k)} = d_{(t)}q_n^{(k-1)}. \qquad (2.15)$$

Theorem 2. The secondary characteristic polynomials Q's satisfy a descent equation

$$d_{(x)}Q_n^{(k)}(A^o,\ldots,A^k;\Delta^k) = \sum_{j=o}^{k}(-1)^j Q_n^{(k-1)}(A^o,\ldots,\hat{A}^j,\ldots,A^k;\Delta^{k-1}_{(j)}).$$

$$\Delta^{k-1}_{(j)} \in \partial\Delta^k . \qquad (2.16)$$

where the caret denotes omission, $\partial\Delta^k$ the boundary of Δ^k.

Theorem 3. The characteristic invariants η's satisfy

$$\sum_{j=o}^{k}(-1)^j\eta_n^{(k-1)}(A^o,\ldots,\hat{A}^j,\ldots,A^k;\Delta^{k-1}_{(j)},M^{2n-k+1})$$
$$= \eta_n^{(k)}(A^o,\ldots,A^k;\Delta^k;\partial M^{2n-k+1}). \qquad (2.17)$$

where ∂M^{2n-k+1} is the boundary of M^{2n-k+1}.

We sketch out how to prove these theorems[1-3]. First, the exterior differentials of both sides of (2.13) give us

$$d_{(x)}q_n^{(k)} = \frac{n!}{k!(n-k)!} d_{(x)}P(H_o^k, F_{o,t^1\ldots t^k}^{n-k}).$$

Make use of the Bianchi identity and the following relations

$$DH_o = d_{(x)}H_o + A_{o,t^1\ldots t^k}H_o + H_oA_{o,t^1\ldots t^k} = d_{(x)}F_{o,t^1\ldots t^k},$$

$$d_{(t)}H_o = 0,$$

the above equation becomes

$$d_{(x)}q_n^{(k)} = \frac{n!}{(k-1)!(n-k+1)!} d_{(t)}P(H_o^{k-1}, F_{o,t^1\ldots t^k}^{n-k+1}) .$$

By definition (2.10), this completes the proof of the theorem 1.

Then take the integral of (2.15) over a k-simplex Δ^k whose orientation is taken to be $[\partial/\partial t^1, \ldots, \partial/\partial t^k]$, we get

$$d_{(x)}Q_n^{(k)} = \int_{\Delta^k} d_{(x)}q_n^{(k-1)} .$$

By Stokes' theorem and the relation between the orientation of Δ^k and that of $\Delta_{(j)}^{k-1}$'s, $\bigcup_j \Delta_{(j)}^{k-1} = \partial\Delta^k$, 1) we get the descent equation (2.16).

Finally, take the integral of (2.16) over the submanifold M^{2n-k+1} M and make use of Stokes' theorem again, we get the theorems 3.

In the cases of k=0,1 the theorems give us

$$d_{(x)}P(F^n) = 0, \tag{2.18}$$

and

$$P(F(A^1)^n) - P(F(A^o)^n) = d_{(x)}Q_n^{(1)}(A^o, A^1; \Delta^1),$$

$$Q_n^{(1)}(A^o, A^1; \Delta^1) = -n\int_o^1 dt^1 P(A^1 - A^o, F_{o,t^1}^{n-1}) ; \tag{2.19}$$

that is, the closeness of $P(F^n)$, the transgression formula between $P(F^n)$ and $Q_n^{(1)}(A^o, A^1; \Delta^1)$, and the definition of the Chern-Simons secondary classes, respectively. In general cases, these theorems have very clear and definite cohomological meaning which we will explain in the next section.

It should be pointed out that these theorems can be applied to the case in which some ghosts are involved. To see this, let us consider a principal bundle $P(M,G)$, and let $d_{(u)}$, $d_{(x)}$, $d_{(s)}$ denote the exterior differential operators on the bundle P, base M, and standard fibre $F \simeq G$. Obviously, we have

$$d_{(u)} = d_{(x)} + d_{(s)}, \quad d_{(x)}d_{(s)} + d_{(s)}d_{(x)} = 0, \text{ etc.} \tag{2.20}$$

Denote

$$A^i(x,s) = g^{-1}(x,s)A^i(x)g(x,s) + g^{-1}(x,s)d_{(u)}g(x,s), \quad i=0,1,\dots,k.$$
$$v(x,s) = g^{-1}(x,s)d_{(s)}g(x,s). \tag{2.21}$$

Then it is easy to see that

$$\begin{aligned} d_{(s)}v &= -v^2, \quad d_{(u)}v + v^2 = d_{(x)}v \equiv F(v), \\ d_{(s)}A(x,s) &= -dv - vA - Av - v^2 . \end{aligned} \tag{2.22}$$

Take ${}^gA^o = 0$, the interpolation then is

$$A_{o,t^1\dots t^k} = A^o + \sum_{j=1}^{k} t^j(A^j - A^o) = v + \sum_{j=1}^{k} t^j\,{}^gA^j(x,s) \tag{2.23}$$

$$F_{o,t^1\dots t^k} = d_{(u)}A_{o,t^1\dots t^k} + A^2_{o,t^1\dots t^k}$$

where ${}^gA = g^{-1}Ag + g^{-1}d_{(x)}g$. Introduce the generalized secondary polynomials with respect to $A^o,\dots,A^k$, as follows

$$Q^{(k)}_n(A^o,\dots,A^k;\ \Delta^k) = \frac{n!}{k!(n-k)!}\int_{\Delta^k} P(H^k, F^{n-k}_{o,t^1\dots t^k}), \tag{2.24}$$

then it is easy to show that such kind of $Q^{(k)}_n$'s satisfy the descent equation in the theorem 2.

For the case $k = 1$, we have

$$P(F^n(A^1)) - P((d_{(x)}v)^n) = d_{(u)}n\int_0^1 dt\,P(A^1, (td\,{}^gA^{1\,2} + (1-t)dv)^{n-1}) \tag{2.25}$$

whose expansion with respect to the power of dv gives rise to Zumino's sequence of polynomials with v regarded as ghosts[7].

3. COHOMOLOGIES OF GAUGE GROUPS

In this section, we will first underline some abstract formalism on cohomology algebra[19,20]. Then we will show that the generalized secondary classes and their degenerate forms present certain realizations of the cohomologies of gauge groups which will play important roles in the analyses on the anomalies in the next section.

3.1 Coboundary Operators And Their Cohomologies

Let $R^{(k)}$ be a set of arrays of functions with respect to arbitrary (k+1) ordered objects $(A^o, A^1, \ldots, A^k)$. Define an operator

$$\Delta : R^{(k)} \to R^{(k+1)} \tag{3.1}$$

acting on the sequence

$$0 \xrightarrow{\Delta} R^{(o)} \xrightarrow{\Delta} R^{(1)} \xrightarrow{\Delta} \ldots \tag{3.2}$$

in the following way:

$$(\Delta r^{(k)})(A^o, \ldots, A^{k+1}) = \sum_{j=0}^{k+1} (-1)^j r^{(k)} (A^o, \ldots, \hat{A}^j, \ldots, A^{k+1}),$$

$$\forall\, r^{(k)} \in R^{(k)}. \tag{3.3}$$

It is easy to prove that

$$\Delta^2 = 0, \tag{3.4}$$

that is, Δ can be defined as a coboundary operator which allows for the introduction of the cohomology of $R^{(k)}$. As usual, A k-cochain $r^{(k)}$ is a k-cocycle if $\Delta r^{(k)} = 0$ whereas a k-cochain $r^{(k)} = \Delta s^{(k-1)}$, for some $s^{(k-1)} \in R^{(k-1)}$, is a k-coboundary. Let $Z_\Delta^{(k)} = \{r^{(k)} \mid \Delta r^{(k)} = 0\}$ be the set of k-cocycles, $B_\Delta^{(k)} = \{r^{(k)} \mid r^{(k)} = \Delta s^{(k-1)}, s^{(k-1)} \in R^{(k-1)}\}$ the set of k-coboundaries. The cohomology of $R^{(k)}$ with respect to the operator Δ is defined by

$$H_\Delta^{(k)} = Z_\Delta^{(k)} / B_\Delta^{(k)}, \qquad k = 0,1,2\ldots \tag{3.5}$$

Obviously, $H_\Delta^{(k)}$ is the set of equivalence classes of k-cocycles, $z^{(k)} \in Z_\Delta^{(k)}$, which differ only by coboundaries. The nontrivial element of $H_\Delta^{(k)}$, of course, is a cocycle which is not a coboundary.

One can introduce another operator $\bar{\Delta}$ on $R^{(k)}$ relevant to the coboundary operator Δ as follows

$$\bar{\Delta}r^{(k)} = \Delta r^{(k)} - r^{(k)}\pi , \qquad r^{(k)} \in R^{(k)} \tag{3.6}$$

where π is an exclusion operator which excludes the first object from the ordered ones, for instance

$$\pi: (A^o,A^1,\ldots,A^k) \longrightarrow (A^1,\ldots,A^k). \tag{3.7}$$

It is easy to see that operators Δ, $\bar{\Delta}$, and π satisfy

$$(\bar{\Delta}r^{(k)})\,\pi = -\Delta\,(r^{(k)}\pi). \tag{3.8}$$

From (3.6) and (3.8), it follows that $\bar{\Delta}$ is also a coboundary operator,

$$\bar{\Delta}^2 = 0. \tag{3.9}$$

Therefore, one can introduce another cohomology of $R^{(k)}$ with respect to $\bar{\Delta}$, $H_{\bar{\Delta}}$.

It should be point out, however, that $(\bar{\Delta}r^{(k)})(A^o,\ldots,A^{k+1})$ has one less than $(\Delta r^{(k)})(A^o,\ldots,A^{k+1})$ in the number of terms so that the number of terms in $\bar{\Delta}r^{(k)}$ is equal to that in $\Delta r^{(k-1)}$. This means that the operation of $\bar{\Delta}$ on $r^{(k)}$ is equivalent to the one of Δ on a (k-1)-cochain rather than a k-cochain and in this sense the k-cochain $r^{(k)}$ with respect to the operator Δ is degenerated under the operation of $\bar{\Delta}$. One could call $\bar{\Delta}$ the degenerate coboundary operator and $H_{\bar{\Delta}}$ the cohomology of the degenerate cochains.

It is plain that if those ordered objects, A's, or some of them, are taken to be the connection 1-forms, and those $R^{(k)}$'s some characteristic polynomials, then the cohomologies H_Δ and $H_{\bar{\Delta}}$ become certain cohomologies of gauge groups.

3.2 Cohomology Of Generalized Secondary Classes

The cohomological meaning of the generalized secondary characteristic classes is very clear and definite. In fact, by means of the coboundary operator Δ, the descent equation (2.16) can be rewritten as

$$(\Delta Q_n^{(k-1)})(A^o,\ldots,A^k;\ \partial\Delta^k) = d_{(x)}\dot{Q}_n^{(k)}(A^o,\ldots,A^k;\ \Delta^k) \tag{3.10}$$

This means that the coboundary of the (k-1)-st secondary classes is exactly equal to the exterior differential of a k-th one as pointed out in Ref.1) so that we would have the following sequence

$$0 \xrightarrow{d^{-1}\Delta} Q_n^{(o)} \xrightarrow{d^{-1}\Delta} Q_n^{(1)} \cdots \xrightarrow{d^{-1}\Delta} Q_n^{(n)} \xrightarrow{d^{-1}\Delta} Q_n^{(n+1)} = 0 \ . \tag{3.11}$$

Similarly, (2.17) satisfied by the characteristic invariants, η's can also be rewritten as

$$(\Delta\eta_n^{(k-1)})(A^o,\ldots,A^k;\ \partial\Delta^k,\ M^{2n-k+1}) = \eta_n^{(k)}(A^o,\ldots,A^k;\Delta^k,\partial M^{2n-k+1}), \tag{3.12}$$

which can be abbreviated as

$$\Delta\eta_n^{(k-1)}(M^{2n-k+1}) = \eta_n^{(k)}(\partial M^{2n-k+1}). \tag{3.12'}$$

Therefore, one could introduce such a cohomology of gauge groups realized upon the characteristic invariants of the secondary classes that $Z^k(\eta,\Delta) = \{\eta_n^{(k)} | \Delta\eta_n^{(k)} = 0\}$ be the set of cocycles, $B^k(\eta\ ,\Delta) = \{\eta_n^{(k)} | \eta_n^{(k)} = \Delta\eta_n^{(k-1)}\}$ the set of coboundaries, and

$$H^k(\eta,\ \Delta) = Z^k(\eta,\ \Delta)/B^k(\eta,\ \Delta)\ , \tag{3.13}$$

the set of equivalence classes of the cocycles. Because of the theorem 3 in the preceding section or the formula (3.11), this cohomology is closely related to the topology of the manifolds on which invariants η's are defined. For instance, $\eta_n^{(k)}(M^{2n-k})$ is a nontrivial element of the cohomology group $H^k(\eta,\ \Delta)$ if and only the boundaryless manifold M^{2n-k} can not be the boundary of any manifold M^{2n-k+1} and so on. As a matter of fact, it is easy to show that there exists a homomorphic relation between the cohomology $H^k(\eta,\ \Delta)$ and the homology of the manifolds, $H_k(M,\ \partial)$, which is made to distinguish topologically inequivalent manifolds[14,20]; that is

$$\begin{aligned} f:\quad & H_k(M,\ \partial) \longrightarrow H^k(\eta,\ \Delta)\ , \\ & [\, M^{2n-k}\,] \longmapsto [\, \eta_n^{(k)}(M^{2n-k})\,]\ , \end{aligned} \tag{3.14}$$

where $[\ldots]$ denotes a representative of an equivalence class, ∂ the boundary operator of $H_k(M,\ \partial)$.

In the simplest case of k=0, for instance, we have

$$\eta_n^{(o)}(M^{2n}) = \int_{M^{2n}} P(F^n) \tag{3.15}$$

If M^{2n} is compact and the characteristic polynomial in the curvature, $P(F^n)$, is taken to be the n-th Chern class, then the zero-th cohomology group $H^o(\eta, \Delta)$ is the Chern number of the gauge field on $M^{2n} \subseteq \mathcal{M}$.

3.3 Cohomology Of Degenerate Forms Of Secondary Classes

We now consider the relation between the cohomology with respect to the coboundary operator $\bar{\Delta}$ and the generalized secondary classes. We will show that one can introduce some polynomials denoted by $\bar{Q}^{(k)}$ such that they solve the descent equation for the degenerate coboundary operator $\bar{\Delta}$

$$\bar{\Delta}\,\bar{Q}_n^{(k-1)} = d\bar{Q}_n^{(k)} \ , \tag{3.16}$$

under certain condition

$$\bar{Q}_n^{(o)} = Q_n^{(o)} = P(F^n), \tag{3.17}$$

which implies that there should be some relations between the polynomials $\bar{Q}_n^{(k)}$ and the generalized secondary classes $Q_n^{(k)}$, since they are descended from the same polynomial in the curvature, $P(F^n)$. Naturally, if we find such kind of $\bar{Q}$'s, we would have another sequence

$$0 \xrightarrow{d^{-1}\bar{\Delta}} \bar{Q}^{(o)} \xrightarrow{d^{-1}\bar{\Delta}} \bar{Q}^{(1)} \xrightarrow{d^{-1}\bar{\Delta}} \cdots \ . \tag{3.18}$$

By using the inductive method and the relations (3.6), (3.8) and

$$\begin{aligned} d\bar{\Delta} &= \bar{\Delta} d, \\ d(r^{(k)}\cdot\pi) &= (dr^{(k)})\cdot\pi \ , \end{aligned} \tag{3.19}$$

one can show that

$$\bar{Q}^{(k)} = Q^{(k)} - d^{-1}\big(\bar{Q}^{(k-1)}\cdot\ \pi + \Delta(d^{-1}\bar{Q}^{(k-2)}\pi)\big), \quad k = 1,2,\ldots, \tag{3.20}$$

solve the descent equation (3.16) under the condition (3.17).

In the course of our analyses, we have made use of an antidifferential operator, d^{-1}, which acts on some closed forms only[5]. We should address that the local exact expressions for the closed forms are not unique and the operator d^{-1} can only be defined in some local sense. Therefore, the set of solutions (3.20) is not unique. To see this, let us consider $Q_n^{(o)}\pi(A^o,A^1) = Q_n^{(o)}(A^1) = P(F^n(A^1))$. Since it is closed, by the Poincaré lemma, $Q_n^{(o)}\pi(A^o,A^1)$ can locally be expressed as an exterior differential of something denoted by $d^{-1}Q_n^{(o)}\pi(A^o,A^1)$. On the other hand, however, the local exact expression for $Q_n^{(o)}$ can also be locally written as

$$Q_n^{(o)}(A^1) = dQ_n^{(1)}(0,A^1) = dQ^{(1)}\iota\pi(A^o,A^1), \tag{3.21}$$

where ι is an inclusion operator which puts zero at the first place of some ordered objects on which it acts, for instance,

$$\iota: (A^1,A^2,\ldots,A^k) \longrightarrow (0,A^1,A^2,\ldots,A^k) . \tag{3.22}$$

It is well known that although we have (3.21) in the local sense, but globally it is true only up to addition of a (2n-1) form which has integer-valued integral over R^{2n-1}. Therefore, we should write

$$d^{-1}Q_n^{(o)}\pi(A^o,A^1) = Q_n^{(1)}\iota\pi(A^o,A^1), \quad (\mathrm{mod}\ \mathbb{Z}). \tag{3.23}$$

Notice that the operator d^{-1} satisfies

$$dd^{-1} = 1 , \quad \text{but} \quad d^{-1}d \neq 1 , \tag{3.24}$$

and the inuniqueness of this sort exists for each value of k. As a matter of fact, we can prove that

$$\overline{Q}^{(k)} = Q^{(k)} - Q^{(k)}\iota\pi , \quad k = 1,2,\ldots, \tag{3.25}$$

also solve the descent equation (3.16) under the condition (3.17).

Finally, we would show that the cohomology of gauge groups in Faddeev's approach[5] is the one upon the degenerate forms of the generalized secondary classes. To this end, we take

$$A^o = A , \quad A^1 = 0 , \quad A^2 = g_1 dg_1^{-1},\ldots,$$
$$A^k = (g_1\cdots g_{k-1})d(g_1\cdots g_{k-1})^{-1},\ldots \tag{3.26}$$

in which $g_1 \ldots, g_{k-1}, \ldots$ are elements of the gauge group G. Then it is easy to prove that the solution (3.20) of the descent equation (3.16) under the condition (3.17) gives rise to the cohomology in Ref. 5,7). Similar results have also been given by Song[6].

4. THE HIERARCHY OF GAUGE AND GRAVITATIONAL ANOMALIES

We have already shown that the generalized secondary characteristic classes in the different dimensional submanifolds of M are cohomologically related to each other. On the one hand, since these classes stem from a common characteristic polynomial in the curvature, they must share some topological properties; on the other hand, these classes have definitely topological meaning in their own right as well. It is well-known that the abelian or singlet anomalies in quantum field theories with certain gauge symmetries at classical level are exactly some invariant polynomials in the curvature of the gauge field, therefore, there must exist a hierarchy of anomalous objects corresponding to the generalized secondary classes in a series of submanifolds of M if the singlet anomalies do exist in highest even dimensional submanifold of M. And these anomalous objects have same definitely topological and cohomological meaning as that of corresponding generalized secondary classes as well.

We will start from the analysis on the singlet gauge anomalies in $M^{2n} \subseteq M$. Let U(x) be an element of a Lie group or its non-singular matrix representation which represents some classical background field. Under the action of the gauge group G, U(x) transformed as

$$U(x) \longrightarrow g^{-1}(x)U(x), \quad g(x) \in G, \qquad x \in M^{2n} \subseteq M \tag{4.1}$$

and its covariant derivative is defined by

$$DU = dU + AU \tag{4.2}$$

where A is the gauge potential 1-form valued on the Lie algebra of G. Let J(U,A) be the expectation value of a singlet anomalous current 1-form of some quantized fermion field on the classical background

U(x) and the external gauge field A. The J(U,A) always satisfy the singlet anomalous divergence equation[8)]

$$d{*}J(U,A) = P(F^n), \tag{4.3}$$

where $*$ is the Hodge star operator. This equation can easily be solved be means of the descent formula (3.10) in the cohomology of the gauge group G, H_Δ. To see this, we take

$$\begin{aligned} A^o &= UdU^{-1}, \quad A^1 = A \\ A_t &= UdU^{-1} + tDUU^{-1} \\ F_t &= tF - t(1-t)(dUU^{-1})^2, \end{aligned} \tag{4.4}$$

then the descent formula gives

$$(\Delta Q_n^{(o)})(UdU^{-1},A) = dQ_n^{(1)}(UdU^{-1},A) = P(F^n(A)) \tag{4.5}$$

$$Q^{(1)}(UdU^{-1},A) = n\int_0^1 dtP(DUU^{-1}, F_t^{n-1}). \tag{4.6}$$

Thus, we get the dual of topologically non-trivial part of J

$${*}J(U,A) = Q_n^{(1)}(UdU^{-1}, A) = n\int_0^1 dtP(DUU^{-1}, F_t^{n-1}); \tag{4.7}$$

that is, $*J(U,A)$ is a 1-cochain density with respect to Δ whereas the singlet anomalous divergent is a zero-cochain density, which will become the zero-cocycle density if M^{2n} has no boundary.

We now consider a submanifold $M^{2n-1} \subset \mathcal{M}$, which may have no relation with the submanifold $M^{2n} \subseteq \mathcal{M}$, and the similar kind of quantum field theory is given as before. It is easy to see that 1-cochain density $Q_n^{(1)}(UdU^{-1}, A)$ can be decomposed into a gauge field part, a U-field part and a part contributed from the boundary of M^{2n-1} by means of the descent formula

$$(\Delta Q_n^{(1)})(UdU^{-1}, A, 0) = dQ_n^{(2)}(UdU^{-1}, A, 0) \tag{4.8}$$

$$Q_n^{(2)}(UdU^{-1}, A, 0) = -n(n-1)\int_0^1 dt^1 \int_0^{1-t^1} dt^2 P(DUU^{-1}, dUU^{-1}, F_{t^1t^2}^{n-2}) \tag{4.9}$$

where

$$\begin{aligned} A_{t^1t^2} &= UdU^{-1} + t^1UDU^{-1} + t^2dUU^{-1} \\ F_{t^1t^2} &= dA_{t^1t^2} + A_{t^1t^2}^2. \end{aligned} \tag{4.10}$$

From (4.8), follows the decomposition formula[8,21]:

$$Q_n^{(1)}(UdU^{-1}, A) = Q^{(1)}(UdU^{-1},0) - Q_n^{(1)}(A,0) + dQ_n^{(2)}(UdU^{-1},A,0). \tag{4.11}$$

It is reasonable to take $(\Delta Q^{(1)})(UdU^{-1},A,0)$ as a gauge invariant Euler-Heisenberg effective Lagrangian, which presents the parity violation anomalies on odd dimensions[22,23,24]. If $\partial M^{2n-1} = \phi$, the effective action becomes 1-cocycle. Otherwise, $Q_n^{(2)}$ discribes the boundary effects.

It should be pointed out that in the case $\partial M^{2n-1} = M^{2n-2}$,

$$\tilde{\Gamma}(U,A) = 2\pi \int_{M^{2n-1}} Q^{(1)}(UdU^{-1},0) + 2\pi \int_{M^{2n-2}} Q^{(2)}(UdU^{-1},A,0) \tag{4.12}$$

is just the gauge invariant Wess-Zumino-Witten effective action in M^{2n-2} which should have no boundary and can be dealt with as the boundary of an M^{2n-1}, if the anomaly-free condition holds, which requines the singlet anomalies on M^{2n} being free. Otherwise,

$$\tilde{\Gamma}(U,A) \qquad \text{or} \qquad 2\pi \int_{M^{2n-1}} Q^{(1)}(A,0) \tag{4.13}$$

gives rise to the generating functional of non-abelian anomaly on such an $M^{2n-2} = \partial M^{2n-1}$, and under the gauge transformation of G, like (4.1), they generate the finite form of such an anomaly, which contains global aspects about the anomaly such as so-called non-purturbative anomaly[25,18,21],

$$\mathcal{A}(g(x),A) = 2\pi \int_{M^{2n-1}} Q^{(1)}(gdg^{-1}, 0) + 2\pi \int_{M^{2n-2}} Q^{(2)}(gdg^{-1},A,0), \tag{4.14}$$

whose infinitesmal form corresponds to the purturbative one

$$\mathcal{A}(\varepsilon(x),A) = 2\pi \int_{M^{2n-2}} \varepsilon^a Q_a(A,T^a) \tag{4.15}$$

$$Q_a(A,T^a) = n(n-1) \int_0^1 dt(1-t)P(T^a d(AF(tA)^{n-2})).$$

Note that each term in $\tilde{\Gamma}(U,A)$ and $\mathfrak{A}(g,A)$ has certain topological meaning and both $\tilde{\Gamma}(U,A)$ and $\mathfrak{A}(g,A)$ as a whole has also certain cohomological meaning which will be shown latter.

However, in the cases of either M^{2n-2} has boundary or can not be the boundary of any M^{2n-1}, the above approach does not work, then the question is how to get the Wess-Zumino-Witten term, the generating functionals and the finite forms of the non-abelian anomalies in such kind of manifolds as mentioned before. From the point of view of the generalized secondary classes and relevant cohomologies, this problem could be solved automatically by taking into account the $Q_n^{(2)}$'s directly. In fact, as long as the connection 1-form A^3 is appropriately taken to be relevant to either the boundary or the non-trivial topology of M^{2n-2}, then the descent equation gives

$$(\Delta Q_n^{(2)})(UdU^{-1},A,0,A^3) = dQ_n^{(3)}(UdU^{-1},A,0,A^3). \tag{4.16}$$

The last term on the left-handed side is just $Q_n^{(2)}(UdU^{-1},A,0)$ whose infinitesimal form gives rise to the purturbative anomaly, and the right—handed side gives rise to the contribution from the boundary of M^{2n-2}. If the manifold M^{2n-2} is a non-trivial element of the homology $H_2(M,\partial)$, then the second generalized characteristic invariant, $\eta_n^{(2)}$, is correspondingly a non-trivial element of the cohomology $H^2(\eta,\Delta)$; that is

$$\Delta\eta_n^{(2)}(UdU^{-1},A,0,A^3;M^{2n-2}) = 0, \quad \eta_n^{(2)} \in H^2(\eta,\Delta), \tag{4.17}$$

which presents the Wess-Zumino-Witten term, the non-abelian anomalies on such an M^{2n-2}.

If we go one step further, take

$$A^o = A,\ A^1 = 0,\ A^2 = g_1 d\, g_1^{-1},\ \ldots,\ A^4 = (g_1g_2g_3)d(g_1g_2g_3)^{-1} \tag{4.18}$$

then we have

$$(\Delta Q_n^{(3)})(A,0,g_1dg_1^{-1},\ldots, g_1g_2g_3d(g_1g_2g_3)^{-1}$$

$$= dQ_n^{(4)}(A,0,g_1dg_1^{-1},\ldots, g_1g_2g_3d(g_1g_2g_3)^{-1}) \,, \tag{4.19}$$

$$(\Delta\eta_n^{(3)})(A,0,g_1dg_1^{-1},\ldots, g_1g_2g_3d(g_1g_2g_3)^{-1};\ M^{2n-3})$$

$$= \eta_n^{(4)}(A,0,g_1dg_1^{-1},\ldots, g_1g_2g_3d(g_1g_2g_3)^{-1};\partial M^{2n-3}). \tag{4.20}$$

These relations would be the cohomological conditions for the anomalous Schwinger terms in equal time commutators on space M^{2n-3} $(n\geq 3)$, or the center term in the Kac-Moody algebra on space M^1.

Let us now consider the cohomological meaning of these anomalous objects from the cohomology $H_{\overline{\Delta}}$, the cohomology of the degenerate co-bounday operator $\overline{\Delta}$ and those polynomials $\overline{Q}^{(k)}$ consisting of the generalized secondary classes.

In the case of k=0, we already assume that

$$\overline{Q}_n^{(o)} = Q_n^{(o)} = P(F^n) \,, \tag{4.21}$$

so the singlet anomalies on $M^{2n}\subseteq \mathcal{M}$ is also a density of zero-cochain or a (-1)-degenerate cochain.

In the case of k=1, corresponding to that of 1-cochain density, we now would have

$$\overline{\Delta}\overline{Q}_n^{(1)}(A,0,UdU^{-1}) = d\overline{Q}_n^{(2)}(A,0,UdU^{-1})$$

whose both sides give

$$Q_n^{(1)}(UdU^{-1},A) + Q_n^{(1)}(A,0)$$

$$= d(Q^{(2)}(A,0,UdU^{-1}) - d^{-1}Q^{(1)}(0,UdU^{-1})). \tag{4.22}$$

The integral of (4.22) over an $M^{2n-1} \subset \mathcal{M}$ gives rise to a kind of Euler-Heisenberg effective action. If $\partial M^{2n-1} = \phi$ and M^{2n-1} can be covered by one coordinate neighbourhood, then this action is a 0-degenerate cocycle in $H_{\bar{\Delta}}$ on such an M^{2n-1}.

If $\partial M^{2n-1} \neq \phi$ and the boundary can be considered as a (2n-2)-dimensional spacetime, $M^{2n-2} = \partial M^{2n-1} \subset \mathcal{M}$ then the integral of the right-handed side of (4.22) gives the Wess-Zumino-Witten term $\tilde{\Gamma}(U,A)$ which should be a 1-degenerate cocycle in $H_{\bar{\Delta}}$

$$\tilde{\Gamma}(U,A) = 2\pi \int_{x\in M^{2n-2}} (Q_n^{(2)}(A,0,UdU^{-1}) - d^{-1}Q_n^{(1)}(0,UdU^{-1})), \qquad (4.23)$$

$$\bar{\Delta}\tilde{\Gamma}(U,A) = 0 .$$

Similarly, the finite forms of non-abelian anomalies, $\mathfrak{A}(g,A)$, is also a 1-degenerate cocycle in $H_{\bar{\Delta}}$

$$\mathfrak{A}(g,A) = 2\pi \int_{x\in M^{2n-2}} (Q_n^{(2)}(A,0,gdg^{-1}) - d^{-1}Q_n^{(1)}(0,gdg^{-1})), \qquad (4.24)$$

$$\bar{\Delta}\,\mathfrak{A}(g,A) = 0 .$$

It should be noted, however, that we must take care of the integration region because of the local sense of the degenerate cochains. Furthermore, for k=2 and k=3 we have

$$(\bar{\Delta}\bar{Q}_n^{(2)})(A,0,g_1dg_1^{-1},\ g_1g_2d(g_1g_2)^{-1})$$

$$= d\bar{Q}_n^{(3)}(A,0,g_1dg_1^{-1},\ g_1g_2d(g_1g_2)^{-1}), \qquad (4.25)$$

$$(\bar{\Delta}\bar{Q}_n^{(3)})(A,0,g_1dg_1^{-1},\dots,g_1g_2g_3d(g_1g_2g_3)^{-1})$$

$$= d\bar{Q}_n^{(4)}(A,0,g_1dg_1^{-1},\dots,g_1g_2g_3d(g_1g_2g_3)^{-1}).$$

If the submanifold $M^{2n-3} \subset \mathcal{M}$ has no boundary and can be covered by one coordinate neighbourhood, the integral of $(4.25)_2$ over such an M^{2n-3} gives a 2-degenerate cocycle, $\bar{\eta}^{(3)}_n$,

$$\bar{\Delta}\bar{\eta}^{(3)}_n(M^{2n-3}) = 0. \tag{4.26}$$

For the case of n=3 and the characteristic polynomial $P(F^3)$ being the 3rd Chern class, $\bar{\eta}^{(3)}_3$ it just the 2-cocycle α_2 on R^3 introduced by Faddeev[5] to get the anomalous Schwinger term. For n=2 and $P(F^2)$ being the 2nd Chern class, $\bar{\eta}^{(3)}_2$ is Faddeev's 2-cocycle α_2 on R^1 relevant to the center of the Kac-Moody algebra.

One can deal with the cases in which both left and right handed gauge fields are involved. Let us take the Wess-Zumino-Witten term[8,21] as an example which corresponds to k=1. Take

$$A^o = A_L,\ A^1 = A_R\ ,\ A^2 = {}^{U^{-1}}A_R = UA_RU^{-1} - dUU^{-1}, \tag{4.27}$$

then the descent equation in $H_{\bar{\Delta}}$ gives

$$(\bar{\Delta}\bar{Q}^{(1)}_n)(A_L,A_R,{}^{U^{-1}}A_R) = d(Q^{(2)}_n - d^{-1}Q^{(1)}_n\pi)(A_L,A_R,{}^{U^{-1}}A_R). \tag{4.28}$$

From (4.28), it follows that

$$Q^{(1)}_n(A_L,A_R) - Q^{(1)}_n(A_L,{}^{U^{-1}}A_R) = d(Q^{(2)}_n - d^{-1}Q^{(1)}_n\pi)(A_L,A_R,{}^{U^{-1}}A_R). \tag{4.29}$$

If the submanifold $M^{2n-2}\subset M$ can be regarded as the boundary of another submanifold $M^{2n-1}\subset M$ which can be covered by one coordinate neighbourhood, then by Stokes' formula, we can get the Wess-Zumino-Witten term with both left and right handed gauge fields on $M^{2n-2} = \partial M^{2n-1}$,

$$\tilde{\Gamma}(U,A_L,A_R) = 2\pi \int_{x\in M^{2n-2}} (Q^{(2)}_n - d^{-1}Q^{(1)}_n\pi)\ (A_L,A_R,{}^{U^{-1}}A_R). \tag{4.30}$$

Since

$$2\pi \int_{x\in M^{2n-1}} Q^{(1)}_n(A_L,A_R)$$

contains Bardeen's counter-term R_3 to shift the anomaly from the (L-R) form to the (V-A)-form in 4-dimensional spacetime[26,8), therefore, (4.30) is the Wess-Zumino-Witten term in the (V-A) form. From (4.28), this term is of course a 1-degenerate cocycle in $H_{\bar{\Delta}}$ with respect to the operator $\bar{\Delta}$ and the gauge group $G_L \times G_R$, that is

$$\bar{\Delta}\tilde{\Gamma}(U, A_L, A_R) = 0. \tag{4.31}$$

We should address that the physical implications of the polynomials and their characteristic invariants for $k \geq 4$ in both H_{Δ} are of course intriguing. For instance, in the case n=3, k=4, these quantities may be relevant to the monopole and so on. We would explore these elsewhere.

Finally, we would like to point out our formalism can also be used to deal with the gravitational anomalies including both Einstein and Lorentz anomalies[28-32,21,11). Similar to the analyses on the gauge anomalies, once that the characteristic polynomials in the curvatures of both gauge fields and gravitational fields for the singlet gravitational anomalies are suitablely taken by means of either purturbation calculations[28) or the index theorems[30,31), we would have a corresponding hierarchy of gravitational anomalies in a series of gravitational fields with different dimensions. According to the cohomological analyses based upon the generalized secondary characteristic classes, we will be able to show the concrete forms and the properties of these anomalies not only for some boundaryless gravitational fields with trivial homotopy properties but also for those fields with boundary or non-trivial homotopy properties. We will show these cohomological analyses in some detail elsewhere[4).

ACKNOWLEDGMENTS

The first author (H.Y.G.) would like to thank M.A. Markov for his kind invitation to participate in the 3rd Seminar on Quantum Gravity, Moscow, October 23-25, 1984, and also to thank V.A. Belinskii, V.A.Berezin, A.V. Byalko, I.M. Khalatrikov, L. Kuzin, M.A. Markov, A.M. Prokhorov for their warm hospitality extended to him.

H.Y.G. is also very grateful to L.D. Faddeev and I.M. Gelfand for interesting and useful discussions during the Seminar. It was indicated that Faddeev and his group has also considered similar issues, Gelfand and his coworkers had independently done some works on the secondary classes and their generalization in the case of tanget bundle[15], so the generalized secondary characteristic classes proposed by Guo, Wu, and Wang[1] would be called the characteristic classes of Chern-Simons-Gelfand type.

REFERENCES

1) Guo, H.Y., Wu, K. and Wang, S.K., "Chern-Simons Type Characteristic Classes", Preprint AS-ITP-84-035, (1984).
2) Guo, H.Y., Hou, B.Y., Wang, S.K. and Wu, K., "Anomaly, Cohomology And Chern-Simons Cochain", Preprint AS-ITP-84-039, (1984).
3) Guo, H.Y., Hou B.Y., Wang, S.K. and Wu, K., "Cohomology Of Gauge Groups And Characteristic Classes Of Chern-Simons Type", Preprint AS-ITP-84-041. (1984).
4) Guo, H.Y., Wu, K. and Wang, S.K., "Cohomological Analyses Of Gravitational Anomalies", In preparation.
5) Faddeev, L.D., Phys. Lett <u>145B</u>, 81(1984).
6) Song, X.C., Private communication (1984).
7) Zumino, B., "Cohomology Of Gauge Groups: Cocycles And Schwinger Term", Santa Barbara NSF-ITP Preprint, (1984),
8) Chou, K.C., Guo, H.Y., Wu, K. and Song, X.C., Phys. Lett. <u>134B</u>, 67(1984); Commun. in Theor. Phys. (Beijing) <u>3</u>, 73(1984); Chou, K.C., Guo, H.Y., Li, X.Y., Wu, K. and Song X.C., ibid <u>3</u>,

491(1984).

9) Zumino. B., Les Houches Lectures 1983, to be published by North-Holland, Stora, R. and DeWitt, B. editors.
Zumino, B., Wu, Y.S. and Zee, A., Nucl. Phys. B239, 477(1984).

10) Stora, R., Cargese Lectures (1983).

11) Alvarez-Gaumé, L.and Ginsparg, P., "The Structure Of Gauge And Gravitational Anomalies", Preprint HUTP-84/A016; and the references therein.

12) Chern, S.S. and Simons, J., Proc. Nat. Acad. Sci. USA 68, 791(1971).

13) Chern, S.S., "Complex Manifolds Without Potential Theory", Springer, New York, (1979).

14) Eguchi, T., Gilkey, P.B. and Hanson, A.J., Phys. Reports 66, 213 (1980).

15) Gabrielov, A.M., Gelfand, I.M. and Losik, M.V., Funct. Anal. Appl. 9, 12(1975).

16) See, for example, Goldstone, J. and Jaffe, R.L., Phys. Rev. Lett. 51, 1518(1983).

17) Wess, J. and Zumino, B., Phys. Lett. 37B, 95(1971).

18) Witten, E., Nucl. Phys. B223, 422(1983).

19) Cartan, H. and Eilenberg, S., "Homological Algebra", Princeton University Press, Princeton (1956).

20) Bott, R. and Tu, L., "Differential Forms In Algebraic Geomatery", Springer, New York, (1982).

21) Chou, K.C., Guo, H.Y. and Wu, K. "Anomalies Of Arbitrary Gauge Group And Its Reduction Group, Einstein And Lorentz Anomalies", Preprint AS-ITP-84-030, (1984).

22) Deser, R., Jackiw, R. and Templeton, S., Phys. Rev. Lett. 48, 975 (1982); Ann. Phys. (N.Y.) 140, 372(1982).

23) Neimi, A.J. and Semenoff, G.W., Phys. Rev. Lett. 51, 2077(1983).

24) Redlich, A.N., ibid. 52, 18(1984).

25) Witten, E., Phys. Lett. 117B, 324(1982).

26) Bardeen, W.A., Phys. Rev. 184, 1848(1969).

27) Gross, D.J. and Jackiw, R., Phys. Rev. D6, 477(1972).

28) Alvarez-Gaumé, L. and Witten, E., Nucl. Phys. B234, 269(1984).

29) Bardeen, W.A. and Zumino, B., Preprint UCB-PTH84-12, (1984).

30) Atiyah, M.F. and Singer, I.M., Proc. Nat. Acad. Sci. USA 81, 2597(1984).

31) Alvarez, O., Singer, I.M. and Zumino. B., Preprint UCB-PTH84-9 (1984).

32) Langouche, F., Schücker, T. and Stora, R., Preprint Ref.TH.3938-CERN (1984).

ON QUANTUM THEORY OF MEASUREMENTS OF GRAVITATIONAL FIELD

M.B. Mensky

State Standards Committee
Moscow, USSR

ABSTRACT

Quantum measurement theory for gravitational field is formulated on the basis of Feynman integral in finite limits. Application of this approach to local measurement of curvature gives absolute limits on its precision and localization.

1. INTRODUCTION

In 1933 Bohr and Rosenfeld/1/ analized restrictions on measurability of electromagnetic field due to its quantum properties. Later on an analogous analysis was applied by different authors/2-14/ to the gravitational field. It was shown that in gravitation, in contrast with electrodynamics, absolute limits should exist for measurability of field strength. It may be expressed by the inequality $\Delta\Gamma \gtrsim L^2/\ell^3$ for the error in determination of typical connection coefficient. Here ℓ denotes the size of the measurement domain while L is the Planck length. The other conclusion arrived at in the cited papers is that $\ell \gtrsim L$, so that the gravitational measurement cannot be localized in a sub-Planckian domain.

In the present paper the new approach is proposed to quantum measurements theory of the gravitational field. It contains no explicit description of a measuring device, but indication of information resulting from the measurement. The approach is based upon the Feynman path integral "in finite limits". This means that integral is taken not over all paths (field configurations) but rather over some restricted set of paths (configurations). Earlier this approach has been applied in non-relativistic quantum mechanics and led to quantum theory of continuous measurements[/15-18/].

After general formulation the proposed method is applied to measurement of a curvature, R, in a small space-time region of the size ℓ. It is shown that the character of the measurement process and information about geometry resulting from the measurement depend on relation between the error of classical measuring device, ΔR, and the specific quantum threshold $\Lambda^{-2} = L^2/\ell^4$, where $L = 4(\pi\hbar G/c^3)^{1/2} = 10^{-32}$cm is a length of the order of the Planck length.

If $\Delta R \gtrsim \Lambda^{-2}$, quantum effects are negligible, and the whole measurement noise is due to the errors of the measuring device, $\Delta R_{class} = \Delta R$. This is a classical mode of measurement. But if $\Delta R \ll \Lambda^{-2}$, quantum effects become predominant, and the quantum measurement noise is shown to be infinite, $\Delta R_{quant} = \infty$. This means that the measurement in this quantum mode provides no information at all about space-time geometry in the measurement domain. Thus in any measurement mode the curvature cannot be determined with precision more than $\Lambda^{-2} = L^2/\ell^4$. It is shown that in a sub-Planchkian region, $\ell \ll L$, only quantum mode is possible, so that no information of geometry in such a region is available.

Sect. 2 contains the main principles of quantum theory of continuous measurements. In Sect. 3 the quantum measurement theory for gravity is formulated on this basis. In Sects. 4,5 it is applied to measurement of curvature in a small region. Interpretation of the obtained conclusions by direct analysis of measuring bodies is given in Sect. 6, and Sect. 7 presents some general remarks.

2. QUANTUM THEORY OF CONTINUOUS MEASUREMENTS

The quantum theory of continuous measurements based on the path integration "in finite limits"/15-18/ may be readily illustrated by the example of tracking position of a quantum system, which can be called measurement of a path.

According to Feynman, the amplitude for the system to transit from one point, x, to another, x', can be expressed by the integral over all paths connecting these points. Let however the observer is keeping track of the coordinate of the system during the transition. This can be achieved by measuring the coordinate (with a finite precision) in each time moment. Then the transition amplitude should be taken equal to the integral over the corridor of paths, corresponding to the result of tracking. Denote this corridor by D. Its width is equal to the error in measurement of coordinate in each time, while its form is defined by the results of these measurements. This gives

$$A_D = \int_D d[x] \exp((i/\hbar)S[x])$$

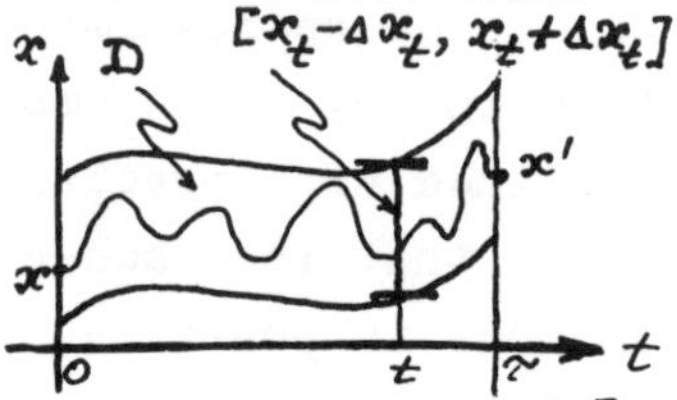

It is convenient to use a weight functional $\rho_D[x]$ instead of the corridor D. Then

$$A_D = \int d[x]\, \rho_D [x] \exp((i/\hbar)S[x]), \qquad (1)$$

provided ρ_D is close to unity in D and decreased outside. The Gaussian weight (an exponential of a quadratic form) is most convenient and physically reasonable.

The equation (1) may be written not only for the measurement of a path, but also for any type of continuous measurement/15-18/. The set of paths, D, should be chosen then to express the concrete result of the measurement. The quantity A_D should be interpreted as a probability amplitude for the measurement to give the result D. The points x, x' fix boundary conditions, i.e. positions of the system before and after measurement.

In practice the integral (1) often can be reduced to a product of one-dimensional integrals of the form

$$A_{z,\Delta z} = \int dz' \ \rho_{z,\Delta z}(z') \exp(i\Lambda^2(z'-z_{class})^2) \qquad (2)$$

where an exponential arises as contribution from $\exp((i/\hbar)S)$ into one-dimensional integral. The amplitude (2) describes measurement of some subsidiary observable z', while Δz and z are correspondingly the error and the result of its measurement. z_{class} stands for the classical value of z', corresponding to the minimum of the action (with given boundary conditions).

With the Gaussian weight $\rho_{z,\Delta z} = \exp(-(z'-z)^2/\Delta z^2)$ the integral (2) gives for the probability density $(P = |A|^2)$

$$P_{z,\Delta z} = \frac{N}{\Delta z_{tot}} \exp(-2 \frac{(z-z_{class})^2}{\Delta z^2_{tot}}) \qquad (3)$$

The most probable result of the measurement coincides

with the classical value $z = z_{class}$, while the width of the interval of comparatively probable results is

$$\Delta z_{tot} = \sqrt{\Delta z^2 + \frac{1}{\Lambda^4 \Delta z^2}} \qquad (4)$$

It contains classical, $\Delta z_{class} = \Delta z$, and quantum, $\Delta z_{quant} = 1 / \Lambda^2 \Delta z$, contributions. The interval $[z - \Delta z_{tot}, z + \Delta z_{tot}]$ may be interpreted as a confidence interval for the measurement result z.

The measurement has a classical nature, when $\Delta z \gg \Lambda^{-1}$. In this case $\Delta z_{tot} \simeq \Delta z$, hence the confidence interval for the measurement result is defined by the errors of the measuring device. This is a classical mode of measurement. The quantum mode corresponds to $\Delta z \ll \Lambda^{-1}$, when the quantum contribution becomes dominant, $\Delta z_{tot} \simeq \simeq 1/\Lambda^2 \Delta z$. It presents a quantum measurement noise, caused by inavoidable influence of a measuring device. The optimal mode, leading to maximum information, is between classical and quantum ones, i.e. for $\Delta z \simeq \Lambda^{-1}$. Then $\Delta z_{tot} \simeq \Lambda^{-1}$. In all modes of the measurement the inequality $\Delta z_{tot} \gtrsim \Lambda^{-1}$ holds, giving an absolute quantum limit, which cannot be overcome by the choice of a measuring apparatus.

3. QUANTUM THEORY OF GRAVITATIONAL MEASUREMENTS

The classical gravitational field may be described by the metric tensor $g_{\mu\nu}(x)$ satisfying the Einstein equation which in turn follows from the Einstein action

$$S[g] = (c^3/16\pi G) \int d^4x \, (g(x))^{1/2} \, R(x)$$

In Feynman integral formulation of quantum gravity all configurations $[g]$ of the field $g_{\mu\nu}$ contribute the amplitudes:

$$A = \int d\,[g] \;\; \exp((i/\hbar)S[g]).$$

The measure $d[g]$ should be actually a measure in the space of equivalence classes of field configurations, equivalence defined by coordinate transformations.

If some characteristic σ' of the gravitational field undergoes measurement with the error $\Delta\sigma$, then the result of the measurement, σ, can be expressed by the set D of field configurations, including those $[g]$, which are characterized by the values $\sigma' \in [\sigma-\Delta\sigma, \sigma+\Delta\sigma]$. Then the probability amplitude for the measurement to give the result D is

$$\begin{aligned} A_D &= \int_D d[g] \; \exp((i/\hbar)S[g]) = \\ &= \int d[g] \; \rho_D[g] \; \exp\left((i/\hbar)S[g]\right), \end{aligned} \qquad (5)$$

with the weight functional ρ_D decreasing outside D. If the measurement in question is localized in the space-time region V, then $[g]$ in (5) have to be field configurations in V, satisfying some conditions at the boundary of V. The action integral S in (5) also should be taken over V.

This treatment of measurement does not include explicit description of a measuring device. The latter is reflected only by indication the characteristic of the field tò be measured as well as the error of this measurement. Back influence of the measuring device on the field is taken into account automatically in the present approach. To be more precise, only minimum possible (inavoidable) influence is accounted in such a way. Therefore, the limitations following from the proposed calculation scheme are common for all devices measuring the same quantity with the same precision (for all devices of the same class).

4. LOCAL MEASUREMENT OF GRAVITATIONAL FIELD

Let us consider measurement of the gravitational field strenght (curvature) in a small space-time region V of the size ℓ. For the sake of simplicity consider a scalar curvature R'as the quantity to be measured, though the main results may be expected to be valid for a typical component of the curvature tensor.

If R be the result of measurement and ΔR be the error of the measuring apparatus, then the set D of field configurations corresponding to this measurement result may be presented by the inequality $R-\Delta R \lesssim R' \lesssim R+\Delta R$. The action integral takes the form

$$S = (\hbar/L^2)\int_V d^4x\, g^{1/2} R' \simeq \hbar\Lambda^2 R', \tag{6}$$

where $L^2 = 16\pi G\hbar/c^3$, $\Lambda^2 = \ell^4/L^2$.

Now one may suppose that integration in $[g]$ may be reduced to integration in R':

$$\begin{aligned} A_{R,\Delta R} &= \int_{R-\Delta R}^{R+\Delta R} d\mu(R') \exp(i\Lambda^2 R') \\ &= \int d\mu(R')\, \rho_{R,\Delta R}(R') \exp(i\Lambda^2 R'). \end{aligned} \tag{7}$$

The simplest choice of the measure, $d\mu(R') = dR'$, proves to be physically senseless. The manifestation of this is that it leads to the probability density not depending of the result of measurement, R. The mathematical origin of this is that the exponent in (7) depends on the integration variable, R', linearly. From the physical point of view this is unacceptable because the action has no extremum in the set of all competing alternatives, so that no analogue exists of a classical trajectory.

The measure $d\mu(R')$ should be chosen so as to avoid this difficulty. For this end the typical integral (2) may serve as a guiding line. Let us introduce a subsidiary variable z' in such a way that the action be quadratic in it, $R' = z'^2 + R_{class}$, and take $d\mu(R') = dz'$. Then the action has a minimum for $z' = 0$, or for $R' = R_{class}$. The latter value plays the role of the classical gravitational field defined by the boundary conditions at the boundry of V.

With this definition (7) takes the form

$$A_{z,\Delta z} = \int dz' \exp\left[i\Lambda^2(z'^2 + R_{class}) - (z'-z)^2/4\Delta z^2\right] \quad (8)$$

where $z = (R - R_{class})^{1/2}$, $2\Delta z = \Delta R/(R-R_{class})^{1/2}$. Then integration gives

$$P_{z,\Delta z} = \frac{N}{\Delta z_{tot}} \exp\left(-2\frac{z^2}{\Delta z_{tot}^2}\right), \quad (9)$$

where

$$\Delta z_{tot}^2 = 4\Delta z^2 + \frac{1}{4\Lambda^4 \Delta z^2}.$$

From (9) one sees that the most probable value of z is $z = 0$, but the values differing from it less than by Δz_{tot} are probable too. This goves $-\Delta z_{tot} \lesssim z \lesssim \Delta z_{tot}$, or $z^2 \lesssim \Delta z_{tot}^2$. In terms of curvature this inequality takes the form

$$R - R_{class} \lesssim \frac{\Delta R^2}{R - R_{class}} + \frac{R - R_{class}}{\Lambda^4 \; \Delta R^2}$$

Solving this inequality in respect of $R-R_{class}$ (and accounting that $R-R_{class}$ cannot be much less than ΔR), one has

$$R - R_{class} \simeq \Delta R \quad \text{for } \Delta R > \Lambda^{-2} \quad \text{(classical mode)},$$
$$\Delta R \lesssim R - R_{class} < \infty \quad \text{for } \Delta R < \Lambda^{-2} \quad \text{(quantum mode)}.$$

Thus $\Delta R = \Lambda^{-2}$ is a quantum threshold separating the classical and quantum modes of measurement. The specific feature of the curvature measurement is that in quantum mode the spread of possible measurement results becomes infinite. Infinite is in this case the confidence interval of any measurement result. Possible interpreration of this is that measurement in quantum mode is impossible.

For better understanding of the result obtained let us derive it in another way. Rewright the probability distribution (9) in terms of the curvature:

$$P_{R,\Delta R} = \frac{N'}{\Delta R_{tot}} \exp\left[-2\,\frac{(R - R_{class})^2}{\Delta R_{tot}^2}\right], \qquad (10)$$

where

$$\Delta R_{tot}^2 = \Delta R^2 + \frac{(R - R_{class})^2}{\Lambda^4 \Delta R^2}. \qquad (11)$$

In the case $\Delta R \gg \Lambda^{-2}$ one can put (10) into the form

$$P^{class}_{R,\Delta R} = \frac{N'}{\Delta R} \exp\left[-2\,\frac{(R-R_{class})^2}{\Delta R^2}\right], \qquad (10')$$

so that the spread of probable measurement results is defined by the inequality $R - R_{class} \lesssim \Delta R$. This is nothing else than a classical mode of measurement. If the opposite condition is fulfilled, $\Delta R \ll \Lambda^{-2}$, (10) may be written as follows:

$$P^{quant}_{R,\Delta R} = \frac{N''}{R - R_{class}}. \qquad (10'')$$

The distribution (10'') is un-normalizable. This means that dispertion of R is infinite in the quantum measurement mode. The quantum measurement noise turns out to be infinite.

Thus, the final conclusion is that in the quantum mode of measurement ($\Delta R \ll \Lambda^{-2}$) the spread of the results becomes infinite. In these conditions confidence interval of the result of measurement is infinite. One may say that measurement is impossible for $\Delta R \ll \Lambda^{-2}$.

5. THE UTMOST LOCALIZATION OF GRAVITATIONAL MEASUREMENT

Up to now no restriction was derived on the size of the measurement domain. Implicitly however some assumptions were accepted which restrict it, leading finally to the inequality $\ell \gtrsim L$. Indeed, replacement of the integral over the field configurations, $[g]$, by integral over the curvature values, R', implies that only those field configurations contribute which are smooth and correspond to Euclidean topology. However, the topology being Euclidean and the geometry being smooth require $|R'| \lesssim \ell^{-2}$. This requirement ought to be implied in the calculation. All R' included into the interval $[R-\Delta R, R+\Delta R]$ should satisfy this condition, hence $\Delta R \lesssim \ell^{-2}$. In the case of sub-Planckian region, $\ell \ll L$, one has $\ell^{-2} \ll \Lambda^{-2}$, so that $\Delta R \lesssim \ell^{-2} \ll \Lambda^{-2}$. This means that only quantum mode of measurement is possible in a sub-Planckian domain. But we saw in the preceding section that the measurement noise in this mode is infinite. This means that the measurement becomes actually impossible in a sub-Planckian domain.

Impossibility of gravitational measurements in sub-Planckian domain was proclaimed in a number of papers (see, for instance, /9,6,11/). The most radical formulation was put forward by Markov[11], who proposed that the fundamental limitation exists in quantum gravity, $\ell > L$, analogous to the limitation $v < c$ in special relativity and $S > \hbar$ in quantum theory. The quantum gravity limitation, $\ell > L$, means, according to Markov, that geometric and topological conceptions become meaningless in a sub-Planckian region. The present investigation supports this statement in the sense that no information is available about the geometry of space-time in sub-Planckian scale.

6. QUALITATIVE ANALYSIS OF THE CURVATURE MEASUREMENT

In Sects, 4,5 above the following conclusions were obtained, in the framework of formal quantum measurement theory, about measurement of curvature:

(i) If $\Delta R \gg L^2/\ell^4$, quantum effects are negligible so that the measurement results disperse only due to the measuring device errors, $\Delta R_{class} = \Delta R$.

(ii) If $\Delta R \ll L^2/\ell^4$, then the quantum effects become predominant, leading to infinite quantum measurement noise, $\Delta R_{quant} = \infty$. The measurement gives then no information about geometry of the measurement domain.

(iii) The measurement carried out in a sub-Planckian domain, $\ell \ll L$, cannot be in the classical mode, so that no information about geometry of such a domain is available.

Now we shall support these statements with the help of physical analysis, considering test bodies explicitly. For the sake of simplicity, the Planck length $L_0 = (\hbar G/c^3)^{1/2}$ will be used in the following, instead of $L = 4\sqrt{\pi} L_0$ used previously. This replacement does not matter in estimations up to the order of value.

Let the test bodies used in the measurement have masses of the order of m. Denote $\lambda = \hbar/mc$ the Compton length of such a body and $r_g = mG/c^2$ its gravitational radius (up to the factor 2). The identity $\lambda r_g = L_0^2$ will be one of the key points in our analysis. The other is that the proper gravitational field of the body of mass m is characterized by the curvature of the order of r_g/ℓ^3 at the distance of the order ℓ. In the measurement of the curvature, fulfilled in the domain of the size ℓ, this proper field inserts the error $\Delta R = r_g/\ell^3 = L_0^2/\lambda\ell^3$.

Now we are in the position to start our analysis. It is necessary for the mass m being a test body that it could be considered to be point-like in the limits of the measurement domain. If $r_g \gg \ell$, then gravitational structure of the body prevents this. Therefore, $r_g \lesssim \ell$ or, equivalently, $\Delta R = r_g/\ell^3 \lesssim \ell^{-2}$, is necessary for the mass m could be a test body. This condition as a purely geometrical requirement was used in the preceding section. Now we see it from the other point of view.

Consider the classic mode of measurement. For the quantum properties of test bodies not being manifested, it is necessary that $\lambda \ll \ell$. This equivalent to $\Delta R = L_0^2/\lambda\ell^3 \gg L_0^2/\ell^4$. Consequently, $\Delta R \gg L_0^2/\ell^4$ is a condition for the measurement to be purely classical. This coincides with the statement (i) formulated at the beginning of the present section, and gives interpretation of this statement. In the classical mode of measurement the spread of the measurement results is defined exclusively by the error of the measuring device, $\Delta R_{class} = \Delta R$, this being the second part of (i).

In the quantum mode of measurement $\lambda \gg \ell$, i.e. the test bodies are essentially quantum in the measurement domain. To observe their position with precision ℓ one can make use of the photons with the wave length of the order of ℓ, and therefore with the energy $E = \hbar c/\ell$. Thus the energy E localized in the region of the size ℓ is needed as an additional measuring body. The gravitational radius of this body is $R_g = EG/c^4$ and satisfies the relation $R_g \ell = L_o^2$. The proper curvature created by this energy at the distance ℓ is equal to $R_g/\ell^3 = L_o^2/\ell^4$. This curvature gives a contribution into the measurement noise ΔR_{quant} which therefore is not less than L_o^2/ℓ^4.

Yet this is not the main origin of the measurement noise in the quantum mode of measurement. Indeed, the movement of the test body, observed with precision $\ell \ll \lambda$, bears no relation to gravitational field. This movement is completely occasional because of quantum fluctuations. Thus the condition $\ell \ll \lambda$ prevents usage of this body as a test body. The quantum mode of measurement turns out to be impossible. The other way to express this thought is that the measurement noise becomes infinite in the quantum mode, $\Delta R_{quant} = \infty$. This confirms the statement (ii) on physical grounds.

As a matter of fact, even more may be said in support of this statement. In the case $\ell \ll \lambda = \hbar/mc$ interaction of the mass m with the photons of energy $E = \hbar c/\ell$ will arise creation of pairs of such masses, so that individuality of the test body will be lost.

In conclusion, consider measurement localized in a sub-Planckian domain, $\ell \ll L_o$. The classical mode of measurement requires then $\lambda \lesssim \ell \ll L_o$. From $\lambda r_g = L_o^2$ it follows that $r_g \gg \ell$. Thus gravitational structure of any body prevents its localization (as a classical body) in the sub-Planckian domain. No material body can therefore serve in such a domain as a point-like test body in a classical mode of measurement. The classical mode of measurement occurs impossible in a sub-Planckian domain. This proves the statement (iii).

7. CONCLUDING REMARKS

In the first sections of the present paper the formal scheme was developed for solving problems of quantum theory of measurements of the gravitational field. This scheme makes use of the Feynman path integral "in finite limits". In the subsequent sections the developed approach was applied to calculation of local measurement of the space-time curvature. A simple approximation was accepted including the only characteristic of the gravitational field, a scalar curvature. (Yet the main results may be expected to maintain for measurement of a typical component of the curvature tensor.)

The absolute limit L^2/ℓ^4 was shown to exist for precision with which the curvature in the space-time region of the size ℓ can be estimated from the measurement result. The measurement in a sub-Planckian region, $\ell \ll L$, is shown to be impossible (or to give no information on geometry in the region).

An interesting specific feature of gravitational measurement occured to be impossibility of the quantum mode of measurement. In fact, in the conditions when quantum effects are essential (for $\Delta R \ll L^2/\ell^4$) they are always lead to infinite measurement noise. In non-relativistic quantum systems investigated earlier /15-18/ quantum measurement noise was always finite though encreased with promotion of measurement precision (because of increased influence of the measuring device on the measured system).

Some general remarks can be made in connection with the proposed approach to quantum theory of gravitational measurements: 1) It can be extended to include not only gravitational but any other (material) fields. 2) It is suitable for treatment of complicated schemes of measurement. 3) It needs no explicit consideration of a measuring apparatus, but fixing the quantities measured and precision of their measurement. 4) It gives the measurability limits which cannot be overcome by the choice of the measuring apparatus. Two latter features of the present path **integral** approach make it a sort of the uncertainty principle for measurements of fields and processes.

Some purely mathematical difficulties necessarily will arise in applications of the present approach, since calculations with path integrals are often arduous. To overcome this one can use the Regge calculus (see/19/) as a regular method for discovering effective approximations. In this connection the paper by Hartle at the present Seminar is of great interest/20/.

The author is extremely obliged to Prof. M.A. Markov for fruitful and stimulating discussions on some of the questions considered here.

REFERENCES

1. Bohr, N. and Rosenfeld, L., Kgl. Danske Videnskab Selskab., Mat.-fys. Medd. 12, 1 (1933).
2. Rosenfeld, L., Report at the Conference on the Role of Gravitation in Physics, Chapel Hill, North Carolina (1957).
3. Peres, A. and Rosen, N., Phys. Rev. 118, 335 (1960).
4. Wheeler, J.A., Ann. Phys. 2 , 604 (1957).
5. Wheeler, J.A., Geometrodynamics and the Issue of the Final State, in "Relativity, Groups and Topology", C.DeWitt and B.DeWitt, eds., Blackie and Son, London and Glasgow, 1964.
6. DeWitt, B., The Quantization of Geometry, in "Gravitation: an Introduction to Current Research", L.Witten, ed., John Wiley and Sons, New York and London, 1962.
7. DeWitt, B., Dynamical Theory of Groups and Fields, in "Relativity, Groups and Topology", C.DeWitt and B.DeWitt, eds., Blackie and Son, London and Glasgow, 1964.
8. Treder, H.-J., Fortschr. Phys. 11, 81 (1963).
9. Regge, T., Nuovo Cimento 7, 215 (1958).
10. Markov, M.A., Progr. Theor. Phys., Suppl. Extra Number (Commemoration issue for 30th Anniversary of the meson theory by dr. H.Wukawa), p. 85 (1965).
11. Markov, M.A., On Quantum Violation of Topology in Small Spatial Regions, preprint, Inst. Nucl. Res., Acad. Sci. USSR, P-0187, 1980.
12. Borzeszkowski, H.-H.v. and Treder, H.-J., Found. Phys. 12, 1113 (1982).
13. Borzeszkowski, H.-H.v. and Treder, H.-J., Annalen der Physik, 40, 287 (1983).
14. Bergmann, P.G. and Smith, G.J., Ben. Rel. Grav. 14, 1131 (1982).
15. Mensky, M.B., Phys. Rev. D20,384 (1979).

16. Menskii, M.B., Sov. Phys. JETP 50, 667 (1979).
17. Mensky, M.B., Quantum Restrictions on Accuracy of Resonant Detectors, in "Abstracts of Contributed Papers, 9th Intern. Conf. on Gen. Rel. and Grav. (Jena, 1980)", Jena, 1980, vol. 2, p. 426.
18. Mensky, M.B., "Group of Paths: Measurements, Fields, Particles", Nauka publishers, Moscow, 1983 (in Russian).
19. Misner, C.W., Thorne, K.S. and Wheeler, J.A., "Gravitation", W.H. Freeman and Co., San Francisco, 1973.
20. Hartle, J.B., Simplicial Quantum Gravity, in the present volume.

THE CONCEPT OF GRAVITONS AND THE MEASUREMENT OF EFFECTS OF QUANTUM GRAVITY

Horst-Heino v. Borzeszkowski
Einstein-Laboratorium für Theoretische Physik
der Akademie der Wissenschaften der DDR
Rosa-Luxemburg-Str. 17a, 1502 Potsdam-Babelsberg
GDR

ABSTRACT

It is demonstrated, first, that there arise limitations in Quantum Gravity which show that General Relativity Theory (GRT) is essentially classical and, second, that the graviton language is accordingly only useful in certain approximations to GRT, in order to harmonise GRT and the quantum theory of matter.

1. INTRODUCTION

Einstein's equations of GRT connect quantized non-gravitational matter described by its energy-momentum tensor $T_{\mu\nu}$ and gravitational fields described by the metric tensor $g_{\mu\nu}$ of a Riemannian space-time. In order to evoid physical and mathematical inconsistencies resulting otherwise from these equations one has to consider quantisation of gravitational fields. The quantum procedure should unify or at least harmonize classical GRT and quantum theory[1]. Accordingly there were developed many methods of quantisation of gravitational fields. Notwithstanding this necessity there is the question to answer whether this quantisation shows the existence of gravitons in the same sense as the physical existence of photons is considered to be verified. To discuss this question we shall consider some features of quantum gravity and ask for their measurable consequences.

2. THE NON-LINEARITY OF EINSTEIN'S EQUATIONS AND THE CONCEPT OF GRAVITONS

To start with, let us cast a short look at one of the usual quantum approaches. We start from the action integral

$$S = \frac{1}{\varkappa^2} \int R \, (-g)^{1/2} \, d^4x + \int \mathcal{L}_M \, d^4x, \qquad (1)$$

where $\varkappa$ is the square root of the gravitational constant, $\mathcal{L}_G = R\,(-g)^{1/2}$ the Lagrange density of Einstein and Hilbert, and $\mathcal{L}_M(\varphi)$ the Lagrange density describing the interaction of gravitation and matter fields φ. Using for conveniency the tensor densities

$$\tilde{g}^{\mu\nu} = (-g)^{1/2} \, g^{\mu\nu} \qquad \text{and} \qquad \tilde{g}_{\mu\nu} = (-g)^{-1/2} \cdot g_{\mu\nu} \qquad (2)$$

as basic fields, $\mathcal{L}_G$ takes the simple Goldberg form

$$\mathcal{L}_G = \left[2\, \tilde{g}^{\rho\sigma} \, \tilde{g}_{\lambda\mu} \, \tilde{g}_{\kappa\nu} - \tilde{g}^{\rho\sigma} \, \tilde{g}_{\mu\kappa} \, \tilde{g}_{\lambda\nu} - 4\, \delta^{\sigma}_{\kappa} \, \delta^{\rho}_{\lambda} \, \tilde{g}_{\mu\nu} \right] \times \tilde{g}^{\mu\kappa}{}_{,\rho} \, \tilde{g}^{\lambda\nu}{}_{,\sigma} \; . \qquad (3)$$

---

[1] This is, at any rate, necessary if one assumes that gravitational fields are basic fields which are classically described by GRT. If one, on the contrary, assumes, as some authors do, that the Newton-Einstein gravitation is only a macroscopic approximation to genuine (quantisable) fields, then the harmonisation by quantising GRT need not be required (cf., for the latter approach, Ref. 1 and the literature cited there).

Just as in Quantum Electrodynamics, one can make now a perturbation theory by adding an appropriate gauge-breaking term and by assuming the <u>ansatz</u>

$$\tilde{g}^{\mu\nu} \rightarrow \delta_{\mu\nu} + \varkappa \, \phi_{\mu\nu} \tag{4}$$

which leads to an infinite series for $\tilde{g}_{\mu\nu}$:

$$\tilde{g}_{\mu\nu} = \delta_{\mu\nu} - \varkappa \, \phi_{\mu\nu} + \varkappa^2 \phi_{\mu\alpha} \phi_{\alpha\nu} + O(\varkappa^3). \tag{5}$$

Substituting (4) and (5) into S , one obtains the series (written symbolically)

$$\begin{aligned} S = & \int \text{"} \phi_{,\mu} \phi_{,\nu} \text{"} + \int \text{"} \varphi_{,\mu} \varphi_{,\nu} \text{"} \\ & + \varkappa \int \text{"} \phi \, \phi_{,\mu} \phi_{,\nu} \text{"} + \varkappa \int \text{"} \phi \, \varphi_{,\mu} \varphi_{,\nu} \text{"} \\ & + \varkappa^2 \int \text{"} \phi^2 \phi_{,\mu} \phi_{,\nu} \text{"} + \varkappa^2 \int \text{"} \phi^2 \varphi_{,\mu} \varphi_{,\nu} \text{"} + \ldots \end{aligned} \tag{6}$$

One can now proceed by discussing the different terms in (6). The parts proportional to $\varkappa^0$ determine the Lagrangian of free gravitational and matter fields. The term proportional to $\varkappa^1$ contains a part with one ϕ and two φ's providing the diagram drawn in Fig. 1a and a part like "$\phi \phi_{,\mu}^2$" implying a definite formula for the interaction of three gravitons, Fig. 1b. In higher-order approximation one obtains diagrams describing gravi-

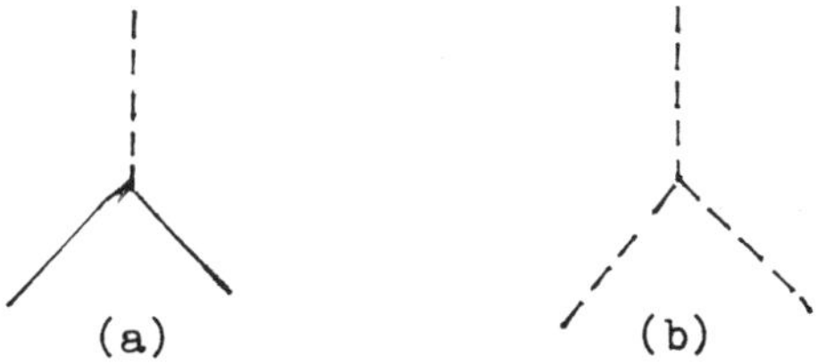

Fig. 1 Feynman diagrams of the $\varkappa$ -order approximation

tational Compton effect and so on; there arise also radiative corrections (closed loops) and hard renormalisation problems related to them.

Following this approach one is nearly automatically led to the language of particle physics. We want however to rediscuss in this paper the question raised by Møller and Rosen, when Feynman[2)] proposed this approach on the 1962 Jablonna Conference on Relativistic Theories of Gravitation: Is this theory really Einstein's theory of gravitation in the sense that if you would have many gravitons the equations would go over into the usual field equations of Einstein?

The answer to this question should not be trivial because the ansatz (4) implies more than the usual weak-field assumption. It confines all considerations to a region of dimension L_0 , where one has a fixed flat background. In other words, one forces Einstein's GRT, as far as possible, to a special-relativistic field theory for a field $\phi_{\mu\nu}$. One asks the queer question, how strongly does GRT differ from usual special-relativistic theories if one forces GRT to behave itself as similar to special-relativistic theories as possible. The field equations arising by this procedure are yet non-linear so that, speaking in the particle language, there arise graviton-graviton interactions in front of a fixed background. This is however not the full non-linearity of Einstein's equations which causes a backreaction of the $\phi_{\mu\nu}$ field on the background and which realizes the strong principle of equivalence.

The above-described approach says therefore only that there are approximations for which one may use the concept of gravitons.[1] There remains to answer the

[1]As long as one assumes a fixed background, this situation does not change essentially if one supposes, instead of a flat Minkowski background, some curved background.

question whether this concept has also a physical sense for regions, where the full non-linearity of GRT has to be considered. To see what happens in this case one should of course consider the full gravitational equations. Because of their complexity we shall however start again with an approximate ansatz, namely with the high-frequency approximation. This approximation[3, 4] considers non-linear effects of GRT in a more essential manner than usual weak-field (low-frequency) methods. Accordingly, it should show at least some features of the non-linearity of GRT.

The high-frequency approximation starts from an ansatz similar to (6). But it is now required that (i) the quantum field, here denoted by $h_{\mu\nu}$, is a high-frequency field of wavelength λ and (ii) the background is an arbitrarily curved Riemannian space-time with the metric $\gamma_{\mu\nu}$, which is not a priori fixed and which changes significantly over a distance $L \gg \lambda$:

$$g_{\mu\nu} = \gamma_{\mu\nu} + \varepsilon h_{\mu\nu}$$
$$\partial\gamma \sim \gamma/L, \quad \partial h \sim h/\lambda, \quad \varepsilon \leq \lambda/L, \quad \lambda \ll L. \tag{7}$$

Substituting (7) into the Ricci tensor one obtains the series

$$R_{\alpha\beta}(\gamma_{\mu\nu} + \varepsilon h_{\mu\nu}) = R^{(0)}_{\alpha\beta}(\gamma_{\mu\nu}) + \varepsilon R^{(1)}_{\alpha\beta}(h_{\mu\nu}) + \varepsilon^2 R^{(2)}_{\alpha\beta}(h_{\mu\nu}) + \dots \tag{8}$$

Here exists, despite the smallness of ε , a backreaction of the h field on γ described by

$$R^{(0)}_{\alpha\beta}(\gamma_{\mu\nu}) = \varepsilon^2 R^{(2)}_{\alpha\beta}(h_{\mu\nu}). \tag{9}$$

Assuming now that the system to quantise has the characteristic length $L_0 \gtrsim \lambda$, one obtains for an arbitrary particle number $n \geq 1$:

$$\rho \approx \frac{\varepsilon^2 c^4}{G \lambda^2} = (n + \tfrac{1}{2}) \frac{\hbar \nu}{L_0^3} \tag{10}$$

(G denotes the Newtonian gravitational constant) leading to

$$\varepsilon = (n + \tfrac{1}{2})^{\frac{1}{2}} \left(\frac{\lambda}{L_0^3} \right)^{\frac{1}{2}}. \tag{11}$$

Together with $\varepsilon \lesssim \lambda / L$ this leads to[5)]

$$\lambda \gtrsim \frac{1}{L_0^3} (\ell_p L)^2 , \tag{12}$$

where $\ell_p = (h\, G/c^3)^{1/2}$. For an optimal field measurement this provides[5, 6)]

$$\lambda \gtrsim (\ell_p L)^{\frac{1}{2}} . \tag{13}$$

Therefore, if one does not assume a fixed background but includes the backreaction due to the non-linearity of Einstein's equations, then there arise limitations for Quantum Gravity, limiting of course the meaning of the term "graviton", too.

Before continuing the discussion of the meaning of these limitations for the gravitational quantum effects, which arise in the above-mentioned series of Feynman diagrams, let us consider the measurement procedure. If one does so, one can in particular reproduce the inequality relations (12) and (13).

3. ON MEASUREMENTS OF EFFECTS IN QUANTUM GRAVITY

In a discussion of electromagnetic field measurements, Bohr and Rosenfeld[7)] formulated requirements on measurement which must be satisfied to have a physically sensible measurement device and procedure. They distinguish strictly between the system to be measured (the electromagnetic field), the measurement device (a test body), and their mutual interaction. As far as the _constitution_ is concerned, one has to consider the measurement body as a classical one; i. e., one has to neglect its atomic structure and to presuppose a rigid body in the sense of classical mechanics carrying a homogeneously distributed charge. For a measurement arrangement one has to realize such bodies with an arbitrarily high accuracy. The main argument for this standpoint is given by the statement that a consideration of the atomic structure of the measurement body is not adequate to a measurement of pure fields because it transfers the discussion to the level of the amalgamation of quantum-electrodynamics and atomic matter theory. If one wants to measure field effects one has to evade such a mixture. Of course, enormeous difficulties can arise in constructing such an artificial test body. But if one intends to measure one has to try its construction. Otherwise, the distinction between measurement apparatus and the physical system whose physical parameters are to measure would be destroyed. Then the concept of measurement loses its sense. - Considering, however, the _interaction_ between test body and electromagnetic field system one has to regard the quantummechanic laws, in particular, the Heisenberg uncertainty relations. They impose an absolute restriction on the displacement of test bodies which cannot be "compensated" by refined mechanisms.

Following Bohr and Rosenfeld, let us consider the measurement of the field strength via a measurement of the momentum change of a test body induced by an electromagnetic and, respectively, a gravitational field. We assume accordingly:

(i) One can realize a measurement body of linear dimension L_0 having a rigid structure in the sense of classical mechanics and carrying a homogeneously distributed charge q.

(ii) Accordingly, the field measurement is carried out so that the time interval $T = t'' - t'$ between the two measurements is much greater than the duration Δt of the momentum measurements at t' and t'' and smaller than the time L_0/c which light needs to cross the measurement body,

$$\Delta t \ll T < L_0/c . \tag{14}$$

(L_0 and T denote the space-time volume over which the field average is calculated and measured.)

(iii) The interaction between the measurement body and the field obeys Heisenberg's uncertainty relation

$$\Delta p_x \, \Delta x \gtrsim \hbar . \tag{15}$$

(We consider, for simplicity, a homogeneous field in x direction.)

On the basis of these assumptions Bohr and Rosenfeld could show that, contrary to the results obtained by Landau and Peierls[8], no absolute limitations on electromagnetic field measurements exist. They obtained, e. g., for the accuracy of the average electric field strength E the relation

$$\Delta E \cdot L_0^2 \gtrsim \frac{\hbar c}{q} . \tag{16}$$

Because q may be chosen, in principle, arbitrarily large the right-hand side of (16) does not mean an absolute limit for the product $\Delta E \cdot L_0^2$.

Transferring the Bohr-Rosenfeld arguments to gravitational fields, it was shown by Rosenfeld[9, 10], Peres and Rosen[11], Wheeler[12], DeWitt[13] and Treder[14] that, in difference to electrodynamics, principle limitations on field measurements arise. In this case, the relation (16) must be replaced by the relations

$$\Delta g \cdot L_0^2 \gtrsim \frac{\hbar G}{c^3} \frac{m_g}{m} \quad (17)$$

and

$$\Delta \Gamma \cdot L_0^3 \gtrsim \frac{\hbar G}{c^3} \frac{m_g}{m}, \quad (18)$$

where Δg denotes the uncertainty $\Delta g_{\mu\nu}$ of the gravitational potential $g_{\mu\nu}$ and $\Delta\Gamma$ the uncertainty of $\Gamma^{\sigma}_{\mu\nu}$. Regarding now that, because of the strong principle of equivalence the gravitational mass m_g is equal to the inertial mass m, one has

$$\Delta g \cdot L_0^2 \gtrsim \frac{\hbar G}{c^3} \quad (19)$$

and

$$\Delta \Gamma \cdot L_0^3 \gtrsim \frac{\hbar G}{c^3}. \quad (20)$$

Assuming the condition $\Delta g < 1$ guaranteeing that the physical space-time structure is not destroyed, (19) provides the well-known absolute limitation on length measurements,

$$L_0 \gtrsim \left(\frac{\hbar G}{c^3}\right)^{\frac{1}{2}}. \quad (21)$$

The measurement situation discussed here in short makes the following point quite clear[15]: Even in Quantum Gravity one has to assume a classical measurement body. The absolute limitations nevertheless arising result from Heisenberg's uncertainty relation together with some fundamental features of gravitational coupling, i. e., from laws on the interaction between field and test body. Therefore, in Quantum Gravity one finds absolute limitations on L_0 and, in general, on all entities having the dimension length, time or mass. Indeed, due to the occurance of the three universal constants $\hbar$, c , G , one may form the Planckian length, time and mass units

$$l_p = \left(\frac{\hbar G}{c^3}\right)^{\frac{1}{2}}, \quad t_p = \left(\frac{\hbar G}{c^5}\right)^{\frac{1}{2}}, \quad m_p = \left(\frac{\hbar c}{G}\right)^{\frac{1}{2}}. \tag{22}$$

They define the dimension of the smallest classical test body and, simultaneously, the biggest elementary particle (cf., for the latter point, the papers of Markov[16, 17]). They define accordingly the border between a measurement body and the system whose entities are measured. This border causes the limitations on measurement which one cannot evade in a full ($\hbar$, c , G) theory.

Considering now the measurement of gravitational fields in front of a curved background, one finds the same limitation (12) and (13) which followed from quantum field formalism[18]. Indeed, if one wants to measure $h_{\mu\nu}$ field effects, then one has to consider perturbations $\delta\Gamma$ of $\Gamma \sim 1/L$ over a characteristic length λ

$$\delta\Gamma \sim \frac{\partial \Gamma}{L}\lambda \sim \frac{1}{L^2}\lambda . \tag{23}$$

The possibility to measure such effects requires that $\delta\Gamma$ be greater than the uncertainty $\Delta\Gamma$ of the measurement,

$$\Delta\Gamma \lesssim \delta\Gamma, \tag{24}$$

where $\Delta\Gamma$ statisfies the relation (20). Therefore, one has

$$\delta\Gamma \cdot L_0^3 \sim \frac{\lambda}{L^2} L_0^3 \gtrsim \Delta\Gamma \cdot L_0^3 \gtrsim l_p^2 \tag{25}$$

and, accordingly again the relation

$$\lambda \gtrsim \frac{1}{L_0^3}\left(l_p L\right)^2 \tag{26}$$

and, for optimal measurement,

$$\lambda \gtrsim (l_p L)^{\frac{1}{2}}. \tag{27}$$

Therefore, we find that the limitations (12) and (13) following from the quantum formalism are in accordance with the limitations resulting from Heisenberg's uncertainty relation for the measurement procedure.

Concluding this section, let us make a remark on the measurement analysis performed by Mensky.[19] He starts from Feynman's path integral

$$A(x',x) = \int_{I(x,x')} d\{x\} \exp\left(\frac{i}{\hbar} S\{x\}\right) \tag{28}$$

giving the total transition amplitude for the transition from the point x to the point x'. ($S\{x\}$ is the action integral calculated along a given path from x to x', and the integral over $\{x\}$ denotes the summation of the amplitudes for the different paths between x and x'.) If one performs now a (contineous) measurement providing some information about the path of transition (measurement value α) then one has not to integrate over all paths $I(x,x')$ but only over a set $I_\alpha(x,x')$. Assuming that this preference of certain paths can be expressed by a weighting function

$$\rho_{\{a\}}\{x\} = \exp\left(-\frac{\langle (x-a)^2\rangle}{\Delta a^2}\right) \tag{29}$$

one may write

$$A_{\{a\}}(x',x) = \int d\{x\}\, \rho_{\{a\}}\{x\} \exp\left(\frac{i}{\hbar} S\{x\}\right). \tag{30}$$

Here $\{a\}$ denotes a path determined by the measurement and Δa the accuracy of the measurement. Analysing $A_{\{a\}}(x',x)$ for an oscillator under the influence of an external force, the uncertainty ΔF of this force was estimated[19]: For an optimal measurement, where

$$\Delta a \approx \left(\frac{2\hbar}{m\tau \, | \Omega^2 - \omega^2 |} \right)^{\frac{1}{2}} \tag{31}$$

it amounts to

$$\Delta F \gtrsim \left(\frac{m\hbar \, | \omega^2 - \Omega^2 |}{\tau} \right)^{\frac{1}{2}} \tag{32}$$

One sees that this relation is equivalent to the Bohr-Rosenfeld relation (16). Indeed, multiplying ΔF by Δa and regarding that

$$\Delta a \approx L_0 \, , \quad \tau \equiv T \lesssim \frac{L_0}{c} \tag{33}$$

and

$$\Delta F \approx q \, \Delta E \tag{34}$$

one obtains

$$\Delta E \cdot \Delta a \approx \Delta E \cdot L_0 \geq \frac{\hbar}{\tau q} \gtrsim \frac{\hbar c}{q L_0} \, . \tag{35}$$

The Bohr-Rosenfeld derivation of this relation has the advantage to make obvious that (32) results mainly from Heisenberg's uncertainty relation (15).

4. ON GRAVITATIONAL QUANTUM EFFECTS

As there was shown in the previous two sections, there arise principle limitation in Quantum Gravity resulting from a contradiction between the quantum principle and the fundamental principle of GRT, namely the principle of equivalence. This was shown by considering the quantum field formalism and the theory of measurement, respectively. In the quantum formalism, this contradiction comes to appearance for high frequencies by means of a contradiction between the quantum requirement (11) and the backreaction implied by the equivalence principle and brought into the calculation by the assumption $\epsilon \lesssim \lambda/L$. The same can be shown by the measurement discussion. Then the contradiction shows up by the incompatibility of the Heisenberg uncertainty relation (15) and the relation $m_g = m$ being also a consequence of the equivalence principle. According to Bohr and Rosenfeld, the coincidence of the uncertainty relations derived from formalism and from measurement, respectively, is necessary to prove the correctness of the quantum field formalism.

Do the relations (12) and (13), however, mean that there should not exist such quantum effects as gravitational Compton effect, Bremsstrahlung, pair creation, and Lamb shift? To answer this question let us return the expression (6) of the action functional .

The term $\propto \varkappa^2$ in (6) (its gravitational part corresponds to the $R^{(3)}_{\alpha\beta}$ term in (8)) provides Feynman diagrams of the form shown in Fig. 2. And the term $\propto \varkappa^3$ in (6) (its gravitational part corresponds

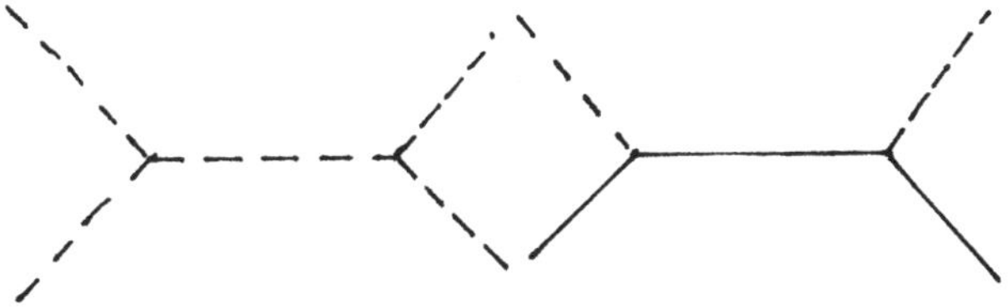

(a) (b)

Fig. 2 Feynman diagrams of the $\varkappa^2$-order approximation

to the $R^{(4)}_{\alpha\beta}$ term in (8)) leads to diagrams of the type drawn in Fig. 3.

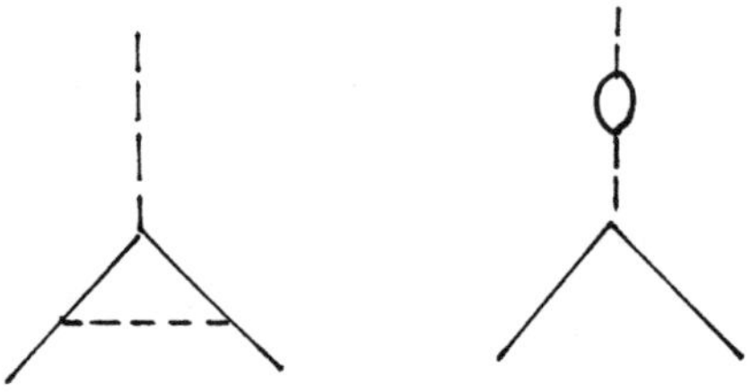

Fig. 3 Feynman diagrams of the $\varkappa^3$-order approximation

One can see that the cut-off length (13)

$$\lambda \gtrsim (l_p L)^{\frac{1}{2}}$$

does not exclude all these effects. It only cuts off the high-frequency part of "graviton-matter" interaction and pure "graviton-graviton" interactions. Indeed, the Compton effect is, e. g., given by the formula

$$\frac{1}{\nu'} = \frac{1}{\nu} + \frac{\hbar}{mc^2}(1 - \cos\theta) . \qquad (36)$$

As there was shown by Heisenberg[20] in Quantum Electrodynamics, the introduction of a cut-off length r_0 restricts the validity of formula (36) to regions satisfying the relation

$$\lambda \Lambda \gg r_0^2 , \qquad (37)$$

where λ is the wave length of the field and $\Lambda = \hbar/mc$ the Compton wavelength of the scattering particle. Assuming now, in accordance with (13), that $r_0 = (l_p L)^{\frac{1}{2}}$, one obtains for $\Lambda \approx L$:

$$\lambda \gg l_p . \qquad (38)$$

That means, such "graviton-matter" Compton effect occurs for a matter-dominated background, where the self-interaction of the gravitational field can be neglected. Assuming however "graviton-graviton" interaction as the dominating process, i. e., assuming $\Lambda \approx \lambda$, then the relation (37) is not satisfied generally. Then $r_0 = (\ell_P L)^{1/2}$ cuts off such effects. - The same can be shown for the higher-order diagrams.

The fact that effects of quantisied GRT are only cut off for high frequencies is due to the method used here. An exact treatment of the theory should show that always when strong non-linearity comes into the calculations, it will not be sensible to speak of gravitons and to look for corresponding effects.[1]

5. CONCLUSIONS

To summarise, the cut-off length arising in Quantum Gravity restricts its validity and, accordingly, the concept of gravitons to the range of weak-field and low-frequency approximation. There are some simularities between the situation in Quantum Gravity and in a theory realizing Heisenberg's programme formulated in his 1938 paper[20)] for Quantum Electrodynamics. Heisenberg introduced there a fundamental length r_0 ad hoc. This length should cut off higher-order approximations of Quantum Electrodynamics to evade the problem of divergencies. There are, however, two differences to Quantum Gravity: First, such a length need not be introduced ad hoc, but arises automatically in Quantum Gravity (this was already repeatedly noticed in literature). Second, the limitations arising here automatically do not signal, as assumed for Quantum Electrodynamics by Heisenberg, the transition to a new type of interaction connected with particles of

---

[1] For the discussion of the Lamb shift, cf. Ref. 21.

mass $M = \hbar / r_0 c$. In Quantum Gravity these particles are planckions which show up a principle limit of the theory[1].

As there was also argued by Treder[22)], this limit may be compared with the role of the vacuum light velocity in SRT. In SRT one cannot overcome this c-border. Relativistic effects can be calculated from SRT mechanics as corrections to classical mechanics signaling the existence of the border c. But one cannot arrive at a physically sensible theory by considering the transition of relativistic mechanics to velocities $v = c$.

In Quantum Gravity one can also calculate quantum effects in certain approximate cases. One cannot however go over to the domain characterized by the dimensions of a planckion. Moreover, the full theory of Quantum Gravity lies in this region characterized by the Planck units (22) what is seen for vacuum Quantum Gravity, because in this case the action of the non-linearity is not suppressed by the matter action.

From this point of view, in GRT the particle language can only be used within the framework of certain approximations describing the interaction of gravitation and matter or, more generally, for some low-frequency approximations to GRT.

---

[1]It was argued by Markov[17, 18)] that the evaporation of black holes leads to a final stable state which is given by the planckion ("maximon"). From our point of view, one can add: This state is stable because there is no interaction beyond ℓ_p according to which a planckion might decay.

6. REFERENCES

1. Adler, S.L., Rev. Mod. Phys. 54, 729 (1982).
2. Feynman, R.P., Acta Phys. Pol. 24, 297 (1963).
3. Brill, D. and Hartle, J.B., Phys. Rev. B 135, 271 (1964).
4. Isaacson, R.A., Phys. Rev. 166, 1263, 1272 (1967).
5. v. Borzeszkowski, H.-H., Found. Phys. 12, 633 (1982).
6. v. Borzeszkowski, H.-H., "On High-Frequency Background Quantization of Gravity", in: M.A. Markov and P.C. West (eds.), Quantum Gravity, New York and London (1984).
7. Bohr, N. and Rosenfeld, L., Det. Kgl. Danske Vidensk. Selskab., Mat.-fys. Medd. 12, 1 (1933).
8. Landau, L. and Peierls, R., Z. Phys. 69, 56 (1931).
9. Rosenfeld, L., "Report of the Conference on the Role of Gravitation in Physics", Chapel Hill, North Carolina (1957).
10. Rosenfeld, L., "Quantentheorie und Gravitation", in: H.-J. Treder (ed.), Entstehung, Entwicklung und Perspektiven der Einsteinschen Gravitationstheorie, Berlin (1966).
11. Peres, A. and Rosen, N., Phys. Rev. 118, 335 (1960).
12. Wheeler, J.A., Ann. Phys. 2, 604 (1957).
13. DeWitt, B., "The Quantization of Geometry", in: L. Witten (ed.), Gravitation, New York and London (1962).

14. Treder, H.-J., Monatsber. Akad. Wiss. DDR 3, 241 (1961).

15. v. Borzeszkowski, H.-H. and Treder, H.-J., Ann. Phys. Leipz. 40, 287 (1983).

16. Markov, M.A., "On the Maximon and the Concept of Elementary Particle", Preprint P-0208. Inst. Nucl. Res. USSR, Moscow (1981).

17. Markov, M.A., "The Problems of General Relativity", in: M.A. Markov and P.C. West (eds.), Quantum Gravity, New York and London (1984).

18. v. Borzeszkowski, H.-H. and Treder, H.-J., Found. Phys. 12, 1113 (1982).

19. Mensky, M.B., Gruppa putei - izmerenie, polya, chastitzy (in Russ.), Moscow (1983).

20. Heisenberg, W., Z. Phys. 110, 251 (1938).

21. Treder, H.-J., "On the Problem of Physical Meaning of Quantization of Gravitational Fields", in: F. de Finis (ed.), Relativity, Quanta and Cosmology, New York etc. (1979).

22. Markov M.A. Preprint P-0I87, Inst.Nucl.Res.USSR, Moscow (1980).

CLASSICAL AND QUANTUM PREGEOMETRY

Dierck-Ekkehard Liebscher

Zentralinstitut für Astrophysik der
Akademie der Wissenschaften der DDR
DDR - 1502 Potsdam-Babelsberg
Rosa-Luxemburg-Straße 17 A

ABSTRACT

The program for theories explaining the metric structure of space-time as dynamical consequence of a pregeometric structure is considered. It is shown that an affine space allows the definition of a metric by dynamics. The character of small deviations from a classical metric structure is explained.

1. INTRODUCTION

Geometry was linked to physics by classical mechanics /1/. General Relativity Theory /2/ transformed geometry into a dynamical object. The representation of this object by a field quantity may not be final, in spite of the success of Einstein's choice to use the metric tensor of a Riemann space-time as the dynamical field /3/. The only use of the metric tensor leads in second-order differential equations to Einstein's GRT of 1916. It is a minimal construction in more than one respect, as all alternative approaches tell. The experimental situation in gravitation is well explained by Einstein's theory. There is evidence now in three of the four gravitation-dominated cases, that is in the weak quasistationary field (solar system tests), in the essentially time-dependent weak field (binary pulsar observations), and in the large-system time-dependent case (cosmology). The evidence in the case of quasistatic strong fields (black holes) is shaky, because we mainly observe accretion discs of compact objects, and do not yet know the signature of alternatives to the black-hole hypothesis in the radiation of this disc. Nevertheless, classical gravitation theory does not get rid of the singularities, which are present in GRT by three physical reasons expressed by the assumptions of the singularity theorems /4/. A space-time with

1) Einstein equations $R_{\mu}^{\nu} = 0$ in vacuo, with

2) positive definite $R_{0}^{0} = T_{0}^{0} - \frac{1}{2} T \geq 0$ in a source, and

3) global causal structure existing,

will exhibit singularities. Alternatives to General Relativity may be classified with respect to these three suppositions of the singularity theorems.

There are many projects to change the gravitational field equations already in the case of vacuum. This implies - if we stick to second-order theories - that in addition to the metric tensor $g_{\mu\nu}$ other field quantities are introduced. If the substitute the metric tensor in its property of dynamical quantity, one may already speak of a kind of pregeometry. However, the label "pregeometry" should depend on the kind of coupling of matter to gravitation, because geometry is defined physically by the motion of matter in the gravitational field b e f o r e we are forced to look into the field equations for gravitation /5,6/. Almost every alternative to GRT in this step is an explicit or implicit bi-metric theory. The metric we observe directly is given macroscopically by the motion of test bodies. The weak equivalence principle tells us that we have to couple non-gravitational fields by substituting covariant derivatives (by the geodesic connection of the metric) for the usual ones in the canonical equations. If we do so, the only freedom of the theory lies in the construction of this metric out of the dynamical quantities and the field equations for these quantities. Now, any second-order field equation for the dynamical quantities defines a second ("back-ground") metric, the theory becomes bi-metric. Bi-metric theories may become very close to General Relativity /7/. This is the reason why we may use approximations of perturbations in a given large-scale ("back-ground") field in GRT, why we may try to quantize these perturbations on the back-ground described classical and so on. We learn from the study of bi-metric theories that there are theories with gauge invariance (now not coinciding with coordinate invariance as in GRT), theories with stable states of cold matter above the LAndau-Oppenheimer-Volkoff limit, and theories allowing for cosmologies with more time for galaxy formation.

The second class of alternatives change the concept of positive definite sources ($T_{0}^{0} - \frac{1}{2} T \geq 0$). This change is obtained by the addition of hypothetical fields, which is in some approaches equivalent to the former concept. The most popular alternative today is to interpret higher-order terms /9/ or the Poincaré pressure /10/ as consequences of a suitably chosen quantum approach. This large and important work is discussed in other contributions to this meeting.

Pregeometry, strictly speaking, concerns the third group of assumptions in the singularity theorems, the a priori existence of a light-cone, or metric, structure.

2. LORENTZ STRUCTURE

Any pregeometric approach has to produce the metric of macroscopic motion as a consequence of the dynamical equations of other fields, but without the construction of a metric coupled to the matter fields a priori. Such a program implies the elimination of the metric tensor from the usual microscopic Lagrangian, just as it was tried by Terazawa and coworkers in their approach to pregeometry /11,12,13/. Of course, to get scalar-density Lagrangians out of vectors or spinors one needs another tensor to form scalars, and they use the Levi-Civita symbol. In the scalar case the action reads

$$S = \int d^4x \sqrt{\det \left(\sum_i \phi^i{}_{,\mu} \phi^i{}_{,\nu} \right)} \cdot F . \qquad (2.1)$$

In such a way, the Lorentz group of the principal bundle seems to be substituted by the centroaffine group, if we consider a chosen field on the background of the others. However, to construct the pregeometric Lagrangian, a non-dynamical Lorentz metric is already used, and from this point of view the proposed pregeometry is just a special kind of bi-metric theory. The sum about the scalar fields hides a pseudo-euclidean metric in the space of the ϕ ,

$$\sum_{i=1}^{N} \phi^i{}_{,\mu} \phi^i{}_{,\nu} = \eta_{ik} \phi^i{}_{,\mu} \phi^k{}_{,\nu} . \qquad (2.2)$$

One scalar field is assumed to be a ghost field, which in the language of geometry means, that its term is substracted, and not added.

$$\eta_{ik} = \begin{cases} 0 & i \neq k \\ 1 & i = k < N \\ -1 & i = k = N . \end{cases} \qquad (2.3)$$

In the minimal case (N=4), $g_{\mu\nu} = \eta_{ik} \phi^i{}_{,\mu} \phi^k{}_{,\nu}$ is the Minkowski metric in non-linear coordinates, and $\phi^i{}_{,\mu}$ a usual (but trivial) tetrad. In the case N$>$4 the tangent space is chosen to be a higher-dimensional Minkowski space, in which the ghost field ϕ^N represents the Minkowski time coordinate. The action (1) is equivalent to

$$S = \int d^4x \sqrt{-g}\left(g^{\mu\nu}\phi^m{}_{,\mu}\phi^n{}_{,\nu}\eta_{mn} - \frac{1}{F}\right), \tag{2.4}$$

where $g_{\mu\nu} = \eta_{mn}\phi^m{}_{,\mu}\phi^n{}_{,\nu}$ is the metric measured by the field equation for ϕ itself.

The spinor pregeometry uses geometry also: As soon as the Dirac matrices or a linear spinor equation

$$\gamma^{\mu A}{}_B \psi^B{}_{,\mu} = M^A{}_B \psi^B \tag{2.5}$$

is written down, a metric

$$g^{\mu\nu} = \frac{1}{4}\gamma^{\mu A}{}_B \gamma^{\nu B}{}_A \tag{2.6}$$

is defined, which is measured as coefficients of the wave equation for ψ, respectively the substitute for this wave equation. Again, the metric it defines, and its existence, are a priori.

Thus, the main problem in the construction of a pregeometry is to get at the Lorentz group for the principal bundle without using this group in the a priori constructed Lagrangian. That means we have to avoid any second rank tensors (in base space or fibre) of Minkowski signature not only in the construction of the gravitation Lagrangian, but also in the Lagrangian of the non-gravitational fields. These constructs may arise dynamically, a posteriori, as effectively metric couplings of the non-gravitational fields to gravitation. In analogy to the use of the unimodular centroaffine invariance in the action, eq. (2.1), we have to begin with a group of invariance in the tangent spaces, which contains the Lorentz group as a subgroup, but does not define it /14/. The reduction to the Lorentz group has to be the effect of dynamics, i.e. a symmetry breakdown.

Without Lorentz structure there will be no time concept a priori. This represents probably the most difficult notion to understand. The difference between time-like and space-like world lines, i.e. local causal structure, depends on the existence of a light cone structure. Such a structure is not defined a priori in a true pregeometric concept. It may not be derived by defini-tion of certain fields, but by their overall dynamics: by cosmology. As a consequence, pregeometry may turn into geometry by action of the surrounding universe. Any pregeometry must have a connection to the relational concepts today mainly labelled by the name of E.Mach /15,16,17,18/.

3. AFFINE MECHANICS

Relational mechanics are the classical counterpart of the pregeometry approach to field theory. The original problem was that of formal mechanics invariant against rotation and acceleration. The solution was given by Lagrangians depending only on distances and distance variations, for instance /19/

$$L = \sum_{A,B} \frac{m_A m_B}{r_{AB}} \sqrt{\sum_{CD} \frac{m_C m_D}{r_{CD}} \dot{r}_{CD}^{2}} , \qquad (3.1)$$

which is invariant against the full kinematical group of euclidean space, and non-linear time reparametrisations. Mechanics of this type realise the desired symmetry breakdown scheme /20/. A telescopic (Planck, /21/) group, the invariance group of the action of the total particle system, is reduced to a local group, the invariance of a subsystem of particles, left unbroken after integrating the Lagrangian over all particles outside the subsystem considered. Of course, the invariance left unbroken depends now on the actual motion of the particles outside the subsystem. In the physical interpretation, the state of the universe determines the locally observed invariance of physical laws. However, in the conventional approach to relational mechanics, the definition of rotation requires the definition of simultaneity, and the unbroken part of the telescopic group may never be the Lorentz group. At most we may expect to get the Galilei group. Now it turns out that one may construct Minkowski-signature metrics in the effective space also on background Galilei symmetry, thus hiding the existing definition of simultaneity from the equations of motion of non-gravitational fields /22/. As for the law of gravitational interaction, in the post-Newtonian approximation, this a priori simultaneity shows up in the coefficient /23/

$$\alpha_2 = \left(\frac{\text{velocity of light propagation}}{\text{velocity of gravitation prop.}} \right)^2 - 1 . \qquad (3.2)$$

In the case of relational mechanics the problem turns into that of getting Poincaré or Lorentz invariance as unbroken residue of the telescopic symmetry also, the consequences being the same as considered above.

In order to get an impression of the task we show an example of point mechanics, using determinants in the Lagrangian as above, but now only determinants, and no hidden metric. That is we try to construct a formal action of point mechanics invariant against the unimodular affine group. We have to consider world lines in a linear affine space. No causal cones, no time are defined, or may be used in constructing this action. Hence, we get in the simplest case at an action of the Fokker-Planck type /24/

$$S = \sum \iint \dots d\lambda_A \, d\lambda_B \dots m_A m_B \dots f\left(\xi^\mu_A \dots \frac{d\xi^\mu_A}{d\lambda_A} \dots\right) \tag{3.3}$$

all sets of particles to form the term	all worldlines of the particles in the term	weights of the different worldlines coming in	function of determinants homogeneous of degree 1 in all tangent vectors

For example, the determinants may be

$$
\begin{aligned}
&\left|\xi^\mu_A(\lambda_A)\,\xi^\nu_B(\lambda_B)\,\xi^\rho_C(\lambda_C)\,\xi^\sigma_D(\lambda_D)\right|, && \text{denoted by} \quad |ABCD| = I_0 ,\\
&\left|\frac{\partial \xi^\mu_A}{\partial \lambda_A}\;\xi^\nu_B\;\xi^\rho_C\;\xi^\sigma_D\right|, && \text{denoted by} \quad |\dot{A}BCD| ,\\
&\left|\frac{\partial \xi^\mu_A}{\partial \lambda_A}\;\frac{\partial \xi^\nu_B}{\partial \lambda_B}\;\xi^\rho_C\;\xi^\sigma_D\right|, && \text{denoted by} \quad |\dot{A}\dot{B}CD| ,\\
&\left|\frac{\partial \xi^\mu_A}{\partial \lambda_A}\;\frac{\partial \xi^\nu_B}{\partial \lambda_B}\;\frac{\partial \xi^\rho_C}{\partial \lambda_C}\;\xi^\sigma_D\right|, && \text{denoted by} \quad |\dot{A}\dot{B}\dot{C}D| ,\\
&\left|\frac{\partial \xi^\mu_A}{\partial \lambda_A}\;\frac{\partial \xi^\nu_B}{\partial \lambda_B}\;\frac{\partial \xi^\rho_C}{\partial \lambda_C}\;\frac{\partial \xi^\sigma_D}{\partial \lambda_D}\right|, && \text{denoted by} \quad |\dot{A}\dot{B}\dot{C}\dot{D}| .
\end{aligned}
\tag{3.4}
$$

Homogeneity in the tangent vectors $\dot{\xi}^\mu_A$, and symmetrisation by the summation of terms suggest to use the symmetrised expressions

$$
\begin{aligned}
I_0 &= |ABCD| ,\\
I_1 &= |\dot{A}BCD|\,|A\dot{B}CD|\,|AB\dot{C}D|\,|ABC\dot{D}| ,\\
I_2 &= |\dot{A}\dot{B}CD|\,|AB\dot{C}\dot{D}|\,|\dot{A}B\dot{C}D|\,|A\dot{B}C\dot{D}|\,|\dot{A}BC\dot{D}|\,|A\dot{B}\dot{C}D| ,\\
I_3 &= |\dot{A}\dot{B}\dot{C}D|\,|\dot{A}\dot{B}C\dot{D}|\,|\dot{A}B\dot{C}\dot{D}|\,|A\dot{B}\dot{C}\dot{D}| ,\\
I_4 &= |\dot{A}\dot{B}\dot{C}\dot{D}| .
\end{aligned}
\tag{3.5}
$$

A monomial action is given by

$$S=\sum_{ABCD}\iiiint d\lambda_A\, d\lambda_B\, d\lambda_C\, d\lambda_D\, m_A m_B m_C m_D\, I_0^{\kappa} I_1^{\lambda} I_2^{\mu} I_3^{\nu} I_4^{\varrho}. \quad (3.6)$$

Reparametrisation invariance requires

$$\lambda + 3\mu + 3\nu + \varrho = 1. \quad (3.7)$$

The action, eq. (3.6), does not assume the notion of time, or causality in the local sense. No metric is presupposed in space or tangent space. Nevertheless, we get the reduction to a smaller invariance group by defining a subsystem and performing the integration about an assumed state of motion of the particles outside this subsystem, supposing a cosmologically acceptable state of motion. If these particles with their world-lines distinguish one of the affine coordinates by their symmetry, it will turn out to be a time coordinate for the subsystem. For instance, if there exists an affine coordinate system, in which the world-lines form the mantle lines of a formal rotation body with axis coordinate t, the action for a particle on the axis is reduced to

$$S = \int dt \left(M_0 \dot{r}^2 + M_1 (\dot{r}^2)^2 + \dots \right), \quad (3.8)$$

M_0, and M_1 being integrals along these mantle lines, and x, y, z, t the special affine coordinates chosen to represent the (4-dim) rotation body in question. The expansion of the action in powers of the velocity begins like in SRT. The state of the universe, the expansion weighted in appropriate (i.e. by the action, eq. (3.6)) way defines the characteristic velocity in the formula for the dependence of mass on velocity. However, no formal proof of Lorentz invariance for the subsystem in special choices of the overall action has been possible, because the integrals turn out to be too difficult to evaluate them analytically. We only may state the possibility to construct actions which suit the abstract scheme in mechanics.

4. FIELDS IN AFFINE SPACES

How field theory will look without metric? The simplest possible action seems to be

$$S = \int d^4x \sqrt{\varepsilon_{ABCD}\, \phi^A_{,\mu}\, \phi^B_{,\nu}\, \phi^C_{,\varrho}\, \phi^D_{,\sigma}\, \varepsilon^{\mu\nu\varrho\sigma}}. \quad (4.1)$$

ϕ^A (A=1,...,4) are four scalars. Terms without derivatives of ϕ^A are trivial. Comparing with the action, eq.

(2.1), we see that also in the tangent space (coordinates ϕ^A) we have chosen only affine notions. Such a Lagrangian yields no acceptable field theory, the field equation only fixes the Lagrangian to be a constant.

If we expect a wave equation for some field quantity Φ, the effective coefficients $g^{\mu\nu}$ in the wave operator $\Box = g^{\mu\nu}\frac{\partial^2}{\partial x^\mu \partial x^\nu}$ have to be constructs from other field quantities ψ, which for themselves cannot be solutions of a wave equation with wave-operator independent of the solution ψ. Therefore, in manifolds without a priori metric tensor field, the effective metric has to be an integral, i.e. a non-local quantity. This is the technical aspect of the epistemological expectation that inertial (in relativity: metric, or causal) properties are to be defined from the global distribution of the fields in the manifold. We show the technical possibility of constructing an effective wave equation from an affinely invariant action.

We choose as field quantity a vector field u^μ, and use invariants like in mechanics, eq.(3.5). The elementary terms are again determinants like

$$\left|\underset{1\alpha}{u}\,\underset{2}{u}\,\underset{3}{u}\,\underset{4}{u}\right| = \varepsilon_{\mu\nu\varrho\sigma}\frac{\partial}{\partial \underset{1}{x}^\alpha}u^\mu(\underset{1}{x})\,u^\nu(\underset{2}{x})\,u^\varrho(\underset{3}{x})\,u^\sigma(\underset{4}{x}) \qquad (4.2)$$

The invariants may be chosen as

$$I_1 = \left|\underset{1\alpha}{u}\,\underset{2}{u}\,\underset{3}{u}\,\underset{4}{u}\right| \left|\underset{1}{u}\,\underset{2\beta}{u}\,\underset{3}{u}\,\underset{4}{u}\right| \left|\underset{1}{u}\,\underset{2}{u}\,\underset{3\gamma}{u}\,\underset{4}{u}\right| \left|\underset{1}{u}\,\underset{2}{u}\,\underset{3}{u}\,\underset{4\delta}{u}\right| \varepsilon^{\alpha\beta\gamma\delta},$$

$$I_2 = \left|\underset{1\alpha}{u}\,\underset{2\beta}{u}\,\underset{3}{u}\,\underset{4}{u}\right| \left|\underset{1\gamma}{u}\,\underset{2}{u}\,\underset{3\delta}{u}\,\underset{4}{u}\right| \left|\underset{1\varepsilon}{u}\,\underset{2}{u}\,\underset{3}{u}\,\underset{4\zeta}{u}\right| \left|\underset{1}{u}\,\underset{2\eta}{u}\,\underset{3\theta}{u}\,\underset{4}{u}\right|$$
$$\left|\underset{1}{u}\,\underset{2\kappa}{u}\,\underset{3}{u}\,\underset{4\lambda}{u}\right| \left|\underset{1}{u}\,\underset{2}{u}\,\underset{3\mu}{u}\,\underset{4\nu}{u}\right| \varepsilon^{\alpha\beta\mu\nu}\varepsilon^{\gamma\delta\kappa\lambda}\varepsilon^{\varepsilon\zeta\eta\theta},$$

$$I_3 = \left|\underset{1\alpha}{u}\,\underset{2\beta}{u}\,\underset{3\gamma}{u}\,\underset{4}{u}\right| \left|\underset{1\delta}{u}\,\underset{2\varepsilon}{u}\,\underset{3}{u}\,\underset{4\zeta}{u}\right| \left|\underset{1\eta}{u}\,\underset{2}{u}\,\underset{3\theta}{u}\,\underset{4\kappa}{u}\right|$$
$$\left|\underset{1}{u}\,\underset{2\lambda}{u}\,\underset{3\mu}{u}\,\underset{4\nu}{u}\right| \varepsilon^{\alpha\delta\theta\nu}\varepsilon^{\beta\zeta\eta\mu}\varepsilon^{\gamma\varepsilon\kappa\lambda},$$

$$I_4 = \left|\underset{1\alpha}{u}\,\underset{2\beta}{u}\,\underset{3\gamma}{u}\,\underset{4\delta}{u}\right| \varepsilon^{\alpha\beta\gamma\delta},$$

the action being

$$S = \iiiint d^4\underset{1}{x}\,d^4\underset{2}{x}\,d^4\underset{3}{x}\,d^4\underset{4}{x}\;\Pi\,,\quad \Pi = I_1^{\kappa_1} I_2^{\kappa_2} I_3^{\kappa_3} I_4^{\kappa_4}. \qquad (4.3)$$

The manifold is a linear affine space, the invariance being given by

$$u^\mu(x) := A^\mu{}_\nu u^\nu(x)\,,\quad A^\mu{}_\nu \text{ regular}. \qquad (4.4)$$

The first variation of the action, eq.(4.3), produces an expression of the form

$$\frac{\delta S}{\delta \underset{1}{u}^{\mu}} = \iiint d^4\underset{2}{x}\, d^4\underset{3}{x}\, d^4\underset{4}{x}\, \Pi \, *$$

$$\Big[\sum_i \frac{\kappa_i \partial_{u_1^\mu} I_i}{I_i} - \sum_{ij} \frac{\kappa_i \partial_{u^\mu_\nu} I_i \kappa_j \partial_{u^\rho} I_j}{I_i I_j} \underset{1}{u}^{\rho}{}_{,\nu} \tag{4.5}$$

$$+ \sum_i \frac{\kappa_i \partial_{u^\mu_\nu} I_i \partial_{u^\rho} I_i}{I_i^2} \underset{1}{u}^{\rho}{}_{,\nu} - \sum_i \frac{\kappa_i \partial^2_{u^\mu_\nu u^\rho} I_i}{I_i} \underset{1}{u}^{\rho}{}_{,\nu}$$

$$- \Big\{ \sum_{ij} \frac{\kappa_i \kappa_j \partial_{u^\mu_\nu} I_i \partial_{u^\rho_\sigma} I_j}{I_i I_j} + \sum_i \frac{-\kappa_i \partial_{u^\mu_\nu} I_i \partial_{u^\rho_\sigma} I_i + \kappa_i \partial^2_{u^\mu_\nu u^\rho_\sigma} I_i \cdot I_i}{I_i^2} \Big\} \underset{1}{u}^{\rho}{}_{,\sigma\nu} \Big]$$

We get an equation of the form

$$A_{\lambda\mu}{}^{\sigma\nu}\, u^{\mu}{}_{,\sigma\nu} = \ldots \tag{4.6}$$

with

$$A_{\lambda\mu}{}^{\sigma\nu} = \iiint d^4\underset{2}{x}\, d^4\underset{3}{x}\, d^4\underset{4}{x}\, \Pi \, *$$

$$\Big[-\sum_{mn} \frac{\kappa_m \kappa_n}{I_m I_n} \partial_{u^\lambda_\nu} I_m \partial_{u^\mu_\sigma} I_n \tag{4.7}$$

$$+ \sum_m \frac{\kappa_m}{I_m^2} \partial_{u^\lambda_\nu} \partial_{u^\mu_\sigma} I_m - \sum_m \frac{\kappa_m}{I_m} \partial^2_{u^\lambda_\nu u^\mu_\sigma} I_m \Big]$$

If we can show, that for some field configuration u^μ the integral $A_{\lambda\mu}{}^{\sigma\nu}$ decomposes into a product $a_{\lambda\mu}\, g^{\sigma\nu}$, we got an effective (contravariant) metric $g^{\sigma\nu}$ induced by the field itself and "measuring" the quantity of perturbations. This again is expected to work for the velocity field

$$u^{\mu} = \begin{pmatrix} u^0(t) \\ R(t) \cdot \begin{matrix} x \\ y \\ z \end{matrix} \end{pmatrix}. \tag{4.8}$$

Of course, the action, eq.(4.3), defines no applicable physical field. However, the formal construction already shows, in which direction deviations from the usual picture of relativistic field theories are to be expected. An a posteriori definition of wave equations implies, that the wave equation is only approximately separable for the different components of the field, and the finiteness of the potentials of the matter distribution in the universe will allow for small deviations from the usual wave operator. These deviations may be at least of order 10^{-40} (Dirac's number), at most of order 10^{-6} (Newtonian potential of the galaxy). In our action, eq.(4.3), deviations are of the form

$$a_{\lambda\mu} g^{\sigma\nu} u^{\mu}{}_{,\sigma\nu} + \varepsilon \beta_{\lambda\mu}{}^{\sigma\nu} u^{\mu}{}_{,\sigma\nu} = \ldots \tag{4.9}$$

$\varepsilon \beta_{\lambda\mu}{}^{\sigma\nu}$ is the small correction to the factorisable part of $A_{\lambda\mu}{}^{\sigma\nu}$. This structure is a known feature of spinor equations, as we will show now.

Let us assume a linear spinor equation,

$$\gamma^{\mu A}{}_{B} \psi^{B}{}_{,\mu} = M^{A}{}_{B} \psi^{B}. \tag{4.10}$$

Squaring the operator we get

$$E^{A}{}_{B}{}^{\mu\nu} \psi^{B}{}_{,\nu\mu} = Q(\psi, \psi_{,\rho}) \tag{4.11}$$

with

$$E^{A}{}_{B}{}^{\mu\nu} = \gamma^{\mu A}{}_{C} \gamma^{\nu C}{}_{B}.$$

The analogy to our operator, eq. (4.6), is evident. In classical Dirac theory, the spin matrices γ are defined by

$$E^{A}{}_{B}{}^{\mu\nu} + E^{A}{}_{B}{}^{\nu\mu} = 2\delta^{A}{}_{B} g^{\mu\nu}, \tag{4.12}$$

and this approach is possible, because $g^{\mu\nu}$ is already given or supposed to exist. Deviation from metric space-time means that the γ are not object of the c o n d i t i o n (4.12), but d e f i n e , the other way round, an effective metric by eq. (4.6),

$$g^{\mu\nu} = \frac{1}{4} E^{A}{}_{A}{}^{\mu\nu}. \tag{4.13}$$

This metric in its turn does not define in full the coefficients of the wave equation (4.11), and may only be regarded as defining the leading terms.

The deviations from Lorentz invariance by our symmetry breakdown scheme are now mapped onto deviations from Dirac equation by assuming additional outer fields, i.e. the part of the coefficients not defined by the metric, eq. (4.13). The character of these deviations corresponds to additional mixing between the components of a multicomponent quantity not implied by special relativity. The first example of possible effects is the spin-dependence of the energy levels of the free electron. If the plane wave $\psi^A = \psi^A_0 e^{ikx}$ is substituted in eq.(4.10), the energy levels are now given by

$$E^A{}_B{}^{\mu\nu} k_\mu k_\nu \psi^B_0 = O\left(k \text{ linear}\right). \tag{4.14}$$

Only the decomposition

$$E^A{}_B{}^{\mu\nu} = a^A{}_B\, g^{\mu\nu} \tag{4.15}$$

provides for a quadratic form of the four-momentum, which is independent of spin. If one looks for limits of special relativity under the assumption of a mono-metric structure, one should look for deviations from the decomposition property considered.

4. LATTICE APPROACHES TO PREGEOMETRY

We only shortly mention this approach to derive macroscopic geometry. Since the euclidean interpretation of space-time by use of time as imaginary coordinate, euclidean and proto-euclidean spaces have been studied as candidates for pregeometry, i.e. as candidates, whose euclidean laws produce in some way the distinction of one coordinate as time and the causal properties known from the wave equation. Of course, the program of pregeometry requires, that the "Wick rotation" is not put in by hand in an euclidean symmetric situation, but that the coordinate chosen for time is distinguished by protodynamics. Lattice theories today mostly do not worry about this question. It is considered, for instance, by C.Lanczos /25/, H.Nielsen et al./26/, and E.Cohen /27/.

5. REFERENCES

/1/ Helmholtz, H.: Thatsächliche Grundlagen der Geometrie Göttinger Nachr. (1868).
/2/ Einstein, A.: Sber.Preuss.Akad.Wiss. (1914), 1030.
/3/ Einstein, A.: Sber.Preuss.Akad.Wiss. (1915), 844.
/4/ Hawking, S.W., and G.F.R.Ellis: The large-scale structure of space-time, Cambridge 1973.
/5/ Treder, H.-J.(ed.): Gravitationstheorie und Äquivalenzprinzip, Berlin 1971.
/6/ Ehlers, J., F.A.E.Pirani, and A.Schild: The geometry of free fall and light propagation, in: J.L.Synge Festschrift, Oxford 1972.
/7/ Liebscher, D.-E.: Ann.d.Physik (Lpz.) 34 (1977), 402.
/8/ Liebscher, D.-E., and J.Mücket: Astron.Nachr. 302 (1981),143.
/9/ Weyl, H.: Ann.d.Physik (Lpz.) 59 (1919), 101.
Bach, R.: Math.Zeitschr. 9 (1921), 110.
/10/ Poincaré, H.: RC.Circ.mat.Palermo 21 (1906), 129.
Einstein, A.: Sber.Preuss.Akad.Wiss. (1919), 349.
/11/ Akama, K., Y.Chikasige, T.Matsuki, and H.Terazawa: Progr.Theor.Phys.60 (1978), 868,1900.
/12/ Akama, K., and H.Terazawa: GRG Journal 15 (1983), 201
/13/ Terazawa, H., and K.Akama: Physics Letters 96B (1980), 276, 97B (1980), 81.
/14/ Liebscher, D.-E.: GR 10 abstracts, Padova 1983, I, 559.
/15/ Treder, H.-J.: Die Relativität der Trägheit, Berlin 1972.
/16/ Treder, H.-J.: Die Prinzipien der Dynamik bei Mach, Einstein, Hertz und Poincaré, Berlin 1972.
/17/ Barbour, J.B.: Relational concepts of space and time, British Soc.Phil.Sci.London 1980.
/18/ Barbour, J.B., and B.Bertotti: Proc.Roy.Soc. London A 382 (1982), 295.
/19/ Barbour, J.B., and B.Bertotti: Nuovo Cim. 26B (1975), 16.
/20/ Liebscher, D.-E., and W.Yourgrau: Ann.d.Physik (Lpz.) 36 (1979), 20.
/21/ Planck, M.: Das Prinzip der Erhaltung der Energie, 3.Aufl. Leipzig 1913, p. 272,
Neumann, C.: Über die Principien der Galilei-Newtonschen Theorie, Leipzig 1870.
/22/ Liebscher, D.-E.: Astron.Nachr. 302 (1981), 137.
/23/ Kasper, U., and D.-E.Liebscher: Astron.Nachr. 295 (1974), 11.
/24/ Liebscher, D.-E.: Inertia-free mechanics and Lorentz invariance: a simple example, to appear in Ann.d.Physik (Lpz.) 41 (1984).

/25/ Lanczos, C.: Ann.Math. 39 (1938), 842,
and in: Entstehung, Entwicklung und Perspektiven der Einsteinschen Gravitationstheorie, Akademie-Verlag Berlin 1966.

/26/ Nielsen, H.B.: Physics Letters 94B (1980), 135,
Lehto, M., H.B.Nielsen, and M.Ninomiya: Geometry from a pregeometric quantum lattice, Nordita pr. 1984

/27/ Cohen, E.: Random Lattices as Theories of Space-Time, Harvard pr. HUTP 84/A 014 (1984)

SUPERGRAND UNIFIED COMPOSITE MODEL IN PREGEOMETRY

Hidezumi Terazawa

Institute for Nuclear Study, University of Tokyo
Midori-cho, Tanashi, Tokyo 188
JAPAN

ABSTRACT

The supergrand unified theory of composite particles and fields in pregeometry is reviewed. It effectively reproduces not only unified gauge theories for the strong and electroweak forces of leptons and quarks but also general relativity for gravity at low energies and predicts the fundamental relations of the type $e^2 = 16\pi GM^2$, where e, G and M are the gauge coupling constant, the Newtonian gravitational constant and the subquark mass, which indicates $M \cong 10^{18}$GeV. Some other consequences of this theory are also discussed.

1. INTRODUCTION

It is a final goal in theoretical physics to construct the fundamental theory which can not only describe but explain all physical phenomena. Historically, many attempts have been made to construct such a final theory in physics. Notable among them are the unified field theory of Einstein, Weyl, Kaluza and Klein,[1)] the unified spinor theory of Heisenberg, Ivanenko and Nambu-Jona-Lasinio,[2)] the unified gauge theory of Schwinger, Glashow, Salam and Weinberg[3)] and the grand unified gauge theory of Pati-Salam and Georgi-Glashow.[4)] The first two of these have become obsolete in their original forms for obvious reasons while the last two are incomplete as the final theory since they can neither describe gravity nor explain all fundamental physical quantities such as the gauge coupling constants and the masses of the fundamental fermions (leptons and quarks) and the gauge bosons ($W^{\pm}$ and Z).

Recently, we[5] have proposed the supergrand unified pregauge[6] and pregeometric[7] subquark model[8] in which not only the fundamental fermions but the gauge bosons and the space-time metric (as well as the Higgs scalars, if any) are all composites of subquarks, the more fundamental and probably ultimate particles in nature. In this talk, I shall review this supergrand unified subquark model, which is a candidate for the fundamental theory of composite particles and fields, based on the two principles (of relativity and of quantum) and the two hypotheses (of the fundamental length and of compositeness).

The theory effectively reproduces not only gauge theories for the strong and electroweak forces of the fundamental fermions but general relativity for gravity at low energies (or low temperature). At extremely high energies (or high temperature), it predicts an infinite series of nonlinear interactions of the fundamental fermions, the gauge bosons and the space-time metric. It also predicts the relations of the type

$$e^2 = 16\pi G M^2 \tag{1}$$

where e, G and M are the gauge coupling constant, the Newtonian gravitational constant and the subquark mass. These relations indicate that the subquark mass may be as large as 10^{18}GeV.

2. THE TWO PRINCIPLES AND THE TWO HYPOTHESES

Let us start with introducing the two principles and the two hypotheses:

1) Relativity Principle

Among others, this requires that a theory be invariant under the Lorentz and general coordinate transformations. The velocity of a massless particle in the vacuum (which is the maximum speed), c ($\cong 3\times10^{10}$cm/s), is set equal to unity for convenience as usual.

2) Quantum Principle

Among others, this requires that the theory be of quantum and that fields be quantized canonically. The Planck constant divided by 2π (a half of which is the minimum action), $\hbar$ ($\cong 1\times10^{-27}$ erg·s), is set equal to unity for convenience as usual.

3) Fundamental-Length Hypothesis

This requires that either one of the above two principles and the microscopic causality may break down if the distance between two particles becomes smaller than the fundamental length, ℓ, which may be as small as the Planck length, $G^{1/2}$ ($\cong 1\times10^{-33}$ cm).

4) Compositeness Hypothesis

This asserts that not only the fundamental fermions (leptons and quarks) but the gauge bosons and the space-time metric (as well as the Higgs scalars, if any) are all composites of subquarks, the more fundamental and probably ultimate particles in nature.

3. THE SIMPLEST POSSIBLE MODEL

Now I am ready to present the simplest possible model Lagrangian that satisfies these four conditions:

$$\begin{aligned} L' = {} & \bar{w}[e^{k\mu}\gamma_k(iD_\mu+\tau^i A^i_\mu)-\phi]w \\ & +g^{\mu\nu}(\partial_\mu - i\lambda^a G^a_\mu)C^\dagger(\partial_\nu - i\lambda^b G^b_\nu)C-\phi^2 C^\dagger C \\ & +\bar{f}wC+C^\dagger\bar{w}f \end{aligned} \tag{2}$$

where w and C are, respectively, the spinor and scalar subquarks ("wakems" and "chroms") which are the most fundamental particles. The auxiliary fields of f, A^i_μ, G^a_μ, ϕ, $e^{k\mu}$ and $g^{\mu\nu}$ are, respectively, the fundamental fermions, the gauge fields, the Higgs scalar, the vierbein and the space-time metric which represent various composite states (or collective motions) of subquarks. For definiteness and simplicity, the spinor and scalar subquarks are, respectively, assumed to form a fundamental N_W-plet of $SU(N_W)$ and a fundamental N_C-plet of $SU(N_C)$

(whose matrices are denoted by τ^i and λ^a for $i=1,2,3,\cdots,N_W^2-1$ and $a=1,2,3,\cdots,N_C^2-1$). Notice that the model Lagrangian is invariant not only under gauge transformation but under scale transformation.

4. THE EFFECTIVE LAGRANGIAN

Define the effective Lagrangian for composite fields by the path-integral over the subquark fields:

$$\exp[i\int d^4x\sqrt{-g}L_{eff}] = \int dwd\bar{w}dCdC^\dagger \exp[i\int d^4x\sqrt{-g}L'] \tag{3}$$

where $g = \det g_{\mu\nu}$. The path-integration can be formally performed to give

$$\begin{aligned}\int d^4x\sqrt{-g}L_{eff} &= -itr\ell n[e^{k\mu}\gamma_k(iD_\mu+\tau^iA_\mu^i)-\phi]\\ &+itr\ell n[-(\sqrt{-g})^{-1}(\partial_\mu-i\lambda^aG_\mu^a)\sqrt{-g}g^{\mu\nu}(\partial_\nu-i\lambda^bG_\nu^b)-\phi^2]\\ &-itr\ell n\{1+\bar{f}[e^{k\mu}\gamma_k(iD_\mu+\tau^iA_\mu^i)-\phi]^{-1}f[(\sqrt{-g})^{-1}(\partial_\nu-i\lambda^aG_\nu^a)\sqrt{-g}g^{\nu\kappa}(\partial_\kappa-i\lambda^bG_\kappa^b)\\ &\qquad+\phi^2]^{-1}\} .\end{aligned} \tag{4}$$

For further simplicity, the Higgs scalar ϕ will be replaced by its vacuum expectation value M which can be identified with the subquark mass and which is arbitrary because of the scale invariance. Then, expand the effective Lagrangian into an infinite series in the inverse power of M to obtain

$$\begin{aligned}\int d^4x\sqrt{-g}L_{eff} &= i\sum_{n=1}^{\infty}\frac{1}{nM^n}\,tr[e^{k\mu}\gamma_k(iD_\mu+\tau^iA_\mu^i)]^n\\ &+i\sum_{n=1}^{\infty}\frac{(-1)^{n-1}}{nM^{2n}}\,tr[(\sqrt{-g})^{-1}(\partial_\mu-i\lambda^aG_\mu^a)\sqrt{-g}g^{\mu\nu}(\partial_\nu-i\lambda^bG_\nu^b)]^n\end{aligned}$$

$$+i\sum_{n=1}^{\infty}\sum_{\ell=0}^{\infty}\sum_{m=0}^{\infty}\frac{(-1)^m}{nM^{\ell+2m+3n}}$$

$$\times\mathrm{tr}\{\bar{f}[e^{k\mu}\gamma_k(iD_\mu+\tau^i A^i_\mu)]^\ell f[(\sqrt{-g})^{-1}(\partial_\nu-i\lambda^a G^a_\nu)\sqrt{-g}g^{\nu\kappa}(\partial_\kappa-i\lambda^b G^b_\kappa)]^m\}^n. \quad (5)$$

Every term in this infinite series would be divergent and, therefore, ill-defined if there were no regularization in nature. According to the fundamental-length hypothesis, however, it must be finite. Suppose that invariance under both gauge and general coordinate transformations is kept tight even near the fundamental length. Then, I shall introduce a sofisticated way of regularization as illustrated as

$$\mathrm{tr}[(\sqrt{-g})^{-1}\partial_\mu\sqrt{-g}g^{\mu\nu}\partial_\nu]$$

$$=\lim_{x'\to x}\int d^4x d^4y(\sqrt{-g_y})^{-1}\partial_{y\mu}\delta^4(y-x')[\exp(\int_x^y\Gamma^\alpha_{\beta\gamma}dz^\gamma)]^{\mu\nu}\sqrt{-g_x}\partial_{x\nu}\delta^4(x-y)$$

$$=-\frac{\delta^4(0)}{\sqrt{-g_0}}\int d^4x\sqrt{-g}R \quad (6)$$

and

$$\mathrm{tr}[(\partial_\mu+A_\mu)(\partial_\nu+A_\nu)]$$

$$=\lim_{x'\to x}\int d^4x d^4y\,\mathrm{tr}[(\partial_{y\mu}-A_{y\mu})\delta^4(y-x')\exp(-\int_x^y A_\alpha dz^\alpha)(\partial_{x\nu}+A_{x\nu})\delta^4(x-y)]$$

$$=\delta^4(0)\int d^4x\,\mathrm{tr}(\partial_\mu A_\nu-\partial_\nu A_\mu+[A_\mu,A_\nu]) \quad (7)$$

where $A_\mu=-i\tau^i A^i_\mu$, $\Gamma^\alpha_{\beta\gamma}$ and R are the Christoffel symbol and the Riemann scalar curvature, respectively, and $\delta^4(0)$ and g_0 are their values at the fundamental length.

More precisely, our regularization at the fundamental length consists of the following procedures: 1) Introduce the point-splitting in the trace operation with respect to the space-time coordinates as A(x)B(x) is replaced by $\lim_{x'\to x} A(x')B(x)$. 2) Insert the exponential factors such as $[\exp(\int_x^y \Gamma^\alpha_{\beta\gamma} dz^\gamma)]^{\mu\nu}$ and $\exp(-\int_x^y A_\alpha dz^\alpha)$ between the operator products so that they may be kept invariant under gauge and general coordinate transformations. 3) Perform the integration over the space-time coordinates in the trace operations. Then, there will appear the factor $\delta^4(0)$ defined by $\lim_{x'\to x} \delta^4(x'-x)$ which would be a divergent quantity in a theory without the fundamental length. 4) Take the factor $\delta^4(0)$ as a finite quantity of order ℓ^{-4}, according to the fundamental length hypothesis of $|x'-x| \gtrsim \ell$.

A careful calculation leads to the following simple result for the first several terms in the infinite series:

$$\int d^4x\sqrt{-g}L_{eff} = \frac{i\delta^4(0)}{\sqrt{-g_0}} \int d^4x\sqrt{-g}\ [\frac{2N_W+N_C}{M^2} R - \frac{1}{8M^4} g^{\mu\nu}g^{\kappa\lambda}(4N_W F^i_{\mu\kappa}F^i_{\nu\lambda}+N_C G^a_{\mu\kappa}G^a_{\nu\lambda}) + \frac{1}{M^4} \bar{f}e^{k\mu}\gamma_k(iD_\mu+\tau^i A^i_\mu+\lambda^a G^a_\mu)f+\cdots] \tag{8}$$

with

$$F^i_{\mu\nu} = \partial_\mu A^i_\nu - \partial_\nu A^i_\mu + \varepsilon^{ijk}A^j_\mu A^k_\nu$$

and

$$G^a_{\mu\nu} = \partial_\mu G^a_\nu - \partial_\nu G^a_\mu + f^{abc}G^b_\mu G^c_\nu \tag{9}$$

where ε^{ijk} and f^{abc} are the structure constants of $SU(N_W)$ and $SU(N_C)$, respectively.

5. THE FUNDAMENTAL RELATIONS

At low energies, this effective Lagrangian reproduces not only gauge theories for the strong and electroweak interactions of the fundamental fermions but general relativity for gravity if the

Newtonian gravitational constant and the SU(N_W) and SU(N_C) gauge coupling constants, g and f, are identified as

$$\frac{i\delta^4(0)}{\sqrt{-g_0}}\,\frac{2N_W+N_C}{M^2} = \frac{1}{16\pi G}\,, \qquad (10)$$

$$\frac{i\delta^4(0)}{\sqrt{-g_0}}\,\frac{N_W}{2M^4} = \frac{1}{4g^2} \quad \text{and} \quad \frac{i\delta^4(0)}{\sqrt{-g_0}}\,\frac{N_C}{8M^4} = \frac{1}{4f^2}\,. \qquad (11)$$

These 3 identifications lead to the fundamental relations

$$g^2 = f^2\,\frac{N_C}{4N_W} = 16\pi GM^2\left(1+\frac{N_C}{2N_W}\right)\,, \qquad (12)$$

which indicate that the subquark mass is as large as 10^{18}GeV.

They also indicate that the space-time at the fundamental length must be space-like, i.e. $g_0>0$, so that $i\delta^4(0)/\sqrt{-g_0} > 0$. Furthermore, since it is natural to suppose $i\delta^4(0)/\sqrt{-g_0}$ is of the order of ℓ^{-4}, the fundamental length ℓ must be of the order of $(16\sqrt{2}\pi G/e)^{1/2}$ ($\simeq 10^{-18}$ GeV^{-1}). Notice that this result is similar to the relation in generalized Kaluza-Klein theories, $\ell = 2\pi(16\pi G/e^2)^{1/2}$ ($\simeq 10^{-17}$GeV^{-1}) for the size ℓ of the extra dimensions, which has been carefully discussed by Freund[9] although the physical features involved are clearly different.

This theory also predicts at extremely high energies an infinite series of nonlinear interactions of the fundamental fermions, the gauge bosons and the space-time metric, which can be seen in the expression (5). The possible phase transition between the geometric phase (where $g_{\mu\nu} \neq 0$ and $|g_{\mu\nu}|<\infty$) and the pregeometric one (where $g_{\mu\nu} = 0$) which Akama and myself[10] have discussed to present a possible explanation for the origin of the big bang of our Universe is one of the consequences of such nonlinear interactions at high temperature.

There seem to be many other consequences, which are subjects for future investigations.

6. CONCLUSION

In conclusion, I would like to emphasize that the possibility of superheavy subquarks as "maximons",[11] which are called by Professor Markov for the heaviest possible elementary particles in nature whose masses are of order of the Planck mass ($G^{-1/2} \cong 1\times10^{19}$GeV), seems to be one of the most theoretically attractive and noble ones although the possibility of relatively light subquarks whose masses are of order of the Fermi mass ($G_F^{-1/2} \cong 300$GeV where G_F is the Fermi weak coupling constant) seems to be the most experimentally attractive and exciting one that may explain all the anomalous events recently found by the UA1 and UA2 groups in the CERN SPS $\bar{p}p$ Collider experiments.[8]

REFERENCES

1) See for example, Einstein, A., The Meaning of Relativity, 3rd ed. (Princeton Univ. Press, Princeton, N.J., 1950), Appendix II; Weyl, H., Raum, Zeit und Materie (Springer Verlag, Berlin, Heidelberg, N.Y., 1920); Kaluza, T., Sitzker. Preuss. Akad. Wiss. Kl, 966 (1921); Klein, O., Z. Physik 37, 895 (1926).

2) See for example, Heisenberg, W., Z. Naturforsch. 14, 441 (1959), Introduction to the Elementary Particles (John Wiley & Sons, Ltd., London, N.Y., Sydney, 1966) and the earlier references therein; Ivanenko, D., Sov. Phys. 13, 141 (1938) and in Centenario di Einstein (Giunti-Barbara, Firenze, 1979), p.131 [Albert Einstein 1879-1979 (Johnston Reprint Corp., N.Y., 1979), p.295]; Nambu, Y. and Jona-Lasinio, G., Phys. Rev. 122, 345 (1961).

3) See for example, Schwinger, J., Ann. Phys. (N.Y.) 2, 407 (1957); Glashow, S.L., Nucl. Phys. 22, 579 (1961); Salam, A., in Elementary Particle Physics, edited by N. Svartholm (Almqvist and Wiksell, Stockholm, 1968), p.367; Weinberg, S., Phys. Rev. Lett. 19, 1264 (1967).

4) See for example, Pati, J.C. and Salam, A., Phys. Rev. D10, 275 (1974); Georgi, H. and Glashow, S.L., Phys. Rev. Lett. 32, 438 (1974).

5) Terazawa, H., Phys. Rev. D22, 184 (1980). See also Terazawa, H., Chikashige, Y. and Akama, K., Phys. Rev. D15, 480 (1977); Terazawa, H., Chikashige, Y., Akama, K. and Matsuki, T., Phys. Rev. D15, 1181 (1977); Terazawa, H., Phys. Rev. D16, 2373 (1977), D22, 1037 (1980) and D22, 2921 (1980); Terazawa, H. and Akama, K., Phys. Lett. 96B, 276 (1980) and 97B, 81 (1980); Terazawa, H., Phys. Lett. 101B, 43 (1981) and 133B, 57 (1983). For reviews, see for example, Terazawa, H., in Proc. INS Int. Symp. on New Particles and the Structure of Hadrons, Tokyo, July 12-14, 1977 and France-Japan Joint-Seminar on New Particles and Neutral Currents, Tokyo and Kyoto, July 14-16 and 18, 1977, edited by K. Fujikawa et al. (INS, Univ. of Tokyo, Tokyo, 1978), p.231 and p.579; in Proc. XIX Int. Conf. on High Energy Physics, Tokyo, Aug., 23-30, 1978, edited by S. Homma et al. (Phys. Soc. Japan, Tokyo, 1979), p.617, in Proc. Second Marcel Grossmann Meeting on the Recent Developments of General Relativity, ICTP, Trieste, 5-11 July, 1979, edited by R. Ruffini (North-Holland Pub. Co., Amsterdam, New York, Oxford, 1982), p.519; in Proc. Japan-Italy Symposium on Fundamental Physics, Tokyo, Jan. 27-30, 1981, edited by S. Fukui and T. Toyoda (Nagoya Univ., Nagoya, 1981), p.65; in Proc. 1981 INS Symp. on Quark and Lepton Physics, Tokyo, June 25-27, 1981, edited by K. Fujikawa et al. (INS, Univ. of Tokyo, Tokyo, 1981), p.296; in Proc. XXI Int. Conf. on High Energy Physics, Paris, July 26-31, 1982, edited by P. Petiau and M. Porneuf, J. de Phys. C3-289 (1982); in Proc. Third Marcel Grossmann Meeting on the Recent Developments of General Relativity, Shanghai, Aug. 30-Sept. 3, 1982, edited by Hu Ning (Science Press and North-Holland Pub. Co., Beijing, 1983), p.239; in Proc. Topical Symposium on High Energy Physics, Tokyo, Sept. 7-11, 1982, edited by T. Eguchi and Y. Yamaguchi (World Scientific Pub. Co., Singapore, 1983), p.173; in Proc.

Europhysics Conf. on Flavor Mixing in Weak Interactions, Erice, March 4-11, 1984, edited by L.-L. Chau (Plenum Pub. Co., New York, 1984), to be published; in Proc. XXII Int. Conf. on High Energy Physics, Leipzig, July 19-25, 1984, edited by A. Meyer and E. Wieczorek, to be published.

6) Bjorken, J.D., Ann. Phys. (N.Y.) 24, 174 (1963); Terazawa, H., Chikashige, Y. and Akama, K., in Ref. 5; Terazawa, H. and Akama, K., in Ref. 5; Terazawa, H., in Ref. 5.

7) Sakharov, A.D., Doklady Akad. Nauk SSSR 177, 70 (1967) [Sov. Phys. JETP 12, 1040 (1968)] and Teor. Mat. Fiz. 23, 178 (1975) [English translation (Plenum Pub. Co., New York, 1979), p.435]; Terazawa, H., Chikashige, Y., Akama, K. and Matsuki, in Ref. 5; Akama, K., Chikashige, Y., Matsuki, T. and Terazawa, H., Prog. Theor. Phys. 60, 868 (1978); Terazawa, H. and Akama, K., in Ref. 5; Terazawa, H., in Ref. 5; Akama, K. and Terazawa, H., Gen. Rel. Gravit. 15, 201 (1983). For reviews, see for example, Terazawa, H., in Proc. Second Seminar "Quantum Gravity", Academy of Sciences of the USSR, Moscow, Oct. 13-15, 1981, edited by M.A. Markov and P.C. West (Plenum Pub. Co., London, New York, 1984), p.47 and reviews in Ref. 5. See also Adler, S.L., Rev. Mod. Phys. 54, 729 (1982).

8) For the latest review, see Terazawa, H., in Ref. 5.

9) Freund, P.G.O., Phys. Lett. 120B, 335 (1983). See also Cho, Y.M. and Freund, P.G.O., Phys. Rev. D12, 1711 (1975).

10) Akama, K. and Terazawa, H., in Ref. 7. For reviews, see Terazawa, H., reviews in Refs. 5 and 7.

11) Markov, M.A., Prog. Theor. Phys. Suppl., Extra Number, 85 (1965) and ZhETF 51, 878 (1966) [Sov. Phys. JETP 24, 584 (1967)].

THERMODYNAMICS AND FIELD STATISTICS OF A RELATIVISTIC SUPERFLUID*

Werner Israel
Theoretical Physics Institute, Department of Physics,
University of Alberta, Edmonton, Alberta, T6G 2J1, Canada

1. INTRODUCTION

Ideas that invoke spontaneous symmetry breaking now permeate much of elementary particle physics and cosmology[1]. As long ago as 1947, Bogoliubov[2] constructed a prototype model for such ideas with his (non-relativistic) theory of Bose-Einstein condensation in a gas with weak interactions. Indeed, in present usage, spontaneous symmetry breaking and Bose-Einstein condensation are almost interchangeable terms for phase transitions in which scalar fields acquire nonvanishing expectation values[3].

Even in its narrower and more traditional sense, Bose-Einstein condensation appears relevant to several areas of current interest, notably to speculations concerning the existence of a pion or quark superfluid in neutron stars[4], and the possibility that quark confinement is a consequence of gluon condensation at energies less than 200 MeV. If the photon has a finite rest-mass, then, as Kuzmin and Shaposhnikov[5] have noted, photon condensation in the very early universe would have observable consequences. One can view the instability of hot flat space[6] in quantum gravity as due to a Bose-Einstein condensation of gravitons, which acquire an imaginary effective mass $m_{grav} \sim iT^2/m_{P\ell}$ when coupled to thermally excited matter.

Landau's two-fluid model and its field-theoretical validation by Bogoliubov stands as a classic instance of how the phenomenological and

* Work partially supported by the Natural Sciences and Engineering Research Council of Canada.

microscopic descriptions of a complex phenomenon can complement and illuminate one another. To revisit this archetype in a relativistic setting was a primary motivation for the present work. While there have been many studies of relativistic Bose-Einstein condensation in interacting[3] and noninteracting[7] systems, and several dealing with the thermodynamics of relativistic superfluids[8-13], the interplay between the microscopic and phenomenological aspects has not previously been treated from an integrated point of view.

Of particular interest to us will be the form of the equation of the state, especially as regards its dependence on the relative velocity of the normal and superfluid components. This issue is intimately related to the question[13] whether it is justifiable to assume that the superfluid component carries no entropy.

We shall first outline the covariant (nonequilibrium) phenomenological theory (Secs. 2,3), and then proceed to its microscopic justification by studying the statistical thermodynamics of the superfluid in the case of thermal equilibrium, using a Euclidean path-integral formalism. In relativistic quantum field theory a chemical potential is necessarily associated with a conserved charge, so it will be assumed from the outset that we are dealing with a charged superfluid in an Einstein-Maxwell field.

2. PHENOMENOLOGICAL THEORY

As Ho and Mermin[14] first noted in a nonrelativistic context, superfluid thermodynamics assumes its clearest and most elegant form when distinct chemical potentials μ_n, μ_s are assigned to the normal and superfluid components. The thermodynamical formulae are then disencumbered of terms, involving the relative velocity of the components, that clutter up the traditional formalism. Dixon[9] and Israel[10] have developed covariant formulations along these lines. It needs to be stressed that the use of two chemical potentials is a purely formal convenience that does not of itself restrict the physics in any way. Thus, the alternative relativistic formulation of Khalatnikov and Lebedev[8], which follows the tradition of employing a single potential μ ($\equiv\mu_n$), is physically equivalent to the two-potential formalism[12].

The following sketch is a slight extension, to the case where the superfluid is charged, of the covariant formulation of the Landau two-fluid model that I have given previously[10].

We consider an arbitrary (nonequilibrium) state of the medium in an Einstein-Maxwell field. The currents J^{μ}_{A} and stress-energy tensors $T^{\lambda\mu}_{A}$ of the normal and superfluid components (A=n,s) are subject to the overall conservation laws

$$\sum_A J^{\mu}_{A|\mu} = 0, \qquad \sum_A T^{\lambda\mu}_{A\ |\mu} = F^{\lambda}_{\ \ \mu} \sum_A J^{\mu}_{A} \,. \tag{1}$$

The requirement that the superfluid flow be frictionless is captured in the assumptions

$$J^{\mu}_{s} = eN^{\mu}_{s} = en_s u^{\mu}_{s} \,, \qquad T^{\lambda\mu}_{s} = \rho_s u^{\lambda}_{s} u^{\mu}_{s} + P_s \Delta^{\lambda\mu}_{s} \tag{2}$$

$$T^{\lambda\mu}_{s\ |\mu} - F^{\lambda}_{\ \mu} J^{\mu}_{s} \propto u^{\lambda}_{s} \tag{3}$$

where $\Delta^{\lambda\mu}_{s} = g^{\lambda\mu} + u^{\lambda}_{s} u^{\mu}_{s}$ and u^{λ}_{s} is the normalized superfluid velocity ($u^{\lambda}_{s} u_{s\lambda} = -1$). Condition (3) means that the frictionless component is free to exchange particles (and hence energy) with its environment, but it cannot transfer 3-momentum to the normal component or the walls.

As in ref. [10], I shall impose the additional postulate that $S_s = 0$, i.e. the entropy of the superfluid component vanishes identically. Then the equation of state $\rho_s = \rho_s(n_s)$ of the superfluid component involves only a single independent parameter, and the Gibbs-Duhem relations reduce to

$$\frac{d\rho_s}{dn_s} = \frac{\rho_s + P_s}{n_s} = \mu_s \Rightarrow dP_s = n_s d\mu_s \tag{4}$$

This encapsulates the commonly-held notion that thermal excitations are associated exclusively with the normal component. However, this postulate is arguably dispensable*, and, in fact, there would be no difficulty in extending the present scheme to the case of a frictionless yet entropy-carrying superfluid component that still satisfies (2), (3) but has a two-parameter equation of state $\rho_s = \rho_s(n_s, S_s)$. This

* I am indebted to Professor I.M. Khalatnikov for an exchange of letters in late 1982 which helped bring into focus the distinctive role of the assumption $S_s \equiv 0$ made in ref. [10].

possibility is admitted in the theories of Dixon[9] and Khalatnikov and Lebedev[11], and has been advocated and pursued in the recent work of Fomin and Shadura[11]. However, the postulate $S_s \equiv 0$ will be adhered to in this work, and it will be part of our business to explore its experimental implications and theoretical foundations.

Insertion of (2), (4) into (3) yields the explicit form of the proportionality coefficient

$$T_s^{\lambda\mu}{}_{|\mu} = F^{\lambda}{}_{\mu} J_s^{\mu} + \mu_s N_s^{\mu}{}_{|\mu} u_s^{\lambda} \tag{5}$$

as well as the superfluid acceleration

$$\mu_s \dot{u}_s^{\kappa} = eF^{\kappa}{}_{\lambda} u_s^{\lambda} - \Delta_s^{\kappa\lambda} \partial_{\lambda} \mu_s \ . \tag{6}$$

The last equation can be re-expressed as

$$(\mu_s u_{s[\lambda} + eA_{[\lambda})_{|\mu]} u_s^{\mu} = 0 \tag{7}$$

which implies conservation of the superfluid circulation

$$\mathcal{L}_{u_s} \oint (\mu_s u_{s\lambda} + eA_{\lambda}) dx^{\lambda} = 0 \ , \tag{8}$$

i.e. the line integral is constant for a closed loop convected with the superfluid component.

Equation (8) guarantees the consistency of imposing the Onsager-Feynman flux quantization condition

$$\oint (\mu_s u_{s\lambda} + eA_{\lambda}) dx^{\lambda} = nh \ . \tag{9}$$

Since the integral must be invariant under continuous deformations of the loop, it is necessary, and, by (7), dynamically consistent, to replace (7) by the stronger condition

$$(\mu_s u_{s[\lambda})_{|\mu]} = \frac{1}{2} eF_{\lambda\mu} \tag{10}$$

which must hold everywhere except on vortex lines.

The normal component can be handled by the standard methods of covariant irreversible thermodynamics[15]. The covariant Gibbs-Duhem relations are

$$S^{\mu}_{n} = P_{n}(\alpha,\beta)\beta^{\mu} - \alpha N^{\mu}_{n} - \beta_{\lambda}T^{\lambda\mu}_{n} \tag{11}$$

$$d(P_{n}\beta^{\mu}) = N^{\mu}_{(0)n}d\alpha + T_{(0)n\lambda}{}^{\mu}\, d\beta^{\lambda} \tag{12}$$

for a state $(N^{\mu}_{n}, T^{\lambda\mu}_{n}, S^{\mu}_{n})$ close to some equilibrium state $(N^{\mu}_{(0)n},\dots)$ specified by the inverse-temperature 4-vector $\beta^{\mu}=\beta u^{\mu}_{n}=T^{-1}u^{\mu}_{n}$, thermal potential $\alpha=T^{-1}\mu_{n}$ and pressure given by the equilibrium equation of state $P_{n}=P_{n}(\alpha_{n},\beta)$. Omission of the "normal" suffix from β^{μ} and T causes no ambiguity, since the temperature of the superfluid component is undefined[10]. When applied to an off-equilibrium state, (11) automatically entails the standard linear relation between entropy flux and heat flux.

The entropy production is obtained by taking the divergence of (11) and invoking (1), (5) and (12). To avoid complications irrelevant to our purpose I shall neglect ohmic effects by assuming the currents to be purely convective: $J^{\mu}_{A} = eN^{\mu}_{A}$. Then

$$0 \le S^{\mu}_{n|\mu} = -N^{\mu}_{(1)n}(\partial_{\mu}\alpha + \beta^{\lambda}eF_{\lambda\mu}) -$$

$$- T^{\lambda\mu}_{(1)n}\,\beta_{\lambda|\mu} + \beta(\mu_{n} - \gamma\mu_{s})N^{\mu}_{s|\mu}\ , \tag{13}$$

where $\gamma = -u^{\mu}_{n}u_{s\mu} = (1-v^{2}/c^{2})^{-\frac{1}{2}}$ is the relativistic dilation factor associated with the relative speed v of the components and $N^{\mu}_{(1)n}$, $T^{\lambda\mu}_{(1)n}$ are deviations from the equilibrium values.

From (13), it is straightforward to read off the covariant versions[10] of the Khalatnikov phenomenological relations if the Curie principle is valid to a sufficient approximation, or of the Clark equations[16] in the general case. Here I shall note only the necessary conditions for thermodynamical equilibrium that follow from (13):

$$\beta_{(\lambda|\mu)} = 0, \quad \partial_{\mu}\alpha + \beta^{\lambda}eF_{\lambda\mu} = 0, \quad \mu_{n} = \gamma\mu_{s}\ . \tag{14}$$

I interpolate here a remark which is worth stressing: the equations of the theory permit arbitrary interactions between the normal and superfluid components even in the absence of dissipation. If the viscosity and thermal conductivity of the normal component are negligible, so that we can set $N^{\mu}_{(1)n}=T^{\lambda\mu}_{(1)n}=0$, then (13) still allows exchange of particles between the two components (i.e., $N^{\mu}_{s|\mu}$, hence $u^{\mu}_{s|\mu}$, $\dot{n}_s$, $\dot{\mu}_s$, ... can take arbitrary, nonvanishing values), and this exchange will be isentropic if $\gamma\mu_s=\mu_n$. (Arbitrariness of $\dot{\mu}_s$ does not conflict with equations (10), which are degenerate and do not determine $\dot{\mu}_s$.) Thus, as already emphasized, a formalism that assigns separate chemical potentials and equations of state to the components in no way constrains the interactions between them.

3. NONRELATIVISTIC LIMIT

The assumption that the superfluid component carries no entropy implies definite and - at least in principle - measurable restrictions on the velocity-dependence of the equation of state.

This is brought out most clearly in the nonrelativistic limit, in which the equilibrium condition $\mu_n=\gamma\mu_s$ becomes

$$\mu_n = \mu_s + \frac{1}{2}mv^2 , \tag{15}$$

where μ_n, μ_s now denote the nonrelativistic potentials ($\mu^{nonrel}=\mu^{rel}-mc^2$). (The physical interpretation is, of course, that a mass-element $(\delta N)m$, reversibly transferred from the superfluid component, from which it extracts the energy $(\delta N)\mu_s$, must have vanishing momentum with respect to that component, so that the energy $(\delta N)\mu_n$ injected into the normal component is augmented by the kinetic energy of relative motion.)

The equation of state

$$P = P_s(\mu_s) + P_n(\mu_n,T), \quad n_s = \frac{dP_s}{d\mu_s}, \quad n_n = \left(\frac{\partial P_n}{\partial \mu_n}\right)_T \tag{16}$$

is determined by the specification of one function of one variable and one function of two variables. If an entropy were assigned to the superfluid component, two functions of two variables would be needed: $P=P_s(\mu_s,T)+P_n(\mu_n,T)$. Even less specific is the form $P=P(\mu,T,v)$ that traditionally appears in the literature[16,17].

An elementary calculation using (15), (16) yields a definite form for the velocity-dependence of the normal/superfluid mass-ratio:

$$\lambda \equiv \frac{\partial(n_n/n_s)}{\partial(\tfrac{1}{2}v^2)} = \frac{n_n^2}{n_n+n_s}\left\{\frac{1}{n_n}\frac{\partial(n_n m)}{\partial P_n} + \frac{1}{n_s}\frac{\partial(n_s m)}{\partial P_s}\right\} \tag{17}$$

where the partial derivatives are taken at fixed P and T. In this form, the result holds whether the superfluid carries entropy or not; in the latter case, the second term in braces should be independent of T. Generally, (17) gives the order-of-magnitude estimate

$$\lambda \sim \frac{1}{v_{sound}^2} = 2\times 10^{-9}(\text{cm/sec})^{-2} \tag{18}$$

for HeII. Experimentally, very little seems to be known about λ. Putterman[16] cites persistent-current measurements by Kojima which gave an upper limit

$$\lambda < 1.2\times 10^{-7}(\text{cm/sec})^{-2} . \tag{19}$$

The gap between (18) and (19) is not so vast as to preclude all hope of a detailed experimental check of the prediction (17); for the present, one can at least say that it is not incompatible with experiment.

4. STATISTICAL THERMODYNAMICS

We now proceed to correlate the covariant phenomenology developed in Sec. 2 with a simple microscopic model, a covariant version of Bogoliubov's theory[2] of Bose-Einstein condensation in a gas of interacting bosons.

The statistical mechanics of quantum fields now exists in a variety of formulations, in particular the Thermofield Dynamics of Takahahsi and Umezawa[18]. (A recent paper by Niemi and Semenoff[19] includes a brief historical survey and critical comparison of the various approaches.) Here we shall be concerned only with thermal equilibrium, for which the relatively straightforward imaginary-time formalism of Matsubara is adequate.

Microscopic models of a nonrelativistic superfluid are often based on a $\lambda\phi^4$-interaction (e.g. Ichiyanagi[19], Toyoda[20]). This is the form obtained when the realistic two-body potential $V(\underset{\sim}{r},\underset{\sim}{r}')$ is replaced in the nonrelativistic Lagrangian by a pseudo-potential

$V_{pseudo}(\underset{\sim}{r},\underset{\sim}{r}') \propto \delta^3(\underset{\sim}{r}-\underset{\sim}{r}')$. Here we shall just extend this simple model to the relativistic regime. The subject of discussion, therefore, is the thermal equilibrium of a self-interacting charged scalar field, possessing a condensed phase, and under the influence of a stationary Einstein-Maxwell background field.

We suppose, accordingly, that classical fields

$$g_{\lambda\mu}(x), \quad F_{\lambda\mu}(x), \quad \alpha(x), \quad \beta^\lambda(x) \tag{20}$$

have been prescribed, satisfying the stationary equilibrium constraints [cf. (14)]

$$\beta_{(\lambda|\mu)} \equiv \frac{1}{2}\underset{\beta}{\mathcal{L}}\, g_{\lambda\mu} = 0, \quad \underset{\beta}{\mathcal{L}}\, F_{\lambda\mu} = 0, \quad \partial_\mu\alpha + \beta^\lambda e F_{\lambda\mu} = 0 . \tag{21}$$

One can choose a special class of "stationary" gauges, characterized by $\underset{\beta}{\mathcal{L}}(A_\mu\beta^\mu)=0$, i.e. by time-independence of A_0 in co-ordinates adapted to the timelike vector β^μ [see (24)]. In these gauges, the last two conditions in (21) specialize to

$$\underset{\beta}{\mathcal{L}}\, A_\mu = 0, \quad \partial_\mu(\alpha - e\beta^\lambda A_\lambda) = 0 . \tag{22}$$

By a further restriction of the gauge, one could even arrange

$$\alpha - e\beta^\lambda A_\lambda = 0 , \tag{23}$$

and thereby eliminate the chemical potential altogether. However, this elimination would not generally be possible for a mixture of several charged elements. For this reason, and also to preserve as much manifest gauge invariance as is conveniently possible, we shall forgo the simplification (23), although we shall later find it expedient to adopt (22).

For a self-interacting complex scalar field $\tilde{\phi}$, the (Lorentzian-signature) action is iS, where

$$S[\tilde{\phi}] = \int_\Omega d^4x\, g^{\frac{1}{2}}\{g^{\mu\nu}(\nabla^\dagger_\mu\tilde{\phi}^*)(\nabla_\nu\tilde{\phi}) + m^2\tilde{\phi}^*\tilde{\phi} + \frac{1}{2}\lambda(\tilde{\phi}^*\tilde{\phi})^2\}, \tag{24}$$

$$\nabla_\mu = \partial_\mu - ieA_\mu , \qquad \nabla^\dagger_\mu = \partial_\mu + ieA_\mu , \tag{25}$$

and we select $\arg g^{\frac{1}{2}} = +\frac{\pi}{2}$ on the Lorentzian sector.

We shall work throughout in real stationary co-ordinates $x^\lambda = (\tau, x^a)$, defined as any solutions of

$$\beta^\mu \partial_\mu x^\lambda = \beta^\tau \delta^\lambda_0 \; , \tag{26}$$

where β^τ is a constant, equal on the Lorentzian sector to an arbitrary real positive number β^t. (No generality would be lost in taking $\beta^t = 1$.)

Since we are dealing with complex fields in a general-relativistic setting, it is helpful to formulate the machinery of Wick rotation in a form that obviates the additional complication of complex co-ordinates. It is elementary but essential to stress that Wick rotation is an algebraic process that basically has no connection with either complex co-ordinate transformations or analytic continuation of field amplitudes. (The highly nonanalytic character of path-integration variables is well-known.)

For practical purposes, a Wick rotation, through an angle χ $(0 \leq \chi \leq \frac{\pi}{2})$, of the action integral (24), is straightforwardly described, in a fixed system of real stationary co-ordinates x^λ, as a deformation in which the field amplitudes $\tilde{\phi}(x)$ are left unchanged, while the tensor components

$$\Phi(x) \equiv \{g_{\lambda\mu}(\underset{\sim}{x}), A_\mu(x), \alpha(\underset{\sim}{x}), \beta^\mu\} \tag{27}$$

deviate from their Lorentzian values $\Phi_L(x)$ according to the "Lie-Wick transformation"

$$\Phi(x) = \Lambda\begin{pmatrix} x \\ x_L \end{pmatrix} \Phi_L(x) \; , \tag{28}$$

where $\Phi(x) = \Lambda\begin{pmatrix} x \\ x' \end{pmatrix}\Phi'(x')$ designates the transformation law of Φ under the co-ordinate group, and

$$x^\mu = (\tau, x^a), \quad x^\mu_L = (t, x^a), \quad t = \tau e^{i\chi} \; (\tau = \tau^*). \tag{29}$$

Thus, all spatial components are unaffected by a Wick rotation, whereas, e.g.,

$$A_\tau(\tau, \underset{\sim}{x}) = \frac{\partial t}{\partial \tau} A_t(\tau, \underset{\sim}{x}) = e^{i\chi} A_t(\tau, \underset{\sim}{x}) \tag{30}$$

where A_t denotes the (real) value of the time-component on the Lorentzian sector (χ=0). It is evident that a Wick rotation is <u>not</u> a co-ordinate transformation, since (for example) it leaves the co-ordinates x^μ and the gradients $\partial_\mu \tilde{\phi}(x)$ untouched.*

Artificial as the foregoing recipe might seem, it has a simple and natural geometrical basis. To see this, consider the flat extension of the stationary space-time manifold into the complex time-plane. The external fields $\Phi(x)$ in (27) and the field amplitudes $\tilde{\phi}(x)$ are initially defined only on the real time-axis, i.e. the Lorentzian sector, and it is necessary to specify how they are to be extended. The external field components $\Phi(x)$ are constant on the real axis (this is effectively true also of $A_\mu(\underset{\sim}{x})$, whose time-dependence is pure gauge), and it is therefore natural to extend them by requiring that they be numerically (hence also covariantly) constant over the plane. Thus,

$$\Phi'(x'^0, x^a) = \Phi'_L(x^a) \ , \tag{31}$$

independent of x'^0, in a <u>fixed</u> system of stationary co-ordinates x'^μ that is real on the Lorentzian sector, $x'^0_L=\tau$. A similar rule cannot be used to extend the field amplitudes $\tilde{\phi}$, which have an essential (and nonanalytic) time-dependence $\tilde{\phi}_L(\tau, x^a)$ on the Lorentzian sector. For that purpose one introduces the group of Wick rotations. A Wick rotation through angle χ moves the point $x'^\mu_L=(\tau, x^a)$ on the Lorentz sector to a new point $x'^\mu(\chi)=(t, x^a)$, where $t=\tau e^{i\chi}$. The field amplitudes $\tilde{\phi}$ are extended off the Lorentzian sector by <u>Lie-Wick transport</u>, i.e. $\tilde{\phi}$ is required to be constant along each circular orbit, so that $\tilde{\phi}(t, \underset{\sim}{x}) = \tilde{\phi}_L(\tau, \underset{\sim}{x})$. For a given value of χ, one can choose new ("convected") co-ordinates x^μ, with $x^0=x'^0 e^{-i\chi}$, that are real on the sector of given χ: $x^0(\chi)=x'^0_L=\tau$. In these co-ordinates, $\tilde{\phi}(\tau, \underset{\sim}{x}) = \tilde{\phi}_L(\tau, \underset{\sim}{x})$ is functionally and numerically invariant, whereas the external fields transform according to the co-ordinate group

$$\Phi(x) = \Lambda\begin{pmatrix} x \\ x' \end{pmatrix}\Phi'(x') = \Lambda\begin{pmatrix} x \\ x' \end{pmatrix}\Phi'_L(x'_L) = \Lambda\begin{pmatrix} x \\ x' \end{pmatrix}\Phi'_L(x) \tag{32}$$

* Wick rotations nonetheless do enjoy a restricted but very useful covariance property: any scalar concomitant of the tensors $\Phi(x)$ in (27) and a <u>time-independent</u> scalar field $\Psi(\underset{\sim}{x})$ and its gradients is invariant under a Wick rotation. Thus, the Ginzburg-Pitaevski equation - see (53), (54) below - can be transcribed without change from the Euclidean to the Lorentzian sector.

where we have made use of (31) in the second step. This is precisely the prescription (28).

Following the program of Matsubara[20], one conveniently derives the grand partition function from the Euclidean extension of the Feynman persistence amplitude. We consider the path integral

$$Z = \int d[\tilde{\phi}]d[\tilde{\phi}^*]e^{-S[\tilde{\phi}]/\hbar} \tag{33}$$

with the integral (24) evaluated for an arbitrary value of χ, over an oblique cylindrical segment $\Omega = V \times J$; V is a bounded portion of the 3-surface $x^0=0$, and J corresponds to a displacement $\beta^\lambda e^{i\chi}$ (which lies in the χ-sector of the complex time-plane), i.e. to an interval $0<\tau<\beta^t$ in terms of our co-ordinate $x^0 \equiv \tau$ that is real on the χ-sector.

The functional integration in (33) extends over the class of complex functions $\tilde{\phi}(x)$ satisfying the quasi-periodicity condition

$$\exp[i\hbar^{-1}\int_{x^\lambda}^{x^\lambda+\beta^\lambda e^{i\chi}} dx'^\lambda \hat{p}_{\lambda'}]\tilde{\phi}(x) = \tilde{\phi}(x) \tag{34}$$

where

$$\hat{p}_\lambda = -i\hbar\partial_\lambda - eA_\lambda + \mu\partial_\lambda t\ , \quad t = x^0 e^{i\chi}, \tag{35}$$

$$\mu(\underset{\sim}{x}) = \alpha(\underset{\sim}{x})/\beta^t\ . \tag{36}$$

Equivalently,

$$\tilde{\phi}(\tau+\beta^t,\underset{\sim}{x}) = e^{-iB(x)}\tilde{\phi}(\tau,\underset{\sim}{x})\ , \tag{37}$$

$$B(x) = \hbar^{-1}\int_{x^\lambda}^{x^\lambda+\beta^\lambda e^{i\chi}} (\mu\partial_{\lambda'}t - eA_{\lambda'})dx^{\lambda'}\ . \tag{38}$$

The line-integrals in (38) and (34) are independent of the connecting route by virtue of (21).

For $\chi = \frac{\pi}{2}$, Ω corresponds to the Euclidean sector, and (33), (37) and (24) give the grand partition function $Z[\alpha,\beta^\lambda,g_{\mu\nu},A_\lambda]$. This prescription can be ratified by taking the Hamiltonian form of the path-integral as a point of departure[3].

In this formulation, Z is invariant under (i) the local gauge group

$$eA_\mu(x) \to eA_\mu(x) + \hbar\partial_\mu\vartheta(x), \quad \tilde{\phi} \to e^{i\vartheta(x)}\tilde{\phi} \tag{39}$$

and (ii) the global "U(1) symmetry" (which is genuinely U(1) only on the Lorentzian sector)

$$\mu(\underset{\sim}{x}) \to \mu(\underset{\sim}{x}) + C, \quad \tilde{\phi} \to e^{-iCt/\hbar}\tilde{\phi}, \quad C = C^* = \text{const} . \tag{40}$$

In Bose-Einstein condensation, it is precisely this U(1) symmetry that is broken by the ground state[21].

In place of $\tilde{\phi}$, it is more convenient to work with integration variables that are strictly periodic. However, this is only possible in the restricted class of stationary gauges (22), since a periodicity condition would not be preserved by time-dependent phase transformations $e^{i\vartheta(\underset{\sim}{x},t)}$. From here on, we therefore limit ourselves to stationary gauges, in which A_μ becomes time-independent and B(x) reduces to a constant:

$$B = e^{i\chi}(\alpha - e\beta^\mu A_\mu) = e^{i\chi}\beta^t(\mu - eA_t) = \text{const} . \tag{41}$$

The residual gauge freedom is given by (40) and (39) with $\partial_0\vartheta=0$. One can now introduce a new field

$$\phi(x) = e^{i(\mu-eA_t)t/\hbar}\tilde{\phi}(x) \tag{42}$$

that is strictly periodic:

$$\phi(\underset{\sim}{x},\tau+\beta^t) = \phi(x,\tau) . \tag{43}$$

The action integral (24) assumes the form

$$S[\phi] = \int_\Omega d^4x\, g^{\frac{1}{2}}\{g^{\mu\nu}(D^\dagger_\mu\phi^*)(D_\nu\phi) + m^2\phi^*\phi + \frac{1}{2}\lambda(\phi^*\phi)^2\}, \tag{44}$$

involving the new covariant derivative

$$D_\mu = \partial_\mu - i\Gamma_\mu, \quad D^\dagger_\mu = \partial_\mu + i\Gamma_\mu, \quad \Gamma_\mu = [eA_\mu + (\mu-eA_t)\partial_\mu t]/\hbar . \tag{45}$$

Explicitly,

$$\hbar\Gamma_a(\underset{\sim}{x}) = eA_a(\underset{\sim}{x}), \quad \hbar\Gamma_0(\underset{\sim}{x}) = \mu(\underset{\sim}{x})\partial_0 t . \tag{46}$$

The gauge group is

$$\phi \to e^{i\vartheta(x)}\phi, \quad \Gamma_\mu \to \Gamma_\mu + \partial_\mu\vartheta, \quad \vartheta(x) = \vartheta(\underset{\sim}{x}) + Ct/\hbar \tag{47}$$

for real $\vartheta(\underset{\sim}{x})$ and C.

In summary, the grand partition function for the equilibrium state specified by (20), (21) is obtainable from

$$Z = e^{-W/\hbar} = \int d[\phi]d[\phi^*]e^{-S[\phi]/\hbar} \tag{48}$$

with $\phi(x)$ subject to the periodicity condition (37), and S given by (44), (46) and (28) with $\chi = \frac{\pi}{2}$. We note that

$$g^{\frac{1}{2}} = ie^{-i\chi}(-g_L)^{\frac{1}{2}} \tag{49}$$

so that the volume element $g^{\frac{1}{2}}d^4x$ is real and positive on the Euclidean sector.

5. CONDENSATE. MACROSCOPIC WAVE FUNCTION. LOOP EXPANSION.

The superfluid phase, in which Bose-Einstein condensation has developed, is characterized by a nontrivial ground state, in which the Euclidean action S is minimized for some nonvanishing $\phi(x)$. The properties of the condensate are then described by a "macroscopic wave function" $\Psi(\underset{\sim}{x})$ which, at least approximately, agrees with the minimizing function. For the nonrelativistic case, elegant functional treatments along these lines have been given by a number of authors in particular Wiegel and Jalickee[22], Ichiyanagi[23] and Toyoda[24], and our covariant analysis runs parallel to these.

In (48), we set

$$\phi(x) = \Psi(x) + \hbar^{\frac{1}{2}}\phi_{New}(x) \tag{50}$$

and imagine S expanded in powers of $\hbar^{\frac{1}{2}}$ (loop expansion[24,25]). The Euclidean effective action W is then given by

$$W = S[\Psi] + \Delta W \tag{51}$$

where one may think of $S[\Psi]$ as referring to "classical" properties of the ground state, and ΔW, defined by

$$e^{-\Delta W/\hbar} = \int d[\phi]d[\phi^*]e^{-\Delta S/\hbar}, \quad \Delta S = S[\Psi+\hbar^{\frac{1}{2}}\phi] - S[\Psi] , \tag{52}$$

as containing the effects of quantum fluctuations and thermal excitations.

We define $\Psi(x)$ as a solution of the classical field equation, moduls an unspecified function f of one-loop order, $\hbar$:

$$-\hbar f(|\Psi|^2)\Psi = -\frac{\delta S[\Psi,\Psi^*]}{\delta\Psi^*} = (D^\mu D_\mu - m^2 - \lambda|\Psi|^2)\Psi \tag{53}$$

(covariant Gross-Pitaevski equation)*. One ensures that the conjugate equation is also satisfied by imposing the further restrictions

$$\partial_0\Psi(x) = 0 \ , \quad f = f^* \ . \tag{54}$$

Expansion of ΔS, using (52), (44) and (53), yields

$$\Delta S = \hbar S_1[\Psi,\phi] + \hbar^{3/2}\, H[\Psi,\phi] \ , \tag{55}$$

where

$$S_1[\Psi,\phi] = \int_\Omega d^N x\ g^{\frac{1}{2}}\{\phi^*(x)(-D^\mu D_\mu + m^2 + 2\lambda|\Psi|^2)\phi(x) + \\ + \frac{1}{2}\lambda(\Psi^{*2}\phi^2 + \Psi^2\phi^{*2})\} \tag{56}$$

quadratic in ϕ, ϕ^*, contains the one-loop corrections, and

$$H[\Psi,\phi] = \int_\Omega d^N x\ g^{\frac{1}{2}}\{f(\Psi^*\phi + \Psi\phi^*) + O(\phi^3)\} \tag{57}$$

the higher-loop terms. Anticipating dimensional regularization, we have liberated spacetime from 4 to N dimensions.

6. ONE-LOOP CONTRIBUTION TO EUCLIDEAN EFFECTIVE ACTION

To evaluate the one-loop contribution

$$e^{-W_1} = \int d[\phi]d[\phi^*]e^{-S_1[\Psi,\phi]} \tag{58}$$

to the path integral (48), we begin with two simple transformations of the functional integration variables.

First, the phase transformation $\phi \to \phi e^{i\arg\Psi(\underset{\sim}{x})}$, accompanied by the appropriate spatial gauge transformation $\underset{\sim}{A} \to \underset{\sim}{A} - (\hbar/e)\underset{\sim}{\nabla}\arg\Psi$, simplifies the last term of (56) to $\frac{1}{2}\lambda|\Psi|^2(\phi^2 + \phi^{*2})$ without affecting the remaining terms. A second transformation $\phi = 2^{-\frac{1}{2}}(\phi_1 + i\phi_2)$ converts

* In the present context one should interpret $D^\mu = g^{-\frac{1}{2}}\, D_\nu g^{\frac{1}{2}} g^{\mu\nu}$.

(56) to the form

$$S_1[\Psi,\phi] = \frac{1}{2}\int_\Omega d^N x\ g^{\frac{1}{2}}\phi^T(x) L_x \phi(x) \ , \tag{59}$$

in which $\phi(x)$ now represents the column vector $\begin{pmatrix}\phi_1\\ \phi_2\end{pmatrix}$, and L_x the 2×2 matrix differential operator

$$L_x = \begin{pmatrix} -\partial^\mu\partial_\mu+\Gamma^\mu\Gamma_\mu+\kappa+\varepsilon & -2\Gamma^\mu\partial_\mu \\ 2\Gamma^\mu\partial_\mu & -\partial^\mu\partial_\mu+\Gamma^\mu\Gamma_\mu+\kappa-\varepsilon \end{pmatrix} \tag{60}$$

$$\kappa(x) = m^2 + 2\lambda|\Psi|^2 \ , \qquad \varepsilon(x) = \lambda|\Psi|^2 \ . \tag{61}$$

This operator is self-adjoint: if ϕ,ψ are column vectors satisfying appropriate boundary conditions,

$$\int d^N x\ g^{\frac{1}{2}}\ \psi^T L_x \phi = \int d^N x\ g^{\frac{1}{2}}(L_x\psi)^T\phi. \tag{62}$$

The one-loop effective action W_1 is conveniently expressed in terms of the Euclidean Green's function[26]. We define the 2×2 matrix function

$$G(x,x') = \langle\phi(x)\phi^T(x')\rangle = e^{W_1}\int d[\phi]\phi(x)\phi^T(x')e^{-S_1[\Psi,\phi]} \tag{63}$$

(functional integration over ϕ_1, ϕ_2 is now indicated simply as $\int d[\phi]$). This is periodic in Euclidean time $\tau_2-\tau_1$ with period β^t, by virtue of (43), and satisfies

$$L_x G(x,x') = g^{-\frac{1}{2}}\delta^N(x-x')1_{2\times 2} \ , \tag{64}$$

as follows from (noting (62))

$$L_x\phi(x) = g^{-\frac{1}{2}}\delta S_1/\delta\phi^T(x) \ . \tag{65}$$

The identity

$$\delta S_1/\delta\kappa(x) = \frac{1}{2}\,g^{\frac{1}{2}}\ \phi^T(x)\phi(x) \tag{66}$$

then leads to the desired relation between W_1 and G:

$$\delta W_1/\delta\kappa(x) = -\frac{1}{2}\,g^{\frac{1}{2}}\ \mathrm{tr}G(x,x). \tag{67}$$

In principle, therefore, one merely needs to integrate (64) for $G(x,x')$, subject to the periodic boundary conditions, and then functionally integrate (67) to obtain W_1.

This formal scheme leads in four dimensions to a divergent W_1, the divergence arising from the divergence of $G(x,x')$ in the coincidence limit $x' \to x$ when $N \geq 2$. It is not hard to see that this is essentially the only source of divergence, in the sense that higher terms in the loop expansion do not produce divergences of a fundamentally new type. Since the singular behaviour of $G(x,x')$ is independent of boundary conditions, it is clear that the divergences of the formalism are independent of β^t. This simply reflects the new well-known possibility[27,18] of renormalizing finite-temperature quantum field theory in accordance with a temperature-independent scheme.

7. ONE-LOOP EFFECTIVE POTENTIAL

In the effective potential approximation to W_1, one neglects the spatial variation of $g_{\mu\nu}$, A_μ, μ and $|\Psi|^2$, and hence of the coefficients κ, ε, Γ_μ in (60), over the domain of integration Ω. (However, these variations are not neglected in the zeroth-order contribution $S[\Psi]$ in (51).)

It is then trivial to solve (64) by the Fourier ansatz

$$G(x,x') = \int g^{-\frac{1}{2}} \frac{d^N k}{(2\pi)^N} e^{-ik_\mu (x-x')^\mu} G(k), \tag{68}$$

and it turns out that

$$(A^2 + B^2 - \varepsilon^2) G(k) = \begin{pmatrix} A-\varepsilon & B \\ -B & A+\varepsilon \end{pmatrix}, \tag{69}$$

where

$$A = g^{\mu\nu}(k_\mu k_\nu + \Gamma_\mu \Gamma_\nu) + \kappa, \qquad B = -2i\Gamma^\mu k_\mu \ . \tag{70}$$

To accommodate the periodicity condition, the Fourier integral over k_0 in (68) has to be re-interpreted as a Fourier series:

$$\int \frac{dk_0}{2\pi} \to \frac{1}{\beta^t} \sum_{n=-\infty}^{\infty} , \qquad k_0 = \frac{2\pi n}{\beta^t} \ . \tag{71}$$

According to (67), and (68), the one-loop contribution to the effective potential is now obtainable from

$$\partial W_1/\partial\kappa = -\frac{1}{2}(2\pi)^{-N}\int d^Nx\, d^Nk\, \mathrm{tr}G(k). \tag{72}$$

We consider first the discrete summation:

$$\int \frac{dk_o}{2\pi}\,\mathrm{tr}G(k) = \frac{1}{\beta^t}\sum_{n=-\infty}^{\infty}\left(\frac{1}{A+i\sqrt{B^2-\varepsilon^2}} + \frac{1}{A-i\sqrt{B^2-\varepsilon^2}}\right) . \tag{73}$$

This can be reduced to a transparent closed form in the weak-coupling approximation. To first order in λ (i.e., in ε), we have

$$A \pm i\sqrt{B^2-\varepsilon^2} = \kappa + (k\pm\Gamma)^2 , \tag{74}$$

and the two terms of (73) are equal:

$$\int \frac{dk_0}{2\pi}\,\mathrm{tr}G(k) = \frac{2\beta^t}{g^{00}(2\pi)^2}\sum_{n=-\infty}^{\infty}\frac{1}{\left(n+\frac{ia}{2\pi}\right)^2+\left(\frac{b}{2\pi}\right)^2} . \tag{75}$$

We have defined

$$a=-i\beta^t(\Gamma_0+g^{0a}p_a/g^{00}), \quad b = (g^{00})^{-\frac{1}{2}}E\beta^t \tag{76}$$

$$E^2 = {}^{(n-1)}g^{ab}p_ap_b + \kappa, \qquad p_a = k_a \pm \Gamma_a . \tag{77}$$

(Either sign can be chosen in (77) without affecting the value of (75).)

The Euler summation formula gives the identity

$$\frac{1}{(2\pi)^2}\sum_{n=-\infty}^{\infty}\frac{1}{\left(n+\frac{ia}{2\pi}\right)^2+\left(\frac{b}{2\pi}\right)^2} = \frac{1}{b}\,\frac{\sinh b}{\cosh b-\cosh a}$$

$$= \frac{1}{b}\left\{1+\frac{1}{e^{b-a}-1}+\frac{1}{e^{b+a}-1}\right\} . \tag{78}$$

From this, the general character of our final result is already apparent. The second and third terms of (78) correspond to the Bose-Einstein distribution functions for the particle and anti-particle thermal excitations. The first term gives rise, after momentum integration, to a divergence proportional to (β^t/b) (temperature-independent), which will require renormalization. An explicit integration of this term may be performed using the general identity

$$\int {}^{(N-1)}g^{-\frac{1}{2}} \frac{d^{N-1}p}{(2\pi)^{N-1}} \left\{ {}^{(N-1}g^{ab} p_a p_b + \kappa \right\}^{\frac{1}{2}} = -\Gamma(-\tfrac{N}{2})(\tfrac{\kappa}{4\pi})^{N/2}, \tag{79}$$

which isolates the divergence as a simple pole at N=4.

8. SUMMARY: ONE-LOOP EFFECTIVE POTENTIAL IN WEAK-COUPLING APPROXIMATION

Collecting our results, we have obtained the partition function

$$Z = e^{-W/\hbar}, \qquad W = \int d^N x \; g^{\frac{1}{2}} \mathcal{L} \tag{80}$$

for a self-interacting charged scalar field, governed by the action (24), in the presence of a condensate at finite temperature. To one-loop order, to linear order in the self-coupling constant λ, and in the effective-potential approximation where field gradients are neglected in the one-loop corrections, the result for the effective Lagrangian $\mathcal{L}$ can be expressed as

$$\mathcal{L} = \mathcal{L}_{class} + \hbar \mathcal{L}_{quant.} + \hbar \mathcal{L}_{therm} . \tag{81}$$

Here,

$$\mathcal{L}_{class} = g^{\mu\nu}(D^{\dagger}_{\mu}\Psi^*)(D_{\nu}\Psi) + m^2|\Psi|^2 + \frac{1}{2}\lambda|\Psi|^4 , \tag{82}$$

with D_μ defined by (45),

$$\mathcal{L}_{quant.} = \Gamma(-\tfrac{N}{2})(\tfrac{\kappa}{4\pi})^{N/2} , \qquad \kappa = m^2 + 2\lambda|\Psi|^2 \tag{83}$$

$$\mathcal{L}_{therm} = -\frac{1}{\beta^t}\int g^{-\frac{1}{2}} \frac{d^{N-1}p}{(2\pi)^{N-1}} \{\ell n(1-e^{-\alpha+\beta^\mu p_\mu}) + \ell n(1-e^{\alpha+\beta^\mu p_\mu})\} . \tag{84}$$

In the last equation, the integration is over the covariant components $p_1,\ldots,p_{N-1}$ of spatial momentum; p_0 is <u>defined</u> (for arbitrary χ) as the negative solution of

$$-g^{\mu\nu}p_\mu p_\nu = \kappa = m^2 + 2\lambda|\Psi|^2 \tag{85}$$

(cf (76), (77)), where $E=-(g^{00})^{\frac{1}{2}}p_0$, $b=-\beta^{\mu}p_{\mu}$).* The condensate wave function $\Psi(\underset{\sim}{x})$ is time-independent and satisfies the Gross-Pitaevski equation

$$\{D^{\mu}D_{\mu} - m^2 - \lambda|\Psi|^2 + \hbar f(|\Psi|^2)\}\Psi = 0 \tag{86}$$

and its complex conjugate. These results were derived in the Euclidean sector, but remain valid in the Lorentzian sector, since the function $A(\alpha,\beta^{\mu},\Psi,g_{\lambda\mu},A_{\lambda})$ is invariant under Wick rotations (see footnote, p.10).

9. RENORMALIZATION

The next task is to absorb the pole (83) that appears in the effective Lagrangian at the physical dimension N=4 through redefinitions of the mass and coupling constant. To one-loop order the condensate wave function Ψ does not require renormalization, because Ψ is sufficiently close to a minimum of the classical action (see (53)) that a change of order $\hbar$ in the definition of Ψ would modify the form of the effective action only through terms of order $\hbar^2$.

The chemical potential (equivalently, Γ_0 in (46)) also requires no adjustment to one-loop, if one adopts the definition

$$g^{0\mu}i(\Gamma_{\mu})_R \equiv \frac{\partial^2 \mathcal{L}}{\partial\Psi^*\partial(\partial_0\Psi)} = g^{0\mu}i\Gamma_{\mu} \;. \tag{87}$$

It is natural to define the renormalized coupling constant and mass by

$$\lambda_R \equiv \left(\frac{\partial^2\mathcal{L}}{\partial(|\Psi|^2)^2}\right)_{\substack{\Psi=0\\ \beta=\infty}} \quad ; \quad m_R^2 + \Gamma^{\mu}\Gamma_{\mu} \equiv \left(\frac{\partial\mathcal{L}}{\partial|\Psi|^2}\right)_{\substack{\Psi=0\\ \beta=\infty}} \;. \tag{88}$$

For our purposes, it is crucial that these renormalization conditions be imposed at zero temperature, $\beta=\infty$. It ensures that λ_R, m_R, and hence also the renormalized form of the Gross-Pitaevski equation (86), are free of contamination by temperature-dependent terms that would otherwise arise from (84), (85). It should be clear that this is a general feature, independent of the idealizations and approximations

* The derivation of these results makes use of the identity $g^{\mu\nu}p_{\mu}p_{\nu} = g^{00}(p_0 + g^{0a}p_a/g^{00})^2 + {}^{(N-1)}g^{ab}p_ap_b$.

of our particular model: imposition of renormalization conditions at zero temperature ensures that the renormalized form of the Gross-Pitaevski equation is temperature-independent. Since it is the GP equation that controls the properties of the condensate, this gives a general microscopic justification for the phenomenological postulate $S_s \equiv 0$ introduced in Sec. 2. Of course, to renormalize at zero temperature is already a wide-spread custom. What the present argument shows is that this choice is actually mandatory if one wishes to make manifest the separability of thermodynamical properties of the normal and superfluid components.

With $\mathcal{L}$ given by (81)-(84), the definitions (88) yield

$$m_R^2 = m^2[1+\hbar\lambda p(N)] \ , \qquad \lambda_R = \lambda[1+\hbar\lambda(N-2)p(N)] \tag{89}$$

where

$$p(N) = -\Gamma(-\tfrac{N}{2})(4\pi)^{-N/2} N m^{N-4} \ . \tag{90}$$

By solving (89) for m^2 and λ (to first order in $\hbar$), we can re-express the effective Lagrangian $\mathcal{L}$ in terms of the renormalized quantities. The resulting form, in which a pole term no longer appears explicitly, is (dropping the subscript 'R' and setting N=4)

$$\mathcal{L} = \mathcal{L}_{\text{class}} + \hbar\mathcal{L}_{\text{fluct.}} + \hbar\mathcal{L}_{\text{therm.}} \tag{91}$$

to one-loop, where

$$\mathcal{L}_{\text{fluct.}} = -\frac{1}{32\pi^2}\{2\lambda m^2|\Psi|^2 + 6\lambda^2|\Psi|^4 - \\ - (m^2+2\lambda|\Psi|^2)^2 \ln(1+2\lambda|\Psi|^2/m^2)\} \tag{92}$$

and the other two terms have the same form as before, except, of course, that it is actually the renormalized parameters which now appear in (82), (84) and (85).

The interpretation of this result is perfectly evident from (80):

$$\ln Z_{\text{therm}} = -\int d^4x \, g^{\frac{1}{2}} \mathcal{L}_{\text{therm}} \\ = -V\int {}^{(N-1)}g^{-\frac{1}{2}} \frac{d^{N-1}p}{(2\pi)^{N-1}} \sum_{\varepsilon=\pm 1} \ln(1-e^{\varepsilon\alpha+\beta^\mu p_\mu}) \tag{93}$$

is the standard partition function[15] for a hot gas of free particles and anti-particles with masses equal to $\sqrt{\kappa} = (m^2+2\lambda|\Psi|^2)^{\frac{1}{2}}$. (These quasi-particles appear to be non-interacting because the thermal contributions to Z have been evaluated only in the weak-coupling limit, linear in λ.) The other two terms in (91) are temperature-independent and describe the condensate. This is the relativistic finite temperature generalization of Bogoliubov's famous result[2]. It should be noted that for $\varepsilon = +1$ in (93), the argument of the logarithm becomes negative unless one imposes the inequality $\mu < \sqrt{\kappa}$ on the chemical potential.

A natural choice for the disposable one-loop term $\hbar f(|\Psi|^2)$ in the Gross-Pitaevski equation (53) now suggests itself. It should be chosen so that the renormalized GP equation is equivalent to

$$\frac{\delta(S_{\text{class}} + S_{\text{fluct.}})}{\delta\Psi^*} = 0 \ . \tag{94}$$

From (82) and (92), the explicit form of this is easily written down for our model.

10. SUPERFLUID HYDRODYNAMICS AND THE GROSS-PITAEVSKI EQUATION

In conclusion, I shall demonstrate the connection and compatibility of the covariant GP equation with the phenomenological theory of the condensate developed in Sec. 2.

Our starting point is the (renormalized) Gross-Pitaevski equation in the general form

$$g^{\mu\nu}D_\mu D_\nu\Psi - m^2\Psi - F(|\Psi|^2)\Psi = 0 \tag{95}$$

where F is an unspecified real function, $\partial_0\Psi = 0$, and, in the Lorentzian sector,

$$D_\mu = \partial_\mu - i\Gamma_\mu, \quad D^\dagger_\mu = \partial_\mu + i\Gamma_\mu \ ,$$

$$\hbar\Gamma_a = eA_a, \hbar\Gamma_0 = \mu = \alpha/\beta^t, \quad \mu - eA_0 = \text{const.} \tag{96}$$

A conservation law follows from (95):

$$\partial^\mu\{g^{\frac{1}{2}}(\Psi D^\dagger_\mu\Psi^* - \Psi^* D_\mu\Psi)\} = 0 \ . \tag{97}$$

Accordingly, define μ_s and a normalized 4-velocity u_s^μ ($u_{s\mu}u_s^\mu = -1$) by

$$\hbar^{-1}\mu_s u_{s\mu} = \frac{1}{2}\, i(\Psi D_\mu^\dagger \Psi^* - \Psi^* D_\mu \Psi)/\Psi\Psi^* \tag{98}$$

$$= \partial_\mu \Phi - \Gamma_\mu \tag{99}$$

where we have set

$$\Psi(\underset{\sim}{x}) = \sqrt{n(\underset{\sim}{x})}\, e^{i\Phi(\underset{\sim}{x})} \;. \tag{100}$$

Then

$$(n_s u_s^\mu)_{|\mu} = 0 \;, \qquad n_s \equiv n\mu_s \;. \tag{101}$$

From (99) and (96) it follows that

$$\oint(\mu_s u_{s\mu} + eA_\mu)dx^\mu = \hbar\oint d\Phi = 2\pi n\hbar = nh \;. \tag{102}$$

Now define u_n^μ, β, μ_n by

$$\beta^\mu = \beta u_n^\mu, \quad u_{n\mu}u_n^\mu = -1, \quad \mu_n = \alpha/\beta = (-g^{00})^{-\frac{1}{2}}\mu \,. \tag{103}$$

Multiplication of (99) by u_n^μ, noting $\partial_0\Phi = 0$, yields

$$\gamma\mu_s = \mu_n, \quad \gamma \equiv -u_{s\mu}u_n^\mu \;. \tag{104}$$

Finally, (95) and (100) yield the equation of state of the superfluid component,

$$\mu_s^2 = m^2 + F(n) - h^2 n^{-\frac{1}{2}}\nabla^2(n^{\frac{1}{2}}), \tag{105}$$

which agrees (aside from the nonlocal term) with the general (one-parameter, temperature-independent) form $n_s = n_s(\mu_s)$ postulated in Sec. 2.

The microscopic theory thus fully confirms, at least for thermodynamical equilibrium, the phenomenological equations of Sec. 2.

ACKNOWLEDGEMENTS

I should like to express thanks to Professor I.M. Khalatnikov for some highly stimulating correspondence; to Drs. M. Ichiyanagi and

T. Toyoda for inducting me into the elements of microscopic superfluid theory through their publications and through personal discussions; to Drs. M.C.J. Leermakers and Ch.G. van Weert for hospitality and discussions at the University of Amsterdam and for calling reference [3] to my attention; and to Professor M.A. Markov, Dr. V.P. Frolov and other members of the Organizing Committee for the very warm hospitality that helped to make the Third Moscow Seminar on Quantum Gravity such a memorable occasion.

REFERENCES

1. Linde, A.D., Rep. Prog. Phys. 42, 389 (1979).
2. Bogoliubov, N.N., J. Phys. (Moscow) 11, 23 (1947).
3. Kapusta, J.I., Phys. Rev. D24, 426 (1981);
   Haber, H.E. and Weldon, H.A. Phys. Rev. D25, 502 (1982).
4. Baym, G. and Pethick, C., Ann. Rev. Astron. Astrophys. 17, 415 (1979).
5. Kuzmin, V.A. and Shaposhnikov, M.E., Phys. Lett. 69A, 462 (1979).
6. Gross, D.J., Perry, M.J. and Yaffe, L.G., Phys. Rev. D25, 330 (1982);
   Kikuchi, Y., Moriya, T. and Tsukahara, H., Phys. Rev. D29, 2220 (1984).
7. Landsberg, P.T. and Dunning-Davies, J., Phys. Rev. 138, A1049 (1965);
   Bechmann, R., Karsch, F. and Miller, D.E., Phys. Rev. Lett. 43, 1277 (1979); Phys. Rev. A25, 561 (1982);
   Singh, S. and Pathria, R.K., Phys. Rev. A30, 442 (1984).
8. Rothen, F., Helv. Phys. Acta. 41, 591 (1968).
9. Dixon, W.G., Arch. Rat. Mech. Anal. 80, 159 (1982).
10. Israel, W., Phys. Lett. 86A, 79 (1981).
11. Khalatnikov, I.M. and Lebedev, V.V., Phys. Lett. 91A, 70 (1982);
    Lebedev, V.V. and Khalatnikov, I.M., Zh. Eksp. Teor. Fiz. 56, 1601 (1982) [transl. as Sov. Phys. JETP 56, 923 (1982)].
12. Israel, W., Phys. Lett. 92A, 77 (1982).
13. Fomin, P.I. and Shadura, V.N., Relativistic thermodynamics and hydrodynamics of systems with superfluidity. Preprint, Inst. for Theoretical Physics, Kiev (1983).

14. Ho, T.L. and Mermin, N.D., Phys. Rev. B21, 5190 (1980).

15. Israel, W. and Stewart, J.M., Ann. Phys. 118, 341 (1979); Israel, W., Physica 106A, 204 (1981).

16. Putterman, S.J., Superfluid Hydrodynamics (North-Holland, Amsterdam 1974), pp. 414, 425.

17. Landau, L. and Lifshitz, E.M., Fluid Mechanics (Addison-Wesley, Reading, Mass. 1958), Chap. 16.

18. Takahashi, Y. and Umezawa, H. Collective Phenomena 2, 55 (1975); Matsumoto, H., Ojima, I. and Umezawa, H., Ann. Phys. 152, 348 (1984).

19. Niemi, A.J. and Semenoff, G.W., Nucl. Phys. B230, 181 (1984); Ann. Phys. 152, 105 (1984).

20. Matsubara, T., Prog. Theor. Phys. 14, 351 (1955).

21. Aitchison, I.J.R., An Informal Introduction to Gauge Field Theories (Cambridge Univ. Press 1982), p. 79.

22. Wiegel, F.W. and Jalickee, Physica 57, 317 (1972); Wiegel, F.W., Phys. Rep. 16C, 57 (1975).

23. Ichiyanagi, M., Prog. Theor. Phys. 66, 1 (1981).

24. Toyoda, T., Ann. Phys. 141, 154 (1982); 147, 244 (1983).

25. Coleman, S. and Weinberg, E., Phys. Rev. D7, 1883 (1973).

26. Brown, M.R. and Duff, M.J., Phys. Rev. D11, 2124 (1975).

27. Freedman, B.A. and McLerran, L.D., Phys. Rev. D16, 1130 (1977).

QUANTUM GRAVITY EFFECTS IN PARTICLE PHYSICS AND COSMOLOGY

V.A.Kuzmin
Institute for Nuclear Research
of the Academy of Sciences of the USSR
60th October Anniversary Prospect 7a
Moscow 117312, USSR

ABSTRACT

We estimate possible violations, at Planck scales, of baryon number conservation, CPT and Pauli principle. We find that in some cases these violations, although small, are within the possibility of observation by means of present experimental techniques. Special attention has to be paid to searches for Pauli principle violation in nuclear transitions as the most sensitive tests. It is found that the observed baryon asymmetry of the Universe could be explained by the possible CPT violation due to quantum gravity.

1. INTRODUCTION

It has been conjectured[1)] (see also[2-4)]) that fluctuations of metrics at Planck scale $r \sim M_{Pl}^{-1}$, $M_{Pl} \approx 10^{19}$GeV, could result in non-conservation of a number of quantities, say, baryon or lepton numbers, or even of otherwise strictly conserved (at least in the framework of local quantum field theory) quantities such as CPT or quantum coherence. It is by no means our purpose in this paper to discuss the origin or the dynamics of these quantum gravity induced violations. Rather, we would like to overview and to seach for any possible observational manifestations in particle physics, as well as in cosmology of these violations at small distances. So, our line of reasoning is as follows. Let us suppose from the very beginning that there is something going wrong at the Planck scale, i.e. that quantum gravity does not respect any conservation law up to possible exceptions. We consider in this paper three possible consequences of this hypothesis: (i) baryon number non-conservation, (ii) CPT violation, and (iii) violation of the Pauli principle. One more effect, that of loss of quantum coherence[2)], has recently been considered in refs.[5,6)].

There is no doubt that the effects under discussion are very small at the usual particle physics scales. Throughout this paper, we shall parametrize all of them in the same way to be able to compare the corresponding experimental sensitivities,

$$(\text{violation}) \sim (m/M_{Pl})^a, \tag{1}$$

where m is some characteristic scale of a process and a is an (unknown) exponent; we shall test the values a = = 2,3,4 for the rates and a/2 for the contributions to amplitudes. In this paper we estimate the expected magnitude of effects and compare them with experimental possibilities.

The paper is organized as follows. In sect.2 we estimate the effects of baryon number non-conservation for both $|\Delta B| = 1$ and $|\Delta B| = 2$, in sect.3 we consider possible CPT violation effects, paying special attention to the production of the baryon asymmetry of the Universe, and in sect.4 we estimate the Pauli principle violations in atoms and nuclei and possible experimental tests.

2. BARYON NUMBER NON-CONSERVATION

Here we briefly discuss possible baryon number non-conservation effects due to quantum gravity. Observational manifestations of effects of this kind are, as usual, proton decay (in the $|\Delta B| = 1$ case) and $n\bar{n}$ oscillations (in the $\Delta B = 2$ case). It is interesting to compare the magnitudes of the effects one could expect to emerge due to quantum gravity with the present experimental possibilities of observation. It is known[7] that the proton lifetime in any case exceed the value

$$\tau_p > 3\cdot 10^{31}\mathrm{y},\ \Gamma_p \sim \tau_p^{-1} < 10^{-63}\ \mathrm{GeV}, \tag{2}$$

while the constraints on the $n\bar{n}$ oscillation time coming from matter stability experiments[8] and direct searches of $n\bar{n}$ transitions in experiments with free neutrons[9] are, respectively,

$$\tau_{n\bar{n}} > 10^{8}\ \mathrm{s},\ \varepsilon \sim \tau_{n\bar{n}}^{-1} < 10^{-32}\ \mathrm{GeV} \tag{3}$$

and

$$\tau_{n\bar{n}} > 3\cdot 10^{6}\ \mathrm{s},\quad \varepsilon < 3\cdot 10^{-31}\ \mathrm{GeV}. \tag{4}$$

Let us now estimate what one could expect with respect to these searches in the case of quantum gravity induced violations. Consider first the $\Delta B = 1$ case. Taking as an ansatz

$$\Gamma \sim m_N (m_N/M_{Pl})^{a}, \tag{5}$$

one immediately obtains:

$$\Gamma \sim 10^{-57}\ \mathrm{GeV\ for}\ a = 3,$$

$$\Gamma \sim 10^{-76}\ \mathrm{GeV\ for}\ a = 4. \tag{6}$$

(These value would turn out to be $\Gamma \sim 10^{-61}$ GeV and $\Gamma \sim$ $\sim 10^{-81}$ GeV for a = 3,4, respectively, if one substituted m_π instead of m_N into eq.(5)). Thus, one can conclude that, taking into account the extreme crudeness of these estimates, the sensitivity of the present proton decay experiments can be considered as being close to that one needs to test the a = 3 case of quantum gravity. The a =4 case is, on the contrary, too far away from being possible to test.

We turn now to consideration of $n\bar{n}$ oscillation experiments in the same context. Now taking

$$\varepsilon \sim m_N (m_N/M_{Pl})^{a/2}, \tag{7}$$

where ε is the quantum gravity induced mixing of n and $\bar{n}$ states, one obtains

$$\varepsilon \sim 3 \cdot 10^{-29} \text{ GeV for a = 3,}$$

$$\varepsilon \sim 10^{-38} \text{ GeV for a = 4,} \tag{8}$$

(or, alternatively, $\varepsilon \sim 10^{-31}$ GeV and $\varepsilon \sim 10^{-42}$ GeV with m_π in eq.(7)). The same conclusions as before could be made here too. One remark has to be made, however. While the proton decay rate would not be affected by a possible mass splitting between particle and antiparticle due to possible CPT violation by quantum gravity (see sect.3) this would not be the case for $n\bar{n}$ oscillations[10]. Namely, if CPT violation induced $n\bar{n}$ mass splitting Δm exceeds the inverse time of flight of free neutrons, then free $n\bar{n}$ oscillations are supressed[10], while $n\bar{n}$ oscillations inside nuclei are not, since the CPT mass splitting does not exceed inverse nuclear time ($\Delta m(CPT) \ll 1$ GeV). It is very difficult, however, to distinguich experimentally between $n\bar{n}$ oscillations inside nuclei and the processes like $nn \longrightarrow$ pions. So, observations of the free $n\bar{n}$ oscillations would indeed provide the best constraints on possible CPT (as well as other) mass splitting, since the time of flight used in the experiments[9] is of order 10^{-2}s - 10^{-1} s, so that the discovery of $n\bar{n}$ oscillations would imply $\Delta m < 10^{-23}$ GeV. To conclude this discussion, we would like to note that it seems unlikely that the CPT non-conserving amplitude is larger than the baryon number non-conserving one. One could expect that all quantum gravity induced violations are of the same order of magnitude; in that case, $n\bar{n}$ oscillations would not be affected by $n\bar{n}$ mass splitting.

3. VIOLATIONS OF CPT

One of the main stones in the foundations of the beautiful building of modern quantum field theory is the assumption of the locality of all known fundamental interactions of elementary particles, which provides the exact CPT conservation. However, it has been conjectured by Penrose[1] that the fluctuations of metrics at Planck scale could give rise to non-locality effects and, presumably, to the CPT non-conservation as well. If so, we lose the last reason to be sure that masses (and total decay rates) of a particle and antiparticle are equal. Thus, just in the line of this paper, we suppose that there is indeed a small particle-antiparticle mass splitting, provided CPT (and C) are not conserved at Planck distances. Again, our question is what are possible, if any, observable manifestations of this assumption? Parametrizing again the corresponding mass splitting as follows (it is also an ansatz)

$$\Delta m \sim m(m/M_{Pl})^{a/2}, \qquad (9)$$

(m being the particle mass), we proceed to evaluating the magnitudes of the effects.

3.1. Direct Measurements and Oscillation Phenomena

The effects under discussion are obviously extremely small for any ordinary particle. For exemple, for nucleons one can expect

$$\Delta m/m \sim (10^{-19})^{a/2}, \tag{10}$$

which is too small even for a = 2 in comparison with the precisions of the direct measurements, since the latter do not exceed at the moment the value $\Delta m/m < 10^{-4} - 10^{-6}$. Much more sensitive to any extra origin (CPT in this context) of mass splitting are the oscillation phenomena such as $K^0 \leftrightarrow \bar{K}^0$, $n \leftrightarrow \bar{n}$, etc. It has been noticed in ref.[10] that the very existence of $K^0 \leftrightarrow \bar{K}^0$ oscillations proves that the CPT-induced mass splitting in $K^0 \leftrightarrow \bar{K}^0$ systems does not exceed

$$\Delta m/m < 10^{-18},$$

which is very close to what one could expect in the quantum gravity case with a = 2. Other kinds of oscillation phenomena, such as $n \leftrightarrow \bar{n}$ [11], could in principle restrict possible CPT mass splitting even further, say to $\Delta m/m < 10^{-23}$, provided that such a phenomenon is observed[10], We have already discussed this point in sect.2.

3.2. Is the Baryon Asymmetry of the Universe a Result of CPT Non-Conservation?

The magnitude of effects could become not at all inconsequential in the case of superheavy particles. Indeed, taking $m_x \sim 10^{15}$ GeV as a generic value of leptoquark masses in GUT's based on simple unifying gauge symmetry groups, such as SU(5), SO(10), etc., one immediately obtains

$$\Delta m_x/m_x \sim (m_x/M_{Pl})^{a/2} \sim 10^{-2a}, \tag{11}$$

which is of order 10^{-8} in the case a = 4. This could drastically change the standard scenario of the generation of the baryon asymmetry of the Universe (BAU)[11,12,13,14]. Starting from the first papers[12,11,13] it is usually believed that one of the three necessary conditions for the BAU generation is the non-equilibrium at some stage of cosmological plasma with respect to interactions violating baryon number (the other two conditions are the existence of baryon number non-conservation and that of CP (and C) non-conserving interactions). The baryon excess is usually expected to be produced during the non-equilibrium stage of the expansion of the Universe. This is due to the fact, first supposed in refs.[12,11,13] and then proved in ref.[15] that, given that CPT is valid, no non-conserved quantum number (such as baryon number) could take a non-zero value when averaged over a statistical ensemble at thermodynamic equilibrium. Now we assume, however, that CPT is not valid and, furthermore, that masses of particles and antiparticles are slightly different. One might expect that there could exist some excess in the number density of particles over that of antiparticles (or vice versa) in plasma even at thermodynamic equilibrium. This could in turn result in the non-vanishing average macroscopic value of the baryon number in the plasma at equilibrium. We find that this is indeed the case. Let us consider this in more detail.

Let us discuss what happens in the cosmological plasma at thermodynamic equilibrium at high temperatures in the framework of some grand unified model, if mass of leptoquarks and anti-leptoquarks are different, $M_X = m_X \pm \Delta m_X$. By equilibrium we mean that the collision rate greatly exceeds the expansion rate of the Universe. We suppose that particle distribution functions are still given by the usual equilibrium expressions; for example, for massive charged leptoquarks X at $T < M_X$, the number densities are

$$n_X(\bar{n}_X) \sim (M_X T)^{3/2} \exp\left[-(M_X \pm \mu(X))/T\right], \qquad (12)$$

$\mu(X)$ being the chemical potential of X. Let us consider the particular case of the SU(5) model, which we regard as a prototype model. We shall be most interested in the temperature interval $M_X > T \geq T_f$, where T_f is the 'freezing-out' temperature (typically $T_f \sim 0.1\ M_X$) at which heavy leptoquarks decouple. So for both X(Q = -4/3) and Y (Q= -1/3) vector leptoquarks of the SU(5) model (Q being the electric charge) we shall use, in this temperature range, the expression in (12). In what follows we shall neglect, for the sake of simplicity, any contributions from scalar leptoquarks, since taking them into account does not change the results significantly.

Let us find the particle number densities for all species in the plasma at $T < M_X$ (we take $M_X = M_y$), i.e. find all chemical potentials, assuming that

$$M_X = M_Y = m_X + \Delta m_X,$$

$$M_{\bar{X}} = M_{\bar{Y}} = m_X - \Delta m_X$$

for leptoquarks and anti-leptoquarks, respectively. All other particles of the SU(5) content are massless at $M_X > T \geq T_f$. In order to evaluate the chemical potentials, one has obviously to solve the constraints coming from the conservation of electric charge, color, weak isospin, and B - L (B and L being baryon and lepton numbers, respectively), i.e. from the electric, color, weak isospin and B - L neutrality of the plasma under consideration. It can be shown that the conditions of the color neutrality and weak isospin are fulfilled automatically, so here we shall write explicitly only the conditions of the electric and B - L neutrality. These are

$$Q = -(4/3)(n_X - \bar{n}_X) - \tfrac{1}{3}(n_Y - \bar{n}_Y) + 2(n_u - \bar{n}_u) - (n_d - \bar{n}_d)$$
$$-3(n_e - \bar{n}_e) - (n_W - \bar{n}_W) - (n_H - \bar{n}_H) = 0 \quad , \qquad (13)$$

$$B\text{-}L = -(2/3)(n_X - \bar{n}_X) - (2/3)(n_Y - \bar{n}_Y) + (n_u - \bar{n}_u) + (n_d - \bar{n}_d)$$
$$-3(n_e - \bar{n}_e) - 3(n_\nu - \bar{n}_\nu) = 0 \qquad (14)$$

where the subscripts u, d, e, W, H and ν denote u-quark, d-quark, electron, W-boson, Higgs boson of the Weinberg - Salam model and neutrino, respectively. We have taken into account that there exist three generations of fermions and considered only one Higgs doublet. It is worth noting that the fermion number densities n_i appearing in eqs.(13) and (14) sums of the number densities of right-handed (n_{iR}) and left-handed (n_{iL}) components (except for neutrino); in general, $n_{iR} = n_{iL}$.

Now we should take into account all reactions between particles in the SU(5) model. These are

$$X \to \bar{u}_L u_R,\ e_L d_R,\ e_R d_L$$

$$W \to e_L \bar{\nu},\ \bar{u}_L d_R,\ X\bar{Y}$$

$$H^0 \to u_R \bar{u}_R,\ e_L \bar{e}_L,\ \bar{d}_L d_L$$

$$Z^0 \to \nu\bar{\nu},\ e_L \bar{e}_R,\ e_R \bar{e}_L,\ u_L \bar{u}_R,\ \text{etc};\ X\bar{X},\ \overline{YY}$$

$$\gamma \to e_L \bar{e}_R,\ e_R \bar{e}_L,\ u_L \bar{u}_R,\ \text{etc};\ W\bar{W},\ X\bar{X},\ YY,\ H^+\bar{H}^+$$

$$H^0 \to H^0 2Z^0. \qquad (15)$$

Other reactions are related to the above ones by the weak $SU(2)_L$ symmetry. These reactions lead to the following set of the relations between corresponding chemical potentials,

$$\mu(\text{particle}) = -\mu(\text{antiparticle}),$$

$$\mu(H^0) = -2\mu(u_L) - \mu(X)$$

$$\mu(e_L) = \mu(u_L) + 2\mu(X)$$
$$\mu(W) = \mu(Z^0) = \mu(\gamma) = 0$$
$$\mu(e_R) = -\mu(u_L) + \mu(X)$$
$$\mu(d_R) = -\mu(u_L) - \mu(X)$$
$$\mu(u_R) = -\mu(u_L) - \mu(X) \,. \tag{16}$$

Furthermore, the weak isospin symmetry makes equal the chemical potentials of the different components of the $SU(2)_L$ multiplets, i.e. $\mu(X) = \mu(Y)$, etc. From eqs.(13), (14) and (16) it follows that

$$\frac{5\cdot 1440}{\sqrt{2}\,\pi^{7/2}}\, C\left(\mu_X - \Delta m + \frac{3}{2}\frac{\Delta m}{m}T\right) - 32\mu(u_L) - 100\mu(X) = 0 \tag{17}$$

$$\frac{4\cdot 1440}{\sqrt{2}\,\pi^{7/2}}\, C\left(\mu_X - \Delta m + \frac{3}{2}\frac{\Delta m}{m}T\right) - 21\mu(u_L) - 147\mu(X) = 0 \tag{18}$$

$$C = (m_X/T)^{3/2} \exp(-\,T/m_X) \,,$$

up to higher orders of μ/T and $\Delta m_X/T$. One obtains at $T = T_f \approx 0.1\, m_X$

$$\mu(u_L) = 14.6\,\mu(X)$$
$$\mu(X) = 2\cdot 10^{-3}\,\Delta m_X \tag{19}$$

We see that the chemical potentials for a number of species in the plasma under consideration do not vanish and the plasma is arranged in a very peculiar way. It is seen from eqs.(16) and (19) that the number densities of the right-handed states exceed those of the left-handed ones. However, here we are interested in a different point, namely, we would like to discuss the differences between the number densities of leptoquarks and anti-leptoquarks, as well as quarks (both left- and right-handed) and antiquarks. This is necessary for finding out whether this plasma carries the non-vanishing macroscopic baryon number (either hidden in leptoquarks until the temperature falls to T_f or contained in quarks). We find, using eqs.(16) and (19),

$$\frac{\Delta B_X}{n} = \frac{1}{6n} \Delta n_X = - \frac{0.85 \cdot 15}{\sqrt{2}\, \pi^{3/2}} \, C \cdot \frac{\Delta m_X}{T} \tag{20}$$

$$\frac{\Delta B_q}{n_\gamma} = \frac{1}{n} \left(\Delta n_u - \Delta n_d \right) = - \frac{7\pi^2}{8} \frac{\mu(X)}{T} = -0.017 \frac{\Delta m_X}{m_X} \tag{21}$$

where ΔB_X and ΔB_q are the macroscopic baryon numbers carried by leptoquarks and quarks, respectively, the factor 1/6 in eq.(20) being an average baryon number excess in decays of a single leptoquark, We get the overall baryon asymmetry of the Universe

$$\delta = \frac{1}{N} \left(\frac{\Delta B_X}{n} - \frac{\Delta B_q}{n} \right) \quad , \tag{22}$$

where ΔB_X and ΔB_q are to be taken at $T = T_f$, N is the total number of degrees of freedom, $N \sim 10^2$. We finally obtain

$$-\left(\frac{\Delta B_X}{n} + \frac{\Delta B_q}{n} \right) \approx 0.15 \frac{\Delta m_X}{m_X} \tag{23}$$

and

$$|\delta| \sim 1.5 \cdot 10^{-3} \frac{\Delta m_X}{m_X} \tag{24}$$

This is by no means a small number. It has to be compared with the observational value

$$\delta \sim 10^{-8} - 10^{-10}$$

so that one has to require

$$\Delta m_X/m_X \sim 10^{-5} - 10^{-7} . \quad (25)$$

It is seen from eq.(25) that even in the case a = 4 in eq. (11), one may attribute fairly well the baryon asymmetry generation to the CPT violation, provided that leptoquark masses are of order $m_X \sim 3\cdot10^{15}- 3\cdot10^{16}$ GeV. The effect would clearly be more significant at $a<4$ or $m_X>3\cdot10^{16}$GeV.

Let us now make the following remarks.

(1) If there exists a spectrum of masses of leptoquarks, then only the lightest ones will contribute to the process of the cosmological baryon excess production, since the concentrations of heavy particles decrease exponentially when temperature is decreasing.

(2) In a picture of first-order phase transitions in the framework of GUT's, the CPT mass splitting under discussion is significant during a short period of time only; nevertheless, it gives rise to the appearance of the baryon asymmetry of the Universe. This can be understood as follows. Before the hphase transition, leptoquarks are massless; the phase transition lasts $\Delta t \sim m_X^{-1}$ and leptoquarks acquire masses during the latter period. The system is usually rehated up to temperature $T^* \sim 0.2\ m_X$ 16), while at temperature $T_f \sim 0.1\ m_X$, the decoupling and decay of leptoquarks take place (see fig.1). So, the BAU

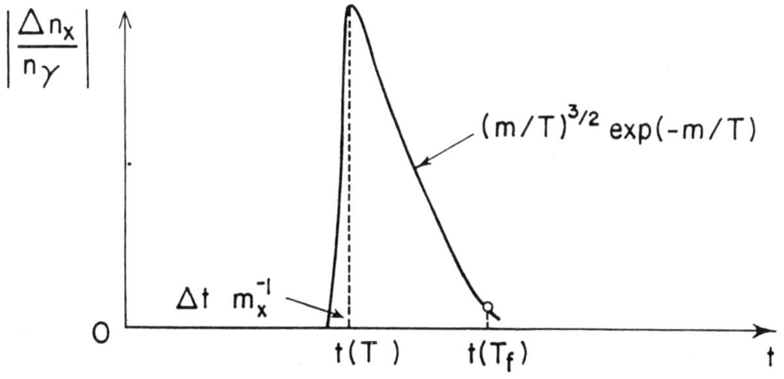

Fig.1

is generated during the period of time in which the temperature decreases from 0.2 m_X to 0.1 m_X. Of course, this is a schematic picture; in fact the process is smoother in time. (The problem of the emergence of equilibrium in such a short period of time is not specific in this context - one encounters it in the standard scenario of the BAU production if the first-order phase transition takes place.)

4. VIOLATION OF THE PAULI PRINCIPLE

We know that fermions obey the Pauli exclusion principle. One may wonder, however, to what extent they really do. And what is the proper way to express a measure of possible violations of the Pauli principle?

We adopt here the following point of view. For our particular purposes of estimating possible magnitudes of effects, which could be induced by the 'wrong' behavior of quantum gravity at small distances, let us take as an ansatz that the width Γ of a transition of an absolutely stable (in the usual sense) atom or nucleus to the Pauli-forbidden state with more than two electrons (protons, neutrons) in the same state is

$$\Gamma \sim \beta^2 \Gamma_{em}^{-1} , \qquad (26)$$

where Γ_{em} is the standard electromagnetic width of the transition, in the case when there is a normal vacancy, and β measures the mixing between the normal and the wrong states. This is shown for an Li atom as an example in fig.2.

In fig.2, the function ψ_1 represents the normal wave function of an Li atom in the ground state, while ψ_2 is the wrong wave function with a vacancy in the K-shell, ψ_1 and ψ_2 describing two states with the same energy.

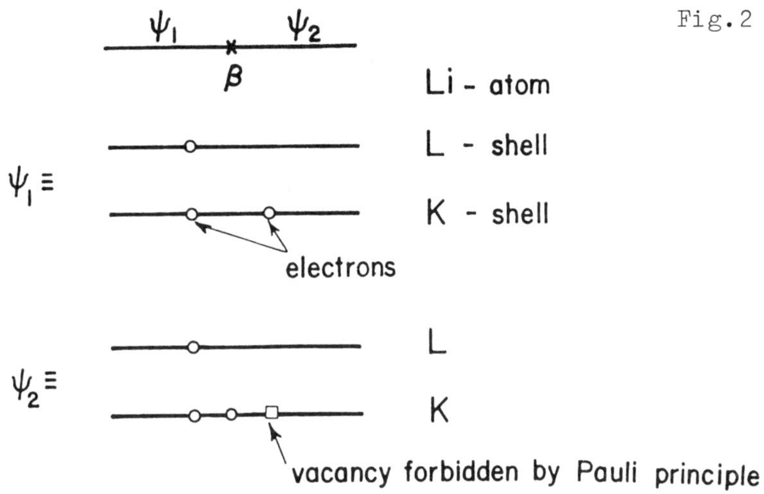

Fig.2

Of course, one could invent [17] many other conjectures as a way of introducing the violation of the Pauli principle. For example, one could imagine that in a time interval between collisions there is a (rising with time) admixture of a wrong wave function ψ' to the normal one ψ_1,

$$\psi \sim c_1 \psi_1 + c_2 \psi' , \tag{27}$$

c_1 and c_2 being given by

$$c_1 \sim \exp(-\beta t) \text{ or } \sim \cos \beta t,$$

$$c_2 \sim 1 - \exp(-\beta t) \text{ or } -\sin \beta t ,$$

and $H_{gr} \sim \beta$, H_{gr} being the wrong 'gravitional' interaction at small distances. Moreover, one could assume

$$\langle \psi' | H_{gr} H_{em} | \psi_0 \rangle \neq 0 . \tag{28}$$

Figure 3 explains the meaning of the symbol ψ'. We shall, however, concentrate in this paper neither on the construction of a consistent quantum mechanical model with the

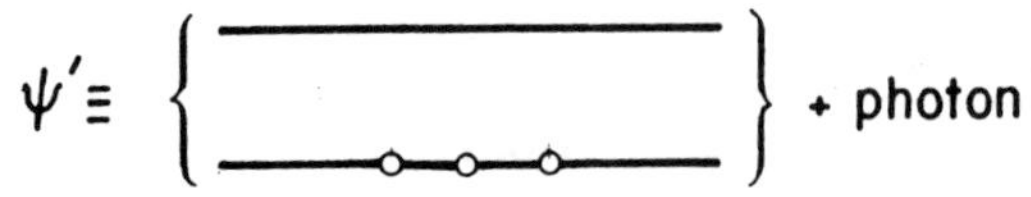

Fig.3

tiny violation of the Pauli principle violation built in, nor on (even more subtle) the question of cosmological consequences of our conjecture, i.e. the problem of overproduction of wrong atoms and nuclei at the epoch of recombination in the early Universe. Both these issues will be discussed in a forthcoming paper [19].

Now, let us assume again that β is given by

$$\beta^2 \sim m^2 \, (m/M_{Pl})^a, \tag{29}$$

and proceed to estimate the sensitivity of atomic and nuclear experiments to possible violations of the Pauli principle.

4.1. Violation of the Pauli Principle in Atoms

A number of experiments on electron stability and electric charge non-conservation have been carried out with an NaJ(Tl) scintillator as a tested substance and a detector at the same time [18]. The result is the limit on the rate of Roentgen γ-ray emission in transitions from L-shell to K-shell of these stable atoms[18].

$$\Gamma < 10^{-23}\ \mathrm{yr}^{-1} \quad 10^{-54}\ \mathrm{GeV}. \tag{30}$$

This Γ is nothing but the rate of Pauli-forbidden transitions in atoms. Thus, using the value $\Gamma_{em} \sim 10^{-13}$ GeV we obtain from [26]

$$\beta^2 < 10^{-67}\ \mathrm{GeV}^2 . \tag{31}$$

This bound looks, at least at first sight, as a very severe constraint on the violation of the Pauli principle.

And indeed it is, in the general sense, but it is presumably much less restrictive than constraints which could come from the study of nuclear transitions. Indeed, let us get a feeling of the sensitivity of the atomic experiments to the 'gravity-induced' violations of the Pauli principle. First of all, we have to decide again what value of m is to be inserted into eq.(29). One can argue that the natural choice is the mass of the electron, since the conjectured 'wrong' interaction at Planck scales makes the electron 'bad'. However, it might happen that it is more adequate to insert into eq.(29) the binding energy ΔE determining the size of the electron orbit.

Further, we do not know what is the value of a, since we know nothing about the new 'wrong' interaction. We may conjecture, however, that the rate of such transitions could be proportional to the ratio of the relevant volumes, namely the Planck volume to to Compton volume of electrons (or nucleon volume in nuclear transitions). It is unclear though what volumes are to be compared: 3 volumes or 4 volumes. The latter conjecture seems to be preferable, although it is by no means conclusive.

We get the following estimates for the rates of transitions under discussion:

$$a=2 \quad \beta^2 \sim 10^{-50} (10^{-59}) \text{ GeV}^2, \; \tau \sim 10^6 \text{ yr } (10^{15} \text{ yr})$$

$$a=3 \quad \beta^2 \sim 10^{-72} (10^{-83}) \text{ GeV}^2, \; \tau \sim 10^{28} \text{ yr } (10^{39} \text{ yr})$$

$$a=4 \quad \beta^2 \sim 10^{-94} \text{ GeV}^2, \; \tau \sim 10^{50} \text{ yr.}$$

The quoted values correspond to $m = m_e$ in eq.(29), while those in brackets are obtained with $m = \Delta E \sim 1$ keV.

Comparing these estimates with the experimental constraints (30), we see that even the case of a = 3 requires the improvement of the experimental sensitivity by 5 orders of magnitude. (It has to be reminded that these estimates are very crude.)

4.2. Violation of the Pauli Principle in Nuclei

Inserting now $m \sim m_N$ and $\Gamma_{em} \sim 0.01 - 1$ eV, as the typical values for nuclear transitions, into eq.(29) we obtain crude estimates of the rates of nuclear rearrangement transitions to states forbidden by the Pauli principle:

$$a=2 \quad \beta^2 \sim 10^{-38}(10^{-46})\ \mathrm{GeV}^2, \quad \tau \sim 10^{-4} - 10^{-2}(10^4 - 10^6)\mathrm{yr}$$

$$a=3 \quad \beta^2 \sim 10^{-57}(10^{-66})\ \mathrm{GeV}^2, \quad \tau \sim 10^{15} - 10^{17}(10^{24} - 10^{26})\mathrm{yr}$$

$$a=4 \quad \beta^2 \sim 10^{-76}(10^{-88})\ \mathrm{GeV}^2, \quad \tau \sim 10^{34} - 10^{36}(10^{46} - 10^{48})\mathrm{yr}.$$

The values given in brackets correspond to $m = \Delta E \sim 10$ MeV substituted into eq.(29).

It is seen that according to our estimates, nuclei are much sensitive to possible violations of the Pauli principle than atoms, at least by 10 orders of magnitude in the transition rates. (This is by no means surprising and follows immediately from eqs.(26) and (29).) Therefore, we propose to search for the Pauli-forbidden nuclear transitions. Clearly, the more detailed analysis concerning the choice of a convenient nucleus and transition has to be carried out.

In table 1 we collect estimates of atomic and nuclear transition rates together with the present cosmological

Table 1

	Lifetimes (yr)		Cosmological abundances of 'wrong'	
	atoms	nuclei	atoms	nuclei
a = 2	$10^6\ (10^{12})$	$10^{-4} - 10^{-2}$	$1\ (10^{-2})$	1
a = 3	$10^{28}(10^{34})$	$10^{15} - 10^{17}\ (10^{24-26})$	$10^{-18}\ (10^{-24})$	$10^{-(5-7)}\ (10^{-(14-16)})$
a = 4	10^{50}	$10^{34-36}\ (10^{46-48})$	10^{-40}	$10^{-(24-26)}\ (10^{-(36-38)})$

abundances of 'wrong' atoms and nuclei produced after recombination, obtained under an assumption that their initial abundances were negligible at the recombination time. It is clear from table 1 that the case a = 2 is excluded by experimental data on both atomic and nuclear transitions and abundance.

5. CONCLUSIONS

The main conclusion we would like to make is that, contrary to the naive expectations, the quantum gravity-induced violations (at least some of them) are not so small as to be unobservable by means of present experimental techniques. In view of the potential importance of (even negative) results, all possible efforts to search for these violations are welcomed. It is amusing that observed baryon excess in the Universe could be attributed to possible CPT violation due to quantum gravity. The preliminary results of this consideration were reported at the Seminar 'Quarks-84' (May 1984, Tbilisi, USSR).

The author is deeply indebted to J.Ambjørn, V.A.Berezin, A.Yu.Ignatiev, V.A.Matveev, H.B.Nielsen, P.Orland, V.A.Rubakov, M.E.Shaposhnikov, A.N.Tavchelidze, I.I.Tkachev and F.V.Tkachov for numerous discussions and comments.

References

1. R.Penrose, in: General Relativity, ed. S.W.Hawking and W.Israel (Cambridge University Press, Cambridge, 1979).
2. S.W.Hawking, Comm. Math. Phys. 43(1975)199; 87(1982) 395.
3. D.N.Page, GRG 14(1982)299.
4. T.Banks, SLAC-PUB-3376 (1984).

5. J.Ellis, J.Hagelin, D.V.Nanopoulos and Srednicki, Nucl. Phys. B241(1984)381.
6. T.Banks, L.Susskind and M.E.Peshkin, Nucl. Phys. B244(1984)125.
7. M.Koshiba, Rapporteur talk at thw XXII Int.Conf. on High Energy Physics, July 1984, Leipzig, GDR.
8. K.G.Chetyrkin, M.V.Kazarnovsky, V.A.Kuzmin and M.E.Shaposhnikov, Phys. Lett. 99B(1981)358.
9. G.Fidecaro, Talk presented at Symposium on Matter-Nonconservation, January 1983, Frascati, Italy.
10. Yu.G.Abov, F.S.Dzheparov and L.B.Okun, Pisma ZhEFT 39(1984)493.
11. V.A.Kuzmin, Pisma ZhEFT 13(1970)335.
12. A.D.Sakharov, Pisma ZhEFT 5(1967)32.
13. A.Yu.Ignatiev, N.V.Krasnikov, V.A.Kuzmin and A.N.Tavkhelidze, Phys. Lett. 76B(1978)436.
14. A.D.Dolgov and Ya.B.Zeldovich, Usp. Fiz. Nauk 130 (1980)559.
15. A.Yu.Ignatiev,V.A.Kuzmin and M.E.Shaposhnikov, Preprint IYaI P-0102 (1978); Sov. Phys. Lebedev Inst. Reports 8(1979)49.
16. V.A.Kuzmin, M.E.Shaposhnikov and I.I.Tkachev, Z.Phys. C12(1982)83.
17. F.Reines, private communication.
18. M.K.Moe and F.Reines, Phys. Rev. B140(1965)992.
19. A.Yu.Ignatiev and V.A.Kuzmin, to be published.

THE USE OF FINITE ENERGY SUM RULES FOR THE CALCULATION OF THE INDUCED GRAVITATIONAL CONSTANT

N.V. Krasnikov and A.A. Pivovarov
Institute for Nuclear Research of the Academy of Sciences of the USSR,
Moscow 117312
USSR

ABSTRACT

We use the finite energy sum rules for the calculation of the induced gravitational constant in gauge theories of the QCD type. The induced gravitational constant is determined by the gluon vacuum condensate. Namely, $(16\pi G_{ind})^{-1} = (11/576\pi)\langle \frac{\alpha}{\pi} G^2 \rangle^{1/2}$ for the QCD with $n_f = 3$ massless quarks.

As is well known the standard classical Hilbert-Einstein Lagrangian for the gravitational field leads to the nonrenormalizable theory at the quantum level (see for example /1/ and refs. quoted therein). Recently Adler/2/ has suggested that gravity can arise as a quantum effect in gauge theories of the QCD type. Radiative corrections in curved space-time induce R-term in the effective action. Adler and Zee /2,3/ have shown that the induced gravitational constant is finite and calculable quantity in the theories with the dynamical breaking of scale invariance. The QCD with massless quarks belongs to this class of theories. Adler and Zee derived also the formula for the induced gravitational constant which has the form/1,3/

$$(16\pi G_{ind})^{-1} = \frac{1}{12}\frac{d}{dQ^2}\Pi(Q^2)\Big|_{Q^2=0} \tag{1}$$

Here $\Pi(Q^2)$ is defined by the relation

$$\Pi(Q^2) = i\int dx\, e^{iqx}\langle T\tilde{\Theta}(x)\tilde{\Theta}(0)\rangle, \quad Q^2=-q^2$$

$$\tilde{\Theta}(x) = \Theta_\nu^\nu(x) - \langle\Theta_\nu^\nu\rangle \tag{2}$$

where $\Theta^{\mu\nu}(x)$ is the energy-momentum tensor.

In this paper we calculate the induced gravitational constant for the QCD type theories. We show that the induced gravitational constant is positive that corresponds to the attractive potential and it is determined by the gluon vacuum condensate $\langle\alpha G^2\rangle$. Our main tool for the calculation of the induced gravitational constant is finite energy sum rules /4,5/. The review of the earlier papers devoted to the calculation of the induced gravitational constant can be found in ref./1/. For the definiteness we shall consider QCD with massless quarks, however, our method is applicable for other gauge theories.

In QCD with massless quarks the trace of the energy-momentum tensor has the form [6]

$$\Theta_\nu^\nu(x) = \frac{\beta(\alpha)}{2\alpha} G^a_{\tau\sigma} G^{a\,\tau\sigma}(x),$$
$$\beta(\alpha) = -\beta_0 \frac{\alpha^2}{4\pi} + O(\alpha^3)\,; \quad \beta_0 = 11 - \frac{2}{3} n_f \qquad (3)$$

n_f is the number of flavours.

The use of the "naive" Källen-Lehmann Representation for the correlator (2)

$$\Pi(Q^2) = \int_0^\infty \frac{\rho(s)\,ds}{s+Q^2}\,, \qquad \rho(s) \geqslant 0 \qquad (4)$$

leads to the negative induced gravitational constant

$$\left(16\pi G_{ind}\right)^{-1} = -\frac{1}{12}\int_0^\infty \frac{\rho(s)\,ds}{s^2} \quad < 0 \qquad (5)$$

However we have to write down the Källen-Lehmann representation with subtractions, namely

$$\Pi(Q^2) = A(\mu^2) + B(\mu^2)Q^2 + C(\mu^2)Q^4$$
$$- \left(Q^2-\mu^2\right)^3 \int_0^\infty \frac{\rho(s)\,ds}{(s+Q^2)(s+\mu^2)^3} \qquad (6)$$

where A, B, and C are some subtraction constants and the number of subtractions is determined by the canonical dimension of the operator $\Theta_\nu^\nu(x)$.

In perturbation theory framework $\Pi(Q^2)$ has the following explicit form for QCD with $n_f = 3$ massless quarks (for instance in the $\overline{MS}$ -scheme) [7]

$$\Pi^{PT}(Q^2) = Q^4 \ln\frac{\mu^2}{Q^2}\left(\frac{\beta(\alpha)}{2\alpha}\right)^2 \frac{2}{\pi^2}\left\{1 + \frac{\alpha}{4\pi}\left(41 + 9\ln\frac{\mu^2}{Q^2}\right)\right. \qquad (7)$$

$$+\left(\frac{\alpha}{4\pi}\right)^2\left(a\ln^2\frac{\mu^2}{Q^2}+b\ln\frac{\mu^2}{Q^2}+c\right)+\ldots\Bigg\} \tag{7}$$

where coefficients a and b can be determined by the renormalization group analysis.

Such a structure is determined by the requirements of generalized dimensional analysis. Really, the renormalization procedure leads only to the logarithmic dependence on the renormalization point μ. Using the spectral representation

$$Q^4\ln^n\frac{Q^2}{\mu^2}=Q^4\int_0^\infty\frac{\rho_n(s)}{s^2}\left(\frac{1}{s+Q^2}-\frac{1}{s+\mu^2}\right)ds \tag{8}$$

where

$$\rho_n(s)=s^2\sum_{k=0}^{\left[\frac{n-1}{2}\right]}(-1)^{k+1}\pi^{2k}\left(\ln s/\mu^2\right)^{n-2k-1} \tag{9}$$

we immediately obtain that in perturbation theory the correlator $\Pi(Q^2)$ has the following spectral representation

$$\Pi^{PT}(Q^2)=Q^4\int_0^\infty\frac{\rho^{PT}(s)}{s^2}\left(\frac{1}{s+Q^2}-\frac{1}{s+\mu^2}\right)ds \tag{10}$$

Here $\rho^{PT}(s)$ is the proper sum of $\rho_n(s)$ and it has the form in the next-to-leading order in α (in $\overline{MS}$ -scheme)

$$\rho^{PT}(s)=s^2\left(\frac{\beta(\alpha)}{2\alpha}\right)^2\frac{2}{\pi^2}\left\{1+\frac{\alpha}{4\pi}\left(41+18\ln\frac{\mu^2}{s}\right)+O(\alpha^2)\right\} \tag{11}$$

The representation (10) leads to zero value for the induced gravitational constant. Therefore in the perturbation theory framework the induced gravitational constant is equal to zero and in order to obtain the nonzero value of $(16\pi G_{ind})^{-1}$ we have to take into account the nonperturbative effects. For this aim let us consider the asymptotic expansion of the correlator $\Pi(Q^2)$ at $Q^2 \to \infty$. This asymptotic behaviour can be found by using the Wilson expansion and it has the form[8]

$$\Pi(Q^2) = \Pi^{PT}(Q^2) + \left(\frac{\beta(\alpha)}{2\alpha}\right)^2 \frac{2}{\pi^2} \left\{ \frac{2\pi^2 \langle \alpha G^2 \rangle}{\alpha} + \frac{48\pi^2 \rho_c^{-2} \langle \alpha G^2 \rangle}{5\alpha Q^2} + \frac{9\pi^3 \langle \alpha G^2 \rangle^2}{4\alpha Q^4} \right\} + O(Q^{-6}) \tag{12}$$

For the difference $\Pi(Q^2) - \Pi^{PT}(Q^2)$ we get at $Q^2 \to \infty$

$$\Pi(Q^2) - \Pi^{PT}(Q^2)\Big|_{Q^2 \to \infty} = \left(\frac{\beta(\alpha)}{2\alpha}\right)^2 \frac{4\langle \alpha G^2 \rangle}{\alpha} + O(Q^{-2}) \tag{13}$$

Using formulae (6,10,13) we can write the Källen-Lehmann representation for this difference with one subtraction only

$$\Pi(Q^2) - \Pi^{PT}(Q^2) = \int_0^\infty [\rho(s) - \rho^{PT}(s)] \left(\frac{1}{s+Q^2} - \frac{1}{s+\mu^2} \right) ds \tag{14}$$

Thus for the induced gravitational constant we obtain the well defined finite expression

$$(16\pi G_{ind})^{-1} = \frac{1}{12} \int_0^\infty \frac{\rho^{PT}(s) - \rho(s)}{s^2} ds \tag{15}$$

To calculate the induced gravitational constant we have to find the physical spectral density $\rho(s)$ /5/. Choosing the simplest ansatz in the form /5/

$$\bar{\rho}(s) = \bar{F} m^4 \delta(s-m^2) + s^2 \Theta(s-s_0)$$
$$\rho(s) = \left(\frac{\beta(\alpha)}{2\alpha}\right)^2 \frac{2}{\pi^2} \bar{\rho}(s) \tag{16}$$

and using the standard technique /5/ we obtain the finite energy sum rules

$$\bar{F} m^2 - \frac{s_0^2}{2} + \frac{2\pi^3}{\alpha}\langle\frac{\alpha}{\pi}G^2\rangle = 0 \tag{17a}$$

$$\bar{F} m^4 - \frac{s_0^3}{3} - \frac{48\pi^3\rho_c^{-2}}{5\alpha}\langle\frac{\alpha}{\pi}G^2\rangle = 0 \tag{17b}$$

$$\bar{F} m^6 - \frac{s_0^4}{4} + \frac{9\pi^5}{4\alpha}\langle\frac{\alpha}{\pi}G^2\rangle^2 = 0 \tag{17c}$$

In these formulae we use for the perturbation theory part of the spectral density K-renormalization scheme (in this scheme all nonleading α corrections are equal to zero, see for example /7/). The nonperturbative corrections to eqs. (17b) and (17c) are very small numerically and can be neglected. Neclecting these corrections we get the following relations for the spectrum parameters $\bar{F}$, s_0, and m

$$m^2 = \frac{3}{4}s_0, \quad \bar{F} = \frac{16}{27}s_0, \quad s_0 = 6\pi\sqrt{\langle G^2\rangle(s_0)} \tag{18}$$

Using eq. (15) we get the following formula for induced gravitational constant

$$(16\pi G_{ind})^{-1} = \left(\frac{\beta(\alpha)}{2\alpha}\right)^2 \frac{2}{\pi^2}\frac{1}{12}\int_0^{s_0} \frac{s^2 - \bar{F} m^4 \delta(s-m^2)}{s^2} ds$$
$$= \left(\frac{\beta(\alpha)}{2\alpha}\right)^2 \frac{2}{\pi^2}\frac{1}{12}(s_0 - \bar{F}) \tag{19}$$

Substituting solution (18) into eq. (19) we finally have

$$\left(16\pi G_{ind}\right)^{-1} = \frac{33\,\alpha^2(s_0)}{64\pi^3}\sqrt{\langle G^2\rangle(s_0)} \tag{20}$$

Formula (20) shows that G_{ind}^{-1} is determined by the essentially nonperturbative parameter of the theory - the gluon vacuum condensate $\langle G^2\rangle$. Note that the gluon vacuum condensate depends on the temperature. The explicit form of this dependence at present is not known but it is reasonable to suggest (by analogy with the superconductivity) that in the deconfinement phase, when free quarks and gluons are genuine asymptotic states, the gluon vacuum condensate is zero. Then we may expect that $G_{ind}^{-1} = 0$ at very high temperatures in the deconfinement phase. This effect can have untrivial consequences for cosmology.

The numerical values of the spectrum parameters at $\langle\alpha G^2\rangle = 0.038\,\mathrm{GeV}^4$ and $\alpha(s_0) = 0.3$ are

$$m^2 = 5.2\,\mathrm{GeV}^2,\quad \bar{F} = 4.2\,\mathrm{GeV}^2,\quad s_0 = 6.9\,\mathrm{GeV}^2 \tag{21}$$

It is natural that we cannot get the right magnitude of $(16\pi G_{ind})^{-1}$ in QCD with a scale $\Lambda_{QCD} = 100 \div 200\,\mathrm{MeV}$ but the considered example shows the possible mechanism of generation of the induced gravitational constant. Really let us suggest that high-energy unified theory can be considered as some effective QCD-like theory with a scale $\Lambda_{eff} = K\cdot\Lambda_{QCD}$. Then if $K \sim 10^{19}$ we can get the right order of magnitude of the induced gravitational constant. Namely, dimensional analysis gives

$$\langle\alpha G^2\rangle_{eff} = K^4\langle\alpha G^2\rangle,\quad \text{where}\ \langle\alpha G^2\rangle = 0.038\,\mathrm{GeV}^4$$

is the low energetic numerical value of the gluon vacuum

condensate parameter in QCD.

Note that there is low energy theorem for $\Pi(Q^2)$ of the form /8/

$$\Pi(0) = \frac{\beta_0}{2}\langle \frac{\alpha}{\pi} G^2 \rangle \tag{22}$$

in the leading order in α. The quantity $\Pi(0)$ is not defined uniquely since one has to subtract the ultraviolet divergences. The subtraction procedure in this case is fixed by resting the low energy contribution only, and high energy contribution determined by the perturbation theory is subtracted. Thus we define

$$\Pi(0) = \int_0^{S_0} \frac{\rho(s)\,ds}{s}$$

Taking into account eqs. (18) we get

$$\frac{\beta_0}{2}\langle \frac{\alpha}{\pi} G^2 \rangle = \bar{F} m^2 \left(\frac{\beta(\alpha)}{2\alpha}\right)^2 \frac{2}{\pi^2}$$

or

$$\alpha(S_0) = \frac{\pi}{\beta_0} = 0.35 \tag{23}$$

The obtained value $\alpha(S_0) = 0.35$ of the effective coupling constant is closed to the effective coupling constant $\alpha(S_0) = 0.3$ which has been used in eqs. (17, 19, 20) so we see that our finite energy sum rules reproduce the low energy theorem (22).

Let us obtain now the low energy theorem (22) using the method of the effective action. Suppose that $\alpha G^2(x) = \pi A\, \sigma^4(x)$ where the quantity A fixes the normalization of the scalar field $\sigma(x)$. The trace of the energy-momentum tensor in the lowest approximation on α is

$$\Theta_\nu^\nu(x) = -\frac{\beta_0}{8} A \sigma^4(x) \tag{24}$$

The effective Lagrangian

$$\mathcal{L} = \frac{1}{2} \partial_\mu \sigma \partial^\mu \sigma - V(\sigma)$$

$$V(\sigma) = \frac{\beta_0}{8} A \sigma^4 \ln \frac{\sigma}{\mu} \tag{25}$$

reproduces formula (24) for the trace of the energy-momentum tensor. The σ -meson mass is

$$m_\sigma^2 = \frac{\beta_0 A}{2} \sigma_0^2 = \frac{\beta_0 \langle \frac{\alpha}{\pi} G^2 \rangle}{\sigma_0^2}$$

$$\sigma_0 = \mu \exp(1/4) \tag{26}$$

In the linear approximation on $\sigma' = \sigma - \sigma_0$ we have

$$\Theta_\nu^\nu(x) = -m_\sigma^2 \sigma_0 \sigma'(x) + const$$

In the one-resonance approximation we have

$$i \int dx e^{iqx} \langle T \tilde{\Theta}_\mu^\mu(x) \tilde{\Theta}_\nu^\nu(0) \rangle \Big|_{q=0} = m_\sigma^2 \sigma_0^2 \tag{27}$$

Eq. (27 coinsides with the low energy theorem (22).

Neglecting the small nonperturbative corrections to eqs. (17b, 17c) we obtain the relation between the mass and the residue of the σ -meson in the form

$$m_\sigma^2 = \frac{81}{64} \bar{F} \tag{28}$$

Using the relations (27, 28) and effective Lagrangian (25) we obtain

$$m_\sigma^2 \sigma_0^2 = \frac{\beta_0}{2} \left\langle \frac{\alpha}{\pi} G^2 \right\rangle$$
$$\sigma_0 = \left(\frac{\beta_0 \alpha}{\pi} \right)^2 \frac{2}{81\pi^2} m_\sigma^2 \qquad (29)$$

or numerically for $n_f = 3$

$$\sigma_0 = 100\,MeV, \quad m_\sigma = 2.3\,GeV$$

So we see that the finite energy sum rules approach (17) and the effective Lagrangian one (25) give similar results.

In conclusion we would like to stress our main result, the induced gravitational constant for the QCD type theories has the positive sign and it is determined by the gluon vacuum condensate.

We are indebted to V.A. Matveev and A.N. Tavkhelidze for very interesting discussions.

References

1. As a review see for example: Adler, S.L., Rev. Mod. Phys. 54, 729 (1982).
2. Adler, S.L., Phys. Rev. Lett. 44, 1567 (1980); Phys. Lett. 95B, 241 (1980).
3. Zee, A., Phys. Rev. D23, 858 (1981).
4. Logunov, A.A., Soloviev, L.D., Tavkhelidze, A.N., Phys. Lett. 24B, 181 (1967); Chetyrkin, K.G., Krasnikov, N.V., Tavkhelidze, A.N., Phys. Lett. 76B, 83 (1978).
5. Krasnikov, N.V., Pivovarov, A.A., Tavkhelidze, A.N., Z. Phys. C19, 301 (1983).
6. Crewther, R., Phys. Rev. Lett. 28, 1412 (1972); Chanowitz, M. and Ellis, J., Phys. Lett. 40B, 397(1972).
7. Kataev, A.L., Krasnikov, N.V., Pivovarov, A.A., Nucl. Phys. B198, 508 (1982).
8. Novikov, V.A. et al., Nucl. Phys. B165, 67 (1980).

SECTION II

SUPERGRAVITY

PRESENT STATE OF QUANTUM SUPERGRAVITY

E.S.Fradkin and A.A.Tseytlin

Department of Theoretical Physics, P.N.Lebedev Physical Institute, Leninsky pr. 53, Moscow 117924 USSR

This is a brief summary of the review talk given at the Seminar. The accent was made on the results of the last three years. The talk consisted of the four parts:

1. Ordinary $N \leq 8$ supergravities in D=4 dimensions
2. Higher dimensional $D \leq 11$ supergravities
3. Conformal (N = 4, D=4) supergravities
4. Superstring theory (D= 10)

The main conclusion of the first part was that in spite of a number of interesting quantum properties (cancellation of one- and two-loop infinities on the mass shell [1], cancellation of one-loop off-shell infinitions in N=8 theory[2] cancellation of quartic and quadratic infinities [2],etc.) ordinary (gauged) supergravities cannot be seriously considered as candidates for a fundamental unified theory valid at arbitrarily small distances. The reason is in the dimensional nature of gravitational coupling constant,i.e. in the presence of the infinite number of invariants which may formally appear as infinities at three and higher loops. It is very unlikely that the infinite number of coefficients of these invariants in the effective action may appear to be vanishing in a theory with a finite number of fields and finite dimensional symmetry group.

The second part of the talk contained the known information about ultraviolet properties of D $\leq$ 11 supergravities. It was noted that the leading power divergences ($\Lambda^{11}, \Lambda^{9}, \ldots , \Lambda^{5}$) cancel on shell in the D=11 theory [3]. At the same time it was shown that there is a $\Lambda^{3} \times (R^{\cdot}_{\ldots})^{4}$ - infinity in this theory which occurs with a non-vanishing coefficient [Λ^{3} and Λ -infinities may be absent on particular conformally-flat backgrounds like $(adS)_4 \times S^7$ [4]]. The presence of a non-trivial Λ^{3} -infinity as well as the presence of invariants - candidates for $L \geqslant 2$ - loop infinities indicates that the D = 11 supergravity is unlikely to be a finite theory. Analogous conclusion is true also for D $<$ 11 theories [some of them (maximal supergravities in D = 5,6,7) are one-loop finite on shell [3]]. Thus ordinary D $\leq$ 11 supergravities may be considered only as low-energy approximations to some more fundamental finite theory. In this second part we also described the method [3] which makes possible to relate the one-loop infinities of a higher dimensional theory and of its lower dimensional reduction. Using this method one can prove that D = 4 supersymmetric theories which follow via reduction from non-maximal D $>$ 4 supergravities fail to be one-loop finite [5].

Conformal supergravities (see [6] for a review and references) are gauge theories of the superconformal group $SU(2,2|N), N\leq 4$. Their actions are supersymmetric extensions of the Weyl ($C^2_{\ldots}$) action which naturally unify the C^2 -gravity with $SU(N)$ Yang-Mills fields. These theories contain only one dimensionless coupling constant and are naturally "tied up" to four dimensions where they are power counting renormalizable. The important property of conformal supergravities is that the condition of their formal consistency (absence of gauge superconformal anomalies) is equivalent to the condition of their ultraviolet finiteness. There are two presently known candidates for a finite superconformal theories, both being based on N=4 conformal supergravity [6] - [9] [Super-space non-renormalization theorem implies that $N \geqslant 2$ conformal supergravities are $L \geqslant 2$ - loop finite [10]; one-loop finiteness is true only for the "non-minimal" N=4 theory and for the "minimal" N = 4 conformal supergravity coupled to $SU(2)\times U(1)$ N = 4 super Yang-Mills [6], [8], [9]]. A presence of conformal scalars in these theories opens a principal possibility to establish a low-energy correspondence with the Einstein (super) gravity.

The main problem of the finite conformal supergravities remains the problem of unitarity. We discussed several proposals for its solution based on non-perturbative approaches (see, e.g., [11]). It was also stressed that conformal supergravities have a number of "phenomenological" advantages as compared to ordinary supergravities: unification with gauge fields without introducing a cosmological constant, chiral $SU(N)$ representations of spinor fields which occur in the spectrum etc.

Another possibility to circumvent the ultraviolet difficulty of ordinary supergravities is connected with a theory of extended objects - (super)strings [12]. String theory may be considered as a theory of an infinite number of fields unifying a finite number of massless fields (graviton, antisymmetric tensor, ...) with the infinite set of the massive "higher spin" ones. An improvement of ultraviolet behaviour of such a theory can be understood as being due to higher derivative character of higher spin field couplings ($\varphi_s \Box \varphi_s + \varphi_s \partial^s \varphi_s \varphi_s + \ldots$) so that summation over all the fields leads to a "form-factor" ("$\exp(-\alpha' p^2)$") which effectively cut off the momentum space integrals. The viable candidates for a consistent string theory are the superstring models in D=10 dimensions. The first (N = 1 supersymmetric) model includes open as well as closed strings and in the $\alpha' \to 0$ (zero string "size") limit reduces to N = 1, D = 10 supergravity interacting with N = 1, D = 10 super Yang-Mills theory with gauge group G . The condition of cancellation of gauge and gravitational anomalies uniquely fixes G to be SO (32) 13/. The second superstring model has the $\alpha' \to 0$ limit being one of the two versions of N= 2, D = 10 supergravity . Both models are likely to be ultraviolet finite 12/, 14/ and certainly are better behaved than their $\alpha' \to 0$ limiting D = 10 field theories which are non-finite. [both N=1 and N = 2 D = 10 supergravities have Λ^2 and $\ln \Lambda^2$ infinities already at the one-loop level] . The important problem of (super)string theories is how to establish a covariant off-shell formalism for a study of low-energy correspondence with the "phenomenological" D=4 theories [a study of compactification of extra six dimensions should be based on an effective action for fields, corresponding to string modes] A way to solution of this problem is suggested in ref. 15/.

Our conclusion is that most interesting candidates for a fundamental theory consistently incorporating quantized gravity are at present the finite conformal supergravities in four dimensions and superstring theories in ten dimensions.

REFERENCES

1. van Nieuwenhuizen, P., Phys. Rept. 68,189 (1981).
Duff, M.J., in: Supergravity 81, ed. by S.Ferrara and J.G.Taylor (Cambridge V.P., 1982).
2. Fradkin, E.S. and Tseytlin, A.A., Phys. Lett. 117B, 303 (1982).
3. Fradkin, E.S. and Tseytlin, A.A., Phys. Lett. 123B, 231 (1983); Nucl. Phys. B227, 252 (1983).
4. Gibbons, G.W. and Nicolai, H., Phys. Lett. 143B, 108 (1984).
Inami, T. and Yamagishi, K., Phys. Lett. 143B, 115 (1982).
5. Fradkin , E.S. and Tseytlin, A.A., Phys. Lett. 137B, 357 (1984).
6. Fradkin, E.S. and Tseytlin, A.A., Conformal Supergravity, Phys. Rept. (1985).
7. Fradkin, E.S. and Tseytlin, A.A., Phys. Lett. 110B, 117 (1982).
(E: 126B, 506 (1983)); Nucl. Phys. B203,157 (1982).
8. Fradkin, E.S. and Tseytlin, A.A., Phys. Lett. 134B, 187 (1984).
9. Fradkin, E.S. and Tseytlin, A.A., Phys. Lett. 134B, 307 (1984).
10. Howe, P.S., Stelle, K.S. and Townsend, P.K., Nucl. Phys. B236,125 (1984).
11. Kaku, M., Phys. Rev. D27,2809 (1983).
Pisarski, R.D., Phys. Rev. D28,2547 (1983).
Tomboulis, E., Phys. Rev. Lett. 52, 1173 (1984).
12. Schwarz, J.H., Phys. Rept. 89, 223 (1982).
Green, M.B., Surveys in H.E. Phys. 3, 127 (1983).
13. Green, M.B. and Schwarz, J.H., Anomaly cancellation in supersymmetric D = 10 gauge theory require SO(32), preprint CALT-68-1182 (1984).
14. Green, M.B. and Schwarz, J.H., Infinity cancellation in SO(32) superstring theory, preprint CALT-68-1194 (1984).
15. Fradkin, E.S. and Tseytlin, A.A., Effective field theory from quantized strings, Phys. Lett. B (1985); contribution to this volume.

GEOMETRICAL ORIGIN OF NEW UNCONSTRAINED SUPERFIELDS

A.A.Rosly, A.S.Schwarz
Moscow Physical Engineering Institute
Kashirskoe Shosse 1, Moscow M-409
USSR

ABSTRACT

We show how the unconstrained formulations of N=2 and N=3 super gauge theories suggested in[1,2] can be obtained from geometrical considerations. An analogous formulation of supersymmetric hyper-Kähler sigma model is given.

1. INTRODUCTION

In papers[1,2] unconstrained formulations were found for N=2 and N=3 supersymmetric gauge theories and for N=2 hypermultiplet. These formulations make use of a superspace with auxiliary dimensions suggested in ref.[3]. The papers[1,2] do not use, however, the construction given in[3] which converts the superfield constraints for these theories to equations in a new extended superspace. Here we will formulate the results and outline the proofs of[3]; we shall show how using this to obtain the results of refs.[1,2]. It is worth mentioning that although the derivation will be simple technically, it uses some important ideas contained in[1,2].

The results of the paper[3] are given there in terms of so-called CR-bundles. Here we shall use a language which is closer to field theory, namely the language of partial gauge fields. A usual gauge field yields lengthening (covariantization) of derivatives and, hence, of all first order differential operators. We shall say that one deals with partial gauge fields if one considers lengthening of only a part of first order differential operators. Thus we assume that a set, $\partial_1, \dots, \partial_n$, of first order differential operators is fixed. To define a partial gauge field means to relate to every ∂_i an operator $\nabla_i = \partial_i + \mathcal{A}_i$, where $\mathcal{A}_i$ are functions with values in the Lie algebra of a gauge group. One can define in a conventional manner the gauge equivalence of such fields ($\mathcal{A}_i \sim g^{-1}\mathcal{A}_i g + g^{-1}\partial_i g$) and the field strengths. It must be pointed out, however, that for partial gauge fields the condition of being gauge equivalent to zero is much stronger than the condition of having field strength equal to zero. Let us mention that for usual gauge fields it is oftenly useful to consider lengthening not of derivatives, but of the differential. Then the gauge field is represented by a 1-form. An analogous treatment will be useful also in the case of partial gauge fields.

An important for us example of a partial gauge field can be found if on a complex manifold with coordinate $z^1, \dots, z^m$ one takes lengthening of the derivatives $\partial/\partial \bar{z}^s$ only, or, equivalently, of the differential $\bar{\partial} = d\bar{z}^s\, \partial/\partial \bar{z}^s$. Such fields are related with holomorphic bundles*/. A slightly more general kind of partial gauge fields is related with so-called CR-bundles[4,5]. In the present work the notion of a CR-manifold and CR-bundle will appear at the very end only.

Let us notice, that the constructions studied here are analogs of some twistor constructions. So the con-

*/ In mathematics, such a partial gauge field is called $\bar{\partial}$-connection. It is known, that every $\bar{\partial}$-connection with vanishing curvature (i.e. field strength) defines a holomorphic vector bundle and that gauge equivalent $\bar{\partial}$-connections correspond to holomorphically equivalent bundles[4].

struction used for super gauge theories is analogous to Ward transformation[6], while the approach to supersymmetric hyper-Kähler sigma model resembles a reasoning due to Penrose[7].

2. SUPER GAUGE THEORIES

Let us start from an observation that the usual superfield constraints[5,6] in N=2 and N=3 super gauge theories can be represented in the following form[3]

$$\{\nabla_\alpha, \nabla_\beta\} = \{\nabla_{\dot\alpha}, \nabla_{\dot\beta}\} = \{\nabla_\alpha, \nabla_{\dot\beta}\} = 0 , \tag{1}$$

where

$$\nabla_\alpha = p^i \nabla_{\alpha i} , \quad \nabla_{\dot\alpha} = u_i \nabla_{\dot\alpha}{}^i . \tag{2}$$

Here p^i and u_i ($i = 1,\ldots, N$) are arbitrary complex constants constrained by the relation $p^i u_i = 0$, while $\nabla_{\alpha i} = D_{\alpha i} + \mathcal{A}_{\alpha i}$, $\nabla_{\dot\alpha}{}^i = D_{\dot\alpha}{}^i + \mathcal{A}_{\dot\alpha}{}^i$ are super gauge covariant derivatives in the N extended Minkowski superspace M. For N=2 the constraints (1) are equivalent to the usual kinematical constraints, whereas for N=3 the constraints (1) are equivalent to the field equations.

As we have said, the constraints (1) are parametrized by vectors p^i , u_i obeying $p^i u_i = 0$. Let us notice that under the transformation $p^i \to \lambda p^i$, $u_i \to \mu u_i$ eqs. (1) are transformed into equivalent ones. This allows to consider eqs. (1) as parametrized by points of a manifold Q on which p^i , u_i constrained by $p^i u_i = 0$ serve as homogeneous coordinates. For N=3, Q is a three-dimensional compact complex manifold which can be identified with $SU(3)/U(1)\times U(1)$. For N=2 we have $Q = \mathbb{C}P^1 = SU(2)/U(1)$. (This can be seen from that one can assume that $u_i = \varepsilon_{ij} p^j$). The remarks above suggest to consider the spaces $P = M \times Q$. The covariant derivatives ∇_α , $\nabla_{\dot\alpha}$ of eqs. (2) define a partial gauge field on the space P,

$$\nabla_\alpha = D_\alpha + \mathcal{A}_\alpha , \quad \nabla_{\dot\alpha} = D_{\dot\alpha} + \mathcal{A}_{\dot\alpha} , \tag{3}$$

where

$$D_\alpha = p^i D_{\alpha i} \; , \quad D_{\dot\alpha} = u_i D_{\dot\alpha}{}^i \; , \qquad (4)$$

$$\mathcal{A}_\alpha = p^i \mathcal{A}_{\alpha i} \; , \quad \mathcal{A}_{\dot\alpha} = u_i \mathcal{A}_{\dot\alpha}{}^i \; . \qquad (5)$$

The condition (1) means that this partial gauge field has vanishing strength.

Now we are ready to describe the convertion of constraints in super gauge theories to some equations on the space P. Let us consider on P a partial gauge field

$$\nabla_\alpha = D_\alpha + \mathcal{A}_\alpha (x, \theta, \bar\theta, z, \bar z) \, ,$$

$$\nabla_{\dot\alpha} = D_{\dot\alpha} + \mathcal{A}_{\dot\alpha} (x, \theta, \bar\theta, z, \bar z) \, ,$$

$$\nabla_s = \frac{\partial}{\partial \bar z^s} + \mathcal{A}_s (x, \theta, \bar\theta, z, \bar z) \, ,$$

which corresponds to lengthening of differential operators $D_\alpha, D_{\dot\alpha}, \partial/\partial \bar z^s$. (Here $\bar z^s$ are some local complex coordinates*/ on the manifold Q ; for N=3 s=1,2,3, while for N=2 s=1). Let us show, that if the field strength of such a partial gauge field vanishes, then in the generic case this field corresponds to a gauge field on Minkowski superspace obeying the constraints (1). As it is known in mathematics, if the fields $\mathcal{A}_s$ lengthening the derivatives $\partial/\partial \bar z^s$ on Q are sufficiently small and if their field strength vanishes, then they can be gauge transformed to zero. After such a transformation the fields $\mathcal{A}_\alpha(x,\theta,\bar\theta,z,\bar z)$, $\mathcal{A}_{\dot\alpha}(x,\theta,\bar\theta,z,\bar z)$ become holomorphic (i.e. not depending on $\bar z^s$). This follows from the

---

*/ As we shall see, it is important to consider the manifold Q in the whole. For this purpose one has to use either a number of local coordinate charts, or homogeneous coordinates p^i, u_i . Note that, although it is often not mentioned explicitly, we are using both ways of defining coordinates; in particular, the operators D_α , $D_{\dot\alpha}$ are defined in eqs. (4) in terms of homogeneous coordinates, while $\partial/\partial \bar z^s$ — in terms of local coordinates.

condition of vanishing field strength which implies that $[\partial/\partial\bar{z}^s, \nabla_\alpha] = 0$, $[\partial/\partial\bar{z}^s, \nabla_{\dot\alpha}] = 0$ */. If expressed in terms of homogeneous coordinates, p^i, u_i, the fields $\mathcal{A}_\alpha$ become holomorphic functions of p^i and u_i which have the degree of homogeneity with respect to these variables equal to one and zero respectively, or (1,0) for short. Similarly, the degree of homogeneity of the fields $\mathcal{A}_{\dot\alpha}$ turns out to be (0,1). This implies that the fields $\mathcal{A}_\alpha$ are linear in p^i and do not depend on u_i, that is $\mathcal{A}_\alpha = p^i \mathcal{A}_{\alpha i}(x,\theta,\bar\theta)$. Analogously, we have $\mathcal{A}_{\dot\alpha} = u_i \mathcal{A}_{\dot\alpha}{}^i(x,\theta,\bar\theta)$. (We have used the fact that any homogeneous of degree k holomorphic function is a polynomial of degree k). The resulting fields $\mathcal{A}_{\alpha i}, \mathcal{A}_{\dot\alpha}{}^i$ can be considered as gauge fields on Minkowski superspace. The part of the condition of vanishing field strength that concerns $\mathcal{A}_\alpha$, $\mathcal{A}_{\dot\alpha}$ implies then the relations (1) and, hence, the usual constraints on superfields $\mathcal{A}_{\alpha i}$, $\mathcal{A}_{\dot\alpha}{}^i$ in Minkowski superspace M. It is easy to see that under a gauge transformation of the fields ($\mathcal{A}_\alpha, \mathcal{A}_{\dot\alpha}, \mathcal{A}_s$) on P the corresponding fields ($\mathcal{A}_{\alpha i}, \mathcal{A}_{\dot\alpha}{}^i$) on M undergo a gauge transformation as well. This implies that there is a one-to-one correspondence between gauge equivalence classes of the fields ($\mathcal{A}_\alpha$, $\mathcal{A}_{\dot\alpha}$, $\mathcal{A}_s$) having zero strengths on P and gauge equivalence classes of the fields ($\mathcal{A}_{\alpha i}, \mathcal{A}_{\dot\alpha}{}^i$) obeying the constraints (1) on M. The arguments above are perfectly correct if the gauge fields take values in a complex Lie algebra. If a gauge field on Minkowski superspace takes values in a real Lie algebra, the corresponding field on P takes values in the complexification of this algebra and satisfies certain reality condition[1,2,5].

*/ For these relations to be accurate on has to express the operators entering it in the same coordinate system.

Let us show how to come with help of the above results to unconstrained formulations of N=2 and N=3 super gauge theories. For this purpose we notice that by means of a gauge transformation on P one can gauge out the fields $\mathcal{A}_\alpha$, $\mathcal{A}_{\dot\alpha}$. After that the partial gauge field is determined by the components $\mathcal{A}_s$ only. The condition of vanishing field strength implies then that

$$D_\alpha \mathcal{A}_s = 0\,, \qquad D_{\dot\alpha} \mathcal{A}_s = 0 \tag{6}$$

$$\bar\partial_s \mathcal{A}_t - \bar\partial_t \mathcal{A}_s + [\mathcal{A}_s, \mathcal{A}_t] = 0\,. \tag{7}$$

The restriction (6) for N=2, as well as for N=3, is analogous to chirality condition of superfields and can be solved in a similar manner. For this reason, we shall call superfields satisfying (6) isochiral superfields. Thus an isochiral superfield corresponds to an unconstrained superfield defined in terms of some new coordinates. The space with these new coordinates will be denoted by $\mathcal{P}$. In the case of N=2, when $\mathcal{A}_s$ has a single component $\mathcal{A}_1 \equiv \mathcal{A}$, the condition (7) is satisfied identically. This means that $\mathcal{A}$ can be regarded as an unconstrained superfield. In the case of N=3, it turns out to be possible to construct an action functional for which eqs. (7) are just the equations of motion. The Lagrangian can be written as follows

$$\varepsilon^{rst}\, \mathrm{tr}\left(\mathcal{A}_r \bar\partial_s \mathcal{A}_t + \tfrac{2}{3} \mathcal{A}_r \mathcal{A}_s \mathcal{A}_t \right), \tag{8}$$

where $\mathcal{A}_1, \mathcal{A}_2, \mathcal{A}_3$ are arbitrary isochiral superfields. Let us notice that the field $\mathcal{A}$ for N=2, as well as the fields $\mathcal{A}_1, \mathcal{A}_2, \mathcal{A}_3$ for N=3, were defined in terms of local coordinates z^s. This means that we have either to use a number of coordinate charts, or, if we choose a single chart which covers almost the whole manifold, to impose certain boundary condition on the fields considered. The transformation law of the fields for passing from one chart to another can be read off from the fact that the expression $\mathcal{A}_s d\Xi^s$ can be considered as a differential 1-form. It may be more convenient, however, to pass to homogeneous coordinates and expand differential forms in the following basis of 1-forms

$$\omega = |p|^{-4} \varepsilon^{ij} \bar{p}_j d\bar{p}_i \ ,$$

for N=2, and

$$\begin{aligned} \omega^1 &= |p|^{-4} \varepsilon^{ijk} \bar{p}_i u_j d\bar{p}_k \ , \\ \omega^2 &= |u|^{-4} \varepsilon_{ijk} p^i \bar{u}^j d\bar{u}^k \ , \\ \omega^3 &= |u|^{-2} |p|^{-2} \bar{u}^i d\bar{p}_i \ , \end{aligned} \tag{9}$$

for N=3. (Here $|p|^2 = p^i \bar{p}_i$, $|u|^2 = u_i \bar{u}^i$). Then for N=2 the field $\mathcal{B}$ defined via $\mathcal{A} d\Xi = \mathcal{B}\omega$ is a function homogeneous of degree two in the variables p^1, p^2. The field $\mathcal{B}$ can be used conveniently as an unconstrained superfield for N=2 gauge theory. It coincides with an analogous superfield introduced in[1]. For N=3, defining $\mathcal{B}_1$, $\mathcal{B}_2$, $\mathcal{B}_3$ as coefficients of decomposition of $\mathcal{A}_s d\Xi^s$ in the basis of $\omega^1, \omega^2, \omega^3$ we obtain functions which have with respect to the variables p^i, u_i the degrees of homogeneity (2,-1), (-1,2), (1,1) respectively. The fields $\mathcal{B}_s$ represent unconstrained superfields for N=3 and do not differ essentially from unconstrained superfields of ref.[2].

In order to characterize the measure appropriate for integrating the Lagrangian (8), let us remark on integration over a homogeneous manifold $\mathcal{N} = G/H$. For H, the isotropy subgroup of a point $p \in \mathcal{N}$, one can define a one-dimensional representation V of this group by relating to every element of H the determinant of the corresponding linear transformation of the tangent space at p. (In the case of superspace, of course, the determinant must be replaced by the Berezinian). An invariant measure on G/H can obviously be defined only if this representation is trivial. If this is not the case, however, and one cannot define an invariant integral for scalar functions over $\mathcal{N}$, one can still define it for the fields that transform according to a representation of H contragradient to V. It can be said that there is in G/H a G invariant integration with a measure which transforms according to the representation V of the group H. It will be important for us to consider the case, when H is a direct product of r copies of the group $U(1)$, and a part of coordinates in the tangent space are taken to be complex*/. In such a case the measure is complex as well, while the representation V is determined by an integer vector $m = (m_1, \ldots, m_r)$ called the weight of the representation. It is easy to see, that $m = \sum_\alpha v_\alpha - \sum_\beta v_\beta$, where v_α and v_β are the weights of respectively even and odd one-dimensional subrepresentations in the decomposition of the representation of H in the tangent superspace.

The superspace $\mathcal{P}$ on which the unconstrained superfields live can be thought of as a homogeneous manifold with respect to the action of the direct product of $SU(N)$ and complexified super Poincaré group. To solve the constraints (6) in case N=2 one has to choose coordinates $y^a, \Theta^\alpha, \Theta^{\dot\alpha}, \tilde\Theta^\alpha, \tilde\Theta^{\dot\alpha}, p^i$ in $M^c \times \mathbb{C}P^1$ (M^c being the complexified Minkowski superspace) in such a way that the operators D_α and $D_{\dot\alpha}$ of eqs. (4) become $\partial/\partial\tilde\Theta^\alpha$ and $\partial/\partial\tilde\Theta^{\dot\alpha}$ respectively. (Explicit expressions which can be found in[1,7]) will not be used here). Then the unconstrained superfield $\mathcal{B}$ can be regarded as depending only on $y^a, \Theta^\alpha, \Theta^{\dot\alpha}, p^i, \bar p_i$. These coordinates are to be considered as coordinates in $\mathcal{P}$. Near the point $x^a = \theta^{\alpha i} = \bar\theta^{\dot\alpha}{}_i = 0$, $(p^1, p^2) = (0,1)$ in the space $M^c \times \mathbb{C}P^1$, the operators $D_\alpha, D_{\dot\alpha}$ look like $\partial/\partial\theta^{\alpha 2} + \ldots$,

*/Strictly speaking, this means that we are dealing with a CR-structure on the manifold, see, e.g.[7].

$\partial/\partial\bar\theta^{\dot\alpha}_1 + \ldots$. Therefore, near this point, one can choose $y^a = x^a + \ldots$, $\Theta^\alpha = \theta^{\alpha 1} + \ldots$, $\Theta^{\dot\alpha} = \bar\theta^{\dot\alpha}_2 + \ldots$, $\tilde\Theta^\alpha = \theta^{\alpha 2} + \ldots$, $\tilde\Theta^{\dot\alpha} = \bar\theta^{\dot\alpha}_1 + \ldots$. The isotropy subgroup of the point in view contains a factor $U(1)$, which is the subgroup of $SU(2)$ that leaves the vector $(p^1, p^2) = (0, 1)$ invariant. One can see that every coordinate among in tangent space has the weight equal to one with respect to this $U(1)$ group. This implies that we are able to integrate the functions on $\mathcal{P}$ of weight four, that is the functions which are homogeneous of degree four. In other words, the weight of the measure equals minus four. This measure will be used later on.

Analogously, for N=3, to solve eqs. (6) one has to choose coordinates $y^a, \Theta^\alpha, \Theta^{\dot\alpha}, \Phi^\alpha, \Phi^{\dot\alpha}, \tilde\Theta^\alpha, \tilde\Theta^{\dot\alpha}, p^i, u_i$ in $M^c \times Q$ such that the operators $D_\alpha, D_{\dot\alpha}$ become $\partial/\partial\tilde\Theta^\alpha, \partial/\partial\tilde\Theta^{\dot\alpha}$. Near the point $x^a = \theta^{\alpha i} = \bar\theta^{\dot\alpha}_i = 0$, $(p^i) = (0,0,1)$, $(u_i) = (1,0,0)$ one can choose $y^a = x^a + \ldots$, $\Theta^\alpha = \theta^{\alpha 1} + \ldots$, $\Theta^{\dot\alpha} = \bar\theta^{\dot\alpha}_3 + \ldots$, $\Phi^\alpha = \theta^{\alpha 2} + \ldots$, $\Phi^{\dot\alpha} = \bar\theta^{\dot\alpha}_2 + \ldots$, $\tilde\Theta^\alpha = \theta^{\alpha 3} + \ldots$, $\tilde\Theta^{\dot\alpha} = \bar\theta^{\dot\alpha}_1 + \ldots$. The weights of coordinates $\Theta^\alpha, \Theta^{\dot\alpha}, \Phi^\alpha, \Phi^{\dot\alpha}, \tilde\Theta^\alpha, \tilde\Theta^{\dot\alpha}$ with respect to a $U(1) \times U(1)$ subgroup of $SU(3)$ preserving $(p^i) = (0,0,1)$ and $(u_i) = (1,0,0)$ are respectively as follows (0,1), (1,0), (1,-1), (-1,1), (-1,0), (0,-1). Therefore one is able to define an invariant integral for functions which depend on the variables $y^a, \Theta^\alpha, \Theta^{\dot\alpha}, \Phi^\alpha$, $\Phi^{\dot\alpha}, p^i, \bar p_i, u_i, \bar u^i$ (i.e. for functions on $\mathcal{P}$) and have the weight (2,2) with respect to the isotropy subgroup $U(1) \times U(1)$. That is to say, a measure, $d\mu$, of weight (-2,-2) is defined on $\mathcal{P}$.

The object to be integrated with this measure in order to obtain the action functional can be given by the expression

$$\mathrm{tr}\,(\mathcal{A} \wedge \bar{\mathcal{D}}\mathcal{A} + \tfrac{2}{3}\,\mathcal{A} \wedge \mathcal{A} \wedge \mathcal{A}) \equiv \varepsilon^{rst}\,\mathrm{tr}\,(\mathcal{A}_r \bar{\mathcal{D}}_s \mathcal{A}_t + \tfrac{2}{3}\,\mathcal{A}_r \mathcal{A}_s \mathcal{A}_t)\, d\bar z^1 d\bar z^2 d\bar z^3, \tag{10}$$

where $\mathcal{A} = \mathcal{A}_s d\bar{z}^s$. The operator $\bar{\mathcal{D}} = d\bar{z}^s \bar{\mathcal{D}}_s$ in (10) is the operator, that was obtained from $\bar{\partial} = d\bar{z}^s \partial/\partial\bar{z}_s$ after passing from isochiral superfields on P to unconstrained superfields on $\mathcal{P}$. The expression (10) can be viewed on as a function of weight (2,2) and, hence, can be integrated with the measure of weight (-2,-2). This follows from the fact that the representation of the group $U(1) \times U(1)$ in the space of forms on Q depending linearly on $d\bar{z}^s$ can be decomposed into one-dimensional representations of weights (2,-1), (-1,2) and (1,1). Consequently, the form $\hat{\mathcal{L}} = \mathcal{L}\, d\bar{z}^1 d\bar{z}^2 d\bar{z}^3$ can be regarded as a field of weight (2,-1)+(-1,2)+(1,1)=(2,2). In other words, one can associate to $\hat{\mathcal{L}}$ a function $\tilde{\mathcal{L}}$, which has the degree of homogeneity equal to (2,2), and then write the action as

$$S = \int d\mu\, \tilde{\mathcal{L}} + \text{c.c.}$$

This function $\tilde{\mathcal{L}}$ of homogeneity (2,2) can be found from the relation $\hat{\mathcal{L}} = \mathcal{L}\, d\bar{z}^1 d\bar{z}^2 d\bar{z}^3 = \tilde{\mathcal{L}}\, \omega^1 \omega^2 \omega^3$, where ω^s are the 1-forms defined in eqs. (9). This form of the Lagrangian, $\tilde{\mathcal{L}}$, can be conveniently expressed in terms of unconstrained superfields $\mathcal{B}_s$ introduced above (the explicit expression can be found in refs. [2,5]).

To conclude, let us point out that the conversion of N=3 super gauge theory to a theory in superspace with auxiliary dimensions remains an equivalence only in perturbation theory. Indeed, to prove such equivalence we used strongly the fact that a field on Q satisfying eqs.(7) can be gauged out. As we mentioned, this is true for sufficiently small fields. It can be proved, however, that there exist on Q such fields satisfying (7) that are not pure gauges. This takes place for N=2, as well as for N=3. In the case of N=3, however, a stronger statement is valid: there exists such a field $\mathcal{A}_s$ satisfying (7) on Q that there are no fields sufficiently near to it that can be gauged out[5].

3. HYPERMULTIPLET AND HYPER-KÄHLER SIGMA MODELS

It was pointed out in ref.[3] that the equations of motion on the superfield ϕ^i, $i=1,2$, corresponding to free N=2 hypermultiplet can be written in the following form

$$D_\alpha \Phi = 0 , \quad D_{\dot\alpha} \Phi = 0 , \tag{11}$$

where $\Phi = p_i \phi^i$, and the operators $D_\alpha, D_{\dot\alpha}$ are the same as in eqs. (4). Consequently, the hypermultiplet can be described on shell by a superfield on the superspace P which satisfies eqs. (11), is holomorphic in p^1, p^2 and has the degree of homogeneity equal to one with respect to these variables (as every homogeneous holomorphic function of degree one in p^1, p^2 is linear).

In ref.[1] it was shown that using this conversion of the hypermultiplet into a superfield on P one can construct an unconstrained off-shell formulation for this theory. In fact, as we have already mentioned, the fields satisfying (11) (i.e. isochiral superfields) can be interpreted as unconstrained superfields on P. Furthermore, the condition on the superfield $\phi(x^a, \theta^{\alpha i}, \bar\theta^{\dot\alpha}_i, p^i, \bar p_i)$ to be holomorphic with respect to p^1, p^2 can be represented by the equation

$$D\Phi = 0 , \tag{12}$$

where

$$D = |p|^2 \varepsilon_{ij} p^j \frac{\partial}{\partial \bar p_i} . \tag{13}$$

Lastly, eq. (11) can be derived as an equation of motion from the action[1]

$$S = \int d\mu \, \tilde\Phi \mathcal{D} \Phi + \text{c.c.} \tag{14}$$

where we have passed to superfields on $\mathcal{P}$ and $d\mu$ is the measure of weight minus four on $\mathcal{P}$ described in the preceding section. Correspondingly, the operations

$\phi \to \mathcal{D}\phi$ and $\phi \to \tilde{\phi}$ for superfields on $\mathcal{P}$ result from the following operations on isochiral superfields on P, namely, from $\phi \to D\phi$ as defined by eq.(13) and from an operation which consists in making complex conjugation of ϕ and simultaneously substituting $(p^1, p^2) \to (\bar{p}_2, -\bar{p}_1)$. It is easy to check that the integrand in eq.(14) has the weight equal to four as required.

Alternatively, in formulating the superfield theory for the hypermultiplet one could start from two isochiral superfields ϕ^1, ϕ^2 with degree of homogeneity equal to one. Then one could write the Lagrangian as $\phi^2\mathcal{D}\phi^1 - \phi^1\mathcal{D}\phi^2$ and require the fields to satisfy a reality condition, $\widetilde{\phi^1} = \phi^2$. Such a reformulation, although trivial, is convenient, as it allows to pass to fields of homogeneity zero (i.e. to ordinary functions) and then to consider the linear hypermultiplet as a simplest hyper-Kähler sigma model.

Let us proceed now to a general supersymmetric hyper-Kähler sigma model. To this end, let us remind that for this theory one can consider the superfields in N=2 Minkowsky superspace which take values in a hyper-Kähler manifold. The hyper-Kähler manifold is a 4n-dimensional Riemannian manifold $\mathcal{H}$, the holonomy group of which is contained in $Sp(n)$. On the hyper-Kähler manifold $\mathcal{H}$ with coordinates φ^m, $m = 1, \ldots, 4n$, one can choose a vielbein field $e^m_{ia}(\varphi)$, $i = 1, 2$; $a = 1, \ldots, 2n$, in such a way, that $e^m_{ia} = \varepsilon_{ij}\Omega_{ab}(e^m_{jb})^*$ and the Riemann connection satisfies: $\omega_{m\,ia}^{\;\;jb} = \omega_{ma}^{\;\;b}\,\delta^j_i$. Here Ω_{ab} is some constant non-degenerate antisymmetric matrix, while $\omega_{ma}^{\;\;b}$ represent for each m a matrix which belongs to the Lie algebra of $Sp(n)$, the subgroup of $SU(2n)$ preserving Ω_{ab}. The equations of motion for this sigma model can be written in terms of superfields φ^m on N=2 Minkowski superspace as follows[8])

$$e^{ia}_m(\varphi) D_{\alpha j} \varphi^m - \frac{1}{2}\delta^i_j e^{ka}_m D_{\alpha k} \varphi^m = 0 ,$$
$$e^{ia}_m(\varphi) D_{\dot\alpha}{}^j \varphi^m + e^{ja}_m(\varphi) D_{\dot\alpha}{}^i \varphi^m = 0 , \tag{15}$$

where e^{ia}_m is the inverse vielbein.
These equations can be obviously rewritten in the following form (using the notation $p_i = \varepsilon_{ij} p^j$, while $\bar{p}_i \equiv (p^i)^*$)

$$p_i e^{ia}_m(\varphi) D_\alpha \varphi^m = 0 ,$$
$$p_i e^{ia}_m(\varphi) D_{\dot\alpha} \varphi^m = 0 . \tag{16}$$

Till now we considered the field φ^m as a field on Minkowsky superspace M. From now on let us assume that the field φ^m is defined on $P = M \times \mathbb{C}P^1$, the parameters p^i being the homogeneous coordinates in $\mathbb{C}P^1$. (That is to say, φ^m can depend now not only on $x, \theta, \bar\theta$, but also on a point which runs over $\mathbb{C}P^1$.) In doing so, let us add to equations (16) the following equations

$$p_i e^{ia}_m(\varphi) D \varphi^m = 0 \tag{17}$$

and a reality condition which requires that φ^m does not change under the substitution $(p^1, p^2) \to (\bar{p}_2, -\bar{p}_1)$. It can be proved that eqs. (17) together with the above reality condition imply that φ^m does not depend on $p^i, \bar{p}_i$ (This is true for those fields at least which are sufficiently close to functions constant along $\mathbb{C}P^1$.) That is to say, the equations (16), (17) supplied with the reality condition on P are equivalent (in perturbation theory at least) to the original field equations (15) on M.

In analogy with the case of a linear hypermultiplet, let us consider eqs. (16) and reality condition as kinematical constraints on P. Then eqs. (17) can be seen to be equations of motion corresponding to an action functional. The kinematical constraints (16) can be given the form of isochirality conditions after a coordinate change in $P \times \mathcal{H}$. In fact, the real coordinates

φ^m, $m=1,\dots,4n$, must be replaced by complex coordinates $\phi^a=\phi^a(\varphi^m, p^i, \bar{p}_i)$, $a=1,\dots,2n$, which can be found as solutions of the system

$$p^i e^m_{ib}(\varphi)\frac{\partial\phi^a}{\partial\varphi^m} = 0\ . \qquad (18)$$

It follows from the properties of hyper-Kähler manifolds, that the operators $p^i e^m_{ib}(\varphi)\partial/\partial\varphi^m$ form for each (p^i) a closed algebra. Consequently, by virtue of the Frobenius theorem, the system of differential equations (18) has just $2n$ independent complex solutions. It is straightforward to see that after passing to coordinates ϕ^a equations (16) acquire the form of $D_\alpha\phi^a = D_{\dot\alpha}\phi^a = 0$. Thus the fields ϕ^a can be considered as unconstrained superfields.

It remains to specify the action functional for which the Euler-Lagrange equations are equivalent to eqs.(17). Let us look for such a functional starting from the condition that its variation satisfies

$$\delta S = \int d\mu\ \delta\phi^a R_{ab} L^b + \text{c.c.} \qquad (19)$$

where R_{ab} are some functions, while L^b is the left hand side of equations resulting from eqs.(17) rewritten in terms of the new variables ϕ^a . The functions R_{ab} in (19) must be found from the requirement that the variation (19) can be integrated to a functional S. The expression (19) can be interpreted as a 1-form on the infinite dimensional space of fields. For a (may be multivalued) functional S to exist, this 1-form must be closed. In order to pick up appropriate functions R_{ab}, let us define the functions Ω_{mn} as follows

$$\Omega_{mn} = p_i p_j e^{ia}_m e^{jb}_n \Omega_{ab}\ . \qquad (20)$$

It is easy to verify that for each (p^i) the functions $\Omega_{mn}(\varphi,p)$ define on $\mathcal{H}$ a closed 2-form, $\Omega = \Omega_{mn}\, d\varphi^m d\varphi^n$. Using this one can check that the expression

$$\delta S = \int d\mu \, \delta\varphi^m \Omega_{mn} D\varphi^n + c.c. \quad (21)$$

is a closed 1-form on the space of the fields φ^m satisfying (16), that is on the space of isochiral superfields ϕ^a. The integrand in eq.(21) has the degree of homogeneity equal to four and can be, thus, integrated with the measure $d\mu$ of weight minus four. To be more precise, before integrating with $d\mu$ one has to pass to the space $\mathcal{P}$ as required by the solution of isochirality conditions.

After the variation of the action is known one can restore the whole functional by integrating along a curve in the space of fields. The result will be independent of the choice of this curve if the 1-form is exact. Otherwise, if this 1-form is only closed, the resulting functional may be multivalued. Thus, we assume that the field φ^m depends also on an auxiliary parameter τ, $0 \leq \tau \leq 1$, provided that $\varphi^m|_{\tau=0}$ is kept fixed. Then we have

$$S = \int d\mu \, d\tau \, \dot{\varphi}^m \Omega_{mn} D\varphi^n + c.c. \quad (22)$$

Let us remind, that the fields under consideration must satisfy the reality condition. Finally, the action (22) generalizes the free action (14) of the linear hypermultiplet.

Let us describe briefly some geometrical features of the above considerations. The crucial point is that every hyper-Kähler manifold $\mathcal{H}$ possesses a family of complex structures parametrized by the points of $\mathbb{C}P^1$. Using this, one can define a natural complex structure on $\mathbb{C}P^1 \times \mathcal{H}$. On the other hand, there is in the space P a natural CR-structure defined by the operators $D_\alpha, D_{\dot\alpha}, D$. Using this CR-structure and complex structure in $\mathbb{C}P^1 \times \mathcal{H}$, one can introduce a CR-structure in $P \times \mathcal{H}$, thus making it a CR-manifold. With respect to this structure the projection $P \times \mathcal{H} \to P$ turns out to be a (non-linear)

CR-bundle. The field φ^m subjected to eqs.(13) and (14) can be considered then as a CR-section of this bundle. Thus the analysis of fields satisfying the equations of motion amounts to the analysis of CR-sections obeying certain reality condition. To make contact with the description of the linear hypermultiplet let us note that the corresponding field entering eqs.(11) and (12) had the degree of homogeneity equal to one on P. Such functions on P can be interpreted naturally as sections of a one-dimensional vector CR-bundle E over the CR-manifold P. Then the superfields satisfying (11) and (12) correspond to CR-sections of E[3,5] analogously to the case of a general hyper-Kähler sigma model.

REFERENCES

1. Galperin, A., Ivanov, E., Kalitzin, S., Ogievetsky,V. and Sokatchev, E., Classical and Quantum Gravity 1, 469 (1984).

2. Galperin, A., Ivanov,E., Kalitzin,S., Ogievetsky,V. and Sokatchev,E., JINR preprint E2-84-441, Dubna (1984).

3. Rosly,A., in: Group Theoretical Methods in Physics. Proc. Int.Seminar, Zvenigorod 1982, Ed. M.A.Markov, Moscow: Nauka, 1983, vol. 1, p.263.

4. Malgrange,B., Lectures on the theory of several complex variables. Bombay, 1958.

5. Rosly,A. and Schwarz,A., ITEP preprint (1985).

6. Ward,R.S., Phys. Lett. 61A, 81 (1977).

7. Penrose,R., Gen.Relat. Grav. 7, 31 (1976).

8. Sierra, G. and Townsend, P.K., Nucl. Phys. B233, 289 (1984).

GEOMETRY OF 11-DIMENSIONAL SUPERGRAVITY

R.E Kallosh
P.N. Lebedev Physics Institute

The new geometrical form of d=11 supergravity Lagrangian is suggested. This Lagrangian depends on vielbein e_μ^m , on supercovariant spin connection $\hat{\omega}_\mu{}^{mn}$ and on gravitino ψ_μ^α. The 3-index gauge potential $A_{\mu\nu\lambda}(e,\hat{\omega},\psi)$ in this theory is a supercovariant torsion tensor. A comment to the Bars-Higuchi recent treatment of the same problem is presented.

The 11-dimensional supergravity [1] attracts much attention since there is some hope [2,3] that the fundamental theory is connected with the spontaneously compactified theory [1].

The formulation of 11-dimensional supergravity which will be presented, is motivated by the desire to understand better the geometry of this theory and, in particular, the origin of non-Riemannian geometry with torsion on S_7 discovered by Englert [3]. Usually d=11 supergravity is described by Cremmer-Julia-Scherk [1] (CJS) Lagrangian, which depends on vielbein e_μ^m, gravitino ψ_μ^α and antisymmetric three-index tensor $A_{\mu\nu\lambda}$, all three variables being independent. The geometrical role of vielbein is clear, the field ψ_μ^α also becomes a geometrical object - the component of the vielbein- when the theory is formulated in superspace [4]. However, the geometrical meaning of $A_{\mu\nu\lambda}$ in the CJS theory, as well as in superspace formulation [4], where $A_{\mu\nu\lambda}$ is some component of the superspace 3-form $A = dx^M dx^N dx^L A_{MNL}$ $(M=\mu_1 \ldots \mu_{11}, \alpha_1 \ldots \alpha_{32})$ remained unclear. Therefore the main purpose of geometrical formulation of d=11 supergravity which will be presented[6], is to clear up the geometrical meaning of $A_{\mu\nu\lambda}(x)$, as it was already done by Bars and MacDowell [5] in the bosonic sector of supergravity.

1. We begin by the presentation of the main result [6]: the geometrical Lagrangian of d=11 supergravity and its symmetries.

The Lagrangian $\mathcal{L}_{d=11}$ which depends on vielbein $e_\mu{}^m$, on supercovariant spin connection $\hat{\omega}_\mu{}^{mn}$ and on gravitino can be presented as follows [x)]

$$\mathcal{L}_{d=11}(e,\hat{\omega},\psi) = \mathcal{L}_{CJS}(e,\hat{\omega},\psi, A_{\mu\nu\lambda} \equiv \hat{T}_{[\mu\nu,\lambda]}(e,\hat{\omega},\psi)) + \frac{e}{2}\hat{T}^2_{[\mu\nu,\lambda]}(e,\hat{\omega},\psi) \quad , \tag{1}$$

where $\hat{T}_{\mu\nu,\lambda}$ is the supercovariant torsion. The torsion $T_{\mu\nu}{}^m$, which by definition depends on vielbein and connection, has the following form

$$T_{\mu\nu}{}^m(e,\hat{\omega}) \equiv \partial_{[\mu} e_{\nu]}{}^m + \hat{\omega}_{[\mu}{}^m{}_n e^n{}_{\nu]} = \mathcal{D}_{[\mu}(\hat{\omega}) e_{\nu]}{}^m . \tag{2}$$

The supercovariant torsion differs from (2) by additional non-linear terms depending on gravitino

$$\hat{T}_{\mu\nu}{}^m(e,\hat{\omega},\psi) = T_{\mu\nu}{}^m(e,\hat{\omega}) - \frac{1}{4}\bar{\psi}_{[\mu}\gamma^m\psi_{\nu]} \ . \tag{3}$$

The antisymmetric tensor $A_{\mu\nu\lambda}(e,\hat{\omega},\psi)$ coincides with the antisymmetrized torsion.

$$A_{\mu\nu\lambda}(e,\hat{\omega},\psi) \equiv \hat{T}_{[\mu\nu,\lambda]}(e,\hat{\omega},\psi) = \hat{T}_{[\mu\nu}{}^m(e,\hat{\omega},\psi) e_{m\lambda]} . \tag{4}$$

The first term in the r.h.s. of eq. (1) is the CJS Lagrangian in the somewhat unusual form. In the paper [1], the

---

x) Our notation are that of [7].

Lagrangian is presented as some function $\mathcal{L}(e,\omega,\hat{\omega},\psi,A)$, where the connection $\omega_\mu{}^{mn}$ and the supercovariant connection $\hat{\omega}_\mu{}^{mn} = \omega_\mu{}^{mn} + \frac{1}{8}\bar{\psi}_\xi \Gamma^{\xi\mu mn\lambda}\psi_\lambda$ should be taken as the functions of e and ψ, according to the equation, which shows that the supercovariant torsion of the theory is equal to zero

$$\hat{T}_{\mu\nu}{}^{m}(e,\hat{\omega},\psi) = 0 . \tag{5}$$

The solution of this equation is

$$\hat{\omega}_{\mu mn}(e,\psi) = \overset{\circ}{\omega}_{\mu mn}(e) + \frac{1}{4}\left(\bar{\psi}_\mu \gamma_m \psi_n - \bar{\psi}_\mu \gamma_n \psi_m + \bar{\psi}_m \gamma_\mu \psi_n\right). \tag{5'}$$

Instead of that in our approach one should take the original expression $\mathcal{L}(e,\omega,\hat{\omega},\psi,A)$ from [1], represent this expression as a function of e, $\hat{\omega}$, ψ, i.e. to replace ω by $\omega = \hat{\omega} - \frac{1}{8}\bar{\psi}\Gamma^{(5)}\psi$ and to replace an independent variable $A_{\mu\nu\lambda}$ by the function $\hat{T}_{[\mu\nu,\lambda]}(e,\hat{\omega},\psi)$, $\hat{\omega}$ being an independent variable.

The explicit form of $\mathcal{L}_{d=11}(e,\hat{\omega};\psi)$ is the following

$$\begin{aligned}
\mathcal{L}_{d=11}(e,\hat{\omega},\psi) = & -\frac{e}{2}R(e,\hat{\omega}) + \frac{e}{2}\hat{T}^2_{[\mu\nu,\lambda]}(e,\hat{\omega},\psi) - \\
& -\frac{e}{2}\bar{\psi}_\mu \Gamma^{\mu\rho\sigma}\mathcal{D}_\rho(\hat{\omega})\psi_\sigma + \frac{e}{8}\hat{T}_{\mu\nu,\lambda}(e,\hat{\omega},\psi)\bar{\psi}_\rho \Gamma^{\rho\mu\nu\lambda\sigma}\psi_\sigma + \\
& +\frac{e}{64}\bar{\psi}_\mu \Gamma^{\mu\nu\lambda\sigma\eta}\psi_\eta \bar{\psi}_\nu \gamma_\lambda \psi_\sigma - \frac{e}{48}F^2_{\mu\nu\lambda\delta}(e,\hat{\omega},\psi) - \\
& -\frac{i\sqrt{2}}{36\cdot 96}\varepsilon^{\mu_1\ldots\mu_{11}}F_{\mu_1\ldots\mu_4}F_{\mu_5\ldots\mu_8}\hat{T}_{\mu_9\mu_{10}\mu_{11}}(e,\hat{\omega},\psi) - \\
& -\frac{\sqrt{2}e}{384}\left(\bar{\psi}_\rho \Gamma^{\rho\mu\nu\lambda\delta\xi}\psi_\xi + 12\,\bar{\psi}^\mu \Gamma^{\nu\lambda}\psi^\delta\right)\left(F_{\mu\nu\lambda\delta} + \hat{F}_{\mu\nu\lambda\delta}\right)(e,\hat{\omega},\psi),
\end{aligned} \tag{I'}$$

where

$$F_{\mu\nu\lambda\delta}(e,\hat{\omega},\psi) = \partial_\mu \hat{T}_{\nu\lambda,\delta} + 23\ \text{terms} ,$$

$$\hat{F}_{\mu\nu\lambda\delta}(e,\hat{\omega},\psi) = F_{\mu\nu\lambda\delta} + \frac{3}{2}\,\bar{\psi}_{[\mu}\,\Gamma_{\nu\lambda}\,\psi_{\delta]} \,. \tag{6}$$

The bosonic part of this Lagrangian coincides with the Bars-MacDowell one [5]. The action (1), (1') is invariant under the following transformations. The standard general covariance is present; the vielbein $e_\mu{}^m$, $e_m{}^\mu$ transform tangent space indices $(m, n, \ldots)$ to coordinate basis indices $(\mu, \nu, \ldots)$ and vice versa. The Lorentz structure group acts as usual in the tangent space. An extra gauge transformation of the spin connection, which mixes tangent and base indices, has the form [5]

$$\delta_\Sigma e_\mu{}^m = \delta_\Sigma \psi_\mu{}^\alpha = 0 \,, \quad \delta\hat{\omega}_\mu{}^{mn} = e^{\nu m} e^{\lambda n}\,\partial_{[\mu} \Sigma_{\nu\lambda]} \,. \tag{7}$$

The torsion and curvature, which are tensors in the tangent space under the Lorentz transformations, are not tensors under the transformations (7). However, $F_{mnet} = e_m{}^\mu \ldots e_t{}^\delta F_{\mu\ldots\delta} + \text{sym.}$ has the tensor properties under the Lorentz transformations as well as under the extra gauge transformations (7).

The supersymmetry transformation laws are given by:

$$\delta_\varepsilon e_\mu{}^m = \frac{1}{2}\,\bar{\varepsilon}\,\Gamma^m \psi_\mu \,, \tag{8a}$$

$$\delta_\varepsilon \psi_\mu = \mathcal{D}_\mu(\hat{\omega})\bar{\varepsilon} - \frac{1}{4}\hat{T}_{[\mu\nu,\lambda]}(e,\hat{\omega},\psi)\,\Gamma^{\nu\lambda}\bar{\varepsilon} + T^{\lambda\delta\eta\xi}{}_\mu\,\bar{\varepsilon}\,\hat{F}_{\lambda\delta\eta\xi}(e,\hat{\omega},\psi) , \tag{8b}$$

$$\delta_\varepsilon \hat{\omega}_\mu{}^{ab} e_{\nu a} e_{\lambda b} = \bar{\varepsilon}\, g_{\mu[\nu}\Gamma_{\lambda]}{}^{\tau\sigma}\hat{\psi}_{\tau\sigma} - \frac{1}{18}\bar{\varepsilon}\, g_{\mu[\nu}\Gamma_{\lambda]}{}^{\rho\eta\tau\sigma} \times T_{\rho\eta\tau}(e,\hat{\omega})\psi_\sigma + \bar{\varepsilon}\, S_{\nu\lambda}{}^{abcd}\,\psi_\mu \hat{F}_{abcd} +$$

$$+\bar{\varepsilon}\left(\Phi_{\nu\lambda\mu} - \Phi_{[\nu\lambda\mu]} + \frac{1}{2}\bar{\varepsilon}\,\Gamma_{[\nu}\hat{\psi}_{\lambda\mu]} + \right.$$

$$+\frac{1}{2}\hat{T}_{[\mu m\nu]}(e,\hat{\omega},\psi)\bar{\varepsilon}\Gamma^{m}\psi_{\lambda} - \frac{1}{2}\hat{T}_{[\mu m\lambda]}\cdot \qquad (8c)$$

$$\cdot\,\bar{\varepsilon}\Gamma^{m}\psi_{\nu} + \frac{\sqrt{2}}{8}\bar{\varepsilon}\,\Gamma_{[\nu\lambda}\psi_{\mu]}$$

where the following notations are introduced:

$$\hat{\psi}_{[\tau\sigma]} = \mathcal{D}_{[\tau}(\hat{\omega})\psi_{\sigma]} - \frac{1}{4}\hat{T}_{[\tau\lambda,\delta]}\Gamma^{\lambda\delta}\psi_{\sigma]}$$

$$+ T^{\lambda\delta\eta\xi}{}_{[\tau}\psi_{\sigma]}\hat{F}_{\lambda\delta\eta\xi},$$

$$\Phi_{\nu\lambda\mu} = \Gamma^{\sigma}{}_{\nu\lambda}\hat{\psi}_{[\mu\sigma]} - \frac{1}{2}T_{\eta\tau[\nu}(e,\hat{\omega})\Gamma_{\lambda}{}^{\eta\tau\sigma}{}_{\mu]}\psi_{\sigma}$$

$$T^{\lambda\delta\eta\xi}{}_{\mu} = \frac{\sqrt{2}}{288}\left(\Gamma^{\lambda\delta\eta\xi}{}_{\mu} - 8\,\delta^{[\lambda}_{\mu}\Gamma^{\delta\eta\xi]}\right)$$

$$S_{\mu\nu}{}^{\lambda\delta\eta\xi} = -\frac{\sqrt{2}}{288}\left(\Gamma_{\mu\nu}{}^{\lambda\delta\eta\xi} + 24\,\delta^{[\lambda\delta}_{\mu\nu}\Gamma^{\eta\xi]}\right)$$

The chain rule transformation of the supercovariant torsion takes the following form:

$$\delta_{\varepsilon}A_{\mu\nu\lambda}(e,\hat{\omega},\psi) = -\frac{\sqrt{2}}{8}\bar{\varepsilon}\,\Gamma_{[\mu\nu}\psi_{\lambda]} \qquad (8d)$$

The alternative form of the supersymmetry transformation rules for the same Lagrangian is x)

$$\tilde{\delta}_\varepsilon e_\mu{}^m = \delta_\varepsilon e_\mu{}^m \tag{9a}$$

$$\tilde{\delta}_\varepsilon \psi_\mu = \delta_\varepsilon \psi_\mu - \frac{1}{4} \frac{\delta S(e,\hat{\omega},\psi)}{\delta \hat{\omega}_{[\mu,\nu\lambda]}} \Gamma_{\nu\lambda} \varepsilon \tag{9b}$$

$$\tilde{\delta}_\varepsilon \hat{\omega}_{\mu,\nu\lambda} = \delta_\varepsilon \omega_{\mu,\nu\lambda} + \frac{1}{4} \frac{\delta S(e,\hat{\omega},\psi)}{\delta \psi_\mu} \Gamma_{\nu\lambda} \varepsilon \tag{9c}$$

In this case the chain rule transformation of the supercovariant torsion is

$$\tilde{\delta}_\varepsilon A_{\mu\nu\lambda}(e,\hat{\omega},\psi) = -\frac{\sqrt{2}}{8} \bar{\varepsilon} \Gamma_{[\mu\nu} \psi_{\lambda]} - \frac{1}{4} \frac{\delta \bar{S}_I(e,\hat{\omega},\psi,A)}{\delta \psi_\mu} \Bigg|_{A=\hat{T}(e,\hat{\omega},\psi)} \Gamma_{\nu\lambda} \varepsilon \tag{9d}$$

The function $\bar{S}_I(e,\hat{\omega},\psi,A)$ will be defined later.

---

x) The last term in (9b) was missed in [6]. Note also the difference between $\frac{\delta S(e,\hat{\omega},\psi)}{\delta \psi_\mu}$ and $\frac{\delta \bar{S}_I(e,\hat{\omega},\psi,A)}{\delta \psi_\mu}$ in (9c),(9d).

2. The proof of the invariance of the theory (1), (1') under the transformations presented above proceeds as follows. We consider d = 11 supergravity in the first order formalism, as given by Castellani, Fre, Giani, Pilch and Nieuwenhuizen [8].

$$\mathcal{L}_I(e,\hat{\omega},\psi,A) = -\frac{e}{2}R(e,\hat{\omega}) - \frac{e}{2}\bar{\psi}_\mu \Gamma^{\mu\rho\sigma}\mathcal{D}_\rho(\hat{\omega})\psi_\sigma -$$
$$-\frac{e}{48}F^2_{\mu\nu\lambda\delta} - \frac{i\sqrt{2}}{36\cdot 96}\varepsilon^{\mu_1\ldots\mu_{11}}F_{\mu_1\ldots\mu_4}F_{\mu_5\ldots\mu_8}A_{\mu_9\mu_{10}\mu_{11}} - \tag{10}$$
$$-\frac{\sqrt{2}e}{384}\left(\bar{\psi}_\rho \Gamma^{\rho\mu\nu\lambda\delta\sigma}\psi_\sigma + 12\bar{\psi}^\mu\Gamma^{\nu\lambda}\psi^\delta\right)\left(F_{\mu\nu\lambda\delta} + \hat{F}_{\mu\nu\lambda\delta}\right) +$$
$$+\frac{e}{8}T_{\mu\nu}{}^{m}(e,\hat{\omega})\bar{\psi}_\rho\Gamma^{\rho\mu m\nu\sigma}\psi_\sigma - \frac{e}{64}\bar{\psi}_\mu\Gamma^{\mu\nu\lambda\delta\sigma}\psi_\sigma\bar{\psi}_\nu\Gamma_\lambda\psi_\delta .$$

The Lagrangian (10) is the function of independent variables e, $\hat{\omega}$, ψ, A and is invariant under symmetry transformations laws given in [8]. The transformation from (10) to the second (or 1.5) order formalism is realized as usual, since the connection $\hat{\omega}$ in (10) satisfies the non-propagating field equation (5). Replacing $\hat{\omega}$ in (10) by $\hat{\omega}(e,\psi)$ from (5') we obtain the usual theory [1] in the second order formalism. The useful observation is the following. The Lagrangian $\mathcal{L}_{CPGPN}(e,\hat{\omega},\psi,A)$ coincides with $\mathcal{L}_{CJS}(e,\hat{\omega},\psi,A)$, i.e. the CJS Lagrangian taken as a function of independent variables e, $\hat{\omega}$, ψ and A (but not ω, as was stressed already) is exactly $\mathcal{L}_I$ (10) of the first order formalism.

Consider further

$$\overline{\mathcal{L}}_I(e,\hat{\omega},\psi,A) = \mathcal{L}_{CJS}(e,\hat{\omega},\psi,A) +$$
$$+\hat{T}_{\mu\nu,\lambda}(e,\hat{\omega},\psi)A_{\mu\nu\lambda} - \frac{e}{2}A^2_{\mu\nu\lambda} . \tag{11}$$

Just this Lagrangian, which is obtained from (10) by the simple change of variables

$$\hat{\omega}_\mu{}^{mn} \longrightarrow \hat{\omega}_\mu{}^{mn} + A_{\mu\nu\lambda} e^{\nu m} e^{\lambda n}, \quad (12)$$

is the appropriate starting point for the geometrical description of 11-dimensional supergravity. All variables are independent in $\bar{\mathcal{L}}_I(e,\hat{\omega},\psi,A)$ and it is invariant under the standard general covariance and Lorentz transformations. Besides, $\bar{\mathcal{L}}_I$ is invariant under the extra gauge transformations

$$\delta_\Sigma A_{\mu\nu\lambda} = \partial_{[\mu} \Sigma_{\nu\lambda]}, \quad \delta_\Sigma e_\mu{}^m = \delta_\Sigma \psi_\mu{}^\alpha = 0, \quad (13)$$
$$\delta_\Sigma \hat{\omega}_\mu{}^{mn} = e^{\nu m} e^{\lambda n} \partial_{[\mu} \Sigma_{\nu\lambda]}.$$

The supersymmetry transformation laws are obtained from the corresponding ones in [8] with account taken of change of variables (12). Varying the antisymmetric part of the connection $\hat{\omega}_{[\mu,\nu\lambda]}$ in (11) we get the equation of motion

$$\frac{\delta \bar{S}_I(e,\hat{\omega},\psi,A)}{\delta \omega_{[\mu,\nu\lambda]}} \equiv \hat{T}_{[\mu\nu,\lambda]}(e,\hat{\omega},\psi) - A_{\mu\nu\lambda} = 0. \quad (14)$$

There exist two possibilities to solve this equation. The first one is to express the antisymmetrized part of the connection through the elementary fields e, ψ, A.

$$\hat{\omega}_{[\mu,\nu\lambda]} = \overset{\circ}{\omega}_{[\mu,\nu\lambda]}(e) + \tfrac{1}{4}\bar{\psi}_{[\mu}\Gamma_\nu \psi_{\lambda]} + A_{\mu\nu\lambda}.$$

With account taken of (12) we are coming back to the conventional theory, depending on e, ψ, A and $\hat{\omega}_\mu{}^{mn} = \hat{\omega}_\mu{}^{mn}(e,\psi)$ (5') (taking into account also the equation $\hat{T}_{\mu(\nu,\lambda)} = 0$), i.e. to the standard [1] 11-dimensional supergravity theory. Another possibility is to use eq. (14) to identify the antisymmetric third rank tensor $A_{\mu\nu\lambda}$ with the torsion, leaving e, $\hat{\omega}$, ψ as independent variables. In this way we come to the theory (1), (1').

$$\overline{\mathcal{L}}_I(e,\hat{\omega},\psi,A)\Big|_{\frac{\delta \bar{S}_I}{\delta \hat{\omega}_{[\mu,\nu\lambda]}}=0} = \mathcal{L}_{d=11}(e,\hat{\omega},\psi) \qquad (15)$$

3. It is well known that the action $S[\varphi^i]$ is always invariant under the transformations of the type

$$\delta\varphi^i = \eta^{ij} S_{,j} \quad , \quad S_{,i} \equiv \frac{\delta S}{\delta \varphi^i} , \qquad (16)$$

since $\delta S = S_{,i}\,\eta^{ij} S_{,j} = 0$

(Here $\eta^{ij} = \pm\eta^{ji}$, + or - for fermions or bosons correspondingly). This property of the theory is quite interesting when we start from the first order formulation $\overline{\mathcal{L}}_I(e,\hat{\omega},\psi,A)$ and come to the theory $\mathcal{L}\Big|_{\frac{\delta \bar{S}_I}{\delta \hat{\omega}_{[\mu,\nu\lambda]}}=0}$ We can modify the supersymmetry transformation rules [8] according to (16) as follows

$$\delta e_\mu{}^m = \delta^I e_\mu{}^m \quad , \quad \delta A_{\mu\nu\lambda} = \delta^I A_{\mu\nu\lambda}$$

$$\delta \hat{\omega}_{[\mu,\nu\lambda]} = \delta^I \omega_{[\mu,\nu\lambda]} + \delta^I \frac{\delta \bar{S}_I(e,\hat{\omega},\psi,A)}{\delta \hat{\omega}_{[\mu,\nu\lambda]}}$$

$$\delta \hat{\omega}_{(\mu,\nu)\lambda} = \delta^I \omega_{(\mu,\nu)\lambda} + \frac{\delta \bar{S}_I(e,\hat{\omega},\psi,A)}{\delta \hat{\omega}_{[\lambda,s\eta]}} \times \qquad (17)$$

$$\times \frac{\delta\, \delta^I \hat{\omega}_{[\lambda,s\eta]}}{\delta\, \omega_{(\mu,\nu)\lambda}}$$

$$\delta \psi_\mu = \delta^I \psi_\mu + \frac{\delta \bar{S}_I(e,\hat{\omega},\psi,A)}{\delta \hat{\omega}_{[\lambda,s\eta]}} \frac{\delta\, \delta^I \psi_\mu}{\delta \hat{\omega}_{[\lambda,s\eta]}}$$

where by δ^{I} we denote the original supersymmetry transformation rules [8] with the change of variables (I2) taken into account.

Using eq. (I4) we get eqs. (8a)-(8d), i.e. the supersymmetry transformation of $A_{\mu\nu\lambda}(e,\hat{\omega},\psi)$ coincides with the supersymmetry transformation of the elementary field $A_{\mu\nu\lambda}$.

The alternative form of the supersymmetry transformations of the same theory $\mathcal{L}_I(e,\hat{\omega},\psi,A)$ in accordance with (I6) is

$$
\begin{aligned}
\tilde{\delta} e_\mu{}^m &= \delta e_\mu{}^m \\
\tilde{\delta} \omega_{\mu,\nu\lambda} &= \delta \omega_{\mu,\nu\lambda} \\
\tilde{\delta} A_{\mu\nu\lambda} &= \delta A_{\mu\nu\lambda} - \frac{1}{4} \frac{\delta \bar{S}(e,\hat{\omega},\psi,A)}{\delta \psi_{[\mu}} \Gamma_{\nu\lambda]} \varepsilon \\
\tilde{\delta} \psi_\mu &= \delta \psi_\mu + \frac{1}{4} \frac{\delta \bar{S}_I(e,\hat{\omega},\psi,A)}{\delta A_{\mu\nu\lambda}} \times \\
&\times \Gamma_{\nu\lambda} \varepsilon
\end{aligned}
\tag{I8}
$$

After using eq. (I4) we get the supersymmetry transformations (9a) - (9d) of the theory (I).

3. We present now the formulation of the bosonic sector of d=11 supergravity (1), (1') in terms of differential forms in x-space. The vielbein, connection, torsion and curvature forms are introduced in the usual way and they have the usual transformation law under the Lorentz group

$$E^m = dx^\mu E_\mu{}^m \,, \quad \omega^m{}_n = dx^\mu \omega_\mu{}^m{}_n \,, \tag{19}$$
$$D = d + \omega^{mn} M_{mn} \,,$$

where M_{mn} is the Lorentz group generator

$$T^m = DE^m \,, \tag{20}$$

$$R^m{}_n = d\omega^m{}_n + [\omega \wedge \omega]^m{}_n \,.$$

Curvature and torsion forms obey the Bianchi identities.

$$DT^m = R^m{}_n \wedge E^n \,, \tag{21}$$

$$DR^m{}_n = 0 \,.$$

The 3-form $A = dx^\mu dx^\nu dx^\lambda A_{\mu\nu\lambda}$ is the function of the already introduced geometrical forms

$$A = T^m \wedge E_m \,. \tag{22}$$

The extra gauge transformation is given by

$$\delta_\Sigma A = \delta_\Sigma (T^m \wedge E_m) = d\Sigma \,, \quad \delta_\Sigma E^m = 0 \,, \tag{23}$$

where Σ is the 2-form $dx^\mu dx^\nu \Sigma_{\mu\nu}$. The transformation (23) means that the connection $\hat{\omega}_\mu{}^{mn}$ transforms as in eq. (7). The 4-form invariant under (23) is given by

$$F = dA = DT^m \wedge E_m + T^m \wedge T_m \,. \tag{24}$$

Taking into account the Bianchi identities (21) we have

$$F = R^m{}_n \wedge E^n \wedge E_m + T^m \wedge T_m \quad , \tag{25}$$

i.e.

$$F_{mnlt} = R_{[mn,lt]} + T_{[mn}{}^{\kappa} T_{lt],\kappa} = \tag{26}$$

$$= D_{[m} T_{nl,t]} + T_{[mn}{}^{\kappa} T_{lt],\kappa} \quad .$$

Thus we see that the torsion and curvature are fundamental objects, as it should be in the geometrical theory, and antisymmetric tensor field strength is a function of them. The Bianchi identity for F

$$dF = 0$$

is the consequence of the definition (24), (25) and of the Bianchi identities (21) for T^m and $R^m{}_n$.

It would be important to construct the d=11 superspace geometry in accordance with the presented above x-space geometry, i.e. preserving eqs. of the type (19), (20). We stress that the already existing on-shell superspace geometry [4] is based on the antisymmetric tensor field strength through which the torsion and curvature tensors are expressed. In this theory [4] the supercurvature tensor $\hat{R}_{mn,lt}(x,\theta)$ has the symmetry properties of Riemannian geometry $\hat{R}_{[mn,l]t}(x,\theta) = 0$ (since at $\theta = 0$ it coincides with the x-space curvature of the CJS theory [1]), as distinct from our formulation, where just the antisymmetrized curvature tensor $\hat{R}_{[mn,lt]}$ is connected with the antisymmetric tensor field strength according to (23).

4. The main property of the presented d=11 supergravity geometry is the non-Riemannian connection, the curvature tensor $\hat{R}_{mn,lt}$ being connected (eq. (26)) with the torsion, the last being a dynamical

variable, satisfying the classical equation

$$\frac{\delta S_{d=11}(e,\hat{\omega},\psi)}{\delta\hat{\omega}_{[\mu,\nu\lambda]}} = \hat{T}_{[\mu\nu,\lambda]}(e,\hat{\omega},\psi) - \tag{27}$$

$$-\hat{T}_{[\mu\nu,\lambda]}(e,\hat{\omega},\psi) + \left.\frac{\delta S_{CJS}(e,\hat{\omega},\psi,A)}{\delta A_{\mu\nu\lambda}}\right|_{A_{\mu\nu\lambda}=\hat{T}_{[\mu\nu,\lambda]}(e,\hat{\omega},\psi)} = 0 .$$

The origin of Englert's solutions [3] is now quite clear. The parallelizability condition of S_7, $\hat{R}_{mn,\ell t} = 0$ at $m,n,\ell,t = 1,\dots,7$, applies just to the curvature tensor of the present formulation of 11-dimensional supergravity. It follows from (26) in particular that $F_{mn\ell p}$ $(m,\dots,p=1,\dots,7)$ is a bilinear function of torsion T_{mn}^{ℓ}, which is a solution of eq. (27), as it takes place in the solutions [3].

There is a hope that the geometrized formulation of 11-dimensional supergravity might be more suitable than the conventional one [1] for relaxing the constraints, i.e. for constructing the off-shell theory, and also for constructing the quantum theory near the classical background, corresponding to the spontaneously compactified solutions [2, 3]. As is well known, these solutions may possess or not possess the nontrivial torsion on 7-sphere. This fact makes it desirable to have such an object as torsion in the theory as a possible <u>property of space</u> even before solving equations of motion.

What kind of torsion will appear in the theory depends on the solution of dynamical classical (or better classical with quantum corrections) equations. Thus, instead of the conventional picture that in the curved space with Riemannian connection in d=11 an antisymmetric tensor field exists, which is required mainly to balance bosonic and fermionic degrees of freedom in eleven dimensions, we come to another picture: The d=11 space is characterized by the curvature and torsion, the antisymmetrized part of the torsion being a dynamical variable.

5. Very recently the paper by Bars and Higuchi "First order formulation and the geometrical interpretation of d=11 supergravity" [9] has been published. At the first sight there is an essential difference between the two approaches.

The basic equation of the Bars-Higuchi theory is the following

$$A_{\mu\nu\lambda}(e,\omega) = T_{\mu\nu,\lambda}(e,\omega) = \partial_{[\mu} e_{\nu}{}^{m} e_{m\lambda]} + \omega_{[\mu,\nu\lambda]}$$

which should be compared with our eq. (14)

$$A_{\mu\nu\lambda}(e,\hat{\omega},\psi) = \partial_{[\mu} e_{\nu}{}^{m} e_{m\lambda]} + \hat{\omega}_{[\mu,\nu\lambda]} - \tfrac{1}{4}\bar{\psi}_{[\mu}\Gamma_{\nu}\psi_{\lambda]}\,.$$

The spin connection $\omega_{\mu,\nu\lambda}$ in [9] is not supercovariant, $\delta\omega_{\mu,\nu\lambda} = \partial_{[\mu}\bar{\varepsilon}\gamma_{\nu}\psi_{\lambda]} + \ldots$ But since we work in the first order formulation, and the spin connection is not the gauge field of supersymmetry, but the gauge field of the Lorentz transformation and of Maxwell-type gauge transformation (7) it is natural to work with supercovariant spin connection $\hat{\omega}_{\mu,\nu\lambda}$. We expect, that the Bars-Higuchi theory reformulated in terms of the supercovariant connection

$$\hat{\omega}_{\mu}{}^{mn} = \omega_{\mu}{}^{mn} + \frac{1}{8}\bar{\psi}_{\xi}\Gamma^{\xi\mu mn\lambda}\psi_{\lambda}$$

will coincide with our formulation up to the non-uniqueness of transformation rules of the type $\delta\varphi^{i} = \eta^{ij}S_{,j}$, discussed in point 3 of the present paper and also in [9].

REFERENCES

1. E.Cremmer, B.Julia and J.Scherk, Phys. Lett. 76B, 409 (1978)

2. See e.g., M.J.Duff, B.E.W.Nilsson and C.N.Pope Nucl. Phys. B233, 433 (1984)
3. F.Englert, Phys. Lett. 119B, 339 (1982)
4. E.Cremmer and S.Ferrara, Phys. Lett. 91B, 61 (1980)
L.Brink and P.Howe, Phys. Lett. 91B, 384 (1980)
5. I.Bars and S.W.MacDowell, Phys. Lett. 129B, 182 (1983)
6. R.Kallosh, Phys. Lett, 143B, 373 (1984)
7. P. van Nieuwenhuizen, Phys. Rep. 68, 189 (1981)
8. L.Castellani, P.Fré, F.Giani, K.Pilch and P. van Nieuwenhuizen, Annals of Physics 146, 35 (1983)
9. I.Bars and A.Higuchi, Phys. Lett. 145B, 329 (1984)

DE SITTER SUPERGRAVITY AND GENERALIZED LIE SUPERALGEBRAS

M.A.Vasiliev
Lebedev Physical Institute, Academy of Sciences, Moscow, USSR

1. INTRODUCTION

Apart from the gravitational constant, at least one more free parameter exists in supergravity theories [1], the cosmological constant Λ. At the first time, this fact was demonstrated for N=2 supergravity by Freedman and Das [2] and by Fradkin and Vasiliev[3]. Namely, in this theory, supersymmetry requires $\Lambda \neq 0$ if gravitino has non-zero electric charge. Somewhat later, Townsend [4] observed that the truncation of N=2 supergravity with $\Lambda \neq 0$ must lead to N=1 supergravity with $\Lambda \neq 0$, and constructed the N=1 theory. It is well-known however that the cosmological constant is not a completely free parameter in supergravity but satisfies the constraint $\Lambda \leq 0$ (more exactly, $\Lambda \leq 0$ in supersymmetric stationary points of an action). This fact essentially differs supergravity as compared with gravity. Its simple explanation can be given in the following way. If $\Lambda \neq 0$, the supergravity action contains the mass-like terms for gravitinos with the massive parameter m related to Λ by the equation

$$\Lambda = -3m^2 \tag{1}$$

Obviously, the parameter m is real in the anti- de Sitter (adS) case $\Lambda \leqslant 0$ only. For the $\Lambda > 0$ de Sitter (dS) case, m is pure imaginary, and one may think that the dS supergravity action is not hermitian.

In this report, we would like to emphasize that a modified conjugation law for Fermi fields exists providing hermiticity of the supergravity action precisely in the dS case $\Lambda > 0$ (and in the limiting flat case $\Lambda = 0$). Unfortunately, as we show, the modified conjugation law cannot be realized in a positive definite Hilbert space. This means that although dS supergravity exists, it possesses ghosts (negative norm states). It seems for us that the presence of ghosts makes dS supergravity of little interest from the point of view of its physical applications. Nevertheless, dS supergravity may be quiet interesting from the more formal point of view. Really, it provides the example of a theory possessing supersymmetry for which there does not correspond any Lie superalgebra. We show however that a generalized Lie superalgebra exists which leads to dS supersymmetry.

2. MODIFIED CONJUGATION FOR FERMION FIELDS

In terms of two-component spinors, the gravitino mass-like term has the form

$$S_m^{3/2} = m \epsilon^{\nu\mu\rho\delta} \int d^4x \left(h_\nu{}^{\alpha}{}_{\dot\beta} h_\mu{}^{\gamma\dot\beta} \psi_{\rho\alpha} \psi_{\delta\gamma} - h_{\nu\gamma}{}^{\dot\beta} h_\mu{}^{\gamma\dot\delta} \varphi_{\rho\dot\beta} \varphi_{\delta\dot\delta} \right), \quad (2)$$

where $h_{\nu\alpha\dot\beta}$ is the tetradic gravitational field, and the gravitino fields $\psi_{\rho\alpha}$ and $\varphi_{\rho\dot\beta}$ are conjugate one to another (the indices μ, ν, ρ, δ take values 0-3; the spinorial indices α, β, γ... ; $\dot\alpha$, $\dot\beta$, $\dot\gamma$... take values 1,2).

In the adS case, the parameter m is real and the expression (2) is hermitian if the fieldsare conjugate in the following way

$$h^{+}_{\nu\alpha\dot{\beta}} = h_{\nu\beta\dot{\alpha}} \ , \quad \psi^{+}_{\rho\alpha} = \varphi_{\rho\dot{\alpha}} \ , \quad \varphi^{+}_{\rho\dot{\beta}} = \psi_{\rho\beta} \ . \tag{3}$$

In the dS case with pure imaginary m , $S^{3/2}_{m}$ is self-conjugate if

$$h^{*}_{\nu\alpha\dot{\beta}} = h_{\nu\beta\dot{\alpha}}, \quad \psi^{*}_{\rho\alpha} = i\,\varphi_{\rho\dot{\alpha}}, \quad \varphi^{*}_{\rho\dot{\alpha}} = -i\,\psi_{\rho\alpha} \ . \tag{4}$$

As it will be shown below, the full dS supergravity action is self-conjugate also if the fields are conjugate by the relation (4).

The operation $*$, by its definition, is antilinear ($i^{*}=-i$) and changes an order of operators (elements of Grassmann algebra) into the inverse one. Nevertheless, this operation cannot be considered as hermitian conjugation since its twice applicatinn changes signs of Fermi fields F (in the our case $F=(\psi,\varphi)$)

$$(F^{*})^{*} = -F \ . \tag{5}$$

For belinear combinations of fermions, i.e. for bosons B , the relation holds

$$(B^{*})^{*} = B \ . \tag{6}$$

It turns out however that, by the use of any given operation $*$ possessing the properties (5), (6), one can construct a hermitian conjugation (an involution) in the following way.

Let us consider fermion (F) and boson (B) fields as operators in a space H of quantum states. The operator K (oftenly called as Klein operator) is defined in H by the relations

$$K|b\rangle = |b\rangle \ , \quad K|f\rangle = -|f\rangle \ . \tag{7}$$

It has the following properties

$$KB=BK\ ,KF=-FK\ ,\quad K^2=I\ ,K^{+}=K\ . \qquad (8)$$

For any given operation $*$, we define the operation $+$ by the relations

$$F^{+}=iKF^{*}\ ,\quad B^{+}=B^{*}\ . \qquad (9)$$

The operation $+$ (9) has all properties of the hermitian conjugation (involution) e.g.,

$$(F^{+})^{+}=(iKF^{*})^{+}=-i\,(F^{*})^{+}K=K(F^{*})^{*}K=F, \qquad (10)$$

$$(F_1F_2)^{+}=(F_1F_2)^{*}=F_2^{*}F_1^{*}=-KF_2^{+}KF_1^{+}=F_2^{+}F_1^{+}. \qquad (11)$$

We stress that, due to the second of the relations (9), it is sufficient to show that $B=B^{*}$ in order to prove hermiticity of any boson quantity (e.g. an action). So, in terms of classical fields which are considered as elements of some Grassmann algebra, one can use the operation $*$ only without explicit reffering to the operation $+$ and Klein operator K which, striktly speaking, is not defined within the ordinary Grassmann algebras (note however that it is not difficult to construct the corresponding generalized Grassman algebras containing an analog of the Klein operator K).

Comparing the relations (4) and (9), we see that the supergravity mass term (2),with the pure imaginary mass parameter m , is hermitian if $\psi^{+}_{\rho\alpha}=-K\varphi_{\rho\dot\alpha}$, $\varphi^{+}_{\rho\dot\alpha}=K\psi_{\rho\alpha}$. Let us consider now the whole actinn for dS supergravity.

3. N = 1 dS SUPERGRAVITY

Among various approaches to N = 1 supergravity, one of the most simple and elegant is the formulation by MacDowell and Mansouri [5] describing the theory in the four-dimensional space-time in terms of curvatures corresponding to the superalgebra Osp(1,4;R). As the first step towards dS supergravity, we consider the curvatures of the complex superalgebra osp(1,4;$\mathbb{C}$) instead of the real one osp(1,4;$\mathbb{R}$).

In the general case, if a Lie superalgebra is given by the relations

$$[e_A, e_B\} = U^C_{AB}\, e_C \quad , \qquad (12)$$

corresponding curvatures and transformation laws are of the form

$$R^A_{\nu\mu} = \partial_\nu A^A_\mu - \partial_\mu A^A_\nu + A^B_\nu A^C_\mu U^A_{BC} \quad , \qquad (13)$$

$$\delta A^A_\nu = \partial_\nu \mathcal{E}^A + A^B_\nu \mathcal{E}^C U^A_{BC} \quad , \qquad (14)$$

where $\mathcal{E}^A$ are the infinitesimal gauge parameters.

By the use of two-component spinors, osp(1,4;$\mathbb{C}$) curvatures can be written in the form

$$R_{\nu\mu\alpha\beta} = [\partial_\nu \omega_{\mu\alpha\beta} + \omega_{\nu\alpha\gamma}\omega_{\mu\beta}{}^{\gamma} + \lambda^2 h_{\nu\alpha\dot\delta} h_{\mu\beta}{}^{\dot\delta} - i\lambda \psi_{\nu\alpha}\psi_{\mu\beta}] - [\nu \leftrightarrow \mu] \quad , \qquad (15)$$

$$r_{\nu\mu\dot\alpha\dot\beta} = [\partial_\nu W_{\mu\dot\alpha\dot\beta} + W_{\nu\dot\alpha\dot\delta} W_{\mu\dot\beta}{}^{\dot\delta} + \lambda^2 h_{\nu\gamma\dot\alpha} h_\mu{}^{\gamma}{}_{\dot\beta} - i\lambda \varphi_{\nu\dot\alpha}\varphi_{\mu\dot\beta}] - [\nu \leftrightarrow \mu] \quad , \qquad (16)$$

$$\rho_{\nu\mu\alpha\dot\beta} = [\partial_\nu h_{\mu\alpha\dot\beta} + \omega_{\nu\alpha\gamma} h_\mu{}^{\gamma}{}_{\dot\beta} + W_{\nu\dot\beta\dot\delta} h_{\mu\alpha}{}^{\dot\delta} - i\psi_{\nu\alpha}\varphi_{\mu\dot\beta}] - [\nu \leftrightarrow \mu] \quad , \qquad (17)$$

$$\chi_{\nu\mu\alpha} = [\partial_\nu \psi_{\mu\alpha} + \omega_{\nu\alpha\gamma}\psi_\mu{}^{\gamma} + \lambda h_{\nu\alpha\dot\delta}\varphi_\mu{}^{\dot\delta}] - [\nu \leftrightarrow \mu] \, , \qquad (18)$$

$$\eta_{\nu\mu\dot\beta} = [\partial_\nu \varphi_{\mu\dot\beta} + W_{\nu\dot\beta\dot\delta}\varphi_\mu{}^{\dot\delta} + \lambda h_{\nu\gamma\dot\beta}\psi_\mu{}^{\gamma}] - [\nu \leftrightarrow \mu] \qquad (19)$$

(the contraction parameter λ is real and proportional to the inverse (a)dS radius).

Following to the method of Ref.5), it is easy to construct the locally supersymmetric functional

$$S^{\mathbb{C}}=\alpha\int d^4x\,\epsilon^{\nu\mu\rho\delta}\left[i\left(R_{\nu\mu\alpha\beta}R_{\rho\delta}{}^{\alpha\beta}-r_{\nu\mu\dot{\alpha}\dot{\beta}}r_{\rho\delta}{}^{\dot{\alpha}\dot{\beta}}\right)\right. \tag{20}$$
$$\left.+2\lambda\left(\chi_{\nu\mu\alpha}\chi_{\rho\delta}{}^{\alpha}-\eta_{\nu\mu\dot{\beta}}\eta_{\rho\delta}{}^{\dot{\beta}}\right)\right],$$

where α is some constant. Really, one can verify that, if the fields ω and W fulfill the equation $\rho_{\nu\mu\alpha\dot{\beta}}=0$, that corresponds to the "1.5 order formalism " 6),3),7), then $S^{\mathbb{C}}$ is invariant under the gauge transformations (14) for the fields h , ψ , φ .

The functional $S^{\mathbb{C}}$ (20) cannot be considered as an action since it depends on the arbitrary complex fields h , ψ , φ (and ω , W , if one uses the first order formalism) and is complex itself. To obtain the supergravity action based on functional $S^{\mathbb{C}}$, it is neccessary to impose some hermiticity conditions guaranteeing hermiticity of $S^{\mathbb{C}}$. Obviously, it is possible only if the hermitian conjugated curvatures $\tilde{R}$, $\tilde{r}$, $\tilde{\chi}$, $\tilde{\eta}$ are expressed in terms of the initial curvatures R , r , χ , η (we use here the simbol $\sim$ to avoid misunderstandings which may be related with the use of the simbols $\dagger$ or $*$ corresponding to the conjugations of two different kinds). In other words, we ask the existence of such hermiticity conditions for the fields $A^A_\nu=(\omega, W, h, \psi, \varphi)$ which are of the form

$$\tilde{A}^A_\nu = a^A{}_B\,A^B{}_\nu \tag{21}$$

and, in accordance with eq. (13), satisfy the relations

$$\tilde{R}^A_{\nu\mu} = a^A{}_B\,R^B_{\nu\mu} \tag{22}$$

where $a^A{}_B$ is some numerical matrix. Taking into account the results of the previous section, we do not require the square of conjugation $\sim$ to be equal to the identity transformation. However, we require the following usual properties to be satisfied

$$\widetilde{(\mu_1 A_1 + \mu_2 A_2)} = \bar{\mu}_1 \tilde{A}_1 + \bar{\mu}_2 \tilde{A}_2 \,, \qquad (23)$$

$$\widetilde{(A_1 A_2)} = \tilde{A}_2 \tilde{A}_1 \quad , \qquad (24)$$

where complex numbers $\bar{\mu}_1$, $\bar{\mu}_2$ are conjugate to μ_1, μ_2 .

In the frameworks of N=1 supergravity,we restrict ourselves with the conjugations for which the relations hold

$$\tilde{\omega}_{\nu\alpha\beta} = W_{\nu\dot{\alpha}\dot{\beta}} \quad , \quad \tilde{W}_{\nu\dot{\alpha}\dot{\beta}} = \omega_{\nu\alpha\beta} \quad , \qquad (25)$$

since precisely this conditions guarantee that the fields ω and W correspond to the Lorentz algebra O(3,1), but not some other real form of O(4; $\mathbb{C}$). Suppose also that

$$\tilde{h}_{\nu\alpha\dot{\beta}} = a\, h_{\nu\beta\dot{\alpha}} \,, \quad \tilde{\Psi}_{\nu\alpha} = b\, \varphi_{\nu\dot{\alpha}} \,, \quad \tilde{\varphi}_{\nu\dot{\beta}} = c\, \Psi_{\nu\beta} \,. \qquad (26)$$

Then, as one can readily verify, application of the relation (22) to the curvatures R , τ , χ , η (15), (16), (18), (19) leads to the following restrictions on the parameters a,b,c

$$a^2 = b^2 = c^2 = a \cdot b \cdot c = 1 \,. \qquad (27)$$

Note, that from eq. (27) it follows that the conditions (22) are satisfied for the curvature ρ (18) also i.e., eq. (22) is satisfied for all curvatures $R^A_{\nu\mu}$ (this means that the conjugation, we have obtained, corresponds to some antiautomorphism of osp (1,4; $\mathbb{C}$).

The system of eqs. (27) has four independent solutions, but only two of them are really inequivalent. To see this, one can note that the change of variables $\omega \to \omega$, $W \to W$, $h \to -h$, $\psi \to \psi$, $\varphi \to -\varphi$, which is accompanied by the corresponding redefinition of the curvatures, does not modify the form of curvatures (15)-(19), (i.e. this change of variables is generated by an automorphysm of Osp (1,4; $\mathbb{C}$)). Hence, it is possible to find such variables for which $b = 1$. As a result, eq. (27) takes the form $a = c$, $a^2 = 1$, and the most general admissible conjugation law of the form (25), (26) is the following one

$$\widetilde{h}_{\nu\alpha\dot{\beta}} = a\, h_{\nu\beta\dot{\alpha}}\,,\quad \widetilde{\psi}_{\nu\alpha} = \varphi_{\nu\dot{\alpha}}\,,\quad \widetilde{\varphi}_{\nu\dot{\beta}} = a\,\psi_{\nu\beta}\,,\quad a^2 = 1\,. \tag{28}$$

Thus, we have shown that there exist two essentially different conjugations with $a = 1$ and $a = -1$, resp. Obviously, in the both cases, the functional (20) is self-conjugate, $\widetilde{S^{\mathbb{C}}} = S^{\mathbb{C}}$.

If $a = 1$, then for all fields $A^A_\nu = (\omega, W, h, \psi, \varphi)$ the relation holds $(\widetilde{A^A_\nu}) = A^A_\nu$, and the conjugation $\sim$ can be identified with the hermitian conjugation $\dagger$. This case corresponds precisely to adS supergravity, which can be obtained if the curvatures of the real superalgebra osp (1,4; $\mathbb{R}$) are used from the very beginning.

The second case, $a = -1$, is of most interest for us since, as we immediately show, it corresponds to dS supergravity. First of all, it is readily to verify that twice application of the operation $\sim$ is equivalent to the identity transformation for bosons and changes signs of fermions ψ and φ. So, if $a = -1$, the conjugation $\sim$ must be identified with the type $*$ conjugation considered in the previous section. Note that the difference in the conjugation laws (4) and (28) is related to the antihermiticity of the tetradic field $h_{\nu\alpha\dot{\beta}}$ in eq. (28) and can be removed by the fields redefinitions considered below. Stress that the very fact of antihermiticity of the tetradic field in the curvatures (15)-(19) means that this curvatures correspond to dS supergravity. Really, if $\psi = \varphi = 0$, the transition from $h_{\nu\alpha\dot{\beta}}$ to $i h_{\nu\alpha\dot{\beta}}$ changes the signs of the terms which are proportional to λ^2 in eqs. (15), (16) without any modifications of all other terms in eqs. (15)-(17). However, if the fields ω and W are conjugate one to another by eq. (25), the field h is hermitian, and the fermions are absent, then the change $\lambda^2 \to -\lambda^2$ in eqs. (15)-(17) corresponds to the transition O(3,2)→ O(4,1) exactly.

Another possible way to establish that α =-1 leads to dS supergravity, is to verify a sign of the cosmological constant. To do it, we note that, after simple transformation including an integration by parts, the functional $S^{\mathcal{C}}$ can be written in the form

$$\begin{aligned} S^{\mathcal{C}} = 4\lambda^2\alpha\int d^4x\, \epsilon^{\nu\mu\rho\delta}\{ & i(h_{\nu\alpha\dot{\delta}} h_{\mu\beta}{}^{\dot{\delta}} R^{0}_{\rho\delta}{}^{\alpha\beta} - h_{\nu\gamma\dot{\alpha}} h_{\mu}{}^{\gamma}{}_{\dot{\beta}}\, \tau^{0}_{\rho\delta}{}^{\dot{\alpha}\dot{\beta}}) \\ & + i\lambda^2 (h_{\nu\alpha\dot{\gamma}} h_{\mu\beta}{}^{\dot{\gamma}} h_{\rho}{}^{\alpha}{}_{\dot{\delta}} h_{\delta}{}^{\beta\dot{\delta}} - h_{\nu\alpha\dot{\gamma}} h_{\mu}{}^{\alpha}{}_{\dot{\delta}} h_{\rho\beta}{}^{\dot{\gamma}} h_{\delta}{}^{\beta\dot{\delta}}) \\ & + 2(h_{\nu\alpha\dot{\delta}} \varphi_{\mu}{}^{\dot{\delta}} \chi^{0}_{\rho\delta}{}^{\alpha} - h_{\nu\gamma\dot{\beta}} \psi_{\mu}{}^{\gamma} \eta^{0}_{\rho\delta}{}^{\dot{\beta}}) \\ & + 2\lambda (h_{\nu\alpha\dot{\delta}} \varphi_{\mu}{}^{\dot{\delta}} h_{\rho}{}^{\alpha}{}_{\dot{\beta}} \varphi_{\delta}{}^{\dot{\beta}} - h_{\nu\gamma\dot{\beta}} \psi_{\mu}{}^{\gamma} h_{\rho\alpha}{}^{\dot{\beta}} \psi_{\delta}{}^{\alpha}) \}, \end{aligned} \qquad (29)$$

where

$$R^{0}_{\nu\mu\alpha\beta} = [\partial_\nu \omega_{\mu\alpha\beta} + \omega_{\nu\alpha\gamma} \omega_{\mu\beta}{}^{\gamma}] - [\nu \leftrightarrow \mu], \qquad (30)$$

$$\tau^{0}_{\nu\mu\dot{\alpha}\dot{\beta}} = [\partial_\nu W_{\mu\dot{\alpha}\dot{\beta}} + W_{\nu\dot{\alpha}\dot{\delta}} W_{\mu\dot{\beta}}{}^{\dot{\delta}}] - [\nu \leftrightarrow \mu], \qquad (31)$$

$$\chi^{o}_{\nu\mu\alpha}=[\partial_\nu\psi_{\mu\alpha}+\omega_{\nu\alpha\gamma}\psi_\mu{}^{\gamma}]-[\nu\leftrightarrow\mu]\ , \qquad (32)$$

$$\eta^{o}_{\nu\mu\dot\beta}=[\partial_\nu\varphi_{\mu\dot\beta}+W_{\nu\dot\beta\dot\delta}\varphi_\mu{}^{\dot\delta}]-[\nu\leftrightarrow\mu]\ . \qquad (33)$$

In the expression (29), the first and the second terms describe the Einstein actinn and the cosmological term, resp., while the third and the fourth terms describe kinetic and mass-like terms for gravitino.

To normalize the action (29) as usual, one must set $\alpha=-\partial/16K^2\lambda^2$. The factor ∂ appears here because if $\partial=-1$, then the tetrad $h_{\nu\alpha\dot\beta}$ is antihermitian and, after transition to the hermitian tetrad $ih_{\nu\alpha\dot\beta}$, the first term in the action (29) changes its sign. As a result, one obtains that the cosmological cnnstant have the form $\Lambda\sim-\partial\lambda^2$. So, it is shown that the case $\partial=1$ corresponds to adS supergravity, while the case $\partial=-1$ corresponds to dS supergravity.

In order to obtain in the case $\partial=-1$ the usual normalization for the gravitino kinetic term with hermitian tetradic field $\hat h_{\nu\alpha\dot\beta}=ih_{\nu\alpha\dot\beta}$, it is sufficient to introduce new fermion fields $\hat\psi_{\nu\alpha}=e^{-\frac{i\pi}{4}}\psi_{\nu\alpha}$, $\hat\varphi_{\nu\dot\beta}=e^{-\frac{i\pi}{4}}\varphi_{\nu\dot\beta}$. This leads to appearence of the additional imaginary unit i in the mass term and to the conjugation law (4) for fermions.

Thus, we have shown that, besides the ordinary adS supergravity theory, the dS supergravity theory exists also. It requires however the essential modification of the conjugation law for fermions. Let us stress that dS supergravity is locally supersymmetric.The supersymmetry parameters $\varepsilon_\alpha(x)$, $\xi_{\dot\beta}(x)$ are related one to another by the same hermiticity relations, as the corresponding gauge fields $\psi_{\nu\alpha}$ and $\varphi_{\nu\dot\beta}$ do.

4. MODIFIED CONJUGATION AND QUANTIZATION

In fact, one must consider the quantization of the free massless spin 3/2 field on the dS background or,as the limiting case, in the flat space (preserving the modified conjugatinn law). Instead, we consider the more simple and clear example with the finite number degrees of freedom possessing however all characteristic features related to the modified conjugating law.

Let us consider the lagrangian

$$L=\frac{1}{2}\left[i(\psi_\alpha\dot\varphi_\beta-\dot\psi_\alpha\varphi_\beta)\delta^{\alpha\beta}+m(\psi_\alpha\psi^\alpha-\varphi_\beta\varphi^\beta)\right], \qquad (34)$$

where $\delta^{\alpha\beta}$ is the Kronecker delta, $\dot\varphi=\frac{\partial\varphi}{\partial t}$, and the fermion "fields" ψ_α and φ_β depend on the time t only. For real m , the lagrangian (34) is hermitian if $\psi_\alpha^+=\varphi_\alpha$, $\varphi_\alpha^+=\psi_\alpha$. This case corresponds to the massive Majorana spinor field at zero momentum. The case of pure imaginary mass $\overline{m}=-m$ is of most interest for us here.* At this case, the lagrangian (34) is self-conjugate if

$$\psi_\alpha^*=i\varphi_\alpha \quad , \quad \varphi_\alpha^*=-i\psi_\alpha \quad , \qquad (35)$$

* Note that the obvious pathology of the lagrangian (34) with pure imaginary mass m is not so evident for the dS supergravity where the parameter m satisfies the relation (1) and hence is zero in the flat limit. A priori, one can think that not only for $\Lambda\leq 0$ but also for $\Lambda>0$ the variety of parameters $3m^2\geqslant-\Lambda$ exists for which at least free massive ($3m^2>-\Lambda$) and massless $3m^2=-\Lambda$ spin $3/2$ fields admit a consistent quantization scheme.

in complete analogy with the gravitino conjugation law (4).

Usual canonical anticommutational relations for the operators φ and ψ have the form

$$\{\psi_\alpha, \varphi_\beta\} = \delta_{\alpha\beta}, \quad \{\psi_\alpha, \psi_\beta\} = \{\varphi_\alpha, \varphi_\beta\} = 0 . \tag{36}$$

These relations are obviously invariant both under the ordinary conjugation and the modified conjugation (35).

Using the hermitian conjugation definition (9), the properties of the operator K (8), and also eqs. (35), (36), we obtain that, for the imaginary mass case, the following relations hold

$$[\psi_\alpha, \psi_\beta^+] = K\delta_{\alpha\beta}, \quad \{\psi_\alpha, \psi_\beta\} = \{\psi_\alpha^+, \psi_\beta^+\} = 0. \tag{37}$$

Note that the left hand side of the first of these relations contains the commutator while the others contain anticommutators.

In complete analogy with the Clifford algebra case corresponding to the ordinary conjugation, one can convice that, since $\{\psi_\alpha, \psi_\beta\} = 0$ and in particular $(\psi_\alpha)^2 = 0$, a "vacuum" state $|\Omega\rangle$ exists which fulfills the relations

$$\psi_\alpha |\Omega\rangle = 0 . \tag{38}$$

Moreover, one can allways choose the vacuum with the definite fermion number $\nu = 0, 1$

$$K|\Omega\rangle = (-1)^\nu |\Omega\rangle . \tag{39}$$

Acting on $|\Omega\rangle$ by the operators ψ_α^+, one obtains the following states

$$|\Omega_\alpha\rangle = \psi_\alpha^+ |\Omega\rangle , \tag{40}$$

$$|\Omega_{\alpha\beta}\rangle = \psi_\alpha^+ \psi_\beta^+ |\Omega\rangle = -|\Omega_{\beta\alpha}\rangle . \qquad (41)$$

Suppose the vacuum has positive norm $\langle\Omega|\Omega\rangle = 1$. Then

$$\langle\Omega_\alpha|\Omega_\beta\rangle = \langle\Omega|\psi_\alpha \psi_\beta^+|\Omega\rangle = (-1)^\nu \delta_{\alpha\beta} . \qquad (42)$$

If $\nu = 0$, the states $|\Omega_\alpha\rangle$ will have positive norm also. However for the state $|\Omega_{12}\rangle$, one obtains

$$\langle\Omega_{12}|\Omega_{12}\rangle = \langle\Omega|\psi_2\psi_1\psi_1^+\psi_2^+|\Omega\rangle = \langle\Omega|\psi_2 K \psi_2^+|\Omega\rangle = -1 . \qquad (43)$$

The indefiniteness of the metric, caused by the conjugation operation of the type $*$, has very general nature if the usual (anti) commutational relations are maintained. For instance, it is not difficult to verify that the dS supergravity theory, suggested in the previous section, has the same defect. To do it, one may consider the flat limit $\lambda \to 0$ keeping the modified conjugation law unchanged and then quantize the free massless spin 3/2 field in the flat space. Following to the well-known Dirac prescription for sistems with constraints, after a complete fixing of gauges, one will obtain Dirac anticommutational relatinnswhich will be equivalent to the relations (36), (37) with the indices α , β supposed including also momentum $\vec{p}$. As a result, there appears an indefinite metric in dS supergravity in the same fashion as in the example considered.

5. MODIFIED CONJUGATION IN GENERAL CASE AND EXTENDED DS SUPERGRAVITY

The possibility to introduce the modified conjugation law leading to the change of the cosmological constant sign is not a peculiarity of supergravity only but turns out to be quiet general. Really,let in the frameworks of gravity, some action $S(q)$ is given which depends on arbitrary set of fields $q^{i}_{\nu_1 \dots \nu_{N_i}, \alpha_1 \dots \alpha_{n_i}, \dot{\beta}_1 \dots \dot{\beta}_{m_i}}$ being ranks N_i tensors with respect to the "world" indices $\nu_1 \dots \nu_{N_i} = 0-3$ and multispinors of ranks n_i and m_i with respect to the spinorial indices $\alpha_1 \dots \alpha_{n_i}, \dot{\beta}_1 \dots \dot{\beta}_{m_i}$ corresponding to the tangent space where the local Lorentz group acts as usual. Let the fields q^i are restricted by hermiticity relations $(q^i)^{+} = a^i{}_j q^j$ where $a^i{}_j$ is some numerical matrix, and $+$ denotes the usual hermitian conjugation which transforms dotted indices into undotted ones and vice versa (i.e., if the numbers $N_i^{+}, n_i^{+}, m_i^{+}$ correspond to the quantity $(q^i)^{+}$, then $N_i^{+} = N_i$, $n_i^{+} = m_i$, $m_i^{+} = n_i$).

If the action $S(q)$ is hermitian, $S = S^{+}$, then it remains self-conjugate with respect to the modified conjugation conditions $(q^i)^{*} = (-1)^{n_i} a^i_j q^j$ This fact is the evident consequence of the local Lorentz covariance of the action, since any its term contains an even number of contracted in pairs spinorial indices of each type, and all additional sign factors $(-1)^{n_i}$ turn out to be compensated. From the other hand, one can readily see that the modified conjugation has the properties of the conjugation operatinn $*$ considered in the previous sections.

The tetradic field $h_{\nu\alpha\dot{\beta}}$ is antihermitian with respect to the conjugation $*$ (if it is hermitian with respect to the conjugation $+$). The transition to the hermitian tetrad $\hat{h}_{\nu\alpha\dot{\beta}} = i h_{\nu\alpha\dot{\beta}}$ is accompanied by the transition from the metric tensor $g_{\nu\mu} = h_{\nu\alpha\dot{\beta}} h_{\mu}{}^{\alpha\dot{\beta}}$ to the metric tensor $\hat{g}_{\nu\mu} = \hat{h}_{\nu\alpha\dot{\beta}} \hat{h}_{\mu}{}^{\alpha\dot{\beta}} = -g_{\nu\mu}$.

Let us consider now the boson fields sector. First of all, we note that, from the very beginning, one can choose the variables in such a way that all boson fields will be "world" tensors without spinorial (tangent) indices i.e., for bosons $n_i = m_i = 0$. In this case, the modified hermiticity conditions do not differ from the initial ones for boson fields. However, after transition from the metric tensor $g_{\nu\mu}$ to $\hat{g}_{\nu\mu} = -g_{\nu\mu}$, there appears an additional sign factor $(-1)^{\ell}$ in any term in the action, where ℓ is the number of the metric tensors that are present in this term. Assuming as usual that a field of integer spin s is described by second order equations of motion, we see that transition to $\hat{g}_{\nu\mu}$ leads to appearence of the additional sign factors $(-1)^{s+1}$ and $(-1)^{s}$ for the kinetic and mass parts of the spin s action, resp. So, if the initial action $S(\varphi)$ has correct sign structure for kinetic and mass terms of all integer spin fields then transition to the modified conjugation law $*$ leads, firstly, to disagreement of the kinetical sign factors for bosons with odd andeven spins and secondly, to the change of the mass square sign for all Bose fields.

To normalize the Einstein part ($s=2$) of the action in appropriate way (after transition to the conjugation $*$), one must change its sign: $S \to S' = -S$. As a result, all odd spins fields, and in particular spin 1 fields, will have wrong kinetical terms signs. Simultaneously, the sign of the cosmological constant will be changed.

As one can readily see, the N = 1 dS supergravity theory, which has been suggested above, is related to ordinary N=1 adS supergravity [4), 5)] by the general procedure described in this section. Obviously, extended dS supergravities can be constructed just in the same way starting from the known adS extended supergravities (see e.g., [1) -3), 7)]). However all spin 1 fields will have the wrong kinetical terms signs in these theories.

To obtain a "good" dS theory with the help of the general procedure described, one must consider an appropriate "bad" adS theory. It seems that in the frameworks of supergravity, such "bad" adS theories do not exist.

6. DS SUPERSYMMETRY AND GENERALIZED LIE SUPERALGEBRAS

Although direct physical applications of dS supergravity are rather awkward due to the presence of ghosts, this theory is very interesting as an example of a theory possessing (local) supersymmetry for which there does not correspond any Lie superalgebra. In this section we show that some generalized Lie superalgebra exists which corresponds to dS supergravity.

To begin with, let us consider the Lie superalgebra osp(1,4;$\mathbb{C}$). We introduce the (super)generators $\{E_A\}=(L_{\alpha\beta}, N_{\dot\alpha\dot\beta}, \mathcal{P}_{\alpha\dot\beta}, e^{-\frac{i\pi}{4}}Q_\alpha, e^{-\frac{i\pi}{4}}R_{\dot\beta})$ corresponding to the gauge fields (ω, W, h, ψ, φ) of section 3. In this terms the superalgebra osp(1,4;$\mathbb{C}$) is given by the relations

$$[L_{\alpha\beta}, E_{\gamma_1\ldots\gamma_n,\dot\delta_1\ldots\dot\delta_m}]=\frac{1}{2}(\epsilon_{\alpha\gamma_1}E_{\beta\gamma_2\ldots\gamma_n,\dot\delta_1\ldots\dot\delta_m}+\ldots +\epsilon_{\alpha\gamma_n}E_{\gamma_1\ldots\gamma_{n-1}\beta,\dot\delta_1\ldots\dot\delta_m})+\frac{1}{2}(\alpha\leftrightarrow\beta), \tag{44}$$

$$[N_{\dot\alpha\dot\beta}, E_{\gamma_1\ldots\gamma_n,\dot\delta_1\ldots\dot\delta_m}]=\frac{1}{2}(\epsilon_{\dot\alpha\dot\delta_1}E_{\gamma_1\ldots\gamma_n,\dot\beta\dot\delta_2\ldots\dot\delta_m}+\ldots +\epsilon_{\dot\alpha\dot\delta_m}E_{\gamma_1\ldots\gamma_n,\dot\delta_1\ldots\dot\delta_{m-1}\dot\beta})+\frac{1}{2}(\dot\alpha\leftrightarrow\dot\beta) \tag{45}$$

($\epsilon_{\alpha\beta}=-\epsilon_{\beta\alpha}=\epsilon_{\dot\alpha\dot\beta}=-\epsilon_{\dot\beta\dot\alpha}$, $\epsilon_{12}=1$ and $E_{\gamma_1\ldots\gamma_n,\dot\delta_1\ldots\dot\delta_m}$ denotes any generator of the set $\{E_A\}$ having n undotted and m dotted spinorial indices) and also by the relations

$$[\mathcal{P}_{\alpha\dot\beta}, \mathcal{P}_{\gamma\dot\delta}]=2\lambda^2(\epsilon_{\alpha\gamma}N_{\dot\beta\dot\delta}+\epsilon_{\dot\beta\dot\delta}L_{\alpha\gamma}), \tag{46}$$

$$[\mathcal{P}_{\alpha\dot\beta}, Q_\gamma] = \lambda \epsilon_{\alpha\gamma} R_{\dot\beta} \quad , \quad [\mathcal{P}_{\alpha\dot\beta}, R_{\dot\delta}] = \lambda \epsilon_{\dot\beta\dot\delta} Q_\alpha \, , \tag{47}$$

$$\{Q_\alpha, Q_\beta\} = 2\lambda L_{\alpha\beta} \, , \quad \{R_{\dot\alpha}, R_{\dot\beta}\} = 2\lambda N_{\dot\alpha\dot\beta} \, , \tag{48}$$

$$\{Q_\alpha, R_{\dot\beta}\} = \mathcal{P}_{\alpha\dot\beta} \tag{49}$$

The generators of osp(1,4; $\mathbb{C}$) can be restricted by the following hermiticity conditions

$$L^{+}_{\alpha\beta} = -N_{\dot\alpha\dot\beta}, \; N^{+}_{\dot\alpha\dot\beta} = -L_{\alpha\beta}, \; \mathcal{P}^{+}_{\alpha\dot\beta} = -\mathcal{P}_{\beta\dot\alpha}, \; Q^{+}_{\alpha} = iR_{\dot\alpha}, R^{+}_{\dot\alpha} = iQ_\alpha \tag{50}$$

that are consistent with eqs. (44)-(49) and correspond precisely to the real form osp(1,4,$\mathbb{R}$) of osp(1,4,$\mathbb{C}$).

In fully accordance with the results of section 3, the different "conjugation" law (more exactly, antiautomorphism) of the type $*$ exists

$$L^{*}_{\alpha\beta} = -N_{\dot\alpha\dot\beta}, \; N^{*}_{\dot\alpha\dot\beta} = -L_{\alpha\beta}, \; \mathcal{P}^{*}_{\alpha\dot\beta} = \mathcal{P}_{\beta\dot\alpha}, \; Q^{*}_{\alpha} = iR_{\dot\alpha}, \; R^{*}_{\dot\alpha} = -iQ_\alpha, \tag{51}$$

which is consistent with eqs. (44)-(49). This conjugation has the property $(E^*_A)^* = (-1)^{\pi(E_A)} E_A$ where $\pi(E_A)=0$ for the even generators ($L, N, \mathcal{P}$) and $\pi(E_A)=1$ for the odd ones (Q, R).

Let us consider some representation of an arbitrary Lie superalgebra. In the representation space, one can introduce the operator K which possesses the following properties

$$K E_A = (-1)^{\pi(E_A)} E_A K \; , \; K^2 = I \, . \tag{52}$$

Really, in the basis where even elements of the superalgebra are realized as the matrices of the form $\begin{pmatrix} A & 0 \\ 0 & B \end{pmatrix}$

and odd ones are realized as $\begin{pmatrix} 0 & C \\ D & 0 \end{pmatrix}$, the operator K has the form $\begin{pmatrix} I & 0 \\ 0 & -I \end{pmatrix}$.

By analogy with the results of section 2, it seems naturally to relate to the conditions (51) the following hermiticity conditions

$$L^{+}_{\alpha\beta}=-N_{\dot\alpha\dot\beta},\ N^{+}_{\dot\alpha\dot\beta}=-L_{\alpha\beta},\ \mathcal{P}^{+}_{\alpha\dot\beta}=\mathcal{P}_{\beta\dot\alpha},\ Q^{+}_{\alpha}=-KR_{\dot\alpha},\ R^{+}_{\dot\alpha}=KQ_{\alpha}. \quad (53)$$

These relations cannot be considered as some hermiticity conditions for Lie superalgebra osp(1,4,$\mathbb{C}$) since the operators KQ_α and $KR_{\dot\beta}$ do not belong to this algebra.

The main point of this section is however the following. If, in artibtrary representation of osp(1,4,$\mathbb{C}$), one introduces the new set of operators $\{e_A\}=(\ell_{\alpha\beta}, n_{\dot\alpha\dot\beta}, p_{\alpha\dot\beta}, q_\alpha, r_{\dot\beta})$ as follows

$$\ell_{\alpha\beta}=L_{\alpha\beta},\ n_{\dot\alpha\dot\beta}=N_{\dot\alpha\dot\beta},\ p_{\alpha\dot\beta}=K\mathcal{P}_{\alpha\dot\beta},\ q_\alpha=Q_\alpha,\ r_{\dot\beta}=KR_{\dot\beta} \quad (54)$$

then, firstly, these operators satisfy the relations, which do not contain the operator K and have the form

$$[p_{\alpha\dot\beta}, p_{\gamma\dot\delta}]=2\lambda^2(\epsilon_{\alpha\gamma}n_{\dot\beta\dot\delta}+\epsilon_{\dot\beta\dot\delta}\ell_{\alpha\gamma}), \quad (55)$$

$$\{p_{\alpha\dot\beta}, q_\gamma\}=\lambda\epsilon_{\alpha\gamma}r_{\dot\beta}, \quad \{p_{\alpha\dot\beta}, r_{\dot\delta}\}=\lambda\epsilon_{\dot\beta\dot\delta}q_\alpha, \quad (56)$$

$$\{q_\alpha, q_\beta\}=2\lambda\ell_{\alpha\beta}, \quad \{r_{\dot\alpha}, r_{\dot\beta}\}=-2\lambda n_{\dot\alpha\dot\beta}, \quad (57)$$

$$[q_\alpha, r_{\dot\beta}]=-p_{\alpha\dot\beta} \quad (58)$$

(the operators ℓ and n satisfy the relations analogous to (44), (45)) and, secondly, the operators $\{e_A\}$ can be restricted by the following hermiticity conditions

$$l^{+}_{\alpha\beta}=-n_{\dot{\alpha}\dot{\beta}},\ n^{+}_{\dot{\alpha}\dot{\beta}}=-l_{\alpha\beta},\ p^{+}_{\alpha\dot{\beta}}=p_{\beta\dot{\alpha}},\ q^{+}_{\alpha}=-r_{\dot{\alpha}},\ r^{+}_{\dot{\alpha}}=-q_{\alpha}, \qquad (59)$$

generated by the relations (53), (54).

By its very construction, the algebra given by eqs. (55)-(58) admits realizations with the brackets $[\ ,\ ]$ and $\{\ ,\ \}$ realized as commutators and anticommutators. However this algebra (in what follows designated as $\widetilde{osp}(1,4,\mathbb{C})$) is not a Lie superalgebra, as it can be easily seen from eqs. (56), (58). One can see also that the hermitian conjugation (more exactly, involution) law (59) corresponds to dS symmetry since the transition from the hermitian operator $p_{\alpha\dot{\beta}}$ to the antihermitian one $i p_{\alpha\dot{\beta}}$ leads to the change of the λ^2 sign in eq. (55) precisely corresponding [*] to the transition from o(3,2) to o(4,1).

In fact, the example of $\widetilde{osp}(1,4,\mathbb{C})$ leads us to some generalized Lie superalgebras. Namely, an algebra A_n (n is non-negative integer) we call degree n Lie superalgebra if it is $\underbrace{Z_2\times Z_2\times\ldots\times Z_2}_{n}$ graded algebra and, for any $a, b, c \in A_n$ possessing definite parities $\pi_i(a)$, $\pi_i(b)$, $\pi_i(c)$ ($i=1\ldots n$, $\pi_i(x)=0,1$) with

---

[*] Nevertheless note that the real form $\widetilde{osp}(1,4,\mathbb{R})$ of $\widetilde{osp}(1,4,\mathbb{C})$, which corresponds to the involution (59), provides an extension of the Lie algebra sp(4) $\sim$ O(3,2), similary to the ordinary osp(1,4, $\mathbb{R}$) case. It should be stressed however that from the physical point of view complex algebras with involution, but not real algebras by itself, are of importance and in this sense algebra $\widetilde{osp}(1,4,\mathbb{C})$ with the involution (59) can be considered as an extension of dS symmetry.

respect to this $(\times Z_2)^n$ grading, a multiplication operation $[a, b\}$ in A_n satisfies (anti) symmetry conditions

$$[a, b\} = -(-1)^{\sum_{i=1}^{n} \pi_i(a)\pi_i(b)} [b, a\} \tag{60}$$

and generalized Jacobi's identities

$$(-1)^{\sum_{i=1}^{n} \pi_i(a)\pi_i(c)} [a, [b, c\}\} + (-1)^{\sum_{i=1}^{n} \pi_i(b)\pi_i(a)} [b, [c, a\}\} + (-1)^{\sum_{i=1}^{n} \pi_i(c)\pi_i(b)} [c, [a, b\}\} = 0. \tag{61}$$

Obviously, this definition is a direct generalization of the definitions for Lie algebras and Lie superalgebras coinciding with them at n = 0,1.

As one can verify, the algebra $\widehat{osp}(1,4,\mathbb{C})$ (55)-(58) represents by itself the degree 2 Lie superalgebra with the parities π_1 and π_2 defined as follows:

$$\pi_1(\ell) = \pi_1(n) = \pi_1(q) = 0\ ,\quad \pi_1(p) = \pi_1(\tau) = 1; \tag{62}$$

$$\pi_2(\ell) = \pi_2(n) = \pi_2(\tau) = 0\ ,\quad \pi_2(p) = \pi_2(q) = 1. \tag{63}$$

Evidently, these parities coincide with the ordinary parities of dotted and undotted spinorial indices numbers. Note, that the usual Bose-Fermi parity π is the sum of π_1 and π_2 .

The algebras of the form (60), (61), and in fact even more general algebras, were considered by Rittenberg and Wyler [8] and by Omote and Kamefuchi [9]. Rittenberg and Wyler had called such algebras as color superalgebras and had studied some of their properties in more details.

Apart from the degree n Lie superalgebras, corresponding generalized degree n Gassmann algebras G_n can be naturally introduced as associative, $(\times Z_2)^n$-graded algebras with elements which satisfy the relations

$g_1 g_2 = (-1)^{\sum_{i=1}^{n} \pi_i(g_1)\pi_i(g_2)} g_2 g_1$. These algebras and some their further generalizations were also considered in Ref. [8] and, especially, in Ref. 9) where an analysis on generalized Grassmann algebras was developed, that may provide a possibility to quantize dinamical systems of fields taking values in these algebras with the help of continual integral approach.

Let us stress that in complete analogy with the ordinary Lie algebras and superalgebras, for any degree n Lie superalgebra given by eq. (12) it is straightforward to introduce the gauge fields taking values in G_n and corresponding curvatures (13) and transformation laws (14). In terms of these quantities, one can construct some objects of physical interest. For example, the action for dS supergravity can be re-written in terms of the curvatures generated by the degree 2 Lie superalgebra $\widehat{osp}(1,4;\mathbb{C})$ directly. In fact, this corresponds to the transition in eqs. (15)-(20) from the set of fields ($\omega, W, h, \psi, \varphi$) to the fields $\omega' = \omega, W' = W, h' = Kh, \psi' = \psi, \varphi' = K\varphi$.

7. CONCLUSIONS

At the conclusion, let us formulate the main results obtained. The locally supersymmetric N=1 dS supergravity action with positive cosmological constant is constructed which is hermitian with respect to the modified conjugation law. It is shown however that the modified conjugation leads to the indefinite metric quantization i.e., although dS supergravity exists, it contains ghosts (negaive norm states).

It is noted that the modified conjugation law, leading to the change of the cosmological constant sign, is admissible not only in supergravity but also in an arbitrary relativistic theory and, in particular, in extended supergravity theories. It is shown that in the general case transition to the modified conjugation leads not only to the change of cosmological constant sign, but also to the change of signs of mass square parameters and to the change of signs of kinetic terms for all odd-integer spins (at the condition that the Einstein action is normalized correctly).

The algebra is found corresponding to N=1 dS supergravity. It is shown that this algebra is not an ordinary Lie superalgebra but belongs to the class of generalized ($Z_2 \times Z_2$ -graded) Lie superalgebras. It seems for us that this result can be considered as an indication on the new interesting possibility of the construction of supersymmetric theories based on generalized Lie superalbras. However, it is not clear for us now whether consistent theories of such type exist which are free from the indefinite metric problem.

I would like to express my thanks to I.A.Batalin, E.S.Fradkin, D.A.Kirzhnits, A.D.Linde, O.V.Ogievetsky, A.A.Tseytlin, I.V.Tyutin, and B.L.Voronov for helpfull discussions.

Note added. After the present paper has been prepaired for publication, we have known that supergravity theories with $\Lambda > 0$ have been considered also in the very recent publications by Lukierski and Nowicki [10] and by Pilch et al [11] (I am grateful to R.E.Kallosh and to A.A.Tseytlin for the attraction of my attention to these works). It should be stressed however that the method suggested here, based on the modified conjugation of the type $\divideontimes$ and on the generalized Lie superalgebras, essentially differs as compaired with the method of Refs. 10) 11), based on ordinary conjugation and usual Lie superalgebras (that does not save however from the indefinite metric problem [11]). This difference is exhibited most evidently in the fact that our method provides the possibility to construct supergravity theories with $\Lambda > 0$ corresponding to N-extended adS theories with arbitrary $1 \leq N \leq 8$ but not only with even N as in Refs. 10) 11).

REFERENCES

1) Van Nieuwenhuizen,P., Phys. Rep., 68, 189 (1981).
2) Freedman D.Z., and Das,A., Nucl. Phys., B 120, 221 (1977).
3) Fradkin E.S., and Vasiliev,M.A., Lebedev Physical Institute Preprint N 197 (1976).
4) Townsend,P.K., Phys. Rev., D15,2808 (1977).
5) MacDowell,S.W., and Mansouri, F., Phys. Rev. Lett., 38, 739 (1977).
6) West,P.C. and Chamseddine,A.H., Nucl. Phys., B129, 39 (1977).
7) Van Nieuwenhuisen, P. and Townsend,P.K., Phys. Lett., 167B, 439 (1977).
8) Rittenberg,V., and Wyler, D., Nucl. Phys., B139, 189 (1978); J. Math. Phys., 19, 2193 (1978).
9) Omote,M., and Kamefuchi, S., IL Nuovo Cim., 50A, 21 (1979); IL Nuovo Cim., 70A,435 (1982); IL Nuovo Cim., 77A, 99 (1983).
10) Lukierski, J. and Nowicki, A., Institute of Theoretical Physics, University of Wroclaw, Preprint, 1984, No 609.
11) Pilch, K., van Nieuwenhuizen, P. and Schius,M.F., Preprint ITP-SB-84-46 (1984).

HIGHER CONSERVATION LAWS FOR SUPERSYMMETRIC GAUGE THEORIES

I.Ya.Aref'eva and I.V.Volovich
Steklov Mathematical Institute, Vavilov street 42, Moscow,GSP-I,II7966,USSR

I. INTRODUCTION

Extended supergravity is, a priori, non-renormalizable in the perturbation theory[I]. It is tempting to speculate that the trouble with non-renormalizability can be circumvented in the following way. i)Some hidden symmetry could forbid ultraviolet quantum divergences, beyond naive symmetry arguments. ii)The theory may have a non-perturbative solution on the quantum level. In the latter case the problem of non-renormalizability would disappear. Strange as it may seem, reasons for such kind of expectations and even for the crazy idea like ii) do exist.They are based on the analogy between the superspace formulation of the N=8 extended supergravity[2,3] and the N=4 supersymmetric Yang Mills (SYM) theory [4,5,6]. In both cases equations of motion follow from constraint equations, which are simple first order differential equations. This implies that the N=4 SYM theory admits inverse scattering formulation in superspace[7]. The inverse scattering formulation usually provides for a detailed investigation of the theory[8]. Namely, it leads to the existence of an infinite number of conserved currents and a hidden symmetry. The properties of classical equations of motion are important for investigation of renormalizability of the theory on mass shell in the background field method (quantization can be performed in Minkowski space).Note also that conserved currents and hidden symmetry usually survive in the quantum case. The inverse scattering formulation in superspace allows us to study the properties of the theory in superspace. It is interesting to what extent these properties are inherited in Minkowski space. To answer this question it is necessary to have an explicit equivalence between the superspace formulation and the description of the theory in Minkowski space.

In the superspace formulation of gauge theory, the main object is a superconnection $A_A(y)$ ($y=x^\mu, \theta^s_\alpha, \bar{\theta}_{\dot\beta t}$, $\mu=0,1,2,3$, $\alpha,\dot\beta=1,2$, $s,t=1,\ldots N$). To exclude nonphysical fields, constraints on the superconnection are introduced. For N = 1,2 the constraints can be solved [4,14,9]. They do not have, as a consequence, any equations in x-space. For N = 3 the constraint equations are

$$F^{(st)}_{\alpha\beta}=0=F_{\dot\alpha(s,\dot\beta t)} \qquad (1.a)$$

$$F^{s}_{\alpha,\dot\beta t}=0 \qquad (1.b)$$

where F_{AB} is the supercurvature. As was pointed out by Sohnius [5] and Witten [6], (1.a),(1.b) lead to differential equations in x-space (put the theory on-shell). This was regarded as a shortcoming and the constraints (1) were considered as unacceptable. On the contrary, it was pointed out in [7] that the fact that the constraint equations (1) for N=3 lead to the equations of motion in x-space is extremely useful for the investigation of N=3,4 SYM theory. The reason for such a judgement is that the constraint equations (1) admit inverse scattering formulation.

We are going to start our report by demonstrating that the equations of motion in x-space for the N=4 SYM theory follow from the constraints (1). Although this fact was taken for granted and, moreover, was intensively used in [6,7],its direct varification is worth to be performing. This part of work was done in collaboration with Ling Lie Chau and Ge Mo Lin. In the Abelian case the proof was given by Sohnius [5] and Witten [6]. It seemed natural to expect the same result for the non-Abelian case.The more non-trivial point is that the non-Lagrangian equations (1.a), (1.b) lead to Lagrangian equations in x-space. Indeed, assuming that it is true, and taking into accou supersymmetry and gauge invariance of the constraints (1) we deduce the desired result from the unique of N=4 (or equivalent N=3) supersymmetric and gauge invariant Lagrangian with dimentionless interaction [10].

The N=4 SYM theory posseses an infinite sequence of conserved currents in superspace [11,12,13]. Then there is a natural question whether exists any connection between the cancellation of ultraviolet divergences for the N=4 SYM theory [14-17] and the existence of an infinite number of conserved currents. To derive concequences of these currents in x-space we ought to have the explicit expressions for the superconnection $A_A(y)$ in terms of physical fields in x-space. In this report we demonstrate that it can be done. In other words we show that a solution of eqs. (1) for N=3 can be written in terms of physical fields satisfying the equations of motion in x-space.

It should also be mentioned that the existence of an infinite number of conserved currents is usually connected with some kind of hidden symmetry. So the natural question arises about the corresponding hidden symmetry for the N=4 SYM theory. There is at present no general way to answer this question. In investigating it via the Riemann-Hilbert transformation we are faced with a reduction problem (cf[20]). At the end of our report we present a new supersymmetric and gauge invariant model (B-model) which overcomes this problem.

2. DERIVATION OF THE N=4 FIELD EQUATIONS FROM THE N=3 CONSTRAINTS

In Minkowski space the N=4 SYM theory [10] describes the Yang-Mills vector field A_μ, spinors fields $\psi^K, \bar{\psi}_{\dot\alpha K}$ and $\bar{\varphi}^{KL}; \varphi_{KL}$, K,L=I,...4. All these fields belong to the adjoint representation of a gauge group G. The fields φ^{KL} and $\bar{\varphi}_{KL}$ are related through: $\bar{\varphi}^{KL} = \frac{I}{2}\varepsilon^{KLMN}\varphi_{MN}$. The equations of motion for these fields follow from the Lagrangian

$$L = Tr\Big\{-\frac{I}{4}F^2_{\mu\nu} - i\bar{\psi}^{\dot\beta K}\mathcal{D}_{\alpha\dot\beta}\psi^\alpha_K + \frac{I}{2}\mathcal{D}_\mu\bar{\varphi}^{KL}\mathcal{D}_\mu\varphi_{KL} + \{\bar{\psi}^{\dot\alpha K}, \bar{\psi}^L_{\dot\alpha}\}\varphi_{KL}$$
$$+\bar{\varphi}^{KL}\{\psi^\alpha_K, \psi_{\alpha L}\} + \frac{I}{4}[\varphi_{MN}, \varphi_{KL}][\bar{\varphi}^{MN}, \bar{\varphi}^{KL}]\Big\} \qquad (2)$$

and have the following form

$$\mathcal{D}_\mu\mathcal{D}^\mu\varphi_{ij} = [\bar{\varphi}^{pq}, [\varphi_{ij}, \varphi_{pq}]] + [\varphi_{pq}, [\varphi_{ij}, \bar{\varphi}^{pq}]]$$
$$+\{\psi^\alpha_i, \psi_{\alpha j}\} - \varepsilon_{ijl}\{\bar{\psi}^{\dot\alpha 4}, \bar{\psi}^l_{\dot\alpha}\}, \quad i,j,l,p,q = I,2,3, \qquad (3.a)$$

$$i\mathcal{D}_{\alpha\dot\beta}\psi^\alpha_K = 2[\bar{\psi}^L_{\dot\beta}, \varphi_{KL}], \quad K,L = I,2,3,4, \qquad (3.b)$$

$$i\mathcal{D}_{\alpha\dot\beta}\bar{\psi}^{K\dot\beta} = 2[\psi_{\alpha L}, \bar{\varphi}^{KL}], \qquad (3.c)$$

$$\mathcal{D}^\mu F_{\mu\nu} = \sigma_{\alpha\dot\beta,\nu}\{\bar{\psi}^{\dot\beta L}, \psi^\alpha_L\} + i([\mathcal{D}_\nu\bar{\varphi}^{ik}, \varphi_{ik}] + [\mathcal{D}_\nu\varphi_{ik}, \bar{\varphi}^{ik}]) \qquad (3.d)$$

where $\mathcal{D}_\mu = \partial_\mu + i[A_\mu,\ ]$, $\mathcal{D}_{\alpha\dot\beta} = \sigma^\mu_{\alpha\dot\beta}\mathcal{D}_\mu$, $\varphi_{i4} = \frac{I}{2}\varepsilon_{i4kl}\bar{\varphi}^{kl}$, $\varepsilon_{I234} = I$.
We also use the following notation:

$$\varepsilon^{I2} = I = -\varepsilon_{I2},\ \varepsilon^{\alpha\beta} = -\varepsilon_{\alpha\beta};\ \psi^\alpha = \varepsilon^{\alpha\beta}\psi_\beta,\ \bar{\psi}^{\dot\alpha} = \varepsilon^{\dot\alpha\dot\beta}\bar{\psi}_{\dot\beta},$$
$$\sigma^0_{\alpha\dot\beta} = \begin{pmatrix} I & 0 \\ 0 & I\end{pmatrix},\ \sigma^I_{\alpha\dot\beta} = \begin{pmatrix} 0 & I \\ I & 0\end{pmatrix},\ \sigma^2_{\alpha\dot\beta} = \begin{pmatrix} 0 & i \\ -i & 0\end{pmatrix},\ \sigma^3_{\alpha\dot\beta} = \begin{pmatrix} I & 0 \\ 0 & -I\end{pmatrix},$$
$$\sigma^\mu_{\alpha\dot\beta}\sigma^{\nu\alpha\dot\beta} = 2\eta^{\mu\nu},\ \eta_{\mu\nu} = (I,-I,-I,-I).$$

Let us recall the notation adopted in the superspace formulation of the supersymmetric gauge treories. Let $C^{4,4N}$ be the complex Minkowski superspace with coordinates

$y^A=(x^\mu,\theta^\alpha_s,\bar\theta^{\dot\beta t}),\mu=0,1,2,3;\ \alpha,\dot\beta=1,2,\ s,t=1,\dots N$, where x^μ are the even space-time variables and $\theta,\bar\theta$ are the odd spinor coordinates. The real superspace corresponds to the sector satisfying $(x^\mu)^*=x^\mu$, $(\theta^\alpha_s)^*=\theta^{\dot\alpha s}$. The superconnection $A_A(y)$ is Lie algebra valued. The covariant derivatives have the form $\mathcal{D}_A=D_A+iA_A$, where

$$D_A=(\partial_\mu,D^s_\alpha,D_{\dot\beta t}),\ \partial_\mu=\partial/\partial x^\mu,\ \partial_{\alpha\dot\beta}=\sigma^\mu_{\alpha\dot\beta}\partial_\mu,$$

$$D^s_\alpha=\frac{\partial}{\partial\theta^\alpha_s}+i\bar\theta^{\dot\beta s}\partial_{\alpha\dot\beta}\quad,\quad D_{\dot\beta t}=-\frac{\partial}{\partial\bar\theta^{\dot\beta t}}-i\theta^\alpha_t\partial_{\alpha\dot\beta}.$$

The supercurvature F_{AB} is defined by the graded commutator of covariant derivatives

$$[\mathcal{D}_A,\mathcal{D}_B]=T^C_{AB}\mathcal{D}_C+iF_{AB},$$

where T^C_{AB} is the torsion. All components of T^C_{AB} are equal to zero, except the following components: $T^{s\ \ \mu}_{\alpha,\dot\beta t}=T^{\ \ s\mu}_{\dot\beta t,\alpha}=-2i\delta^s_t\sigma^\mu_{\alpha\dot\beta}$. The superconnection has to satisfy eqs.(1). These equations are differential equations in superspace. We shall consider the latter within the superanalysis developed in ref.[18].

In the following calculations we shall intensively use the consequences of Bianchi identities, established by Sohnius. Assuming the constraint equations (1), one gets

$$F^{st}_{\alpha\beta}=\varepsilon_{\alpha\beta}\bar W^{st}\quad,\quad \bar W^{st}=-\bar W^{ts} \tag{4.a}$$

$$F_{\dot\alpha s,\dot\beta t}=\varepsilon_{\dot\alpha\dot\beta}W_{st},\quad W_{st}=-W_{ts} \tag{4.b}$$

From the Bianchi identities Sohnius derived the following identities [5]

$$\mathcal{D}^i_\alpha\bar W^{jk}=-\mathcal{D}^j_\alpha\bar W^{ik} \tag{5.a}$$

$$\mathcal{D}_{\dot\alpha i}\bar W^{jk}=\frac{1}{2}(\delta^j_i\mathcal{D}_{\dot\alpha l}\bar W^{lk}-\delta^k_i\mathcal{D}_{\dot\alpha l}\bar W^{lj}) \tag{5.b}$$

$$\mathcal{D}_{\dot\alpha i}W_{kl}=-\mathcal{D}_{\dot\alpha k}W_{il} \tag{5.c}$$

$$\mathcal{D}^i_\alpha W_{jk}=\frac{1}{2}(\delta^i_j\mathcal{D}^l_\alpha W_{lk}-\delta^i_k\mathcal{D}^l_\alpha W_{lj}) \tag{5.d}$$

Together with the equalities

$$F_{\beta\dot\alpha,\gamma}^{\quad\ s}=\frac{i}{4}\varepsilon_{\beta\gamma}\mathcal{D}_{\dot\alpha t}\bar W^{ts} \tag{6.a}$$

$$F_{\alpha\dot\beta,\dot\gamma s}=\frac{i}{4}\varepsilon_{\dot\beta\dot\gamma}\mathcal{D}^t_\alpha W_{ts} \tag{6.b}$$

$$F_{\alpha\dot\beta,\gamma\dot\nu}=-\frac{1}{48}[\varepsilon_{\dot\beta\dot\nu}(\mathcal{D}^t_\alpha\mathcal{D}^s_\gamma W_{st}+(\alpha\leftrightarrow\gamma))+\varepsilon_{\alpha\gamma}(\mathcal{D}_{\dot\beta t}\mathcal{D}_{\dot\nu s}\bar W^{st}+(\dot\beta\leftrightarrow\dot\nu)) \quad (6.c)$$

identities (5.a)-(5.d) make up the set of relations which eqs.(I) and the Bianchi identities implie for the super-curvature.

Let us introduce the notation

$$\lambda_{\alpha i}=\mathcal{D}^j_\alpha W_{ji}\ , \qquad (7.a)$$

$$\bar\lambda^{\ i}_{\dot\alpha}=\mathcal{D}_{\dot\alpha j}\bar W^{ji}\ , \qquad (7.b)$$

$$\lambda_{\beta 4}=\frac{1}{3}\,\varepsilon_{ikl}\mathcal{D}^i_\beta\bar W^{kl}\ , \qquad (8.a)$$

$$\bar\lambda^{\ 4}_{\dot\beta}=\frac{1}{3}\,\varepsilon^{ikl}\mathcal{D}_{\dot\beta i}W_{kl}\ , \qquad (8.b)$$

$$\bar\Phi^{i4}=\frac{1}{2}\,\varepsilon^{ijl}W_{jl}\ , \quad \Phi_{i4}=\frac{1}{2}\,\varepsilon_{ilk}\bar W^{lk}, \qquad (9.a)$$

$$\bar\Phi^{ij}=\bar W^{ij}\ , \quad \Phi_{ij}=W_{ij}\ . \qquad (9.b)$$

Eqs.(5.a),(5.b) say that $\mathcal{D}_{\dot\alpha i}W_{jk}$ and $\mathcal{D}^i_\alpha\bar W^{jk}$ are totally anti-symmertic in (ijk) and thus from (8.a) and (8.b) we get

$$\mathcal{D}^i_\alpha\bar W^{kl}=\frac{1}{2}\,\varepsilon^{ikl}\lambda_{\alpha 4} \qquad (8.a')$$

$$\mathcal{D}_{\dot\beta i}W_{kl}=\frac{1}{2}\,\varepsilon_{ikl}\bar\lambda^{4}_{\dot\beta} \qquad (8.b')$$

Using (5.a)-(5.d) we can derive the following identities

$$\mathcal{D}^i_\alpha\mathcal{D}_{\dot\beta e}\bar W^{ek}=-4i\,\mathcal{D}_{\alpha\dot\beta}\bar W^{ik} \qquad (10)$$

$$\mathcal{D}_{\dot\alpha i}\mathcal{D}^j_\alpha W_{jk}=-4i\mathcal{D}_{\alpha\dot\alpha}W_{ik} \qquad (11.a)$$

$$\mathcal{D}^1_\alpha\mathcal{D}_{\dot\beta i}W_{1k}=-2i\mathcal{D}_{\alpha\dot\beta}W_{1k} \qquad (11.b)$$

$$\mathcal{D}^l_\beta\mathcal{D}^j_\tau W_{jl}=\mathcal{D}^l_\tau\mathcal{D}^j_\beta W_{jl}+i\varepsilon_{\beta\tau}[\bar W^{il},W_{li}] \qquad (12)$$

$$\varepsilon^{\dot\alpha\dot\beta}\mathcal{D}_{\dot\beta j}\mathcal{D}_{\dot\alpha l}\bar W^{lk}=-i\delta^k_j[W_{lq},\bar W^{lq}]+4i[W_{jl},\bar W^{lk}]\ . \qquad (13)$$

We start from spinor equations. Differentiating (11.a) by $\mathcal{D}^k_\beta$ and anticommuting $\mathcal{D}^k_\beta$ with $\mathcal{D}_{\dot\alpha i}$ and also taking into account eq.(6.a) we get

$$-2i(\mathcal{D}_{\beta\dot\alpha}\lambda_{\alpha i}+2\mathcal{D}_{\alpha\dot\alpha}\lambda_{\beta i})=\mathcal{D}_{\dot\alpha i}\mathcal{D}^k_\beta\mathcal{D}^l_\alpha W_{lk}-i\varepsilon_{\alpha\beta}[\mathcal{D}_{\dot\alpha l}\bar W^{lk},W_{ik}]\ .$$

So the antisymmetric part of $\mathcal{D}_{\beta\dot{\alpha}}\lambda_{\alpha i}$ in (α,β) can be written as

$$\mathcal{D}_{\alpha\dot{\alpha}}\lambda_{\beta i}-\mathcal{D}_{\beta\dot{\alpha}}\lambda_{\alpha i}=\frac{\varepsilon_{\alpha\beta}}{2}(\mathcal{D}_{\dot{\alpha}i}[\bar{W}^{kl},W_{lk}]+2[\lambda^{k}_{\dot{\alpha}},W_{ik}]).$$

Using the identity (5.b) and definitions (8.b),(9.a) the first term on the R.H.S. of the previous formula can be written as

$$\mathcal{D}_{\dot{\alpha}i}[\bar{W}^{kl},W_{lk}]=-[\bar{\lambda}^{k}_{\dot{\alpha}},W_{ik}]+\frac{I}{2}\varepsilon_{ilk}[\bar{W}^{kl},\bar{\lambda}^{4}_{\dot{\alpha}}]=$$
$$-[\bar{\lambda}^{k}_{\dot{\alpha}},W_{ik}]+[\lambda^{4}_{\dot{\alpha}},\Phi_{i4}]$$

We have thus obtained the following spinor equation

$$\mathcal{D}^{\beta}_{\ \dot{\alpha}}\lambda_{\beta i}=\frac{I}{2}[\bar{\lambda}^{K}_{\dot{\alpha}},\Phi_{iK}]\ ,\quad i=I,2,3;\quad K=I,2,3,4. \tag{I4}$$

We wish to derive the equation of motion for $\bar{\lambda}^{4}_{\dot{\beta}}$. Acting on the identity (I0) by $\mathcal{D}_{\dot{\beta}i}$ yields

$$-4i\mathcal{D}_{\dot{\beta}i}\mathcal{D}_{\alpha\dot{\alpha}}W_{lk}=\mathcal{D}_{\dot{\beta}i}\mathcal{D}_{\dot{\alpha}l}\mathcal{D}^{j}_{\alpha}W_{jk}\ . \tag{I5}$$

Commuting $\mathcal{D}_{\dot{\beta}i}$ with $\mathcal{D}_{\alpha\dot{\alpha}}$ on the L.H.S. of (I5) and aticommuting $\mathcal{D}_{\dot{\alpha}l}$ with $\mathcal{D}^{j}_{\alpha}$ on the R.H.S., taking into account eq. (6.b) and collecting the terms with space derivatives we obtain

$$\mathcal{D}_{\alpha\dot{\alpha}}\mathcal{D}_{\dot{\beta}i}W_{lk}+\mathcal{D}_{\alpha\dot{\beta}}\mathcal{D}_{\dot{\alpha}l}W_{ik}=\frac{\varepsilon_{\dot{\beta}\dot{\alpha}}}{4}[\lambda_{\alpha i},W_{lk}]+\frac{i}{2}\mathcal{D}^{j}_{\alpha}\mathcal{D}_{\dot{\beta}i}\mathcal{D}_{\dot{\alpha}l}W_{jk}\ .$$

For the antisymmetric part of this formula in $(\dot{\alpha},\dot{\beta})$ one has

$$\mathcal{D}_{\alpha\dot{\alpha}}\mathcal{D}_{\dot{\beta}i}W_{lk}-\mathcal{D}_{\alpha\dot{\beta}}\mathcal{D}_{\dot{\alpha}i}W_{lk}=\frac{\varepsilon_{\dot{\beta}\dot{\alpha}}}{4}[\lambda_{\alpha i},W_{lk}]+\frac{i}{2}\mathcal{D}^{j}_{\alpha}(\mathcal{D}_{\dot{\beta}i}\mathcal{D}_{\dot{\alpha}l}-\mathcal{D}_{\dot{\alpha}i}\mathcal{D}_{\dot{\beta}l})W_{jk}$$

and subsequently the full antisymmetrical part of the latter formula with respect to the indences (i,l,k) after some algebra can be written as

$$\varepsilon^{ilk}(\mathcal{D}_{\alpha\dot{\alpha}}\mathcal{D}_{\dot{\beta}i}W_{lk}-\mathcal{D}_{\alpha\dot{\beta}}\mathcal{D}_{\dot{\alpha}i}W_{lk})=\frac{3}{4}\varepsilon_{\dot{\beta}\dot{\alpha}}\varepsilon^{ilk}[\lambda_{\alpha i},W_{lk}] \tag{I6}$$

Eq.(I6) is merely the equation of motion for the fourth componet of the spinor field:

$$\mathcal{D}_{\alpha}^{\ \dot{\beta}}\bar{\lambda}^{4}_{\dot{\beta}}=-\frac{I}{2}[\lambda_{\alpha i},\bar{\Phi}^{i4}] \tag{I7}$$

Equations of motions for $\bar{\lambda}^{i}_{\dot{\alpha}}$ and $\lambda_{\alpha 4}$ can be obtained by similar calculations.

Let us calculate $\mathcal{D}^{\mu}F_{\mu\nu}$. By (6.c) we get

$$\mathcal{D}^{\mu}F_{\mu\nu}=\frac{I}{4\cdot 8\cdot 3\cdot 2}\,\sigma^{\alpha\dot{\beta}}_{\nu}(J^{(I)}_{\alpha\dot{\beta}}+J^{(2)}_{\alpha\dot{\beta}}), \tag{I8}$$

where

$$J^{(1)}_{\alpha\dot\beta}=\mathcal{D}^{\tau}{}_{\dot\beta}(\mathcal{D}^{i}_{\alpha}\lambda_{\tau i}+\mathcal{D}^{i}_{\tau}\lambda_{\alpha i})\ ;\ J^{(2)}_{\alpha\dot\beta}=\mathcal{D}_{\alpha}{}^{\dot\tau}(\mathcal{D}_{\dot\beta i}\bar\lambda^{i}_{\dot\tau}+\mathcal{D}_{\dot\tau i}\bar\lambda^{i}_{\dot\beta})$$

Taking into account eq.(12.a), commuting the operators $\mathcal{D}^{\tau}{}_{\dot\beta}$ and $\mathcal{D}^{i}_{\alpha}$ and then using the spinor equations of motion (14) we get

$$J^{(1)}_{\alpha\dot\beta}=-\frac{3}{2}\{\bar\lambda^{K}_{\dot\beta},\lambda_{\alpha K}\}+[\mathcal{D}^{i}_{\alpha}\bar\lambda^{k}_{\dot\beta},W_{ik}]-[\bar W^{lk},\mathcal{D}^{i}_{\alpha}\mathcal{D}_{\dot\beta i}W_{lk}]-i\mathcal{D}_{\alpha\dot\beta}[\bar W^{ik},W_{ki}]$$

The second and the third term on the R.H.S. of this formula can be expressed, according to (10),(11.b) through the commutators of W and its space derivatives. So $J^{(1)}_{\alpha\dot\beta}$ is given by

$$J^{(1)}_{\alpha\dot\beta}=-\frac{3}{2}\{\bar\lambda^{L}_{\dot\beta},\lambda_{\alpha L}\}-3i\,[\mathcal{D}_{\alpha\dot\beta}\bar W^{ik},W_{ik}]-3i[\mathcal{D}_{\alpha\dot\beta}W_{ik},\bar W^{ik}]\,. \quad (19)$$

In the similar way one can derive the following relation

$$J^{(2)}_{\alpha\dot\beta}=-\frac{3}{2}\{\lambda_{\alpha L},\bar\lambda^{L}_{\dot\beta}\}-3i[\mathcal{D}_{\alpha\dot\beta}W_{ik},\bar W^{ik}]-3i[\mathcal{D}_{\alpha\dot\beta}\bar W^{kl},W_{kl}]\,. \quad (20)$$

It follows from (18),(19) and (20) that

$$\mathcal{D}^{\mu}F_{\mu\nu}=-\frac{1}{64}\sigma_{\nu}^{\alpha\dot\beta}\{\lambda_{\alpha L},\bar\lambda^{L}_{\dot\beta}\}-\frac{i}{16}([\mathcal{D}_{\nu}W_{ik},\bar W^{ik}]+[\mathcal{D}_{\nu}\bar W^{kl},W_{kl}]), \quad (21)$$

i.e. the equation of motion for the vector field $A_{\mu}(y)$.

We only have to derive the equation of motion for the scalar field W. If we differentiate (11.a) by $\mathcal{D}^{\alpha k}$ we get, after anticommuting $\mathcal{D}^{k}_{\beta}$ with $\mathcal{D}_{\alpha i}$ and commuting $\mathcal{D}^{k}_{\alpha}$ with $\mathcal{D}_{\alpha\dot\alpha}$

$$-\varepsilon^{\alpha\beta}\mathcal{D}_{\dot\alpha i}\mathcal{D}^{k}_{\beta}\mathcal{D}^{l}_{\alpha}W_{lk}+2i\varepsilon^{\alpha\beta}\mathcal{D}_{\beta\dot\alpha}\mathcal{D}^{k}_{\alpha}W_{ik}=2i[\lambda^{k}_{\dot\alpha},W_{ik}]\,. \quad (22)$$

Applying the operator $\mathcal{D}^{\dot\alpha}_{j}$ to the relation (22) and taking the antisymmetric part in (i,j) we find

$$-\frac{1}{2}\varepsilon^{\dot\alpha\dot\beta}\varepsilon^{\alpha\beta}\{\mathcal{D}_{\dot\beta j},\mathcal{D}_{\dot\alpha i}\}\mathcal{D}^{k}_{\beta}\mathcal{D}^{l}_{\alpha}W_{lk}-i\varepsilon^{\alpha\beta}\{\lambda_{\beta[j},\lambda_{\alpha i]}\}-8i\mathcal{D}^{\alpha}{}_{\dot\alpha}\mathcal{D}_{\alpha}{}^{\dot\alpha}W_{ij}=$$
$$2i\varepsilon^{\dot\alpha\dot\beta}\mathcal{D}_{\dot\beta[j}[\bar\lambda^{k}_{\dot\alpha},W_{i]k}]\,. \quad (23)$$

By (4.b) and (12) the first term on the L.H.S. of (23) can be expressed through the commutator of W and $\bar W$, so that it remains to calculate the R.H.S. of (23). We claim that by virtue of eq.(13) the R.H.S. of (23) can be written as

$$2i\mathcal{D}_{\dot\beta[j}[\bar\lambda^{k}_{\dot\alpha},W_{i]k}]=2[[W_{lq},\bar W^{ql}],W_{ij}]-8[[W_{[jl},\bar W^{lk}],W_{i]k}]+$$
$$2i\{\bar\lambda^{k\dot\beta},\mathcal{D}_{\dot\beta j}W_{ik}\}\,.$$

So we obtain

$$16\mathcal{D}^{\mu}\mathcal{D}_{\mu}W_{ij}=3[W_{ij},[\bar W^{kl},W_{lk}]]-8([[W_{[jl},\bar W^{lk}],W_{i]k}]-$$

$$i\{\lambda_i^{\alpha},\lambda_{\alpha j}\}+i\varepsilon_{ijk}\{\bar{\lambda}^{4\dot{\beta}},\bar{\lambda}^{k}_{\dot{\beta}}\}.$$

For N=3 it is possible to prove

$$3[W_{ij},[\bar{W}^{kl},W_{lk}]]-8[[W_{[jl},\bar{W}^{lk}],W_{i]k}]=$$
$$-([\bar{W}^{rs},[W_{ij},W_{rs}]]+[W_{rs},[W_{ij},\bar{W}^{rs}]]).$$

At last we get the equation of motion for W:

$$16\mathcal{D}^{\mu}\mathcal{D}_{\mu}W_{ij}=-([\bar{W}^{rs},[W_{ij},W_{rs}]]+[W_{rs},[W_{ij},\bar{W}^{rs}]]-i\{\lambda_i^{\alpha},\lambda_{\alpha j}\}$$
$$+i\varepsilon_{ijk}\{\bar{\lambda}^{4\dot{\beta}},\bar{\lambda}^{k}_{\dot{\beta}}\}. \qquad (24)$$

Equations (14),(17),(21) and (24) after the substitution

$$W_{ik}=4i\varphi_{ik},\ \bar{W}^{ik}=4i\bar{\varphi}^{ik},\ \lambda_{\alpha I}=8i\psi_{\alpha I},\ \bar{\lambda}^{I}_{\dot{\alpha}}=8i\bar{\psi}^{I}_{\dot{\alpha}} \qquad (25)$$

for $\theta=\bar{\theta}=0$ coincide with eqs.(3.a),(3.b),(3.c) and (3.d). Thus, the dynamical equations for the N=4 supersymmetric Yang-Mills theory in Minkowski space are obtained from the constraint equations (1) for N=3 superspace.

3. RECONSTRUCTION OF SUPERCONNECTION FROM PHYSICAL FIELDS

For the ease of illustration let us first consider the Abelian case. In this case we can suppose that the expansions of the superfields $W,\bar{W}$ in powers of $\theta,\bar{\theta}$ are given, because the expressions $(D)^n(\bar{D})^m W$ for arbitrary (n,m) can be determined according to formulas (5),(6.c),(7),(8),(10)-(13) in terms of physical fields $W|$, $\bar{W}|$, $\lambda|$, $\bar{\lambda}|$, $A_{\mu}|$, their space derivatives and commutators. It is obvious that the reconstructed superfields W, $\bar{W}$ satisfy eqs.(5),(6.c),(10)-(13).

We can find particular solutions of eqs.(4.a),(4.b). $A^{t}_{\beta}=\theta_{\beta k}\bar{C}^{tk}$, $A_{\dot{\beta}s}=\bar{\theta}^{k}_{\dot{\beta}}C_{sk}$, where

$$\bar{C}^{tk}=\frac{1}{2}\bar{W}^{tk}+\frac{1}{6}\varepsilon^{\gamma\rho}\theta_{\gamma r}D^{r}_{\rho}\bar{W}^{tk}\ ,\quad C_{tk}=-\frac{1}{2}W_{tk}+\frac{1}{6}\varepsilon^{\dot{\gamma}\dot{\rho}}\bar{\theta}^{r}_{\dot{\gamma}}D_{\dot{\rho}r}W_{tk}\ .$$

The general solutions of (4.a),(4.b) in some gauge are $A^{t}_{\beta}=\theta_{\beta k}\bar{C}^{tk}$, $A_{\dot{\beta}s}=\bar{\theta}^{k}_{\dot{\beta}}C_{sk}+D_{\dot{\beta}s}X$. It remains to solve eq.(1.b),i.e.

$$D^{s}_{\alpha}D_{\dot{\beta}t}X=Q^{s}_{\alpha,\dot{\beta}t}-2i\delta^{s}_{t}A_{\alpha\dot{\beta}}\ , \qquad (26)$$

where $Q^{s}_{\alpha,\dot{\beta}t}=\bar{\theta}^{k}_{\dot{\beta}}D^{s}_{\alpha}C_{tk}+\theta_{\alpha k}D_{\dot{\beta}t}\bar{C}^{sk}$. About $A_{\alpha\dot{\beta}}$ we have according to (6.a),(6.b) some additional information

$$D^s_\gamma A_{\alpha\dot\beta} = K^s_{\gamma,\alpha\dot\beta} \,, \tag{27}$$

$$D_{\dot\gamma t} A_{\alpha\dot\beta} = \bar K_{\dot\gamma t,\alpha\dot\beta} + D_{\dot\gamma t}\partial_{\alpha\dot\beta} X \,, \tag{28}$$

where $K^s_{\gamma,\alpha\dot\beta} = -\frac{i}{4}\varepsilon_{\alpha\gamma}\bar\lambda^s_{\dot\beta} + \theta_{\gamma k}\partial_{\alpha\dot\beta}\bar C^{sk}$, $\bar K_{\dot\gamma t,\alpha\dot\beta} = -\frac{i}{4}\varepsilon_{\dot\beta\dot\gamma}\lambda_{\alpha t} + \bar\theta^k_{\dot\gamma}\partial_{\alpha\dot\beta}C_{tk}$.
The system of eqs.(26)-(28) is the overdeterminant system of equations for unknown superfields X and $A_{\alpha\dot\beta}$. It is possible to prove that necessary and sufficient conditions for compatibility of the system of eqs.(26)-(28)

$$D^t_\varrho K^s_{\gamma,\alpha\dot\beta} + D^s_\gamma K^t_{\varrho,\alpha\dot\beta} = 0,$$

$$D_{\dot\nu p}\bar K_{\dot\gamma t,\alpha\dot\beta} + D_{\dot\gamma t}\bar K_{\dot\nu p,\alpha\dot\beta} = 0,$$

$$D^p_\nu Q^s_{\alpha,\dot\beta t} + D^s_\alpha Q^p_{\nu,\dot\beta t} - 2i(\delta^s_t K^p_{\nu,\alpha\dot\beta} + \delta^p_t K^s_{\alpha,\nu\dot\beta}) = 0,$$

$$D_{\dot\nu q} Q^s_{\alpha,\dot\beta t} + D_{\dot\beta t} Q^s_{\alpha,\dot\gamma q} - 2i(\delta^s_t \bar K_{\dot\gamma q,\alpha\dot\beta} + \delta^s_q \bar K_{\dot\beta t,\alpha\dot\gamma}) = 0,$$

$$D_{\dot\gamma t} K^s_{\gamma,\alpha\dot\beta} + D^s_\gamma \bar K_{\dot\gamma t,\alpha\dot\beta} + \partial_{\alpha\dot\beta} Q^s_{\gamma,\dot\gamma t} + 2i\delta^s_t F_{\gamma\dot\gamma,\alpha\dot\beta} = 0$$

are true. So for arbitrary $A_{\alpha\dot\beta}|$ and $X(x,\theta,\bar\theta=0)$ the superfields $A_{\alpha\dot\beta}(y)$ and $X(y)$ can be found from (26)-(28).

In non-Abelian case we proceed as in the Abelian case by first trying to find particular solutions of eqs.(4.a), (4.b). The same ansatz $A^s_\alpha = \theta_{\alpha k}\bar C^{ks}$, where $\bar C^{ks}$ satisfies the relation

$$\bar C^{st} = \frac{1}{2}\bar W^{st} + \frac{1}{2}\varepsilon^{\beta\alpha}\theta_{\beta k}D^{[s}_\alpha \bar C^{t]k} + \frac{i}{4}\varepsilon^{\beta\alpha}\theta_{\alpha k}\theta_{\beta l}[\bar C^{sk},\bar C^{tl}]$$

solves eqs.(4.a). $A_{\dot\alpha q} = \bar\theta^k_{\dot\alpha}C_{kq}$ solves eqs.(4.b) if

$$C_{st} = \frac{1}{2}W_{st} + \frac{1}{2}\varepsilon^{\dot\beta\dot\alpha}\bar\theta^k_{\dot\beta}D_{\dot\alpha[s}C_{t]k} + \frac{i}{4}\varepsilon^{\dot\beta\dot\alpha}\bar\theta^k_{\dot\alpha}\bar\theta^l_{\dot\beta}[C_{sk},C_{tl}] \,. \tag{30}$$

The general solution of (4.a),(4.b) in some gauge are given by

$$A^q_\alpha = \theta_{\alpha k}\bar C^{kq} \,, \quad A_{\dot\alpha q} = e^{-iV}\bar\theta^k_{\dot\alpha}\tilde C_{kq}e^{iV} - ie^{-iV}D_{\dot\alpha q}e^{iV}, \tag{31}$$

where $\bar C^{kq}$ is given by (29) and $\tilde C_{kq}$ is determined by (30) in which instead of W_{st} enters $\tilde W_{st} = e^{iV}W_{st}e^{-iV}$. Substituting (31) in eq.(1.b) yields

$$\begin{aligned} iD^s_\alpha(e^{-iV}D_{\dot\alpha t}e^{iV}) = &-\theta_{\alpha k}D_{\dot\alpha t}\bar C^{ks} + D^s_\alpha(e^{-iV}\bar\theta^k_{\dot\alpha}\tilde C_{kt}e^{iV}) + \\ &\{\theta_{\alpha k}\bar C^{sk}, e^{-iV}D_{\dot\alpha t}e^{iV}\} + i\{\theta_{\alpha k}\bar C^{sk}, e^{-iV}\bar\theta^k_{\dot\alpha}\tilde C_{kt}e^{iV}\} + 2i\delta^s_t A_{\alpha\dot\alpha} \,. \end{aligned} \tag{32}$$

In non-Abelian case we can't suppose that we know the superfields $W, \bar W$ from the beginning. In this case it is necessery to reconstruct the superfields $W,\bar{W},A^s_\alpha,A_{\dot\alpha q}$ and $A_{\alpha\dot\alpha}$ by iteration procedure at a time. Supposing $V|=0$ and taking into account (32) one get

$$V=-2iA_{\alpha\dot\beta}|\theta^\alpha_s\bar\theta^{s\dot\beta}+\ldots$$

So the first order terms in the expansion of the $A^s_\alpha,A_{\dot\alpha q}$ in powers of $(\theta,\bar\theta)$ are

$$A^s_\alpha=\frac{I}{2}\theta_{\alpha k}\bar{W}^{ks}|+\ldots\ ,\quad A_{\dot\alpha t}=\frac{I}{2}\bar\theta^k_{\dot\alpha}W_{kt}|-2i\theta^\beta_t A_{\beta\dot\alpha}|+\ldots$$

The following terms can be in principle reconstructed, but here we have not the possibility to go in details. One can see that the reconstruction procedure discribed above is the generalization of the method of solution of the N=2 constraint equations developed by R.Grimm, M.Sohnius and J.Wess[4]. We are informed that similar results was also obtained by J.Harnad et al.

4. Explicit expression for the first non-trivial current.

In this section explicit expression in terms of superconnection for the first non-trivial conserved current will be written (for details see ref.[12]). The currents $J_{(2)s}$ and $J^{(2)s}$ are constructed by the following formulas

$$J_{(2)s}=-iA_{2s}\ ,\quad J^{(2)s}=(D^s_2+iA^s_2)\Psi, \tag{33}$$

where

$$\Psi(x,\theta,\bar\theta)=-i\int A_{2\dot I}(x,\theta,\bar\theta)\Big|_{\theta^I=0=\bar\theta^{\dot I}}dx_{I\dot I}+$$

$$f^s(z)\theta^I_s+(P^tf^s)(z)\frac{\theta^I_t\theta^I_s}{2}+(P^tP^uP^s)(z)\frac{\theta^I_u\theta^I_t\theta^I_s}{3!}\ ,$$

$$z=(\tilde{x},\theta^I_s=0,\theta^2_s,\bar\theta^{\dot\alpha s}),\ \tilde{x}_{I\dot I}=x_{I\dot I},\ \tilde{x}_{I\dot 2}=x_{I\dot 2},\ \tilde{x}_{2\dot I}=x_{2\dot I}+i\theta^2_s\bar\theta^{\dot Is},$$

$$\tilde{x}_{2\dot 2}=x_{2\dot 2}\ ;\ f^s=-iA^s_2-\frac{i}{2}\bar\theta^{\dot Is}D_{\dot Iq}A^q_2\ ,\ P^s=\frac{\partial}{\partial\theta^I_s}+i\bar\theta^{\dot 2s}\partial_{I\dot 2}\,.$$

Here the gauge $A^s_I=A_{\dot It}=A_{I\dot I}=0$ is used.

The currents (33) satisfy the following identities

$$D^{(s}_I J^{(2)t)}=0,\qquad D_{\dot I(s}J_{(2)t)}=0\ ,$$

$$D_{\dot It}J^{(2)s}+D^s_IJ_{(2)t}=\delta^t_s(D_{\dot Iq}J^{(2)q}+D^q_IJ_{(2)q})$$

To express the currents in terms of the physical

fields we are left to find the explicit expression for the gauge transformation from $A^s_I = A_{\dot{I}t} = A_{I\dot{I}} = 0$ gauge to the gauge used in the previous section.

5. B-MODEL

As it was mentioned in Introduction the hidden symmetry for the N=4 SYM theory corresponding to infinite number of conserved currents is unknown.

In this section we shall consider the model described by the superconnection $A_A(y)$ and some additional superfield $B_s(y)$. We impose on these fields the following equations

$$
\begin{aligned}
&F^{(s,t)}_{\alpha\ \beta}=0\ ,\quad F_{\dot{I}t,\dot{I}s}=0=F_{\dot{2}t,\dot{2}s}\ ,\quad F_{\dot{I}(t,\dot{2}s)}+B_{(t}B_{s)}=0\ ,\\
&\mathcal{D}_{\dot{\alpha}(t}B_{s)}=0\ ,\quad F^s_{I,\dot{I}t}=0=F^s_{2,\dot{2}t}\ ,\\
&F^s_{2,\dot{I}t}+\mathcal{D}^s_I B_t=0\ ,\quad F^s_{I,\dot{2}t}+\mathcal{D}^s_2 B_t=0.
\end{aligned}
\tag{34}
$$

This system of equations is non-relativistic one. But it is invariant under the gauge transformation

$$A_A \rightarrow K^{-I}A_A K + K^{-I}D_A K\ ,\qquad B\rightarrow K^{-I}BK\ ,$$

where K is an arbitrary superfield. If B=0 eqs.(34) are the N=3 SYM constraints (I).

Let us consider a set of linear systems

$$
\begin{aligned}
&X^s(\lambda)\Psi(\lambda)=(\nabla^s_I+\lambda\nabla^s_2)\Psi(\lambda)=0,\\
&Y_t(\lambda)\Psi(\lambda)=(\nabla_{\dot{I}t}+\lambda B_t+\lambda^2\nabla_{\dot{2}t})\Psi(\lambda)=0,\\
&Z(\lambda)\Psi(\lambda)=(\nabla_{I\dot{I}}+\lambda\nabla_{2\dot{I}}+\lambda^2\nabla_{I\dot{2}}+\lambda^3\nabla_{2\dot{2}})\Psi(\lambda)=0,
\end{aligned}
\tag{35}
$$

where $\nabla^s_\alpha=D^s_\alpha+iA^s_\alpha$, $\nabla_{\dot{\alpha}t}=D_{\dot{\alpha}t}+iA_{\dot{\alpha}t}$, $\nabla_{\alpha\dot{\beta}}=\partial_{\alpha\dot{\beta}}+iA_{\alpha\dot{\beta}}$, λ is an arbitrary complex parameter.

The integrability conditions of these linear equations are

$$\{X^s,X^t\}=0=\{Y_t,Y_s\}\ ,\quad \{X^s,Y_t\}=-2i\delta^s_t Z\ .$$

They give eqs.(34) on the supercurvature and superfield B. Note that the expansions of X,Y,Z in the powers of λ content the powers of λ without admission. This implies that it is possible to deal with the integrability properties of system (34) by the Riemann-Hilbert transformation method without face with a reduction problem. About the discussion reduction problem see ref.[19]. For the system of eqs.(35) we can use the usual Riemann-Hilbert approach [8] and find the following hidden symmetry transformations:

$$\delta A_2^s=-\varepsilon i D_2^s S(\mu)\ ,\quad \delta A_1^s=0\ ,\ \delta A_{\dot{2}t}=-\varepsilon i D_{\dot{2}t} S(\mu)\ ,\quad \delta A_{\dot{1}t}=0\ ,$$

$$\delta B_t=i\mu\delta A_{\dot{2}t}+\varepsilon\,[B_t,S(\mu)],\ \delta A_{1\dot{1}}=0\ ,\quad \delta A_{2\dot{2}}=-i\varepsilon\mathcal{D}_{22}S(\mu)\ ,$$

$$\delta A_{2\dot{1}}=-i\varepsilon\,(\mathcal{D}_{2\dot{1}}S(\mu)+\mu\mathcal{D}_{1\dot{2}}S(\mu)+\mu^2\mathcal{D}_{2\dot{2}}S(\mu))\ , \qquad (36)$$

$$\delta A_{1\dot{2}}=-i\varepsilon(\mathcal{D}_{1\dot{2}}S(\mu)+\mu\mathcal{D}_{2\dot{2}}S(\mu)),$$

where $S(\mu)$ satisfies the following equations

$$(\mathcal{D}_1^s+\mu\mathcal{D}_2^s)S(\mu)=0\ ,$$

$$(\mathcal{D}_{\dot{1}t}+\mu^2\mathcal{D}_{\dot{2}t})S(\mu)=-\mu[B_t,S(\mu)]\ ,$$

$$(\partial_{1\dot{1}}+\mu\mathcal{D}_{2\dot{1}}+\mu^2\mathcal{D}_{1\dot{2}}+\mu^3\mathcal{D}_{2\dot{2}})S(\mu)=0.$$

The transformations (36) form a loop algebra, which occurs in two-dimensional integrable models such as chiral model [8].

REFERENCES

[1] Gates S.J.Jr., Grisaru M.T., Roček M. and Siegel W. Superspace. The Benjamin/Cummings publishing Company, 1983.
[2] Brink L. and Howe P.S., Phys.Lett.88B,268(1979).
[3] Gates S.J. Jr. and Grimm R., Phys.Lett.133B,192(1983).
[4] Grimm R., Sohnius M. and Wess J., Nucl.Phys.133B,275(1978
[5] Sohnius M. Nucl.Phys.136B,461(1978).
[6] Witten E. Phys.Lett.77B,394(1978).
[7] Volovich I.V. Lett.Math.Phys.7,517(1983).
[8] Solitons, eds. Bullough R.K. and Caudrey P.J., Springer-Verlàg, Berlin-Heidelberg-New York, 1980.
Julia B. in:Frontiers in particle physics '83, eds. Šijački Dj. et al., World Scientific, Singapore, 1984.
L.Dolan, Kac-moody algebras and exact solvability in hadron physics, preprint RU 83/B/63,(1983).
Chau L.L., Ge M.L., Sinha A. and Wu Y.S., Phys.Lett. 121B,391(1983).
Ueno K. and Nakamura Y., Phys.Lett.109B,273(1982).
Devchand C. and Fairlie D.B., Nucl.Phys.B194,232(1982).
Breitenlohner P. and Maison D., Explicit and hidden symmetries of dimensionally reduced (super-)gravity theories, preprint MPI-PAE/PTh 1/84.
[9] Galperin A., Ivanov E., Kalitzin S, Ogievetsky V. and

Sokatchev E., Unconstrainted N=2 matter, Yang-Mills and supergravity theories in harmonic superspace, preprint, IC/84/43, 1984.

[10] Gliozzi F., Scherk J. and Olive D., Nucl.Phys.122B, 253(1977).
Brink L., Schwarz J. and Scherk J.,Nucl.Phys.121B, 77(1977).

[11] Devchand C.,Nucl.Phys. B238,331(1984).

[12] Aref'eva I. and Volovich I.,Phys.Lett.148,500 (1984).

[13] Vladimirov V.S. and Volovich I.V., Consruction of local and non-local conservation laws for non-linear field equations, preprint, IC/84/128, 1984.

[14] Howe P.S., Stell K.S. and Townsend P,Nucl.Phys.236B, 125(1984).

[15] Mandelstam S., Nucl.Phys.213B,149(1983).

[16] Brink L., Lindgren O. and Nillson B.E.W.,Phys.Lett. (1983).

[17] Namazie M.A., Salam A. and Strathdee J., Finiteness of broken N=4 super Yang-Mills theory, preprint, IC/82/230,1982.

[18] Vladimirov V.S. and Volovich I.V., Theor. and Math. Phys. 59,3(1984);60,169(1984).

[19] Bohr H., Ge M.-L. and Volovich I., New hidden symmetries in 2-dimensional models, preprint, IC/84/165, 1984.

[20] Chau L.-L., Ge M.-L. and Popovich Z.,Phys.Rev.Lett.52, 1940(1984).

KALUZA-KLEIN THEORIES AND SPONTANEOUS COMPACTIFICATION MECHANISMS OF EXTRA SPACE DIMENSIONS

D.P.Sorokin, V.I.Tkach, D.V.Volkov
Institute of Physics and Technology
the Ukrainian Academy of Sciences ,
310108, Kharkov,
USSR

1. The extension of supersymmetry and supergravity theories to space-time with $d > 4$ /1/ has led recently to the revival of Kaluza-Klein ideas /2/. If extra dimensions belong to a compact subspace of a small size, their existence does not contradict to the every day experience and exhibits itself through symmetry properties of fields in the physical 4-dimensional space-time. (For a review of modern approaches to the Kaluza-Klein theory see ref./3/.) The existence of compact subspaces must not also contradict to the equations of motion of an original 4+n - dimensional theory. Thus the compactification of extra dimensions into a compact spaces must be spontaneous, i.e., it must be caused by the interaction of gravitation with other fields which are contained in the theory. It is necessary to note that the Einstein gravity itself cannot provide the formation of the desired vacuum state where space-time is the direct product of the 4-dimensional Minkowski space M^4 and n-dimensional compact space M^n.

The main hope of the approach is that multidimesional theory of (super)gravity in which the spontaneous compactification takes place can be one of the candi -

dates for unified field theory of gravitational, strong and electroweak interactions. It seems that d=10 and 11 supergravities are the most probable candidates for the realistic Kaluza-Klein theory. This is because

i) these theories necessarily contain gauge fields which are superpartners to gravitational and Rarita - Schwinger fields and thus the spontaneous compactification can occur naturally as a result of their interactions;

ii) one may hope that these theories can be finite;

iii) it is possible that just a spontaneous compactification of extra dimensions in these theories may result in a definite group structure of internal symmetries, it may also give a clue to an understanding of the consecutive spontaneous breaking of internal symmetries and of spontaneous supersymmetry breaking, thus determining the phenomenology of all known interactions.

2. In this connection much attention has been given to the study of various spontaneous compactification mechanisms in 4+n-dimensional Einstein-Yang-Mills theories and d = 10, 11 supergravities. It seems that all posible mechanisms must give the contribution to the spontaneous compactification of extra dimensions, because all that is not forbided can happen.

The first mechanism of the spontaneous compactification of exstra dimensions was considered by Cremmer and Scherk for compactification into S^n-spheres /4/ and then was generalized by Luciani /5/ in case of symmetric spaces G/H (G is the symmetry group and H is the holonomy group of such spaces). In this mechanism the spontaneous compactification in the 4+n-dimensional Eistein-Yang-Mills theory and the formation of nontrivial vacuum states in which space-time is $M^4 \times G/H$ with a metric

$$g_{MN} = \begin{pmatrix} g_{\mu\nu}(x) & 0 \\ 0 & g_{mn}(y) \end{pmatrix}, \qquad (1)$$

are caused by the interaction of the gravitational field with the background gauge field A_M^α being transformed under group G and having the following components

$$A_\mu^\alpha = 0, \quad A_m^\alpha = K_m^\alpha(y), \qquad (2)$$

where $K_m^\alpha(y)$ are the Killing vectors of the symmetric space G/H. Such an ansatz Satisfies the equations of motion*)

$$R_{MN} - \frac{1}{2} g_{MN}R = -2\varkappa^2 T_{MN} + 2\varkappa^2 \lambda g_{MN}, \qquad (3a)$$

$$(\hat{D}_M F^{MN})^\alpha = (D_M + A_M) F^{MN\alpha} = 0, \qquad (3b)$$

$$T_{MN} = \frac{1}{e^2}(F_{ML}F_N{}^L - \frac{1}{4} g_{MN}F^2),$$

if we introduce into the theory the original cosmological constant λ . The size of the compact space G/H is characteriesed by a mean curvature $K \sim \frac{e^2}{2\varkappa^2}$ ($R_{nm} = \frac{K}{2} g_{mn}$) Which dimension is determined by original gravitational ($\varkappa$) and gauge (e) coupling constants. The field strength $F_{mn}^\alpha = \partial_m A_n^\alpha - \partial_n A_m^\alpha + C_{\beta\gamma}^\alpha A_m^\beta A_n^\gamma$ corresponding to the compactified solution has the form

$$F_{(M)(n)}^\alpha = -KC_{(m)(n)}^A D_A^\alpha(y), \qquad (4)$$

*) Our convention is the following: the indices M,N...=0, ..., 4+n-1 are the curved indices and (M),(N),... are the flat indiced of 4+n-dimensional space-time; Greek indiced μ , ν ,... = 0,...,3 correspond to 4-dimensional space time with coordinates x^μ ; Latin indeces m,n,...=1,.. ., n correspond to the compact space G/H with coordinates y^m; α , β are the indeces of group G and A, B, C,... are the indices of its subgroup H

in orthonormal basis. $D_A^{\alpha}(y)$ the matrix components belonging to the adjoint representation of the group G.

A more economical mechanism of spontaneous compactification has been proposed by Volkov and Tkach /6/ and later by Randjbar-Daemi and Percassi /7/. In this mechanism only the fields of the gauge group H participate in the formation of the symmetric space G/H. To obtain nontrivial background solutions (different from the 4+n-dimentional Minkowski space) of eqs.(3), which must be simple and have a comparatively high symmetry to consider them as possible ground states, it was proposed to impose on the background gravitational and gauge feilds the parallelizability conditions /6/ :

$$D_m R_{enpr} = 0 \tag{5a}$$

$$(\hat{D}_m F_{np})^A = 0 \tag{5b}$$

We put the vacuum values of all other components of R_{LMNQ} and $F_{MN}{}^A$ equal to zero. The first condition can be considered as the definition of the n-dimensional symmetric space G/H which has a maximum symmetry group. The curvature tensor of the G/H-space is constructed in the orthonormal basis with the help of the group G structure constants $C_{(m)(n)}{}^A$:

$$R_{(m)(n)(l)}{}^{(p)} = K\, C_{(m)(n)}{}^A\, C_{A(l)}{}^{(P)} \tag{6}$$

The transition to the curved basis is realized by means of veilbains $e_m^{(m)}(y)$.

If the second condition (5b) is fulfilled, then the Yang-Mills equations are satisfied identically. To derive the solution of eq.(5b) for the n-dimensional symmetric space, let us consider the integrability condition

$$[\hat{D}_e, \hat{D}_m]\, F_{np}^A = 0 \tag{7}$$

The solution of eq.(7) in the orthonormal basis is

$$F_{(m)(n)}{}^{A} = - \varkappa\, C_{(m)(n)}{}^{A} \tag{8}$$

which also satisfies (5b) as a consequence of the Maurer-Cartan structure equations for the G-forms.

Substituting (6) and (8) into eqs.(3) we obtain, just as in the first mechanism, that

$$\lambda = \frac{1}{4e^2} F_{mn}{}^{A} F^{nmA} , \qquad \varkappa = \frac{e^2}{2\varkappa^2} \tag{9}$$

It can be shown /8/ that the two considered mechanisms of spontaneous compactification are equivalent for symmetric spaces with a simple holonomy group H, i.e., the gauge strength tensors of eqs. (4) and (8) and corresponding gauge field are connected with each other by definite gauge group G transformation which is determined by matrix $D_\beta^\alpha(y)$. Thus the number of background gauge fields participating in the Cremmer-Scherk spontaneous compactification mechanism can be reduced and they become equal to the gauge fields of the mechanism of refs. /6,7/.

Further one can try to find compact spaces demanding the least number of the original gauge fields participating in their formation. It occurs that such spaces are symmetric ones having nonsimple holonomy groups, i.e., the groups which are the direct product of several groups ($H = H_1 \times H_2 \times \ldots$). For example, grassmanian spaces

$$\frac{SO(p+q)}{SO(p)\times SO(q)} , \quad \frac{S_p(p+q)}{S_p(p)\times S_p(q)} , \quad \frac{SU(p+q)}{SU(p)\times SU(q)\times U(1)}$$

can be formed as a due to result of spontaneous compactification caused by the interaction of the gravitational field with gauge fields transforming at least under one of the normal subgroups of their holonomy group /9/. The field strength tensor of the background gauge fields participating in the formation of the such spaces has the form of eq. (8), but index A belongs now to one of the normal subgroups of the holonomy group. Note that the Cremmer-Scherk mechanism becomes unequivalent to the second mechanism in case of symmetric spaces with nonsimple holonomy groups, because in spite of the reducibility of the group G to the holonomy group, a further reduction of H to one of its subgroups is impossible. The economy of the considered compactification mechanism in the choice of the gauge fields participating in compactification allows it to hold in d=10, 11 supergravities the supermultiplets of which contain vector and tensor abelian fields (see below).

There are some other mechanism of spontaneous compactification. One of them is based on taking account of matter field Casimir energy arising after the formation of compact subspaces and supporting there existence /10/. Another mechanism takes place when extra dimensions are compactified into parallelizable compact manifolds (i.e. manifolds admitting zero curvature, but in general nonzero torsion) such as group manifolds and seven sphere . In these cases the spontaneous compactification is caused by the interaction of gravitation with background tensor antisymmetric gauge fields A_{mn}, A_{mnl}, which play the role of "torsion" of parallelizable manifolds /11,12/ . It is interesting to note that such background fields satisfy parallelizability conditions analogous to eg.(5b).

3. The most popular mechanism of spontaneous compactification in d=11 supergravity is the Freund-Rubin mechanism/13/. As a result of its action 11-dimensional spacetime is compactified into the direct product of 4-dimensional anti-de-Sitter space (Ad S^4) and 7-dimensional compact space. This happens when the antisymmetric field

A_{MNL} of d=11 supergravity /1/ acquires a nonzero background value

$$F_{\mu\nu\lambda\rho} = \sqrt{-\det g_{\mu\nu}(x)}\ 3m\ \mathcal{E}_{\mu\nu\lambda\rho} \qquad (10)$$

(where $F_{MNLP} = \frac{1}{4}\ \partial_{[M} A_{NLP]}$, m - parameter) and other components are zero. In this case the solutions of the bosonic part of the d=11 supergravity equations of motion (vacuum expectation values of the Schwinger field and their bilinear combinations are supposed to be zero)

$$R_{MN} - \frac{1}{2} g_{MN} R = -\frac{\varkappa^2}{3} (F_{MLPQ} F_N{}^{LPQ} - \frac{1}{8} F^2 g_{MN}) \qquad (11)$$

$$D_M F^{MNLQ} = -\frac{\varkappa}{(4!)^2} \frac{1}{\sqrt{-\det g_{MN}}} \mathcal{E}^{NLQM_1 \ldots M_8} F_{M_1 \ldots M_4} \times F_{M_5 \ldots M_8}$$

are the vacuum states characterised by space-time being the direct product of 4-dimensional and 7-dimensional Einstein space which have the following Ricci tensors

$$R_{\mu\nu}(x) = 12\ m^2\ g_{\mu\nu}(x)\ , \qquad (12a)$$

$$R_{mn}(y) = -\ 6\ m^2\ g_{mn}(y)\ . \qquad (12b)$$

It is seen that we cannot demand a flateness of 4-dimensional background space-time without putting the curvature of a compact space to zero. On the other hand, if we postulate a small size of a compact space, then an enormous cosmological constant appears in the effective 4- dimensional theory. This is one of the important but yet unsolved problems arising in the Kaluza-Klein d=10, 11 supergravity theories as a result of impossiblility of introdusing the original cosmological constant into these theories. One of the possible solutions of this problem was proposed by Duff and Orzalesi /14/ and consists in taking into account nonzero background fermionic bilinear fields contributed to the torsion of compactified space-time.

There are infinitely many Freund-Rubin solutions characterized by egs.(10),(12). The main task is to separate from this infinity probable candidates for actual vacuum configurations of the theory. Note that the complete quantum mechanical vacuum could be an admixture of different vacuum configurations with possible tunnelling from one to another. One of the most convincing criteria for the compactified vacuum-state stability is the existence of residual supersymmetries in the effective 4-dimensional theory the number of which is characterized by the number of covariantly constant 8-component spinors η (so-called Killing spinors) being solutions of the equation.

$$D_m \eta - \frac{m}{2} e_m^{(m)}(y) \Gamma_{(m)} \eta = 0 , \tag{13}$$

Where $\Gamma_{(m)}$ are the d=7 Dirac matrices.

An examples we mention are the round seven-sphere S^7 with a symmetry group SO(8) conserving N=8 supersymmetries in the d=4 theory and corresponding to the gauged N=8 supergravity of de Wit , Nicolai /15/; squashed S^7 (i.e., the manifold which is topologically equivalent to the round S^7 but having a different Einstein metric) with SO(5)×SU(2) as a symmetry group and N=1 supersymmetry*); coset spaces $\frac{SU(3)\times SU(2)\times U(1)}{SU(2)\times U(1)\times U(1)}$ admiting N=2 supersymmetry /17/; $\frac{SU(3)\times U(1)}{U(1)\times U(1)}$ - spaces with N=3,1 symmetries /18/ and some other. A list of 7-dimensional compact spaces being Freund-Rubin solutions of d=11 supergravity and their topologies are considered in ref./19/.

Duff, Nilsson and Pope /20/ have shown that not only vacuum configurations admitting supersymmetry, but

---

*) For comprehensive review of Freund-Rubin and Englert solutions with S^7 topology see Duff et al./16/ and ref. theres.

also the configurations obtained from supersymmetric ones by a reversal of M^7 orientation by sending $e^{(m)}_m(y)$ to $-e^{(m)}_m(y)$ are stable. It was however shown by several authors [20,21] that Freund-Rubin solutions corresponding to the Einstein spaces M^7 which are direct-product spaces ($S^5\times S^2$, $S^4\times S^3$, $CP^2\times S^3$ ect.) are unstable. The stability of Englert-type solution[11] where the background field $A_{mn\ell}$ plays the role of "parallelized torsion"of the compact space M^7 (i.e.,corresponds to the existence of Ricci flattening torsion on M^7) and leads to the breaking of all supersymmetries in d=4, is also not clear because Englert S^7 is unstable against small fluctuations[22].

4. Though much attention is given to the Freund-Rubin and Englert like compactification mechanisms of d=11 supergravity,there are other possibilities of spontaneous compactification in the theory which lead to the compactification of the d=11 space-time into the direct product of a 5-dimensional space-time and 6-dimensional symmetric space $G/H_o\times H_1$ (where H_o is the abelian subgroup)[23]. Note that there are no 7-dimensional symmetric spaces with this property. This compactification takes place by means of a mechanism considered in refs.[6-8] and generalised for the case of antisymmetric gauge field A_{MNL}. Background solutions of eqs.(11) for A_{MNL}, or equivalently for F_{MNLP} corresponding to the spontaneous compactification have nonzero values with internal indices and are built in the orthonormal basis with the help of group G structure constants having one index (o) corresponding to the H_o subgroup and other indices corresponding to $G/H_o\times H_1$:

$$F_{(m)(n)(l)(p)} = \rho\, C_{[(m)(n)}{}^{o}\, C_{(l)(p)]}{}^{o} \quad , \qquad (14)$$

where ρ is a parameter. The examples of possible symmetric spaces are $CP^2\times S^2$, CP^3, $S^2\times S^2\times S^2$, $\frac{SO(5)}{SO(3)\times SO(2)}$.

The vacuum states characterized by eq.(14) have

$SO(4,2) \times G$ as a symmetry group ($SO(4,2)$ being the symmetry group of AdS^5).

The subsequent possibility of compactifying five dimensions into the direct product of 4- dimensional Minkowski space and an S^1-curl is connected with the solution of a cosmological constant problem. It is possible that such spontaneous compactification mechanism can extend the class of stable vacuum configurations.

5. Let us now turn to the consideration of a spontaneous compactification in N=2, d=10 supergravities. There exist two versions of the N=2, d=10 supergravity, both connected with a d=10 superstring theory /24/ but having a different field content.

In case of the so-called chiral N=2, d=10 supergravity /25/ none of interesting compactified solutions has been found yet /26/.

Another (nonchiral) N=2, d=10 theory can be obtained from d=11 supergravity by trivial dimensional reduction of one coordinate (or its compactification into S^1) /27/. As a consequance of such compactification an abelian gauge vector field arises in d=10 dimensions similar to the classical 5-dimensional Kaluza-Klein theory.

As both the A_{MNL} and A_M fields are contained in the theory, a spontaneous compactification of d=10 space-time can be induced by means of a combined Freund-Rubin mechanism (eq.(10)) and the mechanism of ref./6-9/ (eq.(8)). This indeed takes place and several solutions of the equations of motion were obtained /9, 23, 27-29/ corresponding to the direct product of AdS^4 and a compact 6-dimensional space ($M^6 = CP^2 \times S^2$, CP^3, $S^4 \times S^2$, $S^2 \times S^2 \times S^2$, $\frac{SU(3)}{U(1) \times U(1)}$ $\frac{SO(5)}{SO(3) \times SO(2)}$).

6. Now the question arises: if a nonchiral N=2, d=10 supergravity obtained from d=11 by compactification of

one dimension admits a nontrivial spontaneous compactification of six extra dimensions, then can such a compactification be interpreted as a consecutive compactification of the d=11 supergravity, at the first stage of which the S^1-curl and the abelian gauge field A_M arise, which ensures the second stage - a compactification of six coordinates with the help of the background field (10) ? This means that if such a consecutive compactification takes place indeed, then the Freund-Rubin solutions (i.e., compact Eistein manifolds M^7) must be characterized by the metric which has the form of a classical Kaluza-Klein metric

$$g_{mn}(y) = \begin{pmatrix} g_{\hat{m}\hat{n}}(z) + R_o^2 A_{\hat{m}}(z)A_{\hat{n}}(z) & R_o A_{\hat{m}} \\ R_o A_{\hat{m}} & 1 \end{pmatrix} \tag{15}$$

where $g_{\hat{m}\hat{n}}(z)$ is the metric of the compact 6-dimensional space M^6 with the coordinates z^m, R_o is the radius of the S^1-curl. From the differential-geometry point of view the manifolds characterized by the metric of eq.(15) can be interpreted as principal fiber bundles $P(M^6,S^1)$ with a compact space M^6 as a base and a structure group $U(1) = S^1$ as a fiber (see, for example,/29/). The abelian field $A_{\hat{m}}(z)$ plays the role of fiber bundle connection components.

It appears that all known Freund-Rubin solutions of the d=11 supergravity including S^7-spheres and $\frac{SU(3)\times SU(2)\times U(1)}{SU'(2)\times U'(1)\times U''(1)}$ - spaces (except the solution with a $\frac{SU(3)}{SO(3)} \times S^2$-space, which is unstable /21/) can be interpreted as fiber bundle manifolds having metrics of eq. (15) with a proper choice of M^6 and $A_{\hat{m}}$ connections. The results are listed in the Table, where $E(S^4,S^2,SU(2))$ and $E(CP^2,S^2,SU(2))$ - are the associative instanton fiber bundles over the S^4-sphere and CP^2, respectively, with an S^2-fiber and the SU(2) structure group.

Compact manifolds $M^7 = P(M^6, S^1)$		Supersymmetries in d=4	Symmetry groups of vacuum configurations
$T^7 \sim T^6 \times S^1$		$N=8$	$Superpoinc. \times [U(1)]^7$
$S^7 \sim P(CP^3, S^1) \sim U(4)/U(1) \sim P(E(S^4, S^2, SU(2)), S^1) \sim P(S^4, SU(2)) \sim SO(5)/SU(2)$		$N=8$	$OSp(4,8)$
$K3 \times T^3 \sim (K3 \times T^2) \times S^1$		$N=4$	$Superpoinc. \times [U(1)]^3$
$SU(3)/U(1) \sim P(E(CP^2, S^2, SU(2)), S^1) \sim P(CP^2, SU(2)) \sim SU(3) \times SU(2)/SU(2) \times U(1)$		$N=3$	$OSp(4,3) \times SU(3)$
$\frac{SU(3) \times SU(2) \times U(1)}{SU'(2) \times U'(1) \times U''(1)} \sim P(CP^2 \times S^2, S^1)$	$p/q = 1$	$N=2$	$OSp(4,2) \times SU(3) \times SU(2)$
	$p/q \neq 1$	$N=0$	$SO(3,2) \times SU(3) \times SU(2) \times U(1)$
$\frac{SO(5) \times SO(2)}{SO(3) \times SO(2)} \sim P(\frac{SO(5)}{SO(3) \times SO(2)}, S^1)$		$N=2$	$OSp(4,2) \times SO(5)$
$\frac{SU(2) \times SU(2) \times SU(2) \times U(1)}{U'(1) \times U''(1) \times U'''(1)} \sim P(S^2 \times S^2 \times S^2, S^1)$	$p \neq q \neq r$	$N=0$	$SO(3,2) \times [SU(2)]^3 \times U(1)$
	$p = q = r$	$N=2$	$OSp(4,2) \times [SU(2)]^3$
"squashed"-$S^7 \sim P(E(S^4, S^2, SU(2)), S^1) \sim P(S^4, SU(2))$		$N=1$	$OSp(4,1) \times SO(5) \times SU(2)$
"squashed"-$SU(3)/U(1) \sim P(E(CP^2, S^2, SU(2)), S^1) \sim P(CP^2, SU(2))$		$N=1$	$OSp(4,1) \times SU(3) \times SU(2)$
$\frac{SU(3) \times U(1)}{U'(1) \times U''(1)} \sim P(\frac{SU(3)}{U(1) \times U'(1)}, S^1)$		$N=1$	$OSp(4,1) \times SU(3) \times U(1)$
$S^4 \times S^3 \sim P(S^4 \times S^2, S^1)$		$N=0$	$SO(3,2) \times SO(5) \times SO(4)$

It is interesting to discuss in brief this SU(2) - instanton on CP^2. It is invariant up to SU(2) gauge transformations under the action of SU(3)-group transformations, and in the orthonormal basis its field strength tensor has the same form as the classical BPST-instanton solution constructed with the help of t'Hooft tensors.

One can see an interesting analogy between the S^7 and $\frac{SU(3)}{U(1)}$ spaces both interpreted as instanton fiber bundles. The both spaces admit two different Einstein metrics related to each other by continuous deformations. In case of squashed S^7 and "squashed" $\frac{SU(3)}{U(1)}$, N=1 supersymmetry survive in four dimensions.

Thus, we can state that all compactified vacuum solutions of the nonchiral N=2, d=10 supergravity, which can be obtained due to the action of the combined Freund-Rubin mechanism, including the background abelian field A_M, are equivalent to those of the d=11 supergravity represented in the Table. The examples are all known M^6 spaces being the solutions of d=10 supergravity equations of motion /9,27-29/. We can also assume that if the d=11 background solution conserves supersymmetry in d=4, then so does the corresponding solution of the d=10 supergravity. Note that the relation between d=10 backgrounds $AdS^4 \times M^{4(compact)} \times S^2$ and d=11 backgrounds $AdS^4 \times M^{4(compact)} \times S^3$ was considered also by Giani and Pernici /27/. It was mentioned there that though there is a relationship between the backgrounds of two theories, the fluctuations about these two backgrounds are different since the dimensional reduction from d=11 to d=10 selects only a subset of the harmonics on which the fluctuations of the 11-dimensional theory are expanded. An example of such differences is the calculation of $SU(3) \times SU(2) \times U(1)$ gauge coupling constants for the case of the d=10 background space $AdS^4 \times CP^2 \times S^2$ and d=11 backgrounds AdS^4 x

$\times \frac{SU(3)\times SU(2)\times U(1)}{SU'(2)\times U'(1)\times U''(1)}$ /30/. In case of the d=11 theory, the obtained coupling constants are drastically different from strong- and electroweak-interaction strengths. On the other hand, in case of the N=2, d=10 supergravity, the coupling constants are adjustable to the strong, electroweak interaction strengths.

7. The realization of beautiful and tempting Kaluza-Klein ideas involves however a lot of important problems to be solved before the unified theory based on extra dimensions could be called a realistic one. First of all, it's necessary to mention such problems as:

i) An enormous cosmological constant in the effective d=4 theory arising after spontaneous compactification of the d=10 and 11 supergravities. Since the spontaneous supersymmetry breaking leads to the cosmological term of an apposite sign, the general solution of cosmological constant value problem requires the consideration of both bosonic and fermionic excitations.

ii) Identification of symmetries of internal compact spaces with those of strong and electroweak interactions. This problem arises because in the most case the symmetry groups of compactified vacuum solutions are too small to contain the $SU(3)_C\times SU(2)_W\times U(1)_Y$ group of internal symmetries. Even in case of $\frac{SU(3)\times SU(2)\times U(1)}{SU'(2)\times U'(1)\times U''(1)}$ spaces the symmetry group cannot be interpreted as a group of a standard model. (It was shown by Randjibar-Daemi, Salam and Strathdee /31/ using the harmonic analysis on homogeneous space/3/). One of the ways out is to consider the theory obtained after the spontaneous compactification as a preonic one and postulate that some of the gauge bosons and some of the fermions of the standard model are the composites formed from the preons of supergravity.

In this case the relevant hidden symmetry must be found, e.g. the SU(8) symmetry of the N=8 supergravity. (More detailed speculations on this topic can be found in ref. /16/.

iii) The problem of chirality which is also connected with the second one. It was pointed out by Witten /17/ that none of the classical solutions of the d=11 supergravity corresponding to spontaneous compactification could provide the arising of chiral fermions (i.e. the asymmetry between the left- and right-handed fermions) necessary for the d=4 effective theory. It is Worth noting that in the Einstein-Yang-Mills theories this difficulty can be overcome and chiral fermions can appear in d=4 due to monopole- and instanton-indiced compactification /32/.

iv)The problem of C,P,T symmetries and their violation /33/.

v) The problem of defining a consistent criterion of truncation from the complete theory with infinite towers of multiplets to an effective d=4 theory with a finite number of multiplets. In A-de Sitter space the idea of keeping the massless modes and discarding the massive ones wrong. Other criteria are required /34/.

vi) And the most fundamental problem of the finiteness of supergravity theories.

The future will show whether the Kaluza-Klein theory can answer all the questions and become the realistic unified theory of gravitational, strong and electroweak interactions.

REFERENCES

1. Cremmer E., Julia B., Scherk J., Phys.Lett, 76B, 469, (1978).
2. Kaluza Th., Sitzungsber.preuss.Acad.Wiss., 966 (1921); Klein O., Z.Phys., 37, 895 (1926).
3. Salam A., Strathdee J., Ann.of Phys., 141, 316 (1982).
4. Cremmer E., Scherk J., Nucl.Phys., B118, 61, (1977).
5. Luciani J.F., Nucl.Phys., B135, 111 (1978).
6. Volkov D.V., Tkach V.I., Zh.E.T.Fiz.Lett., 32, 681, (1980); Teor.Mat.Fiz., 51, 171 (1982).
7. Randjibar-Daemi S., Percassi R., Phys.Lett., 117B, 378 (1982).
8. Volkov D.V., Sorokin D.P., Tkach V.I., Teor.Mat.Fiz., 56, 171 (1983).
9. Volkov D.V., Sorokin D.P., Tkach V.I., Zh.E.T.Fiz.Lett, 38, 397 (1983); Teor.Mat.Fiz., 61, 241 (1984).
10. Voronov N.A., Kogan Ya.I., Zh.E.T.Fiz.Lett., 38, 262, (1983); Candelas P., Weinberg S., Nucl.Phys., B237, 586 (1984).
11. Englert F., Phys.Lett., 119B, 339 (1982).
12. Tze C.H., Phys.Lett., 128B, 160 (1983).
13. Freund P.G.O., Rubin M.A., Phys.Lett., 97B, 233 (1980).
14. Duff M.J., Orzalesi C.A., Phys.Lett., 122B, 37 (1983)
15. De Wit B., Nicolai H., Phys.Lett., 108B, 285 (1982); Nucl.Phys., B208, 323 (1982).
16. Duff M.J., Nilsson B.E.W., Pope C.N., Nucl.Phys., B233, 433 (1984).
17. Witten E., Nucl.Phys., B186, 412 (1981); Castellani L., D'Auria R., Fre F., Nucl.Phys., B239,610 (1984).
18. Castellani L., Romans L.J., Nucl.Phys., B238, 683 (1984); Volkov D.V., Sorokin D.P., Tkach V.I., Zh. E.T.Fiz.Lett., 40, 356 (1984).

19. Castellani L., Romans L.J., Warner N.P., Nucl.Phys., B241, 429 (1984).
20. Duff M.J., Nilsson B.E.W., Pope C.N., Phys.Lett., 139B, 154 (1984)
21. Ito K., Yasura O., Phys.Rev.Lett., 52, 1849 (1984).
22. De Wit B., Nicolai H., Nucl.Phys., B231, 506 (1984). Biran B., Spindel Ph., Bruxelles Preprint U.L.B, TH 84/002 (1984); Kogan Ya.I., Morozov A.Yu., Zh. E.T.Fiz.Lett., 39, 482 (1984).
23. Volkov D.V., Sorokin D.P., Tkach V.I., Yad.Fiz., 39, 1314 (1984).
24. Schwarz J.H., Phys.Reports, 89, 223 (1982).
25. Hawe P.S., West P.C., Nucl.Phys., B238, 181 (1984).
26. Schwarz J.H., Preprint CALT-68-1049 (1984).
27. Giani F., Pernici M., Phys.Rev., D30, 325 (1984) Campbell I.C.G., West P.C., Nucl.Phys., B243,112(1984)
28. Watamura S., Phys.Lett., 129B, 188 (1983); Chapline G., Slansky R., Los Alamos Preprint, LA-UR-84-939 (1984).
29. Eguci T., Gilkey P.B., Hanson A.J., Phys.Reports, 66, 214 (1980); Percassi R., Randjibar-Daemi S., J.Math.Phys., 24, 807 (1983); Sorokin D.P., Tkach V.I., Ukr.Fiz.Zh., 28, 1605 (1984).
30. Ezawa Z.F., Koh J.G., Phys.Lett., 142B, 153, 157 (1984).
31. Randjibar-Daemi S., Salam A., Strathdee I., Tries Preprint IC/84/15 (1984).
32. Randjibar-Daemi S., Salam A., Strathdee I., Nucl. Phys., B214, 491 (1983); Phys.Lett.,132B, 56 (1983).
33. Wetterich C., Nucl.Phys., B234, 413 (1984).
34. D'Auria R., Fre P., CERN Preprint TH 3860; TH 3861 (1984).

THE CHIRAL ANOMALY IN CONFORMAL AND ORDINARY SIMPLE SUPERGRAVITY IN FUJIKAWA'S APPROACH

by

P. H. Frampton

Department of Physics
University of North Carolina at Chapel Hill
Chapel Hill, North Carolina 27514

D. R. T. Jones

Department of Physics
University of Colorado
Boulder, Colorado 8039

P. van Nieuwenhuizen

Institute for Theoretical Physics
State University of New York at Stony Brook 11794

S. C. Zhang

Institute for Theoretical Physics
State University of New York at Stony Brook 11794

1. INTRODUCTION

In this contribution we shall reobtain the chiral anomaly of simple ordinary supergravity by means of Fujikawa's method[1)] as well as by the Pauli-Villars method. Then we shall present, as a new result, the axial anomaly for simple conformal supergravity.

Axial anomalies have been discussed extensively in recent articles. For supergravity, the issue is, as usual, more subtle than elsewhere, because one must fix gauges and add ghosts for the fermions in the loop. The axial anomal in simple ordinary supergravity has been calculated by various methods, see below. We begin by reobtaining the same result by means of the original Fujikawa method, since it is interesting in itself and will be used to illustrate certain aspects in the conformal computation. We show that using as regulator either the operator which is obtained directly from the classical action plus gauge fixing term, or simply the Dirac operator itself, yields the same result, which agrees with observations made in ref.[2)]. We present the Pauli-Villars computation[3)] because it most clearly shows which regulator should be used for a given anomaly. [As an example, we note that in a theory with only lef-handed spin 1/2 Dirac fermions the regulator is given by

$$\frac{1}{2}\, \not{D}(1+\gamma_5) + \frac{1}{2}\, \not{\partial}(1-\gamma_5) \tag{1}$$

since massive Pauli-Villars fields contains propagating left-and right-handed fields, while, however, only the (left-handed) fermion-loops must be regularized.]

The computations in conformal supergravity are based on the original papers of Kaku and Townsend and van Nieuwenhuizen[4)]. We will also use some important results obtained by Fradkin and Tseytlin, who computed the β-function in N = 1,2,3,4 conformal supergravity[5)]. It would be interesting to study the multiplet structure of the trace, chiral and other anomalies of conformal supergravity.

The gravitational spin 3/2 axial anomaly in four dimensions has been computed by various methods: by determining the eigenvalues of the relevant Hodge-de Rham operators[6)], by a Feynman graph analysis (imposing gravitational conservation, the Adler-Rosenberg method[7)], by zeta-function regularization (determining the a_2 coefficients by the coincidence limit method of Synge and DeWitt)[7], by the point splitting method[7], and by the topological method (determining the index of certain operators involved)[7]. Recently, Alvarez-Gaumé and Witten[2] computed the gravitational spin 3/2 axial anomaly in n dimensions, using a direct Feynman graph method (not by imposing gravitational conservation). They also gave a derivation of their results using a modificaiton of Fujikawa's method (by introducing a one-dimensional quantum-mechanical system whose Hamiltonian is equal to the exponent of Fujikawa's regulator, exp $(\not{D}/M)^2$).

In this article we intend to give yet another derivation of the gravitational spin 3/2 axial anomaly, namely following Fujikawa's original approach for the gravitational spin 1/2 axial anomaly[1)]. For higher dimensions, this method becomes rather complicated, and the modification of ref.[2)] is more suitable. However, the original Fujikawa method is rather simple in principle. As we shall show, no point splitting techniques need be used. We also point out that the method is quite similar to those Pauli-Villar's regularizations of the path-integral whose Jacobian for chiral transformations is unity; however, unlike in the Yang-Mills case, in the gravitational case one needs more than one regulator, due to the derivative couplings of gravity.

2. SPIN 1/2 CASE BY FUJIKAWA'S METHOD

Consider a complex massless spin 1/2 fermion ψ in a gravitational background described by vielbeins $e_\mu^{\ m}$. The generator for connected (i.e., one-particle irreducible, in this case) graphs reads

$$W = \int d\psi\, d\bar{\psi}\, \exp\,[-e\,\bar{\psi}\gamma^\mu D_\mu \psi] \tag{2}$$

Under a chiral transformation of integration variables

$$\psi \to (1 + \alpha\gamma_5)\,\psi, \quad \bar{\psi} \to \bar{\psi}(1 + \alpha\gamma_5) \tag{3}$$

the path-integral does not change. Hence, the Jacobian cancels the variation of the action.

$$-\ \text{"Tr"}(2\alpha\gamma_5) + \langle -e\,\bar{\psi}\gamma^\mu\gamma_5\psi\rangle\,\partial_\mu\alpha = 0 \tag{4}$$

To regularize the trace "Tr" over spacetime points and spinor indices, Fujikawa showed that any function $f(\not{D}^2)$ can be used, provided $f(0) = 1$. The most convenient choice is $f = \exp(\not{D}/M)^2$. Hence, using plane waves in a four-dimensional box

$$\text{"Tr"}(2\alpha\gamma_5) = \Sigma\,\langle\vec{k}|\; e^{(\not{D}/M)^2}\; 2\alpha\gamma_5|\vec{k}\,\rangle$$

$$\equiv \int \frac{d^4k}{(2\pi)^4}\, e^{-ikx}\,[\mathrm{tr}\; e^{(\not{D}/M)^2}\, 2\alpha\gamma_5]\,e^{ikx} \tag{5}$$

where tr denotes the trace over spinor indices. By pulling the plane waves exp ikx to the left, the operator $\not{D}$ is replaced by $\not{D} + i\not{k}$, and one obtains.

$$"\mathrm{Tr}"(2\alpha\gamma_5) = \int \frac{d^4k}{(2\pi)^4}\, \mathrm{tr}(2\alpha\gamma_5) e^{-k^2/M^2} \exp(B/M^2)$$

$$B = 2ik\cdot D + D^2 + \frac{1}{4} R \qquad (6)$$

In deriving this result, we used the cyclic identity for the Riemann tensor

$$\frac{1}{4}[\gamma^\mu, \gamma^\nu][D_\mu, D_\nu] = \frac{1}{8}\, \gamma^{\mu\nu} R_{\mu\nu}{}^{ab} \gamma_{ab} = \frac{1}{4} R \qquad (7)$$

The evaluation of the integral is performed by expanding exp B/M^2, and only retaining terms which do not vanish when M^2 tends to infinity. The relevant terms are

$$"\mathrm{Tr}"(2\alpha\gamma_5) =$$

$$\int\frac{d^4k}{(2\pi)^4}\, e^{-k^2/M^2}\, \mathrm{tr}2\alpha\gamma_5 \times \Big[1 + \frac{1}{1!}\,(D^2 + \frac{1}{4} R)M^{-2}$$

$$+ \frac{1}{2!}\,\{(2ik\cdot D)(2ik\cdot D) + (D^2 + \frac{1}{4} R)(D^2 + \frac{1}{4} R)\}M^{-4}$$

$$+ \frac{1}{3!}\,\{(2ik\cdot D)(2ik\cdot D)(D^2 + \frac{1}{4} R) + (2ik\cdot D)(D^2 + \frac{1}{4} R)(2ik\cdot D)$$

$$+ (D^2 + \frac{1}{4} R)(2ik\cdot D)(2ik\cdot D)\}M^{-6} + \frac{1}{4!}\,\{(2ik\cdot D)^4\}M^{-8}\Big]$$

(8)

Using $\int d^4k = M^4 \pi^2 \int d(k^2/M^2)(k^2/M^2)$ and $\int_0^\infty dy(\exp{-y})y^n = \Gamma(n+1)$, we obtain

$$"\mathrm{Tr}"(2\alpha\gamma_5) =$$

$$\frac{2\alpha}{16\pi^2} \mathrm{tr}\gamma_5 \Big[M^2(D^2) + \frac{1}{2!}\{(-2)M^2D^2 + (D^2 + \frac{1}{4}R)(D^2 + \frac{1}{4}R)\} + \frac{1}{3!}(-2)\{[D^2(D^2 + \frac{1}{4}R) + D_\mu(D^2 + \frac{1}{4}R)D^\mu + (D^2 + \frac{1}{4}R)D^2\} + \frac{1}{4!}\frac{6\times 16}{24}\{D^2D^2 + D_\mu D^2 D^\mu + D_\mu D_\nu D^\mu D^\nu\}\Big] \quad (9)$$

The terms proportional to M^2 contain D^2, which contains four Dirac matrices since $D_\mu = \partial_\mu + \frac{1}{4}\omega_\mu{}^{mn}\gamma_{mn}$, but these terms are seen to cancel. The terms containing the scalar curvature R cancel, too, since they are given by

$$\frac{1}{2}(\frac{1}{4}D^2R + \frac{1}{4}RD^2) - \frac{1}{3}(\frac{1}{4}D^2R + \frac{1}{4}D_\mu RD^\mu + \frac{1}{4}RD^2) = \frac{1}{24}(D^2R) \quad (10)$$

which vanishes since $(D_\mu R) = (\partial_\mu R)$ so that the trace over $\gamma_5(D^2R)$ vanishes. In the remaining terms, those proportional to $(D^2)(D^2)$ cancel, too, and one is left with

$$"\mathrm{Tr}"2\alpha\gamma_5" = \frac{\alpha}{8\pi^2}\mathrm{tr}\gamma_5[-\frac{1}{3}D_\mu D^2 D^\mu + \frac{1}{6}\{D_\mu D^2 D^\mu + D_\mu D_\nu D^\mu D^\nu\}]$$

$$= \frac{\alpha}{96\pi^2}\mathrm{tr}\gamma_5[D_\mu, D_\nu][D^\mu, D^\nu]$$

$$= \frac{\alpha}{96\pi^2}\mathrm{tr}\gamma_5(\frac{1}{4}R_{\mu\nu}{}^{ab}\gamma_a\gamma_b)(\frac{1}{4}R^{\mu\nu cd}\gamma_c\gamma_d)$$

$$= \frac{\alpha}{96\pi^2}(\frac{1}{4}\varepsilon^{abcd}R_{\mu\nu ab}R^{\mu\nu}{}_{cd}) \quad (11)$$

This result as well as some of the intermediate steps are identical to[1], but we have avoided the use of point-splitting techniques in order to simplify the calculations. We will use the intermediate steps of this derivation when we discuss the spin 3/2 anomaly.

3. THE SPIN 3/2 CASE BY FUJIKAWA'S METHOD

Let us now consider the spin 3/2 case. The Rarita-Schwinger action for a real massless spin 3/2 field in an external gravitational field which satisfies the Einstein condition $R_{\mu\nu} = 0$ (necessary in order that the Rarita-Schwinger action be gauge-invariant by itself under $\delta\psi_\mu = D_\mu \varepsilon$) reads, after adding the usual gauge fixing term

$$\mathcal{L} = -\frac{e}{2}\bar{\psi}_\mu \gamma^{\mu\rho\sigma} D_\rho \psi_\sigma + \frac{e}{4}\bar{\psi}\cdot\gamma \not{D} \gamma\cdot\psi$$

$$= \frac{e}{4}\bar{\psi}_\mu \gamma^\sigma \gamma^\rho \gamma^\mu D_\rho \psi_\sigma = \frac{e}{2}\bar{\psi}_\mu (\Gamma^\rho)^{\mu\sigma} D_\rho \psi_\sigma \tag{12}$$

where $(\Gamma^\rho)^{\mu\sigma} \equiv \frac{1}{2}\gamma^\sigma\gamma^\rho\gamma^\mu$. Throughout we will suppress the spinor indices of the gravitino, but we will often explicitly exhibit the vector indices of the gravitino.

$$\text{"Tr"}(2\alpha\gamma_5) =$$

$$\int\frac{d^4k}{(2\pi)^4} e^{-ikx} \mathrm{tr}\, 2\alpha\gamma_5 \exp\left[\frac{(\Gamma^\mu D_\mu)(\Gamma^\nu D_\nu)}{M^2}\right] e^{ikx} \tag{13}$$

Using $\{\Gamma^\mu, \Gamma^\nu\}_{\rho\sigma} = 2g^{\mu\nu} g_{\rho\sigma}$ we get

$$\text{"Tr"}(2\alpha\gamma_5) =$$

$$\int\frac{d^4k}{(2\pi)^4} e^{-ikx} \mathrm{tr}\, 2\alpha\gamma_5 \exp M^{-2}\left(D^2 + \frac{1}{4}[\Gamma^\mu,\Gamma^\nu][D_\mu, D_\nu]\right) e^{ikx}$$

$$= \int\frac{d^4k}{(2\pi)^4} \mathrm{tr}\, 2\alpha\gamma_5 \exp\left(\frac{-k^2}{M^2}\right) \exp(B/M^2) \tag{14}$$

where now

$$B = 2ik\cdot D + D^2 + C, \quad C = \frac{1}{4}\,[\Gamma_\mu, \Gamma_\nu][D^\mu, D^\nu] \tag{15}$$

The evaluation of $[\Gamma^\mu,\Gamma^\nu][D_\mu,D_\nu]$ was given in[7)], but for completeness we will rederive it here.

$$C_{\lambda\rho} = \frac{1}{2}(\Gamma_\mu)_{\lambda\alpha}(\Gamma_\nu)_{\alpha\beta}\left[D_\mu,D_\nu\right]_{\beta\rho} = \frac{1}{8}(\gamma_\alpha\gamma_\mu\gamma_\lambda\gamma_\beta\gamma_\nu\gamma^\alpha)$$

$$\times\ \left[\frac{1}{4}\,R_{\mu\nu}{}^{ab}\gamma_a\gamma_b g_{\beta\rho} + R_{\beta\rho\mu\nu}I_s\right] \tag{16}$$

The symbol I_s denote the unit matrix in spinor space. Elementary Dirac matrix algebra leads to

$$\frac{1}{8}\,\gamma_\alpha\gamma_\mu\gamma_\lambda\gamma_\beta\gamma_\nu\gamma^\alpha = \frac{1}{2}(g_{\mu\lambda}g_{\beta\nu} - g_{\mu\beta}g_{\lambda\nu} + g_{\mu\nu}g_{\lambda\beta} - \varepsilon_{\mu\lambda\beta\nu}\gamma_5) \tag{17}$$

since $\gamma_\alpha\gamma^{\rho\sigma}\gamma^\alpha = 0$ in four dimensions. Hence

$$C_{\lambda\rho} = (g_{\mu\lambda}g_{\beta\nu} - \frac{1}{2}\,\varepsilon_{\mu\lambda\beta\nu}\gamma_5)(\frac{1}{4}\,R_{\mu\nu}{}^{ab}\gamma_{ab}g_{\beta\rho})$$

$$= \frac{1}{4}\,R_{\lambda\rho}{}^{ab}\gamma_{ab} + \frac{1}{4}\,R_{\lambda\rho}{}^{cd}\gamma_{cd} = \frac{1}{2}\,R_{\lambda\rho}{}^{ab}\gamma_{ab} \tag{18}$$

Instead of using $(\Gamma_\mu)_{\rho\sigma} = \frac{1}{2}\,\gamma_\sigma\gamma_\mu\gamma_\rho$ one could have used $(\Gamma_\mu)_{\rho\sigma} = g_{\rho\sigma}\gamma_\mu$ in the regulator. As claimed in ref.[2], "under certain broad assumptions" the result should be the same. In our case it is easy to see that this is true. Namely, not only is $\Gamma_{(\mu}\Gamma_{\nu)} = g_{\mu\nu}$ but also $C_{\lambda\rho}$ comes out the same. To see this, note that

$$C_{\lambda\rho} = \frac{1}{2}\,\gamma^{\mu\nu}g_{\lambda\beta}\left[D_\mu,D_\nu\right]_{\beta\rho} = \frac{1}{2}\,\gamma^{\mu\nu}\left[D_\mu,D_\nu\right]_{\lambda\rho}$$

$$= \frac{1}{2}\,\gamma^{\mu\nu}R_{\mu\nu\lambda\rho} \tag{19}$$

since the first term in $[D_\mu, D_\nu]$, namely $\gamma^{\mu\nu}R_{\mu\nu ab}\gamma^{ab}$, does not contribute, due to the cyclic identity, when $R_{\mu\nu} = 0$.

Thus the only difference in the expression for B for spin 3/2 as compared to spin 1/2 is that the term $\frac{1}{4}$ R is replaces by $\frac{1}{2} R_{\lambda\rho}{}^{ab}\gamma_{ab}$. In the spin 1/2 case, the $\frac{1}{4}$ R terms did not contribute at all. Looking at the expansion in (8), we see that the curvature term in the term with 1! still does not contribute, since it only has two Dirac matrices. The sum of all terms with only one curvature term is still as before, except that in this expression $\frac{1}{4}$ R is again to be replaced by $\frac{1}{2} R_{\lambda\rho}{}^{ab}\gamma_{ab}$. These terms again cancel, because they are equal to $((D^2)_{\lambda\alpha} \frac{1}{2} R_{\alpha\rho}{}^{ab})\gamma_{ab}$ which does not contain enough Dirac matrices to contribute (namely, it contains only the explicitly shown Dirac matrices γ_{ab}). Thus the only modification comes from the R^2 term in "Tr"$(2\gamma_5)$.

One finds for this contribution

$$R^2\text{-term} = \frac{2\alpha}{16\pi^2} \operatorname{tr} \gamma_5 \frac{1}{2!} \left(\frac{1}{2} R_{\mu\nu}{}^{ab}\gamma_{ab}\right)\left(\frac{1}{2} R^{\nu\mu cd}\gamma_{cd}\right)$$

$$= \frac{2\alpha}{16\pi^2} \left(-\frac{1}{2}\right)\left(\varepsilon^{abcd} R_{\mu\nu ab} R^{\mu\nu}{}_{cd}\right) \tag{20}$$

the terms without any curvature arrange themselves as in the spin 1/2 case, except that one must multiply the result by four, since one must trace over the vector indices as well as over the spinor indices of the gravitino. Thus the result from the gravitino plus gauge fixing terms to the spin 3/2 anomaly is

$$\text{"Tr"}\, 2\alpha\gamma_5 = \frac{\alpha}{96\pi^2} \left(\frac{1}{4} \varepsilon^{abcd} R_{\mu\nu ab} R^{\mu\nu}{}_{cd}\right)(4-24) \tag{21}$$

To obtain the complete anomaly, one must add the contribution from the Faddeev-Popov ghost (a complex spin 1/2 ghost with the same chiral weight as the gravitino) and of the Nielsen-Kallosh ghost (a real ghost, coming from the $\not{\partial}$ in the gauge fixing term, whose chiral weight is opposite to that of the gravitino). The sum of these contributions is (-1) and the total result is -21 times the anomaly for a real spin 1/2 field.

4. THE SPIN 1/2 CASE WITH PAULI-VILLARS REGULARIZATION

Let us now rederive the spin 1/2 axial anomaly, using Pauli-Villars regularization. We shall need two Pauli-Villars regulator fields, because, as we shall see, terms with the operator D^2 which vanish in the trace over spinor indices in the Yang-Mills case, do no longer vanish in the gravitational case, and are, moreover, divergent. These divergent terms cancel if one employs two regulator fields. We consider

$$Z = \int (d\psi d\bar{\psi})(d\chi_1 d\bar{\chi}_1)(d\chi_2 d\bar{\chi}_2) \exp S$$

$$S = \int d^4x \Big[-e\bar{\psi}\gamma^\mu (D_\mu + iA_\mu\gamma_5)\psi + \sum_{i=1}^{2} \{ -e\bar{\chi}_i \gamma^\mu (D_\mu + iA_\mu \gamma_5)\chi_i - eM_i \bar{\chi}_i \chi_i \} \Big] \tag{22}$$

We choose the chiral weights of χ_i such that <u>the measure is chirally invariant</u>

$$\delta\psi = \alpha\gamma_5\psi, \quad \delta\chi_i = -\frac{\alpha}{2}\gamma_5\chi_i \tag{23}$$

After a chiral transformation we obtain

$$0 = \int d\psi d\bar{\psi}\, d\chi_1 d\bar{\chi}_1 d\chi_2 d\bar{\chi}_2 \exp\,[S + \Delta S]$$

$$\Delta S = \int d^4x\; -e(\bar{\psi}\gamma^\mu\gamma_5\psi - \frac{1}{2}\bar{\chi}_1\gamma^\mu\gamma_5\chi_1 - \frac{1}{2}\bar{\chi}_2\gamma^\mu\gamma_5\chi_2)\partial_\mu\alpha \tag{24}$$

$$+ e\alpha M_1 \bar{\chi}_1\gamma_5\chi_1 + e\alpha M_2 \bar{\chi}_2\gamma_5\chi_2$$

Integrating over χ_i we obtain the product of the inverses of the determinants of the kinetic operators for χ_1 and χ_2 (which are complex commuting spinors). One finds, expanding to first order α

$$\det{}^{-1}(A_1+B_1)\det{}^{-1}(A_2+B_2) = \prod_{i=1}^{2} (\det A_i)^{-1}\left[1-\mathrm{Tr}A_i^{-1}B_i\right]$$

$$A_i = -e\gamma^\mu(D_\mu + iA\gamma_5)-eM_i + \frac{e}{2}\gamma^\mu\gamma_5\partial_\mu\alpha$$

$$B_i = \alpha e M_i\gamma_5 \qquad (25)$$

Re-exponentiating $(\det A)^{-1}$, one obtains, to first order in α, and dropping the axial vector field

$$\langle -e\bar\psi\gamma^\mu\gamma_5\psi + \frac{e}{2}\bar\chi_1\gamma^\mu\gamma_5\chi_1 + \frac{e}{2}\bar\chi_2\gamma^\mu\gamma_5\chi_2\rangle\partial_\mu\alpha$$

$$= \mathrm{Tr}(A_1^{-1}B_1 + A_2^{-1}B_2) = \sum_{i=1}^{2}\mathrm{Tr}\frac{-1}{(\not{D}+M_i)}(\alpha M_i\gamma_5)$$

$$= \sum_{i=1}^{2}\int\frac{d^4k}{(2\pi)^4}\,\mathrm{tr}\, e^{-ikx}(-\alpha M_i\gamma_5)\frac{1}{(\not{D}+M_i)(\not{D}-M_i)}(\not{D}-M_i)e^{ikx}$$

$$= \sum_{i=1}^{2}\int\frac{d^4k}{(2\pi)^4}\,\mathrm{tr}(-\alpha M_i\gamma_5)\frac{1}{(\not{D}+i\not{k})^2 - M_i^2}(\not{D}+i\not{k}-M_i) \qquad (26)$$

since the denominator contains an even number of Dirac matrices, we can drop the $\not{D} + iK$ in the numerator, and find

$$\sum_{i=1}^{2}\mathrm{Tr}\,A_i^{-1}B_i =$$

$$\sum_{i=1}^{2}\int\frac{d^4k}{(2\pi)^4}\,\mathrm{tr}(\alpha M_i^2\gamma_5)\frac{1}{-(k^2+M_i^2) + 2ik\cdot D + D^2 + \frac{1}{4}R}$$

$$= \sum_{i=1}^{2} \int \frac{d^4k}{(2\pi)^4} \, \mathrm{tr} \, \frac{(-\alpha M_i^2 \gamma_5)}{(k^2+M_i^2)} \Big[1 + (2ik\cdot D + D^2 + \frac{1}{4} R)/(k^2+M_i^2)$$

$$+ \frac{(2ik\cdot D)^2 + (D^2 + \frac{1}{4} R)^2}{(k^2 + M_i^2)^2} + \{ (2ik\cdot D)(2ik\cdot D)(D^2 + \frac{1}{4} R)$$

$$+ (2ik\cdot D)(D^2 + \frac{1}{4} R)(2ik\cdot D) + (D^2 + \frac{1}{4} R)(2ik\cdot D)(2ik\cdot D) \}/(k^2+M_i^2)^3$$

$$+ (2ik\cdot D)^4/(k^2 + M_i^2)^4 \Big] \tag{27}$$

The terms with $(\mathrm{tr}\ \gamma_5 D^2)$ are logarithmically divergent, and the spinor trace does not vanish a priori. However, assuming that $\Sigma M_i^{\ 2} = 0$, these terms are regulated. This is the motivation for using two regulators. The result resembles our previous result for the spin 1/2 case in many respects; however there are some differences: the first wo terms with two D's no longer cancel but add up to $[-D^2 M^4_{\ i}]/(k^2 + M_i^2)^3$.

Using

$$\int \frac{d^4k(k^2)^m}{(k^2 + M^2)^n} = \pi^2 (M^2)^{2+m-n} \frac{\Gamma(n-m-2)\Gamma(m+2)}{\Gamma(n)} \tag{28}$$

we obtain

$$\sum_{i=1}^{2} \mathrm{Tr} A_i^{-1} B_i = \sum_{i=1}^{2} \mathrm{tr}(-\frac{\alpha}{32\pi^2} \gamma_5) \Big[(M_i^2 D^2 + D^2 D^2 + \frac{1}{2} R D^2)$$

$$- \frac{2}{3} (2 D^2 D^2 + \frac{3}{4} R D^2 + D_\mu D^2 D^\mu)$$

$$+ \frac{1}{3} (D^2 D^2 + D_\mu D_\nu D^\mu D^\nu + D_\mu D^2 D^\mu) \Big] \tag{29}$$

As before we have dropped terms with $D_\mu R = \partial_\mu R$ since they do not contain enough Dirac matrices. The terms with RD^2 cancel again, (the numerical coefficients coming from the k-integral are essential for this), and the R-independent terms yield again a double commutator.

$$\sum_{i=1}^{2} \mathrm{Tr} A_i^{-1} B_i = \sum_{i=1}^{2} \mathrm{tr}\left(\frac{\alpha}{32\pi^2}\gamma_5\right)\left\{M_i^2 D^2 + \frac{1}{3}[D_\mu, D_\nu][D^\mu, D^\nu]\right\} \tag{30}$$

If we choose the regulator measses such that $M_1^2 + M_2^2 = 0$, then the M^2D^2 terms cancel, and one obtains

$$\sum_{i=1}^{2} \mathrm{Tr}\ A_i^{-1} B_i = \mathrm{tr}\left(\frac{\alpha}{32\pi^2}\gamma_5\right)\frac{1}{3}[D_\mu, D_\nu][D^\mu, D^\nu] \tag{31}$$

which is the same result as obtained from Fujikawa's method.

5. THE AXIAL ANOMALY IN N=1 CONFORMAL SUPERGRAVITY

In this section we use Fujikawa's method to compute the axial anomaly in simple (N=1) conformal supergravity. This is the first time that this method is used to obtain a new result. Of course, the method has also been used extensively to reobtain in elegant ways results which were previously obtained by laborious techniques such as the Feynman graph computations by Bardeen and Adler-Rosenberg. In our case, the Feynman graph calculations would have been very cumbersome, while Fujikawa's method is quite simple, despite the fact that one is dealing with a higher-derivative theory, due to a lemma (see below) according to which one may replace the regulator $\not{D}^3$ by $\not{D}$ in certain cases.

We begin by using a result derived by Fradkin and Tseytlin[5], who cast that part of the N=1 conformal supergravity action which contributes to one-gravitino-loop graphs, in a simple form. It reads

$$\mathcal{L} = -\bar{\psi}_\rho \not{D}^3 \psi_\rho - \frac{2}{3}\bar{\phi}\not{D}\phi + \frac{1}{2}\bar{\chi}\not{D}^3\chi \tag{32}$$

where the gravitationally covariant derivative D_μ contains both a spin-connection <u>and</u> a Christoffel part, and where further

$$\phi \equiv D^{\mu}\psi_{\mu} + \frac{1}{2}\not{D}\chi \;, \quad \chi \equiv \gamma^{\mu}\psi_{\mu} \tag{33}$$

We have put the chiral gauge field A_{μ} to zero. Moreover, we have assumed that $R_{\mu\nu} = 0$, which is necessary in order that the gravitino action by itself be invariant under the ordinary and conformal supersymmetry transformations, which read

$$\delta_{Q}\psi_{\mu} = D_{\mu}\varepsilon_{Q} \;; \quad \delta_{S}\psi_{\mu} = \gamma_{\mu}\varepsilon_{S} \tag{34}$$

The easiest way to see that the gravitino part is separately invariant, is to note that its variation must cancel the variation of the Weyl action $R_{\nu\mu}{}^{2} - \frac{1}{3}R^{2}$. Since the latter variation vanishes when $R_{\mu\nu} = 0$, so does the former.

We will cancel the last two terms in the gravitino action by adding the following gauge fixing terms

$$\mathcal{L}(\text{fix}) = \frac{2}{3}\bar{\phi}\not{D}\phi - \frac{1}{2}\bar{\chi}\not{D}^{3}\chi \tag{35}$$

Thus we have two gauge fixing terms, namely $F^{\alpha} = (\phi,\chi)$, which will yield the usual Faddeev-Popov ghosts, while we also must take into account that the normalization determinates of $\not{D}$ and $\not{D}^{3}$ will give rise to Nielson-Kallosh ghosts. The Faddeev-Popov ghost action is obtained by varying F^{α} with respect to the (Q,S) supersymmetry transformations with parameters $\xi^{\alpha} = (\varepsilon_{Q},\varepsilon_{S})$, and sandwiching the result with commuting ghosts and antighosts. This leads to

$$\mathcal{L}(\text{FP ghosts}) = \bar{C}_{Q}\left(\frac{3}{2}\,\Box\, C_{Q} + 3DC_{S}\right) + \bar{C}_{S}\left(\not{D}C_{Q} + 4C_{S}\right) \tag{36}$$

The exponentiation of the gauge fixing terms in the Dirac delta functions in the path-integral requires as normalization factors $(\det \not{D})^{-1/2}$ and $(\det \not{D})^{-3/2}$, respectively, since ϕ and χ are anticommuting Majorana spinors. These determinants cannot be exponentiated as they stand, because they would require a <u>commuting</u> Majorana spinor whose Dirac action is a total derivative. The resolution is by now well-known[8]: one replaces $\not{D}^{-1/2}$ by $\not{D}^{-1}\not{D}^{1/2}$ and exponentiates by introducing a <u>complex</u> commuting ghost F and a Majorana anticommuting ghost f. Similar remarks apply to $(\det \not{D})^{-3/2}$ with ghosts G and g. In this way one obtains the following ghosts

$$\mathcal{L}(\text{NK ghosts}) = -\bar{F}\not{D}F - \bar{f}\not{D}f - \bar{G}\not{D}^3 G - \bar{g}\not{D}^3 g. \tag{37}$$

We must now determine the chiral weights of all these ghosts. For the Faddeev-Popov ghosts we follow ref.[7], and consider interaction terms like

$$\bar{C}_Q(D^{\mu}\partial_{\rho}\psi_{\mu}C^{\rho}_{g.c.}) + \bar{C}_S(\gamma^{\mu}\partial_{\rho}\psi_{\rho}C^{\rho}_{g.c.}) + \ldots \tag{38}$$

where $C^{\rho}_{g.c.}$ is the general coordinate ghost. Requiring chiral invariance of the complete Faddeev-Popov action yields the following chiral weights.

$$w(\psi_{\mu}) = +1,\ w(\bar{\psi}_{\mu}) = +1,\ w(\bar{C}_Q) = -1,\ w(\bar{C}_S) = +1$$

$$w(C_Q) = +1,\ w(C_S) = -1. \tag{39}$$

For the determination of the chiral weights of the NK ghosts we follow ref.[9] Namely we require that the part of the chiral current corresponding to the gauge-fixing term is cancelled by the part of the chiral current due to the NK ghosts. This is because in the unweighted gauge neither gauge fixing terms nor NK term are present in the quantum action. This argument shows at once that the NK Q-ghosts and the NK S-ghosts have opposite chiral weights.

We can simplify the quantum action by intergrating in the path integral over $\bar{C}_s$. This yields a Dirac-delta function $\delta(4C_s + \not{D}C_Q)$. which replaces C_s by $-\frac{1}{4}\not{D}C_Q$ after integration over C_s. Hence we finally obtain

$$\mathcal{L}(\text{quantum}) = -\bar{\psi}_{\rho}\not{D}^3\psi_{\rho} + \frac{3}{4}\bar{C}_Q\Box C_Q + \mathcal{L}(\text{NK ghosts}) \tag{40}$$

We now turn to the computation of the chiral anomaly associated with the global chiral invariance of $\mathcal{L}$(quantum). We must evaluate according to Fujikawa's method, the trace of the matrix $\gamma_5 w$ for all fields, where w is the chiral weight of the field considered. Moreover, in the computation one must regularize the ill-defined sum over spacetime points by the same operator as appears in the action. The justification of this procedure can, in our opinion, best be seen by comparing this Fujikasw computation with a Pauli-Villars computation in which the Jacobian is exactly unity but where now the anomaly resides in Pauli-Villars fields (see before). Hence, we use as regulators the operators exp R^2 where R is given by $(\not{D}/M)^3$ for ψ_ρ .

We now argue that the sum of the ghost contributions to the axial anomaly cancels. For the FP ghosts this follows easily from the fact that $\bar{C}_Q$ and C_Q have opposite chiral weights. This means that their contributions cancel in the Jacobian. For the NK ghosts, we make use of a general property of both the Fujikawa and the Pauli-Villars methods[1)], namely that any regulator f(R) with f(0) = 1 and vanishing sufficiently fast at infinity, will give the same result. Hence, instead of $\not{D}^3$ operators we can take simply the $\not{D}$ operator as regulator. Since both F and G, and also both f and g, have the same regulators but opposite chiral weights, their contributions cancel.

All that remains to be computed is the anomaly due to ψ_ρ with regulator $\not{D}^3$, or rather, invoking the arguments presented above, with regulator $\not{D}$. This computation was already performed in section (3). Hence we conclude

chiral anomaly of N=1 conformal supergravity theory = (-20)A

where A is the chiral anomaly of a real anticommuting electron.

We conclude this section with a few comments.

1. The chiral weights can also be determined by keeping the chiral gauge field A_μ in the quantum action, and defining the chiral current by the coupling terms to A_μ . As shown by Fradkin and Tseytlin, A_μ appears as

$$-\bar{\psi}_\rho \not{D}\not{D}^+ \not{D}\psi_\rho - \frac{2}{3}\bar{\phi}\not{D}\phi + \frac{1}{2}\bar{\chi}\not{D}^+\not{D}\not{D}^+\chi \tag{41}$$

where

$$\not{D} = \gamma^\mu(D_\mu - \frac{3i}{4}A_\mu\gamma_5) \text{ and } \not{D}^+ = \gamma^\mu(D_\mu + \frac{3i}{4}A_\mu\gamma_5). \tag{42}$$

Clearly, $\mathcal{L}$(class) is locally chiral invariant. From the exponentiation of $\not{D}^{-1/2}$, it follows that the commuting Majorana spinor (which should really be replaced by F and f as we explained) has the same weight as the gravitino. The chiral weight of G and g follows most easily by writing

$$\det(\not{D}^+\not{D}\not{D}^+)^{-1/2} = (\det\not{D}^+)^{-1/2}(\det\not{D})^{-1/2}(\det\not{D}^+)^{-1/2} \tag{43}$$

and observing that after exponentiation one would have three chiral currents from three commuting Majorana NK ghosts with weights -1, +1 and -1 respectively. Hence, the chiral contributions of the NK ghosts indeed cancel. The FP ghost action contains the following terms linear in A

$$\bar{C}_Q[\frac{3}{4}D^\mu D_\mu - \frac{1}{4}D_\mu\gamma^{\mu\nu}D_\nu]C_Q \tag{44}$$

This yields a chiral current

$$j_\mu = \frac{3}{4}\bar{C}_Q\overleftrightarrow{\partial}_\mu\gamma_5 C_Q - \frac{1}{4}\partial_\nu(\bar{C}_Q\gamma_{\mu\nu}\gamma_5 C_Q) \tag{45}$$

The first type of current has no anomaly while the second current is identically conserved.

2. For higher N models, spin 1/2 will contribute, in addition to N gravitinos. In particular, for N=4, one expects finiteness, since Fradkin and Tseytlin showed that the β function vanishes in this model. Actually, the local guage group is SU(4) in this model, not U(4), and although there is a composite local U(1) gauge invariance, it is not part of the expected anomaly multiplet. Hence, in the N=4 model vanishing of the chiral anomaly is replaced by absence of a chiral current; a rather trivial solution.

3. The N=1 conformal supergravity theory by itslef has anomalies in its coupling of the axial vector field, and thus, this theory is inconsistent. However, one can cancel these anomalies by coupling conformal matter.

6. REFERENCES

1. Fujikawa, K., Phys. Rev. Lett. 42 (1979) 1195; Phys. Rev. D21 (1980) 2848; D22 (1980) 1499(E).
2. Alvarez-Gaumé, L., and Witten, E., Nucl. Phys. B234(1984) 269.
3. Einhorn, M.B.,and Jones, D.R.T., Phys. Rev. D29(1984) 331.
4. Kaku, M., Townsend, P.K. and van Nieuwenhuizen, P., Phys.Rev. D17 (1978) 3179.
5. Fradkin, E.S., and Tseytlin, A.A., Nucl. Phys B203 (1982) 157.
6. Christensen, S., and Duff, M. J., Phys. Lett. 76B, 571 (1978).
7. Nielsen, N.K., Römer, H. Grisaru, M.T., and van Nieuwenhuizen, P. Nucl. Phys. B140 (1978) 477.
8. van Nieuwenhuizen, P., Phys. Rep. 68 (1981) 243.
9. Nielsen, N. K. Nucl. Phys. B140 (1978) 499.

SECTION III

QUANTUM EFFECTS IN BLACK HOLES AND IN ACCELERATED FRAMES

EFFECT OF VACUUM POLARIZATION NEAR BLACK HOLES

V.P.Frolov, and A.I.Zel'nikov

P.N.Lebedev Physical Institute, Moscow
Leninsky Prospect 53, 117924
USSR

1. INTRODUCTION

Hawking[1)] has shown that a black hole formed by collapse spontaneously creates and emites particles as if it were a hot body with a temperature proportional to the surface gravity of the black hole. The action of the gravitational field of a black hole on the virtual vacuum quanta can provide them (and with some probability really provides) with the energy that is sifficient to make these quanta real. A part if these quanta reaches infinity and forms the Hawking radiation of the black hole. It is not, however, the only result of the action of the gravitational field on the vacuum. The states of those virtual vacuum quanta which do not become real are also affected by the gravitational field. This results in the rise of a nonzero vacuum expectation value of the energy-momentum tensor. This effect of the change of the vacuum expectation value of the local observables under the action of the external field is known as vacuum polarization.

Recently a number of papers appeared in which the effect of vacuum polarization in black holes was discussed. One of the final aims of these investigations is to obtain a selfconsistent description of the evaporating black holes. This aim is not reached till now. Nevertheless the results obtained are complete enought to provide us not only with a qualitative picture of the vacuum polarization effect in the strong gravitational field of black holes but also to give us detailed information about its quantitative characteristics

In this paper we describe the main of the obtained results. We hope that the part of them appears to be new.

We use the sign conventions of Misner, Thorne and Wheeler[2)] and Planck's units: $\hbar = c = G = 1$.

In the one-loop approximation the contributions of different physical fields to the vacuum polarization are additive. It is convenient to investigate the contributions of massive and massless fields separately using different methods. If the wavelength $\lambda_m = \hbar / mc$ of a massive (with mass m) field is much smaller than the characteristic radius L of the spacetime curvature, then the contribution of this field to the vacuum polarization is essentially determined by the local properties of the geometry and it can be expanded in the series with respect to the small parameter $\varepsilon = (\lambda_m / L)^2$. The contribution of massless fields to the vacuum polarization is essentially nonlocal and its calculation is much more complicated because of the absence of any small parameters. It is rather surprising that for massless fields one can nevertheless develop rather good approximate analytic methods and obtain exact explicit expressions for quantities describing vacuum polarization near black holes.

In section 2 we present the main formulas for the Green's functions in the spacetime of a black hole. In the following two sections these formulas are used for investigation of the massless fields contribution to the vacuum polarization near black holes. The contribution of massive fields is discussed in section 5.

2. VACUUM POLARIZATION OF MASSLESS FIELDS. MAIN FORMULAS

Let us consider the massless scalar field φ satisfying the equation

$$\Box \varphi - \xi R \varphi = 0 \quad , \tag{2.1}$$

where ξ is an arbitrary parameter (ξ = 1/6 for the conformal field). The complete information about the quantum properties of this field is contained in the Green's function $G(x,x') = i\langle T(\hat{\varphi}(x)\,\hat{\varphi}(x'))\rangle$, which is defined as the solution of the following equation

$$(\Box - \xi R)\, G(x,x') = -|g(x)|^{-1/2}\, \delta(x-x') \, . \tag{2.2}$$

The ambiguity of the choice of boundary conditions which uniquely specify the Green's function of Eq.(2.2) corresponds to the ambiguity of the choice of the quantum state of the field φ . If the Green's function $G(x,x')$ is known then the calculation of the vacuum averages such as $\langle T_{\mu\nu}(x)\rangle^{ren}$ requires the following steps[3-4]: 1. expressing of $\langle T_{\mu\nu'}(x,x')\rangle$ at separated points x and x' in terms of combinations of derivatives of $G(x,x')$; 2. expanding of the obtained expression in powers of $\sigma^{\mu}=\nabla^{\mu}\sigma$ (where $\sigma = S^2/2$, and S is the geodetic interval between the points x and x') and 3. performing the renormalization procedure by subtracting the one-loop divergences and taking the coincidence limit.

The analogous procedure of calculation of vacuum fluctuations of a scalar field is much more simple. For example we have for

$$\langle \varphi^2(x)\rangle^{ren} = -i \lim_{x'\to x}\left[G(x,x') - \frac{i}{8\pi^2\sigma}\left(1 + R_{\alpha\beta}\,\sigma^{\alpha}\sigma^{\beta}\right)\right] . \tag{2.3}$$

The geometry of a stationary charged black hole is described by the Kerr-Newman metric which in the Boyer-Lindquist coordinates is of the form

$$ds^2 = -\left(1-\frac{Z}{\Sigma}\right)dt^2 - \frac{2aZ\sin^2\theta}{\Sigma}\,dt\,d\varphi + \frac{\Sigma}{\Delta}dr^2 +$$

$$+\left(r^2+a^2+\frac{Za^2}{\Sigma}\sin^2\theta\right)\sin^2\theta\,d\varphi^2 + \Sigma\,d\theta^2 \quad , \tag{2.4}$$

$$\Sigma = r^2 + a^2\cos^2\theta \quad , \qquad Z = 2Mr - Q^2 \quad ,$$
$$\Delta = r^2 - 2Mr + a^2 + Q^2 \quad .$$

The wave equation (2.1) in this metric allows the separation of variables. It is convenient to introduce as the basis the following complete set of normalized solutions of this equation

$$v_{\omega\ell m} = \frac{1}{\sqrt{4\pi\omega}}\frac{e^{-i\omega t}}{\sqrt{r^2+a^2}}V_{\omega\ell m}(r)Y_{\ell m}(\theta,\varphi) \, , \quad \omega > 0 \, ;$$

$$q_{\omega\ell m} = \frac{1}{\sqrt{4\pi\tilde{\omega}}}\frac{e^{-i\omega t}}{\sqrt{r^2+a^2}}Q_{\omega\ell m}(r)Y_{\ell m}(\theta,\varphi) \, , \quad \tilde{\omega} \equiv \omega - \Omega m > 0 \, . \tag{2.5}$$

Here $Y_{\ell m}(\theta,\varphi) = e^{im\varphi}S_\ell^m(\cos\theta)/\sqrt{2\pi}$, S_ℓ^m being the spheroidal wave functions normalized by the condition

$$\int_{-1}^{1} S_\ell^m(z)\,S_{\ell'}^m(z)\,dz = \delta_{\ell\ell'} \quad , \tag{2.6}$$

$\Omega = a/(r_+^2 + a^2)$ is the angular velocity of the black hole. The radial functions $V_{\omega\ell m}$ and $Q_{\omega\ell m}$ satisfy the following boundary conditions

$$V_{\omega \ell m} \sim \begin{cases} e^{-i\omega r^*} + A_+(\omega \ell m) e^{i\omega r^*}, & r^* \to \infty; \\ B_+(\omega \ell m) e^{-i\tilde{\omega} r^*}, & r^* \to -\infty; \end{cases}$$

$$Q_{\omega \ell m} \sim \begin{cases} B_-(\omega \ell m) e^{i\omega r^*}, & r^* \to \infty; \\ e^{i\tilde{\omega} r^*} + A_-(\omega \ell m) e^{-i\omega r^*}, & r^* \to -\infty; \end{cases} \quad (2.7)$$

The Green's function $G(x, x')$ can be written in the form

$$G(x,x') = i\left[\theta(x,x') S(x,x') + \theta(x',x) S(x',x)\right], \quad (2.8)$$

where $S(x, x') = \langle \hat{\varphi}(x) \hat{\varphi}(x') \rangle$. For the so called Unruh's $|U\rangle$ [6], Hartle-Hawking's $|H\rangle$ [7], and Boulware's $|B\rangle$ [8] vacuum states one has

$$S_U(x,x') = \sum_{\ell,m} \left\{ \int_0^\infty d\omega\, v_J(x,x') + \int_{-\infty}^{+\infty} \frac{d\omega \frac{\tilde{\omega}}{|\tilde{\omega}|} q_J(x,x')}{1 - \exp(-2\pi\tilde{\omega}/\varkappa)} \right\};$$

$$S_H(x,x') = \sum_{\ell,m} \left\{ \int_{-\infty}^{+\infty} \frac{d\omega \frac{\tilde{\omega}}{|\tilde{\omega}|} \left[v_J(x,x') + q_J(x,x')\right]}{1 - \exp(-2\pi\tilde{\omega}/\varkappa)} - \right.$$

$$\left. - \int_0^{\Omega_m} \frac{d\omega \left[\exp(-2\pi\tilde{\omega}/\varkappa)\, \overline{v_J}(x,x') + v_J(x,x')\right]}{\exp(-2\pi\tilde{\omega}/\varkappa) - 1} \right\}; \quad (2.9)$$

$$S_B(x,x') = \sum_{\ell,m} \left\{ \int_0^\infty d\omega\, v_J(x,x') + \int_{\Omega_m}^\infty d\omega\, q_J(x,x') \right\};$$

where $v_y(x,x') \equiv v_{\omega \ell m}(x)\, \overline{v}_{\omega \ell m}(x')$, and

$q_y(x,x') = q_{\omega \ell m}(x)\, \overline{q}_{\omega \ell m}(x')$.

In the case of nonrotating black holes these representations for Green's functions were obtained by Candelas [9]. Their generalization for the electromagnetic and gravitational perturbations in the Kerr metric can be found in the paper [10].

It should be noted that the difference between the average values of $\hat{\varphi}^2$ in any two states is finite since the divergences, which are removed by the renormalization procedure have a universal form. In particular the following expressions are finite

$$\langle \varphi^2(x) \rangle_U^{ren} - \langle \varphi^2(x) \rangle_H^{ren} =$$

$$= \langle U | \hat{\varphi}^2(x) | U \rangle^{ren} - \langle H | \hat{\varphi}^2(x) | H \rangle^{ren} =$$

$$= -2 \sum_{\ell,m} \frac{d\omega\, |v_y(x)|^2}{\exp(2\pi\tilde{\omega}/\varkappa) - 1} , \tag{2.10}$$

$$\langle \varphi^2(x) \rangle_B^{ren} - \langle \varphi^2(x) \rangle_H^{ren} =$$

$$= \langle B | \hat{\varphi}^2(x) | B \rangle^{ren} - \langle H | \hat{\varphi}^2(x) | H \rangle^{ren} =$$

$$= -2 \sum_{\ell,m} \left\{ \int_0^\infty \frac{d\omega\, |v_y(x)|^2}{\exp(2\pi\tilde{\omega}/\varkappa) - 1} + \int_{\Omega_m}^\infty \frac{d\omega\, |q_y(x)|^2}{\exp(2\pi\tilde{\omega}/\varkappa) - 1} \right\} .$$

Let us describe another important Green's function $G_{H,R}$ corresponding to the equilibrium situation when the black hole is inside the bath with thermal radiation at temperature T_{BH} surrounded by the mirror boundary at $r = R$. We take as the basis the following set of solutions of Eq. (2.1)

$$k_{\omega l m}=\frac{1}{\sqrt{4\pi\tilde{\omega}}}\frac{e^{-i\omega t}}{\sqrt{r^2+a^2}}K_{\omega l m}(r)Y_{lm}(\theta,\varphi)\ ,$$
$$\tilde{\omega}>0\ , \qquad (2.11)$$

satisfying the boundary condition $K_{\omega l m}(R)=0$.
Then we have

$$S_{H,R}(x,x')=\sum_{l,m}\int_{-\infty}^{\infty}\frac{d\omega\,\frac{\tilde{\omega}}{|\tilde{\omega}|}\,k_{\omega l m}(x)\,\overline{k_{\omega l m}(x')}}{1-\exp(-2\pi\tilde{\omega}/\varkappa)}\ . \qquad (2.12)$$

To investigate average values in the Hartle-Hawking vacuum it is convenient to use the analytic continuation to the Euclidean section of a complexified black hole metric. Let $g_{E\mu\nu}$ be the metric obtained by the analytic continuation $\tau=it$, $b=-ia$ of the metric (2.4) then one has [7,11]

$$G_H(x,x')=\left[iG_E(x,x')\right]_{\tau=it,\ b=-ia} \qquad (2.13)$$

The Euclidean Green's function allows the following representation

$$G_E(x,x')=\frac{\varkappa_E}{(2\pi)^2}\sum_{n,m=-\infty}^{\infty}G_{nm}(r,\theta;r'\theta')\times$$
$$\times\exp\{i\tilde{n}\varkappa_E(\tau-\tau')+im(\varphi-\varphi')\}\ , \qquad (2.14)$$

where

$$\varkappa_E=\sqrt{M^2+b^2-Q^2}/(r_E^2-b^2)\ ,$$

$$\tilde{n}=n-m\Omega_E/\varkappa_E\ ,$$

$$\Omega_E = \beta/(r_E^2 - \beta^2) \tag{2.15}$$
$$r_E = M + \sqrt{M^2 + \beta^2 - Q^2}$$

and $G_{n,m}$ satisfies the equation

$$[D - \Sigma_E(g_E^{\tau\tau}\varkappa_E^2\tilde{n}^2 - 2g_E^{\tau\varphi}\varkappa_E\tilde{n}m - g_E^{\varphi\varphi}m^2)] \times$$
$$\times\, G_{nm}(r,\theta; r'\theta') = -\delta(r-r')\delta(\theta-\theta')/\sin\theta \, . \tag{2.16}$$

Here

$$\Sigma_E = r^2 - \beta^2\cos^2\theta \quad ,$$
$$\Delta_E = r^2 - 2Mr - \beta^2 + Q^2 \quad ; \tag{2.17}$$

$$D = \partial_r(\Delta_E\partial_r) + \frac{1}{\sin\theta}\partial_\theta(\sin\theta\,\partial_\theta) \, . \tag{2.18}$$

3. VACUUM POLARIZATION OF MASSLESS FIELDS NEAR AN EVENT HORIZON

This section contains the results concerning the vacuum fluctuations and vacuum polarization at the event horizon of stationary black holes for various choices of the vacuum state.

3.1 $\langle\varphi^2\rangle^{ren}$ For The Massless Scalar Field

3.1.a <u>The value of $\langle\varphi^2\rangle_H^{ren}$.</u> In the Kerr-Newman metric $\langle\varphi^2\rangle_H^{ren}$ at the pole of the event horizon can be calculated exactly [12,13]. This property is connected with the fact that the pole x_0 of the event horizon of the Euclidean black hole is stationary under the isometries with the generators ∂_τ and ∂_φ. Therefore $G_E(x,x_0)$ does not depend on τ and φ coordinates

$$G_E(x,x_0)=\frac{\varkappa_E}{(2\pi)^2}\,G_{00}(r,\theta;\,r_E,0)\quad . \qquad (3.1)$$

The equation for G_{00} in the coordinates

$$\rho=\Delta_E^{1/2}\sin\theta\ ,\qquad z=(r-M)\cos\theta \qquad (3.2)$$

reads

$$\left[\frac{1}{\rho}\partial_\rho(\rho\,\partial_\rho)+\partial_z^2\right]G_{00}(\rho,z;0,z_0)=-\frac{\delta(\rho)\,\delta(z-z_0)}{\rho}\ , \qquad (3.3)$$

where $z_0=\sqrt{M^2+b^2-Q^2}$. This equation can be easily solved. Using this solution one has

$$G_H(x,x_0)=i\varkappa\left[8\pi^2(r-M-\sqrt{M^2-a^2-Q^2}\,\cos\theta\,)\right]^{-1}\ . \qquad (3.4)$$

After the renormalization (2.3) one finally obtaines

$$\langle\varphi^2(x_0)\rangle_H^{ren}=\frac{1}{48\pi^2}\,\frac{r_+^2-3a^2-Q^2}{(r_+^2+a^2)^2}\quad . \qquad (3.5)$$

The properties of $\langle\varphi^2(x_0)\rangle_H^{ren}$ were discussed in details in the work [13]. Note only that for $a=0$ this expression is valid for every point at the horizons $\mathcal{H}^+$ and $\mathcal{H}^-$. For $a=Q=0$ it reproduces the Candelas's result [9].

The described method is applicable for a wide class of stationary axi-symmetric metrics, which have a regular event horizon and allow an analytical continuation to a regular Euclidean section. In the paper [14] it was applied for investigation of the vacuum fluctuations of $\langle\varphi^2(x_0)\rangle_H^{ren}$ in axi-symmetric distorted black holes. In analogous way the problem of the influence of boundary conditions on the vacuum fluctuations near the horizon can be solved. In particular, if the Kerr-Newman black hole is surrounded by the mirror surface at $r = R_0$ at which $\varphi|_{r=R_0}=0$ one has

$$\langle \varphi^2(x_0) \rangle^{ren}_{H,R_0} = \langle \varphi^2(x_0) \rangle^{ren}_{H} - \tag{3.6}$$
$$- \frac{1}{8\pi^2 (r_+^2 + a^2)} \sum_{\ell=0}^{\infty} (2\ell+1) \frac{Q_\ell(\eta_0)}{P_\ell(\eta_0)} ,$$

where $\eta_0 = (R_0 - M)/\sqrt{M^2 - a^2 - Q^2}$.

3.1.b <u>The value of $\langle \varphi^2 \rangle^{ren}_{U}$</u> at the past event horizon $\mathcal{H}^-$ coincides with $\langle \varphi^2 \rangle^{ren}_{H}$. Unfortunately it is not so easy to find an exact analytical expression for $\langle \varphi^2 \rangle^{ren}_{U}$ at the future horizon. Using the relations (2.7) and (2.10) however one can obtain the following expression for $\langle \varphi^2 \rangle^{ren}_{U}$ at $\mathcal{H}^+$

$$\langle \varphi^2 \rangle^{ren}_{U} = \langle \varphi^2(x_0) \rangle^{ren}_{H} - \frac{1}{4\pi^2 (r_+^2 + a^2)} \times \tag{3.7}$$
$$\times \sum_{\ell,m} \int_0^\infty \frac{d\omega \, |S^m_\ell(\cos\theta)|^2}{\exp(2\pi\tilde{\omega}/\varkappa) - 1} |B_+(\omega \ell m)|^2 .$$

The numerical calculation of $\langle \varphi^2 \rangle^{ren}_{U}$ performed by Whiting and Fawcett [15)] for the Schwarzschild black hole gives

$$\langle \varphi^2 \rangle^{ren}_{U} \Big|_{\mathcal{H}^+} \simeq \langle \varphi^2 \rangle^{ren}_{H} \Big|_{\mathcal{H}^+} - 0.123 \, T^2_{BH} . \tag{3.8}$$

3.1.c <u>The value of $\langle\varphi^2\rangle_B^{ren}$</u> diverges both a $\mathcal{H}^+$ and at $\mathcal{H}^-$. One can show that near the horizons $\mathcal{H}^\pm$ $\langle\varphi^2\rangle_B^{ren}$ has the following asymptotic behaviour

$$\langle\varphi^2(r)\rangle_B^{ren} \underset{r\to r_+}{\sim} \frac{1}{r-r_+} . \tag{3.9}$$

3.2 $\langle T_\mu{}^\nu\rangle^{ren}$ at The Event Horizon

3.2.a <u>The value of $\langle T_\mu{}^\nu\rangle_H^{ren}$</u>. The general representation of the conserved energy-momentum tensor in Schwarzschild spacetime was obtained by Christensen and Fulling [24] (see Eqs.(4.12)). At the event horizon the ambiguity of $\langle T_\mu{}^\nu\rangle_H^{ren}$ which is regular at $\mathcal{H}^\pm$ reduces to one constant. One can take $\langle T_\theta{}^\theta\rangle_H^{ren}$ as this constant.

In the case of a conformal scalar field in Schwarzschild spacetime Candelas [9] obtained the following value for $\langle T_\mu{}^\nu\rangle_H^{ren}$ at the bifurcation sphere of the event horizons $\mathcal{H}^+$ and $\mathcal{H}^-$

$$\begin{aligned} \langle T_t^{\ t}\rangle_H^{ren} &= \langle T_r^{\ r}\rangle_H^{ren} = \frac{1}{\pi^2(4M)^4}\left\{-\frac{1}{60}+\beta\right\} , \\ \langle T_\theta^{\ \theta}\rangle_H^{ren} &= \langle T_\varphi^{\ \varphi}\rangle_H^{ren} = \frac{1}{\pi^2(4M)^4}\left\{\frac{1}{20}-\beta\right\} . \end{aligned} \tag{3.10}$$

Numerical calculations give [9] $\beta \simeq 0.1286$.

For the electromagnetic field the value of $\langle T_\mu{}^\nu\rangle_H^{ren}$ at the event horizon can be calculated exactly. In the case of the Schwarzschild black hole this result was obtained by Elster [20] who succeded in summing of the series expansions for $\langle T_\mu{}^\nu\rangle_H^{ren}$. We describe below a rather general method applicable to the vacuum type-D spacetimes which allows one to obtain this result in a more direct way.

The main steps of this approach are: i) One introduces the complex null tetrad $(k^\mu, \ell^\mu, m^\mu, \bar{m}^\mu)$, k^μ and ℓ^μ lying along the degenerated principal null directions of the Weyl tensor. Then each of the following spinor components Φ_i of the electromagnetic field

$$\begin{aligned} \Phi_0 &= F_{\mu\nu}\, k^\mu \ell^\nu , \\ \Phi_1 &= \tfrac{1}{2} F_{\mu\nu} (k^\mu \ell^\nu + m^\mu \bar{m}^\nu) , \\ \Phi_2 &= F_{\mu\nu}\, \bar{m}^\mu \ell^\nu , \end{aligned} \tag{3.11}$$

satisfies the separated second order differential equations [17]. ii) The energy-momentum tensor with splitting points is expressed in terms of the null tetrad and the biscalar propagators $\langle \Phi_i(x)\, \bar{\Phi}_k(x') \rangle$. If one of the points (x') in this propagator is stationary under the action of the group of symmetries of spacetime, then the propagator considered as a function of x is invariant under the action of this group. For example if the point x' lies on the pole of the horizon of the Euclidean section of the complexified Kerr black hole ($x' = x_0$), then the propagators $\langle \Phi_i(x)\, \bar{\Phi}_k(x_0) \rangle_H$ do not depend on τ and φ and satisfy the equations, which do not contain the derivatives ∂_τ and ∂_φ. The solution of these equations allow one to find $\langle T_{\mu\nu'} \rangle_H$ at separated points and after the subtraction of the divergent part to calculate the renormalized energy-momentum tensor of the electromagnetic field

$$\langle T_\mu^\nu \rangle_H^{ren} = \lim_{x' \to x} \left\{ \langle T_\mu^\nu \rangle_H - \langle T_\mu^\nu \rangle^{div} \right\} . \tag{3.12}$$

(The general expression for $\langle T_\mu{}^\nu \rangle^{div}$ was given by Christensen [4]).

Let us describe the application of this method to the calculation of $\langle T_\mu{}^\nu \rangle_H^{ren}$ in Schwarzschild geometry. The appropriate null tetrad in the Schwarzschild coordinates (t, r, θ, φ) is

$$
\begin{aligned}
k^\alpha &= \left\{ \left(1-\frac{2M}{r}\right)^{-1},\ 1\ ,\ 0\ ,\ 0 \right\} ,\\
l^\alpha &= \left\{ \frac{1}{2}\ ,\ -\frac{1}{2}\left(1-\frac{2M}{r}\right),\ 0\ ,\ 0 \right\} ,\\
m^\alpha &= \frac{1}{\sqrt{2}\,r}\left\{ 0\ ,\ 0\ ,\ 1\ ,\ \frac{i}{\sin\theta} \right\} .
\end{aligned}
\qquad (3.13)
$$

For the $\langle T_\theta^{\theta'} \rangle_H$ component with separated points we have

$$
\begin{aligned}
\langle T_\theta{}^{\theta'}(x,x') \rangle_H &= \mathrm{Re}\left\{ \mathcal{F}(x,x') + \mathcal{F}(x',x) \right\} ,\\
\mathcal{F}(x,x') &= \langle \Phi_1(x)\, \overline{\Phi}_1(x') \rangle_H .
\end{aligned}
\qquad (3.14)
$$

The biscalar $\mathcal{F}(x,x')$ can be represented in the form, which is analogous to the Eq.(2.14) with $\Omega_E = 0$. When $x' = x_0$ we find

$$
\mathcal{F}(x,x_0) = \frac{\varkappa_E}{(2\pi)^2}\, \mathcal{F}_{00}(r,\theta;\, r_E, 0) , \qquad (3.15)
$$

$\mathcal{F}_{00}$ satisfying the following equation

$$
\hat{P}\, \mathcal{F}_{00} = -\hat{L}_\alpha \hat{K}^{\alpha'} \delta^\alpha_{\alpha'} \frac{\delta(r-r')\,\delta(\theta-\theta')}{r^2 \sin\theta} , \qquad (3.16)
$$

where the operators $\hat{P}$, $\hat{L}_\alpha$, and $K^{\alpha'}$ are

$$\hat{P} = (1-\frac{2M}{r})\frac{\partial^2}{\partial r^2} + \\ + \frac{4r-6M}{r^2}\frac{\partial}{\partial r} + \frac{2}{r^2} + \frac{1}{r^2 \sin\theta}\frac{\partial}{\partial\theta}(\sin\theta\frac{\partial}{\partial\theta}) ,$$

$$\hat{K}^{\alpha} = \frac{1}{2}\{\delta^{\alpha}_{t}\frac{\partial}{\partial r} - i\delta^{\alpha}_{\varphi}\frac{1}{r^2}\frac{1}{\sin\theta}\frac{\partial}{\partial\theta}\} ,$$

$$\hat{L}_{\alpha} = \frac{1}{2}\{\delta^{t}_{\alpha}\frac{1}{r^2}\frac{\partial}{\partial r}(r^2 - 2Mr) + \\ + \delta^{r}_{\alpha}\frac{1}{r^2}\frac{\partial}{\partial r}r^2 + \delta^{\theta}_{\alpha}\frac{1}{\sin\theta}\frac{\partial}{\partial\theta}\sin\theta - \\ - i\delta^{\varphi}_{\alpha}\frac{1}{\sin\theta}\frac{\partial}{\partial\theta}\sin^2\theta\} . \tag{3.17}$$

Note that the Green's function $G(x,x')$ of the operator $\hat{P}$

$$\hat{P}G = -\delta(r-r')\delta(\theta-\theta')\delta(\varphi-\varphi')/r^2\sin\theta \tag{3.18}$$

can be represented in the form [ж)]

$$G(x,x') = \frac{1}{2}\{\frac{\partial}{\partial r}[\frac{1}{r}\int_{\infty}^{r'}\frac{dr'}{r'-2M}(M + \\ + \frac{\Pi(x,x')}{R(x,x')})] + x \rightleftarrows x'\} , \tag{3.19}$$

where

ж) This representation can be easily obtained if we express $G(x,x')$ in terms of the solution found by Linet [18] and describing the electrostatic potential of a point charge located near a black hole.

$$\Pi(x,x') = (r-M)(r'-M) - M^2\lambda \quad ,$$
$$R(x,x') = \{(r-M)^2+(r'-M)^2-2(r-M)(r'-M)\lambda - M^2(1-\lambda^2)\}^{1/2} ,$$
$$\lambda(x,x') = \cos\theta\cos\theta' + \sin\theta\sin\theta'\cos(\varphi-\varphi') \quad .$$

Using this representation and the relations (3.14) ÷ (3.17) we finally get

$$\langle T_\theta^{\theta}(r,0;2M,0)\rangle_H = \frac{1}{\pi^2(4M)^4}\left\{\frac{1}{\delta^2}-1\right\} . \tag{3.20}$$

The divergent part of $\langle T_\mu{}^\nu\rangle^{div}$ is

$$\begin{aligned}
\langle T_t^{t}\rangle^{div} &= \frac{1}{\pi^2(4M)^4}\left\{\frac{1}{\delta^2}-\frac{2}{\delta}+\frac{191}{30}+O(\delta)\right\} ,\\
\langle T_r^{r}\rangle^{div} &= \frac{1}{\pi^2(4M)^4}\left\{-\frac{3}{\delta^2}+\frac{2}{\delta}-\frac{49}{30}+O(\delta)\right\} ,\\
\langle T_\theta^{\theta}\rangle^{div} = \langle T_\varphi^{\varphi}\rangle^{div} &= \frac{1}{\pi^2(4M)^4}\left\{\frac{1}{\delta^2}-\frac{58}{30}+O(\delta)\right\} ,\\
\langle T_\varepsilon^{\varepsilon}\rangle^{div} = \langle T_\varepsilon^{\varepsilon}\rangle_H^{ren} &= \frac{1}{\pi^2(4M)^4}\left\{\frac{26}{30}+O(\delta)\right\} .
\end{aligned} \tag{3.21}$$

Substituting Eqs.(3.20), (3.21) into Eq.(3.12) gives

$$\begin{aligned}
\langle T_t^{t}\rangle_H^{ren} = \langle T_r^{r}\rangle_H^{ren} &= -\frac{41}{30}\,\frac{1}{\pi^2(4M)^4} ,\\
\langle T_\theta^{\theta}\rangle_H^{ren} = \langle T_\varphi^{\varphi}\rangle_H^{ren} &= \frac{28}{30}\,\frac{1}{\pi^2(4M)^4} .
\end{aligned} \tag{3.22}$$

The above described method admits a natural generalization to the problem of a calculation of $\langle T_{\mu\nu}\rangle_H^{ren}$ at the pole of the event horizon of the Kerr black hole [23)]

3.2.b <u>The value of $\langle T_\mu^\nu \rangle_U^{ren}$</u>. In a regular map covering the event horizon $\langle T_\mu^\nu \rangle_U^{ren}$ is finite on the future event horizon [9,27)] and diverges on the past horizon. This divergence is proportional to the luminosity L of the black hole [9)] $a, b = (t, r)$

$$\langle T_a^b \rangle_U^{ren} \underset{r\to 2M}{\sim} \frac{L}{4\pi r^2} \begin{Bmatrix} (1-2M/r)^{-1} & -1 \\ (1-2M/r)^{-2} & -(1-2M/r)^{-1} \end{Bmatrix} \quad (3.23)$$

3.2.c <u>The value of $\langle T_\mu^\nu \rangle_B^{ren}$</u> diverges both on the future horizon and on the past event horizon. The leading part of the divergences of $\langle T_\mu^\nu \rangle_B^{ren}$ for the conformal scalar field ($S = 0$) and for the electromagnetic field ($S = 1$) was found by Candelas [9)]

$$\langle T_\mu^\nu \rangle_B^{ren} \underset{r\to 2M}{\sim} - \frac{h(s)}{2\pi^2 (1-2M/r)^2} \times \int_0^\infty \frac{d\omega\, \omega(\omega^2 + \varkappa^2 s^2)}{\exp(2\pi\omega/\varkappa) - 1} , \quad (3.24)$$

where $h(s)$ - is the number of helicity states for a massless field of spin S .

4. CONTRIBUTION OF MASSLESS FIELDS TO THE VACUUM POLARIZATION OUTSIDE THE BLACK HOLE

As far as we know the behaviour of $\langle \varphi^2 \rangle$ and $\langle T_\mu^\nu \rangle$ outside the event horizon was investigated only in the Schwarzschild metric.

4.1.a <u>The value of $\langle \varphi^2(r) \rangle_H^{ren}$</u>. In a spherically symmetric space the radial wave functions V and Q in the Eq.(2,5) do not depend on m , that enables one to sum over m in the Eq.(2.9). After straightforward transformations we can rewrite the expression for $\langle\varphi^2\rangle_H^{ren}$ in the following form [9]

$$\langle\varphi^2(r)\rangle_H^{ren} = \langle\varphi^2(r)\rangle_H^{P} + \frac{1}{16\pi^2}\int_0^\infty \frac{d\omega}{\omega}\,\mathrm{cth}\frac{\pi\omega}{\varkappa}\,\mu(r,\omega)\ , \tag{4.1}$$

where

$$\mu(r,\omega) = \frac{1}{r^2}\sum_{l=0}^{\infty}(2l+1)\left[|V_{\omega l}(r)|^2 + |Q_{\omega l}(r)|^2\right] - \frac{4\omega^2}{1-\frac{2M}{r}}\ , \tag{4.2}$$

$$\langle\varphi^2(r)\rangle_H^{P} = \frac{1}{768\pi^2M^2}\,\frac{1-(2M/r)^4}{1-2M/r}\ . \tag{4.3}$$

The remarkable fact, which was discovered by Fawcett and Whiting [15] in their analysis of results of the numerical calculations of $\langle\varphi^2\rangle_H^{ren}$, is that the difference between $\langle\varphi^2\rangle_H^{ren}$ and $\langle\varphi^2\rangle_H^{P}$ is uniformally small outside the black hole. This observation was analytically confirmed by Candelas and Howard [21]. They have shown that

$$\max_{2M\le r<\infty} |\langle\varphi^2(r)\rangle_H^{ren} - \langle\varphi^2(r)\rangle_H^{P}| \lesssim 10^{-2}\ . \tag{4.4}$$

The value (4.3) coincides with the Page's approximation formula [22] . Page has shown that if vacuum states for conformally connected spacetimes $\bar{g}_{\mu\nu}(x)=\Omega^2(x)g_{\mu\nu}(x)$ are conformally related, then the renormalized vacuum average values $\langle\varphi^2\rangle^{ren}$ and $\langle\bar{\varphi}^2\rangle^{ren}$ satisfy the relation

$$|g|^{1/4}\left(\langle\varphi^2\rangle^{ren} + \frac{1}{288\pi^2}R\right) = \\ = |\bar{g}|^{1/4}\left(\langle\bar{\varphi}^2\rangle^{ren} + \frac{1}{288\pi^2}\bar{R}\right) . \qquad (4.5)$$

Using the Gaussian approximation for the propagator $\bar{G}_H$ in the ultrastatic spacetime with the metric

$$d\bar{s}^2 = -dt^2 + dr^{*2} + \frac{r^2}{1-\frac{2M}{r}}(d\theta^2 + \sin^2\theta\, d\varphi^2) \qquad (4.6)$$

one has for the state conformally related to the Hartle-Hawking vacuum

$$\langle\bar{\varphi}^2\rangle_H^P = \frac{1}{768\pi^2 M^2} . \qquad (4.7)$$

This relation together with Eq.(4.5) leads to the expression (4.3) for $\langle\varphi^2\rangle_H^P$, which can be identically re-written in the form

$$\langle\varphi^2(r)\rangle_H^P = \frac{1}{12}\left[T_{loc}^2(r) - T_{acc}^2(r)\right] . \qquad (4.8)$$

Here $T_{loc} = \varkappa|g_{00}(r)|^{-1/2}/2\pi$, $T_{acc} = a(r)/2\pi$, $a = |a^\mu a_\mu|^{1/2}$ and $a^\mu(r)$ is the four-acceleration of a point at rest at the radius r near a black hole. Note that on the pole of the event horizon of the Kerr (as well as the Schwarzschild) black hole Eq.(4.6) coincides with the exact value (3.5).

4.1.b <u>The value of $\langle\varphi^2(x)\rangle_U^{ren}$</u>. $\langle\varphi^2(x)\rangle_U^{ren}$ can be represented in the form [9]

$$\langle\varphi^2(x)\rangle_U^{ren} = \frac{1}{16\pi^2}\int_0^\infty \frac{d\omega}{\omega}\mu_U(r,\omega) - \frac{M^2}{48\pi^2 r^4(1-\frac{2M}{r})} , \qquad (4.9a)$$

where

$$\mu_U(r,\omega)=\frac{1}{r^2}\sum_{\ell=0}^{\infty}(2\ell+1)\{|V_{\omega\ell}(r)|^2+ \\ +\operatorname{cth}\frac{\pi\omega}{\varkappa}|Q_{\omega\ell}(r)|^2\}-\frac{4\omega^2}{1-\frac{2M}{r}} \quad . \tag{4.9b}$$

The results of numerical calculations of $\langle\varphi^2(x)\rangle_U^{ren}$ can be found in Ref. 15.

4.1.c <u>The value of $\langle\varphi^2(x)\rangle_B^{ren}$</u>. In Schwarzschild spacetime $\langle\varphi^2(x)\rangle_B^{ren}$ can be represented as follows

$$\langle\varphi^2(x)\rangle_B^{ren}=\frac{1}{16\pi^2}\int_0^{\infty}\frac{d\omega}{\omega}\,\mu(r,\omega)+\langle\varphi^2(x)\rangle_B^{P} \quad , \tag{4.10a}$$

where

$$\langle\varphi^2(x)\rangle_B^{P}=-\frac{M^2}{48\pi^2 r^4(1-\frac{2M}{r})} \quad . \tag{4.10b}$$

In the ultrastatic space the Gaussian approximation of the Green's function $\overline{G}_B$ corresponding to the Boulware vacuum gives

$$\langle\overline{\varphi}^2(x)\rangle_B^{P}=0 \quad . \tag{4.11}$$

In this approximation $\langle\varphi^2(r)\rangle_B^{ren}$ coincides with $\langle\varphi^2(x)\rangle_B^{P}$ given by Eq.(4.10). It can be considered as the indication that not only integrals in Eqs.(4.1) and (4.9a) are small, but the function $\mu(r,\omega)$ itself is uniformly small.

4.2 Vacuum Energy-Momentum Tensor

Christensen and Fulling [24] have shown that an arbitrary symmetrical conserved ($T_{\mu\;;\nu}^{\nu}=0$) tensor T_μ^ν in Schwarzschild spacetime admits the representation

$$T_\mu{}^\nu = \sum_{i=1}^{4} \underset{i}{t}{}_\mu{}^\nu \quad , \tag{4.12}$$

where $\underset{i}{t}{}_\mu{}^\nu$ in coordinates $(t, r^*, \theta, \varphi)$ are of the following form.

$$\underset{1}{t}{}_\mu{}^\nu = \mathrm{diag}\left\{-\frac{HF}{r^2}+\frac{1}{2}T, \frac{HF}{r^2}, \frac{1}{4}T, \frac{1}{4}T\right\} ,$$

$$\underset{2}{t}{}_\mu{}^\nu = \mathrm{diag}\left\{-\frac{GF}{r^2}-2\Theta, \frac{GF}{r^2}, \Theta, \Theta\right\} ,$$

$$\underset{3}{t}{}_\mu{}^\nu = \frac{KF}{Mr^2}\begin{pmatrix} 1 & -1 & 0 & 0 \\ 1 & -1 & 0 & 0 \\ 0 & 0 & 0 & 0 \\ 0 & 0 & 0 & 0 \end{pmatrix} , \tag{4.13}$$

$$\underset{4}{t}{}_\mu{}^\nu = \mathrm{diag}\left\{\frac{QF}{M^2r^2}, \frac{QF}{M^2r^2}, 0, 0\right\} ,$$

where

$$T(r) = T^\varepsilon{}_\varepsilon(r) , \qquad F(r) = \left(1-\frac{2M}{r}\right)^{-1} ,$$

$$\Theta(r) = T^\theta_\theta(r) - \frac{1}{4}T(r) , \tag{4.14}$$

$$H(r) = \frac{1}{2}\int_{2M}^{r} (r'-M)\,T(r')\,dr' ,$$

$$G(r) = 2\int_{2M}^{r} (r'-3M)\,\Theta(r')\,dr' . \tag{4.15}$$

The tensors $\underset{i}{t}{}_\mu{}^\nu$ satisfy the conservation law $\underset{i}{t}{}_\mu{}^\varepsilon{}_{;\varepsilon} = 0$. Only $\underset{1}{t}{}_\mu{}^\nu$ has a nonzero trace, only $\underset{2}{t}{}_\mu{}^\nu$ has a traceless part $T_\mu{}^\nu - \frac{1}{4}\delta_\mu^\nu T$ with nonzero $\theta\theta$ component, only $\underset{3}{t}{}_\mu{}^\nu$ has off-diagonal terms which describe energy fluxes, and only $\underset{4}{t}{}_{\mu}{}^{\nu}$ diverges on the future horizon $\mathcal{H}^{+}$. In this representation two functions $T(r)$, $\Theta(r)$ and two constants Q and K are arbitrary.

If $T_{\mu}{}^{\nu}$ is the vacuum average of the energy of the energy-momentum tensor of a conformally-invariant massless field, then its trace is unambiguously determined by conformal anomalies and is of the form

$$T = \frac{a}{2880\pi^2} C_{\alpha\beta\gamma\delta} C^{\alpha\beta\gamma\delta} = \frac{a}{60\pi^2} \frac{M^2}{r^6} , \qquad (4.16)$$

where

$$a = h(0) + \frac{7}{8} h(1/2) - \frac{13}{2} h(1) , \qquad (4.17)$$

and $h(s)$ is the number of helicity states for a massless field of spin s. The corresponding function H in Eq.(4.15) is

$$H = \frac{aM^3}{1200\pi^2} \frac{1}{r^5} \left[2 - 5\left(\frac{r}{2M}\right) - 3\left(\frac{r}{2M}\right)^5\right] . \qquad (4.18)$$

4.2.a <u>The value of $\langle T_{\mu}{}^{\nu}\rangle_H^{ren}$</u>. The requirement of the regularity of $\langle T_{\mu}{}^{\nu}\rangle_H$ at the horizon and of the absence of the energy fluxes gives $K = Q = 0$.

An approximate expression for $\langle T_{\mu}{}^{\nu}(x)\rangle_H^{ren}$ was obtained by Page [22]. He used the invariance of the following expression

$$|g|^{1/2}\{\langle T_{\mu}^{\nu}\rangle^{ren} + \alpha[(C^{\alpha\nu}{}_{\beta\mu} \ln|g|)_{;\alpha}{}^{;\beta} +$$

$$+ \frac{1}{2} R^{\beta}_{\alpha} C^{\alpha\nu}{}_{\beta\mu} \ln|g|] + \qquad (4.19)$$

$$\beta [2H_{\mu}^{\nu} - 4R_{\alpha}^{\beta} C^{\alpha\nu}{}_{\beta\mu}] + \frac{1}{6}\gamma \underline{I}_{\mu}^{\nu} \quad \}$$

under the conformal transformation $\bar{g}_{\mu\nu} = \Omega^2(x)\, g_{\mu\nu}$.
Here

$$\alpha = \frac{1}{23040\pi^2}[12h(0) + 18h(1/2) + 72h(1)] \quad ,$$

$$\beta = \frac{1}{23040\pi^2}[-4h(0) - 11h(1/2) - 124h(1)] \quad , \qquad (4.20)$$

$$\gamma = \frac{1}{23040\pi^2}[8h(0) + 12h(1/2) + (48 \text{ or } -72)h(1)] \quad ,$$

$$H_{\mu\nu} = -R_{\mu}^{\alpha} R_{\alpha\nu} + \frac{2}{3} R R_{\mu\nu} + (\frac{1}{2} R_{\beta}^{\alpha} R_{\alpha}^{\beta} - \frac{1}{4} R^2) g_{\mu\nu} \, ,$$

$$\underline{I}_{\mu\nu} = 2R_{;\mu\nu} - 2RR_{\mu\nu} + (\frac{1}{2}R^2 - 2R_{;\alpha}{}^{;\alpha}) g_{\mu\nu} \, . \qquad (4.21)$$

Using the Gaussian approximation for the Green's function of a conformal scalar field in the ultrastatic space (4.6) Page obtained

$$\langle \bar{T}_{\mu}{}^{\nu} \rangle_H^P = \frac{\pi^2}{90} T_{BH}^4 (\delta_{\mu}^{\nu} - 4\delta_0^{\nu}\delta_{\mu}^0) \quad , \qquad (4.22)$$

where $T_{BH} = \varkappa/2\pi = 1/8\pi M$. In physical spacetime this leads to the following expression for $\langle T_{\mu}{}^{\nu} \rangle_H^P$

$$\langle T_{\mu}^{\nu} \rangle_H^P = \frac{\pi^2}{90} T_{BH}^4 \{ [1 - (4 - \frac{6M}{r})^2 (\frac{2M}{r})^6] \times$$

$$\times (\delta_{\mu}^{\nu} - 4\delta_{\mu}^0 \delta_0^{\nu})(1 - \frac{2M}{r})^{-2} + \qquad (4.23)$$

$$+ 24 (\frac{2M}{r})^6 (3\delta_{\mu}^0 \delta_0^{\nu} + \delta_{\mu}^1 \delta_1^{\nu}) \} \, .$$

Howard and Candelas [25] corrected a computational mistake in the work [26] and show that Eq.(4.23) is in a good agreement with the results of numerical calculations of $\langle T_\mu{}^\nu \rangle_H^{ren}$ for all values of r : $2M \leq r < \infty$. The Page's approximation corresponds to the following expression for $\Theta(r)$ in Eq.(4.13)

$$\Theta(r) = \frac{\pi^2}{90} T_{BH}^4 \frac{1-\left(\frac{2M}{r}\right)^6 \left[40-72\frac{2M}{r}+33\left(\frac{2M}{r}\right)^2\right]}{\left(1-\frac{2M}{r}\right)^2} \quad . \tag{4.24}$$

4.2.b <u>The value of $\langle T_\mu{}^\nu \rangle_V^{ren}$</u> was obtained by Elster [27] by means of numerical calculations.

4.2.c <u>The value of $\langle T_\mu{}^\nu \rangle_B^{ren}$</u> can be found by the method, which is analogous to the Page's method. If we use the Gaussian approximation for the Green's function $\bar{G}_B$ in the ultrastatic spacetime, then we obtain $\langle \bar{T}_\mu{}^\nu(x) \rangle_B^P = 0$. Using the conformal invariance property of the expression (4.19) we find for the average value of $\langle T_\mu{}^\nu \rangle_B^P$ in the Boulware vacuum state

$$\langle T_\mu^\nu \rangle_B^P = \frac{M^2}{1440\pi^2 r^6} \left[\frac{\left(2-3\frac{M}{r}\right)^2}{\left(1-\frac{2M}{r}\right)^2} \left(-\delta_\mu^\nu + 4\delta_\mu^0 \delta_0^\nu\right) + 6\left(3\delta_\mu^0 \delta_0^\nu + \delta_\mu^1 \delta_1^\nu\right)\right] \quad . \tag{4.25}$$

The obtained results about the massless fields contribution to the vacuum polarization near black holes are briefly summarized in Table 1. This table contains references and information about the vacuum state choice (H, U, B). The reference to the results obtained in this paper contains symbol *.

	Schwarzschild		Reissner- -Nordström	Kerr- (-Newman)
	at the horizon	out of the horizon		
$\langle\varphi^2\rangle^{ren}$ $s=0$	Candelas [9] H(B,U) Frolov & Garcia [14] H Frolov [37] H,r_o	Fawcett & Whiting [15] H(B,U) Candelas & Howard [21] H Page [22] H Bolashenko & Frolov [29] H Frolov & Zel'nikov [*] B	Zannias [28] H	Frolov [12] H
$\langle T_\mu^\nu\rangle^{ren}$ $s=0$	Candelas [9] H(B,U) Elster [19] H,r_o	Page [22] H Howard & Candelas [25] H Howard [38] H Bolashenko & Frolov [29] H Fawcett [26] H(B,U) Frolov & Zel'nikov [*] B	Zannias [28] H	
$\langle T_\mu^\nu\rangle^{ren}$ $s>0$	Elster [20] H Frolov & Zel'nikov [23,*] H			Candelas, Chrzanowsky, & Howard [10] B

Table 1.

5. CONTRIBUTION OF MASSIVE FIELDS TO THE VACUUM POLARIZATION IN THE GRAVITATIONAL FIELD OF BLACK HOLES

If mass of a black hole is much more than the Planck mass $m_{Pl}=\sqrt{\hbar c/G}$, then the contribution of massive fields to the vacuum polarization is much smaller (by the factor $\varepsilon=(\lambda_m/L)^2$, $\lambda_m=\hbar/mc$, L is a characteristic radius of the spacetime curvature) than the contribution of massless fields. However there are some reasons why the investigation of the vacuum polarization of massive fields is of some interest. First of all it should be noted that for $\varepsilon \ll 1$ it is possible to separate contributions of real particles and vacuum polarization to $\langle T_\mu{}^\nu\rangle_H^{ren}$. In the Hartle-Hawking vacuum state the contribution of real particles to $\langle T_\mu{}^\nu\rangle_H^{ren}$ is exponentially small $\sim \exp(-GMm/\hbar c)$, while the vacuum polarization contribution near the horizon is proportional to $(\hbar c/GMm)^2$. For massive fields the vacuum polarization effect can be described in more details, because there is a possibility to use series expansion in powers of small parameter ε . In particular this method allows one to investigate the influence of the black-hole rotation on the vacuum polarization. This effect appears to be quite general and it can be essential for massless fields also.

We consider massive scalar $(s=0)$, Dirac spinor $(s=1/2)$, and vector $(s=1)$ fields $\varphi^{(s)}$ satisfying the equations

$$(\Box - \xi R - m^2)\,\varphi^{(0)} = 0 \quad , \tag{5.1a}$$

$$(\gamma^\alpha \nabla_\alpha + m)\,\varphi^{(1/2)} = 0 \quad , \tag{5.1b}$$

$$(\Box \delta_\alpha^\beta \nabla^\varepsilon \nabla_\varepsilon - R_\alpha^\beta - m^2 \delta_\alpha^\beta)\,\varphi_\beta^{(1)} = 0 \quad . \tag{5.1c}$$

One can obtain the vacuum average $\langle T_\mu{}^\nu \rangle_H^{ren}$ of the energy-momentum tensor for these fields by functionally differentiating the renormalized effective action $W_{ren}^{(s)}$

$$\langle T_{\mu\nu}^{(s)} \rangle_H^{ren} = -\frac{2}{|g|^{1/2}} \frac{\delta W_{ren}^{(s)}}{\delta g^{\mu\nu}} , \tag{5.2}$$

which in the one-loop approximation reads

$$W_{ren}^{(s)} = \frac{1}{(4\pi m)^2} \frac{1}{18\cdot 7!} \int d^4x \, |g|^{1/2} L^{(s)} + O(\varepsilon^2), \tag{5.3}$$

$$\begin{aligned} L^{(s)} = {} & a_1 R_{\alpha\beta\gamma\delta;\varepsilon} R^{\alpha\beta\gamma\delta;\varepsilon} + a_2 R^\alpha_\beta R_{\alpha\gamma\delta\varepsilon} R^{\beta\gamma\delta\varepsilon} + \\ & + a_3 R_{\alpha\beta}{}^{\gamma\delta} R_{\gamma\delta}{}^{\varepsilon\zeta} R_{\varepsilon\zeta}{}^{\alpha\beta} + a_4 R_{\alpha\beta\gamma\delta} R^{\alpha\beta\gamma\delta} + \\ & + a_5 R R_{\alpha\beta} R^{\alpha\beta} \quad + a_6 R^3 \quad + \\ & + a_7 R_{;\varepsilon} R^{;\varepsilon} \quad + a_8 R_{\alpha\beta;\gamma} R^{\alpha\beta;\gamma} + \\ & + a_9 {}'^{\beta}_{-\alpha} R^\gamma_\beta R^\alpha_\gamma \quad + a_{10} R_{\alpha\beta} R_{\gamma\delta} R^{\alpha\gamma\beta\delta} . \end{aligned} \tag{5.4}$$

The coefficients a_i $(i = 1, \dots, 10)$ for $s = 0, 1/2, 1$ which were obtained by using the general expression found by Gilkey [30] are collected in the following table.

For the metric satisfying the vacuum Einstein's equations $R_{\mu\nu} = 0$ the expression for $W_{ren}^{(s)}$ can be written as follows

$$W_{ren}^{(s)} = B_0 \int d^4x \, |g|^{1/2} \{ \alpha^{(s)} R_{\alpha\beta}{}^{\gamma\delta} R_{\gamma\delta}{}^{\varepsilon\zeta} R_{\varepsilon\zeta}{}^{\alpha\beta} + \\ + \beta^{(s)} R R_{\alpha\beta\gamma\delta} R^{\alpha\beta\gamma\delta} + \dots \} + O(m^{-4}), \tag{5.5}$$

where $B_0=(96\cdot 7!\pi^2 m^2)^{-1}$, and ellipsis denotes the omitted terms which do not contribute to $\langle T_{\mu\nu}\rangle_H^{ren}$ in the Ricci-flat ($R_{\mu\nu}=0$) spacetime and

$$\alpha^{(0)}=1 \quad , \alpha^{(1/2)}=-4 \quad , \alpha^{(1)}=3 \quad ,$$
$$\beta^{(0)}=18-84\xi , \beta^{(1/2)}=12 \quad , \beta^{(1)}=-30 \quad . \tag{5.6}$$

Table 2

	S=0	S=1/2	S=1
a_1	-7	-35	105
a_2	-8	158	3504
a_3	24	93	-306
a_4	$42(1-6\xi)$	-147/2	-756
a_5	$-42(1-6\xi)$	-84	3906
a_6	$35(1-6\xi)^3$	35/2	-525
a_7	$-142+1512\xi-3780\xi^2$	127	834
a_8	-26	-400	-2850
a_9	-36	-360	-6660
a_{10}	20	424	-948

The calculation of $\langle T_\mu{}^\nu\rangle_H^{ren}$ for the algebraically special type-D vacuum metrics can be essentially simplified by using the Newman-Penrose approach. In particular in the Kerr metric one can choose the following null complex tetrad

$$k^\mu=\left(\frac{r^2+a^2}{\Delta}, 1, 0, \frac{a}{\Delta}\right) ,$$
$$\ell^\mu=\frac{\Delta}{2\Sigma}\left(\frac{r^2+a^2}{\Delta}, -1, 0, \frac{a}{\Delta}\right) , \tag{5.7}$$
$$m^\mu=-\frac{\bar{\rho}}{\sqrt{2}}\left(ia\sin\theta, 0, 1, \frac{i}{\sin\theta}\right) .$$

for which the spin coefficients ρ , π , μ , τ and the nonzero spin component $\Psi = \Psi_{0011}$ of the Weyl tensor are

$$\rho = -(r - ia\cos\theta)^{-1}, \qquad \pi = i\frac{a}{\sqrt{2}}\rho^2 \sin\theta \quad ,$$
$$\mu = \rho\Delta/2\Sigma, \qquad \tau = -ia\sin\theta/\sqrt{2}\Sigma, \qquad \Psi = M\rho^3 . \tag{5.8}$$

Then

$$\begin{aligned}\langle T^{(s)\,ren}_{\mu\nu}\rangle_H = 24 B_0 \,\mathrm{Re}\{\alpha^{(s)}[-3g_{\mu\nu}(\Psi^2)_{;\alpha}{}^{;\alpha} + \\ +20 g_{\mu\nu}\Psi^3 + 12\Psi_{,\mu}\Psi_{,\nu} + 144\Psi^2 \times \\ \times(\pi\tau k_{(\mu}l_{\nu)} + \rho\mu m_{(\mu}\bar{m}_{\nu)} - \\ - \tau\mu k_{(\mu}\bar{m}_{\nu)} - \rho\pi l_{(\mu}m_{\nu)})] + \\ +4\beta^{(s)}[(\Psi^2)_{;\mu\nu} - g_{\mu\nu}(\Psi^2)_{;\alpha}{}^{;\alpha}]\} .\end{aligned} \tag{5.9}$$

If we denote

$$z = M/\Sigma \quad , \qquad u = Mr/\Sigma \quad ,$$
$$v = r^2/\Sigma \quad , \qquad w = (r^2 + a^2)/\Sigma \quad , \tag{5.10}$$

then $\langle T^{(s)\,\nu\,ren}_{\mu}\rangle_H$ can be written in the form

$$\langle T^{(s)\,\nu\,ren}_{\mu}\rangle_H = \frac{M^2\,\mathrm{Re}\{\rho^7(\alpha^{(s)}p^{\nu}_{\mu} + 12\beta^{(s)}q^{\nu}_{\mu})\}}{10080\,\pi^2 m^2} \tag{5.11}$$

$$p^t_t = (-45 + 106u)\rho + 8z \, , \qquad p^{\varphi}_t = 0 \quad ,$$
$$p^t_{\varphi} = -36a\rho u\sin^2\theta, \qquad p^{\varphi}_{\varphi} = (-27 + 70u)\rho + 8z \; ,$$
$$p^r_r = (-27 + 36w - 2u)\rho + 8z, \quad p^{\theta}_r = \frac{1}{\Delta}p^r_{\theta} = 36ia\rho\sin\theta/\Sigma \, ,$$
$$p^{\theta}_{\theta} = (9 - 36w + 70u)\rho + 8z \qquad ;$$

$$
\begin{aligned}
q_t^{\ t} &= (-5+12u+2uw)\rho + (2-2v-w+4vw)z \,, \\
q_t^{\ \varphi} &= [2\rho u a + z a(4v-1)]\Sigma^{-1} \,, \\
q_\varphi^{\ t} &= -2\rho u a^3 \sin^4\theta\, \Sigma^{-1} + z w a(4v-1)\sin^2\theta \,, \qquad (5.12) \\
q_\varphi^{\ \varphi} &= (-6+16u-2uw)\rho + (1+2v+w-4vw)z \,, \\
q_r^{\ r} &= (-6+8w)\rho + (1+2v)z, \quad q_r^{\ \theta} = \frac{1}{\Delta} q_\theta^{\ r} = 8ia\rho\sin\theta\,\Sigma^{-1}, \\
q_\theta^{\ \theta} &= (2-8w+14u)\rho + (2-2v)z \,.
\end{aligned}
$$

The plots presenting the dependence of $\langle T_\mu^{\ \nu}\rangle_H^{ren}$ on r and θ for the values of the rotation parameter $a/M=$ 0, 1/2, 1 are given in our paper [32].

The following properties of the solution (5.11) are of the main interest: i. The numerical values of $P_\mu^{\ \nu}$ outside the horizon are much smaller than the corresponding components of $q_\mu^{\ \nu}$ and hence the dependence of $\langle T_\mu^{(s)\nu}\rangle_H^{ren}$ on the spin of a field is mainly determined by the factor $\beta^{(s)}$. ii. For a nonrotating black hole ($a=0$) the expression (5.11) reads

$$\langle T_\mu^{(s)\nu}\rangle_H^{ren} = \frac{M^2[\alpha^{(s)}\tilde{P}_\mu^{\ \nu} + 12\beta^{(s)}\tilde{q}_\mu^{\ \nu}]}{10080\pi^2 m^2 r^8} \,, \qquad (5.13)$$

where nonzero components of $\tilde{P}_\mu^{\ \nu}$ and $\tilde{q}_\mu^{\ \nu}$ are ($x \equiv r/M$)

$$
\begin{aligned}
&\tilde{P}_t^{\ t} = -15+32x \,, \quad \tilde{P}_r^{\ r} = -3+8x \,, \quad \tilde{P}_\theta^{\ \theta} = \tilde{P}_\varphi^{\ \varphi} = 9-22x \,, \\
&\tilde{q}_t^{\ t} = -5+11x \,, \quad \tilde{q}_r^{\ r} = 2-3x \,, \quad \tilde{q}_\theta^{\ \theta} = \tilde{q}_\varphi^{\ \varphi} = -6+14x \,. \qquad (5.14)
\end{aligned}
$$

As well as for the massless case the energy densities $\varepsilon^{(s)} = -\langle T_t^{(s)t}\rangle_H^{ren}$ of massive scalar and vector fields at the horizon have different signes. This sign difference is connected with the additional contribution to the vacuum energy density due to the interaction of

spin of virtual particles with their angular momentum.
iii. The rotation of the black hole leads to the appearance of the circular flux of the vacuum energy density in the surrounding space, which is described by the component $\langle T^{(s)\,t}_{\varphi}\rangle^{ren}_{H} \neq 0$. In the Boyer-Lindquist coordinates $\langle T^{(s)\,\nu}_{\mu}\rangle^{ren}_{H}$ is finite at the horizon both for rotating and nonrotating black holes. The total mass M and the angular momentum J of a black hole measured by a distant observer

$$M = -\frac{1}{8\pi}\int_{S_\infty} \nabla^\alpha \xi^\beta_{(t)}\, dS_{\alpha\beta} , \tag{5.15}$$

$$J = \frac{1}{16\pi}\int_{S_\infty} \nabla^\alpha \xi^\beta_{(\varphi)}\, dS_{\alpha\beta} .$$

($\xi^\varepsilon_{(t)}\partial_\varepsilon = \partial_t$, $\xi^\varepsilon_{(\varphi)}\partial_\varepsilon = \partial_\varphi$, S_∞ being two-sphere at infinity surrounding the black hole) differ from the proper mass M_{BH} and the proper angular momentum J_{BH} of the black hole. The quantities M_{BH} and J_{BH} are determined by Eq.(5.15) where the intergration should be performed over the surface of the black hole S_{BH} instead of S_∞ . When the rotation parameter is small $a \ll M$, we have

$$M - M_{BH} = M\,\frac{1}{7!\pi m^2 M^4}\cdot\frac{7}{6}\begin{cases} 2-9\xi , & s=0 ; \\ 1 , & s=1/2 ; \\ -3 , & s=1 ; \end{cases} \tag{5.16}$$

$$J - J_{BH} = J\,\frac{1}{7!\pi m^2 M^4}\,\frac{1}{8}\begin{cases} 17-84\xi , & s=0 ; \\ 16 , & s=1/2 ; \\ 33 , & s=1 ; \end{cases} \tag{5.17}$$

These equations show that for scalar ($\xi = 1/6$) and spinor fields the mass M and the angular momentum J are greater and for the vector field are smaller than M_{BH} and J_{BH}. The considered effects of screening and antiscreening (in dependence on the spin of the field) of the mass and the angular momentum of a black hole are to be expected in the massless case. These effects seem to be rather important for black holes of masses comparable with the Planckian mass. iv. It is instructive to compare the above described angular momentum polarization effect with the analogous effect of the polarization of the magnetic field in quantum electrodynamics. The contribution of massive fields of spin S to the magnetic susceptibility χ is described by the Nielsen-Hughes formula [33-35]

$$\chi = \frac{1}{48\pi^2}\left[12 S^2 - 1\right](-1)^{2S} \quad . \qquad (5.18)$$

The second term in this expression does not depend on the spin and due to the Landau diamagnetism, while the term proportional to S^2 is due to paramagnetic phenomenon. The common sign minus for fermions is connected with the fact that their contribution to the vacuum energy is negative [36]. Qualitative similarity (up to the sign minus) of the spin dependences of $\chi_{BH} \equiv (J - J_{BH})/J$ and χ attracts attention. The reason of this similarity may become clear if one takes into account the deep analogy between gravitation and electromagnetism. In the framework of this analogy (at least in the weak field approximation) gravitational interaction of masses and angular momenta is analogous to electromagnetic interaction of charges and magnetic moments. The main difference between these two cases lies in an additional minus sign which is present in the gravitational theory. The spin dependence (5.17) of χ_{BH} can be interpreted as a consequence of gravitational version of dia- and paramagnetism. One can speculate that the existence of the effects involved make it possible for the gravitating system composed of particles with integer spins only to possess the total half-integer angular momentum.

REFERENCES

1. Hawking, S.W., Nature (London) 248, 30 (1974); Commun.Math.Phys. 43, 199 (1975).
2. Misner, C.W., Thorne, K.S. and Wheeler, J.A. "Gravitation" (Freeman, San Francisco, 1973).
3. De Witt, B.S. "Dynamical Theory of Groups and Fields", Gordon and Breach, New York, 1965).
4. Christensen, S.M. Phys.Rev. D17, 946 (1978).
5. Christensen, S.M. Phys.Rev. D14, 2490 (1976).
6. Unruh, W.G. Phys.Rev. D14, 870 (1976).
7. Hartle, J.B. and Hawking, S.W. Phys.Rev. D13, 2188 (1976).
8. Boulware, D.G. Phys.Rev. D11, 1404 (1975).
9. Candelas, P. Phys.Rev. D21, 2185 (1980).
10. Candelas, P., Chrzanowsky, P. and Howard, K.M. Phys.Rev. D24, 297 (1981).
11. Hawking, S.W. Commun.Math.Phys. 80, 421 (1981).
12. Frolov, V.P. Phys.Rev. D26, 954 (1982).
13. Frolov,V.P. In "Quantum Gravity", proceedings of the Second Seminar on Quantum Gravity, Moscow, 1981, edited by M.A.Markov and P.West (Plenum, New York, 1983) p.303.
14. Frolov, V.P. and Garcia, A.D. Phys.Lett. 99A, 421 (1983).
15. Fawcett, M.S. and Whiting, B. In: "Quantum Theory of Space and Time", eds. M.J.Duff and C.J.Isham (Cambridge: CUP, 1982).
16. Copson, E.T. Proc.Roy.Soc.London A118, 184 (1928).

17. Bičák, J. , Dvořák L. Czech.J.Phys., B27, 127 (1977).
18. Linet, B. J.Phys. A9, 1081 (1976).
19. Elster, T. J.Phys. A16, 983 (1983).
20. Elster, T. Class. and Quant.Grav. 1, 43, (1984).
21. Candelas, P. and Howard K.W. Phys.Rev. D29, 1618 (1984).
22. Page, D.N. Phys.Rev. D25, 1499 (1982).
23. Frolov, V.P. and Zel'nikov, A.I. to appear (1985).
24. Christensen, S.M. and Fulling, S.A. Phys.Rev. D15, 2088 (1977).
25. Howard, K.W. and Candelas, P. Phys.Rev.Lett. 53, 403 (1984).
26. Fawcett, M.S. Commun.Math.Phys. 89, 103 (1983).
27. Elster, T. Phys.Lett. 94A, 205 (1983).
28. Zannias, T. Phys.Rev. D30, 1161 (1984).
29. Bolashenko, P.A. and Frolov, V.P. 10th Int. Conf. on GRG, Padova, 1983, v.2, 1036.
30. Gilkey, P.B. J.Diff.Geom. 10, 601 (1975).
31. Frolov, V.P. and Zel'nikov, A.I. Phys.Lett, 123B, 197 (1983).
32. Frolov, V.P. and Zel'nikov, A.I. Phys.Rev. D29, 1057 (1984).
33. Hughes, R.J. Phys.Lett. 97B, 246 (1980); Nucl.Phys. B186, 376 (1981).
34. Nielsen, N.K. Am.J.Phys. 49, 1171 (1981).
35. Allcock, G.R. Acta Phys.Pol. B11, 875 (1980).
36. Wald, R. Phys.Rev. D6, 406 (1972).
37. Frolov, V.P. "Quantum Effects in Gravitational Field of Black Holes", In: Proceedings of P.N.Lebedev Physical Institute, 169 (1985).

IMPROVED CHARACTERIZATION OF THE KERR METRIC

Z. PERJÉS

Central Research Institute for Physics
H-1525 Budapest 114, P.O.B.49, Hungary

ABSTRACT

The condition of asymptotic flatness is removed from Simon's characterization of the Kerr metric by vanishing of the complexified Bach tensor. The solution of the stationary vacuum equations of relativity is given for a vanishing Simon tensor. One class of metrics consists of three Ehlers-rotated Levi-Civita metrics which have conformally flat three-spaces. The second class contains Hoffmann's plane-fronted standing wave solutions, and the third includes the three-parameter Kerr-NUT space-time.

1. INTRODUCTION

The past decades have seen the recognition that the Kerr space-time is an all-important model of relativistic rotation. Nevertheless, the precise position of the Kerr metric among the solutions of the stationary vacuum equations of relativity remained puzzling. Various available characterizations involve the algebraic properties of the curvature[1)] or the separability of the Hamilton-Jacobi equations[2)]. Apart from Newman's argument of 'transcendent logic'[3)] and the existence of geodesic eigenrays of the Killing vector[4)], however, no straightforward assumption was known about the gravitational field under which the Kerr solution would follow from the stationary field equations. By way of contrast, the Schwarzschild metric can be characterized either by spherical symmetry or by conformal flatness of the 3-space. The latter is equivalent to a vanishing York (or Bach) tensor,

$(R_{i[k} - \frac{1}{4}g_{i[k}R)_{;\ell]} = 0$ where R_{ik} is the Ricci tensor of the 3-space of time-like Killing trajectories.

Recently, Simon[5] has found a complex generalization of the Bach tensor for stationary vacuum fields, using (a variant of) the Ernst potential. The Simon tensor vanishes for the Kerr space-time. However, the converse, i.e., whether the vanishing of the Simon tensor implies Kerr metric is still a valid question. Assuming asymptotic flatness, and using multipole expansion, the Kerr metric does follow from the condition that the Simon tensor vanishes[5].

The purpose of the present paper is twofold. First, in section 2 it will be shown that the Simon tensor can be defined even when the Ernst potential is nonexistent. The discussion does not rely on the assumption of axial symmetry either. The second purpose of the paper is to solve the vacuum stationary gravitational equations of relativity for a vanishing Simon tensor. In doing so, we will apply new techniques which were developed recently for the treatment of conformally flat spaces[6,7,8]. The essential idea is to separate two degenerate classes of fields, characterized either by a null Killing bivector or by a functional relationship between the real and imaginary parts of the Ernst potential. These special metrics will be discussed in Sec.3. The remaining fields can be coordinatized by the Ernst and complex conjugate potential as new coordinates (Sec.4). It is next shown in Sec.5. that the three-parameter Kerr-NUT space-time belongs to that class. The computation of the components of the Simon tensor is carried out in the Appendix.

2. THE SIMON TENSOR

In a stationary space-time with time-like Killing vector K^μ we can introduce the quantities[9]

$$f = K^\mu K_\mu \; ; \qquad f > 0 \tag{2.1}$$

$$\omega_i = \frac{K_i}{f} \tag{2.2}$$

and the complex 3-vector

$$\underline{G} = (2f)^{-1}(\nabla f + i\, f^2 \nabla \times \underline{\omega}) \tag{2.3}$$

where the notation refers to the metric $g_{ik}dx^i dx^k$ of the 3-space of Killing trajectories. We define the complex tensor

$$C_i^{\ \ell} = 2\epsilon^{jk\ell}(G_{j;i}G_k - g_{ij}g^{rs}G_{[s|;r|}G_{k]}) . \tag{2.4}$$

Specializing further to vacuum, Einstein's equations $R^{(4)}_{\lambda\mu}=0$ hold, and the Simon tensor[5] C_{ijk} is defined by

$$C_{ijk} = \epsilon_{jk\ell}C_i^{\ \ell} . \tag{2.5}$$

This can be seen by noting that the Ernst potential ε exists when the vacuum field equations hold and is defined by [10]

$$\nabla\varepsilon = \nabla f + i\, f^2 \nabla \times \underline{\omega} . \tag{2.6}$$

In terms of the Ernst potential we have

$$\underline{G} = \frac{\nabla\varepsilon}{\varepsilon + \bar{\varepsilon}} . \tag{2.7}$$

(Note that in Simon's work[5] the tensor C_{ijk} is given in terms of the variant of the Ernst potential $w = \frac{1 - \varepsilon}{1 + \varepsilon}$.)

By using the detailed form of the vacuum Einstein equations[3]

$$(\nabla - \underline{G} + \bar{\underline{G}})\cdot\underline{G} = 0 \tag{2.8a}$$

$$(\nabla - \underline{G} + \bar{\underline{G}})\times\underline{G} = 0 \tag{2.8b}$$

$$R_{ik} = -\, G_i\bar{G}_k - \bar{G}_i G_k \tag{2.8c}$$

where R_{ik} is the Ricci tensor of the 3-space, one can easily check that the tensor C_{ik} is symmetric and trace-free:

$$C_{ik} = C_{ki} \; ; \qquad C^{k}_{\;k} = 0 \; . \tag{2.9}$$

Furthermore, $C_{i}^{\;k}$ satisfies the divergence equation

$$(\nabla_k - 2G_k + 2\bar{G}_k)C_{i}^{\;k} = 0 \; . \tag{2.10}$$

In a static space-time $\underline{\omega}$ is a gradient and the vector $\underline{G}$ is real, hence $C_{i}^{\;k}$ becomes real. By use of the field equations we find that $C_{i}^{\;k}$ coincides with the York tensor in a static space time[5].

The components of the tensor $C_{i}^{\;k}$ can be conveniently computed by introducing the complex triad vectors

$$z_o^{\;i} = \ell^i \, , \qquad z_+^{\;i} = m^i, \qquad z_-^{\;i} = \bar{m}^i \tag{2.11}$$

normalized by

$$\left[z_{\underline{m}}^{\;\;i} \; z_{\underline{n}i}\right] \equiv \left[g_{\underline{mn}}\right] = \begin{bmatrix} 1 & 0 & 0 \\ 0 & 0 & 1 \\ 0 & 1 & 0 \end{bmatrix} \; . \tag{2.12}$$

The unit vector $\ell^i = i\epsilon^{ijk} m_j \bar{m}_k$ is real. We introduce the following notation for the Ricci rotation coefficients and scalar derivatives[4]:

$$\begin{aligned} \kappa &= m_{i;k}\ell^i \ell^k \\ \epsilon &= m_{i;k}\bar{m}^i \ell^k \\ \rho &= m_{i;k}\ell^i \bar{m}^k \\ \sigma &= m_{i;k}\ell^i m^k \\ \tau &= m_{i;k}\bar{m}^i \bar{m}^k \end{aligned} \tag{2.13}$$

$$D = \ell^i \partial_i, \qquad \delta = m^i \partial_i, \qquad \bar{\delta} = \bar{m}^i \partial_i \quad . \tag{2.14}$$

The orientation of the triad will be chosen along eigenrays[4],

$$G_+ = \underline{G}\cdot\underline{m} = 0 \tag{2.15}$$

and
$$\in = 0 , \tag{2.16}$$

leaving the only permissible transformation

$$\underline{\ell}' = \underline{\ell} , \quad \underline{m}' = e^{i\chi}\, \underline{m}, \quad D\chi = 0. \tag{2.17}$$

Computation (Appendix) yields the triad components $C_{ik} z^i_{\underline{m}} z^k_{\underline{n}}$ in the detailed form

$$i\, C_{++} = 2\sigma G_o^2 \tag{2.18a}$$

$$i\, C_{o+} = -\sigma G_o G_- \tag{2.18b}$$

$$i\, C_{-+} = 0 \tag{2.18c}$$

$$i\, C_{oo} = -2\kappa G_o G_- \tag{2.18d}$$

$$iC_{o-} = G_o G_-(-3\rho + \bar{G}_o - G_o) + G_o DG_- - \kappa G_-^2 - \bar{\kappa} G_o^2 \tag{2.18e}$$

$$\frac{1}{2}\, i\, C_{--} = -\, G_-\bar{\delta} G_o + G_o \bar{\delta} G_- - \rho G_-^2 + \bar{\sigma} G_o^2 + \tau G_o G_- . \tag{2.18f}$$

These expressions are supplemented by the triad form[4)] of the propagation equations (2.8a) and (2.8b) of the vector $\underline{G}$:

$$DG_o - (2\rho + G_o - \bar{G}_o)G_o + \kappa G_- = 0 \tag{2.19a}$$

$$\bar{\delta} G_o - DG_- + \rho G_- + \bar{\kappa} G_o - \bar{G}_o G_- = 0 \tag{2.19b}$$

$$\delta G_o + \sigma G_- + \kappa G_o + \bar{G}_+ G_o = 0 \tag{2.19c}$$

$$-\delta G_- + (\bar{\rho} - \rho) G_o + \bar{\tau} G_- - \bar{G}_+ G_- = 0 . \tag{2.19d}$$

3. DEGENERATE AND NULL SOLUTIONS

Here we consider two special classes of metrics for which the proceduce to be adopted in subsequent sections would fail. First we examine the kind of degeneracy caused by a functional relationship between the real and imaginary parts of the Ernst potential. For such degenerate fields, the complex vector $\underline{G}$ is parallel to its complex conjugate, i.e., $\underline{G} \times \bar{\underline{G}} = 0$. From $G_+ = 0$ we obtain by complex conjugation that

$$G_- = 0 \ . \qquad (3.1)$$

The condition $C_{++} = 0$ yields, noting that $\underline{G} = 0$ characterizes a flat space-time, and using Eq. (2.18a):

$$\sigma = 0 \ . \qquad (3.2)$$

Similarly, from $C_{o-} = 0$ and Eq. (2.18e) we have

$$\kappa = 0. \qquad (3.3)$$

The remaining components (2.18) of the Simon tensor vanish identically.

Space-times characterized by $\kappa = \sigma = 0$ are degenerate in the Petrov sense[6]. The condition that the Bach and Simon tensors are equal not only holds for static space-times[5] (with the vector $\underline{G}$ real) but also for degenerate space-times. As a consequence, the 3-space of Killing trajectories is conformally flat. The complete solution of the degenerate vacuum problem was given in Ref.6. The solution consists of three Ehlers-rotated Levi-Civita metrics, including the NUT metric.

We continue the discussion with the exceptional class of space-times characterized by the condition that the complex 3-vector $\underline{G}$ is null. In terms of triad components, we have

$$\underline{G}\cdot\underline{G} = 2G_+G_- + G_o^2 \ . \qquad (3.6)$$

Here $G_+ = 0$ by the eigenray condition. Thus for null fields with $\underline{G}\cdot\underline{G}=0$,

$$G_o = 0 \ . \qquad (3.7)$$

Vanishing of the Simon tensor (2.18) yields

$$\kappa = 0 \ , \quad \rho = 0 \ . \tag{3.8}$$

This is the condition that the eigenrays are geodesic and orthogonal to a family of planes. In addition, from field equation (2.19c) we obtain, using (3.7), that the eigenrays are shear-free:

$$\sigma = 0. \tag{3.9}$$

The eigenrays define multiple principal null congruences of the space-time when they are geodesic and shear-free[4]. These principal congruences are orthogonal to a family of null planes when $G_o = \rho = 0$. What we have shown is that the null class of the vacuum stationary space-times with a vanishing Simon tensor consists of standing wave solutions belonging to Kundt's plane-fronted waves[11]. These standing waves have been studied by Hoffmann[12].

4. ERNST COORDINATES

Excluding the degenerate space-times of the previous section from the discussion, the real vector

$$\underline{L} = i\underline{G} \times \bar{\underline{G}} \tag{4.1}$$

is nonvanishing and orthogonal to $\underline{G}$ and $\bar{\underline{G}}$. We can use the basis $\{\underline{L},\underline{G},\bar{\underline{G}}\}$ in the tangent 3-space. The condition that the Simon tensor vanishes, i.e.,

$$C_i{}^{\ell} = 2\epsilon^{jk\ell}(G_{j;i}G_k - g_{ij}g^{rs}G_{[s|;r|}G_{k]}) = 0 \tag{4.2}$$

will be seen from a geometric perspective in the new basis. Contraction with G_ℓ and Eq. (2.8b) yields:

$$g_{ij}\epsilon^{jk\ell}[\tfrac{1}{2}(G^rG_r)_{,k} + \bar{G}_k G_rG^r]\, G_\ell = 0 \ . \tag{4.3}$$

Further contraction with G^i yields an identity, and with $\bar{G}^i$ gives

$$L^k(\underline{G}\cdot\underline{G})_{,k} = 0 \ . \qquad (4.4)$$

By transvecting with $i\bar{G}^i\bar{G}_\ell$ we get the complex equation

$$L^k G_{k;i}\bar{G}^i = 0 \ . \qquad (4.5)$$

The real part, with field equation (2.8b) and using the orthogonality properties, becomes

$$L^k(\underline{G}\cdot\underline{\bar{G}})_{,k} = 0 \ . \qquad (4.6)$$

Similarly, from the imaginary part:

$$\epsilon_{ijk}\, L^{i;j}L^k = 0 \ . \qquad (4.7)$$

The geometric content of Eq. (4.7) is that the vector $\underline{L}$ is orthogonal to a family of surfaces[13] $\varphi(x^i)$ = const.:

$$\underline{L} = \chi\nabla\varphi \qquad (4.8)$$

where χ is some real function of proportionality.

Using this information, one can introduce the coordinates[7]

$$x^1 = \varepsilon \ , \quad x^2 = \bar{\varepsilon} \ , \quad x^3 = \varphi \ . \qquad (4.9)$$

Allowable still are the coordinate transformations corresponding to re-parametrization of the surfaces by

$$\varphi \Rightarrow \xi(\varphi) \qquad (4.10)$$

In these coordinate systems the metric can be written[7]

$$\left[g^{ik}\right] = \begin{bmatrix} \alpha & \beta & 0 \\ \beta & \gamma & 0 \\ 0 & 0 & \rho^{-2} \end{bmatrix} \tag{4.11}$$

where the function $\alpha = (\varepsilon + \bar{\varepsilon})^2(\underline{G}\cdot\underline{G})$ is complex and $\beta = (\varepsilon + \bar{\varepsilon})^2(\underline{G}\cdot\underline{\bar{G}})$. From equations (4.4) and (4.6), the functions α and β are independent of the coordinate x^3. The complex conjugate of α is

$$\gamma = \bar{\alpha} . \tag{4.12}$$

In Ernst coordinates, the Simon tensor has the only remaining components C_{13} and C_{23}. From the assumption that the Simon tensor vanishes we get the equations

$$2(\alpha\partial_1 + \beta\partial_2)\ln\rho + \frac{1}{2}\alpha_1 - \frac{\alpha}{r} = 0 \tag{4.13a}$$

$$\alpha_2 = 0 \tag{4.13b}$$

where we are using the notation $r = \frac{1}{2}(\varepsilon + \bar{\varepsilon})$.
That is to say, $\alpha = \alpha(x^1)$ is an analytic function of the Ernst potential ε. In deriving Eqs. (4.13), use has been made of the vector components

$$(G_i) = (\frac{1}{2r}, 0, 0) \qquad (\bar{G}_i) = (0, \frac{1}{2r}, 0) . \tag{4.14}$$

Field equation (2.8a) takes the form[7)]

$$(\alpha\partial_1 + \beta\partial_2)\ \ln\frac{\rho}{\sqrt{D}} + \alpha_1 + \beta_2 = \frac{\alpha}{r} \tag{4.15}$$

where

$$D \stackrel{\text{def}}{=} \alpha\gamma - \beta^2 \tag{4.16}$$

is a nonvanishing real function since degenerate fields have been excluded. We can rewrite Eq. (4.15) in the form

$$-\frac{1}{2}(\alpha\gamma - 3\beta^2)\alpha_1 + \alpha\beta\gamma_2 - 2\alpha(\beta\beta_1 + \gamma\beta_2) + \frac{D}{r}\alpha = 0 . \tag{4.17}$$

The complex conjugate equation is

$$\gamma\beta\alpha_1 - \frac{1}{2}(\alpha\gamma - 3\beta^2)\gamma_2 - 2\gamma(\alpha\beta_1 + \beta\beta_2) + \frac{D}{r}\gamma = 0 . \tag{4.18}$$

Upon multiplying Eq. (4.17) by β and Eq. (4.18) by $-\alpha$ and taking the sum, we can simplify by the factor D to obtain

$$4\alpha\beta_1 - 3\beta\alpha_1 + \alpha\gamma_2 + 2\frac{\alpha}{r}(\beta - \gamma) = 0 . \tag{4.19}$$

Similarly, multiplying Eq. (4.13a) by γ and the complex conjugate equation by $-\beta$ we can express the derivative of the function ρ:

$$D\,\frac{\rho_1}{\rho} = \beta\gamma_2 - \gamma\beta_2 + \frac{1}{2}(\alpha\gamma_1 - \gamma\alpha_1) + \frac{\gamma}{r}(\alpha - \beta) . \tag{4.20}$$

Eliminating the ∂_2 derivatives by use of Eq. (4.19) and the complex conjugate equation, we have

$$D\,\partial_1 \ln\rho = -\frac{1}{4}\gamma\alpha_1 + \frac{3}{4}\beta^2\frac{\alpha_1}{\alpha} - \beta\beta_1 + \frac{D}{2r} . \tag{4.21}$$

This can be rewritten as

$$(D^{-1/2}\alpha^{3/4}r^{-1}\rho)_{,1} = 0 . \tag{4.22}$$

Since γ is analytic in x^2, we can include a factor $\gamma^{3/4}$ here such that on the left hand side we have the ∂_1 derivative of a real function. Integration yields

$$D^{-1/2}(\alpha\gamma)^{3/4}\frac{\rho}{r} = c \tag{4.23}$$

where c is a real constant.

Equation (4.23) yields ρ as a function of α and β (and $\gamma=\bar{\alpha}$). As α and β are independent of the coordinate x^3, so is the function ρ. The space-time is axially symmetric.

5. ELIMINATION OF SECOND DERIVATIVES

In the remaining part of our procedure we solve Einstein's equations (2.8c) which involve the Ricci tensor R_{ik}. We follow Chap. 2 of Ref. 8 as closely as possible in eliminating the second derivatives.

A number of second derivatives can be expressed in terms of β_{12} and lower-order terms, using (4.19) and the conjugate equation

$$4\gamma\beta_2 - 3\beta\gamma_2 + \gamma\alpha_1 + 2\frac{\gamma}{r}(\beta-\alpha) = 0 . \tag{5.1}$$

Excluding null fields with $\alpha=0$ (section 3), the ∂_2 derivative of (5.1) yields

$$\gamma_{22} = -4\,\beta_{12} + 3\,\frac{\alpha_1}{\alpha}\beta_2 - \frac{2}{r}(\beta_2-\gamma_2) + \frac{1}{r^2}(\beta-\gamma) \tag{5.2}$$

and

$$\alpha_{11} = -4\,\beta_{12} + 3\,\frac{\gamma_2}{\alpha}\,\beta_1 - \frac{2}{r}(\beta_1-\alpha_1) + \frac{1}{r^2}(\beta-\alpha) \quad . \tag{5.3}$$

From the ∂_1 derivative we get

$$4\alpha\beta_{11} = -12\beta\beta_{12} + 2\alpha_1\beta_1 - \alpha_1\gamma_2 + 12\beta_1\beta_2 - \frac{8}{r}\alpha\beta_1 + \frac{2}{r}(\gamma+2\beta)\alpha_1 + 3\,\frac{\beta^2}{r^2} - \frac{\alpha}{r^2}(\gamma+2\beta) \tag{5.4}$$

$$4\gamma\beta_{22} = -12\beta\beta_{12} + 2\gamma_2\beta_2 - \alpha_1\gamma_2 + 12\beta_1\beta_2 - \frac{8}{r}\gamma\beta_2 + \frac{2}{r}(\alpha+2\beta)\gamma_2 + 3\,\frac{\beta^2}{r^2} - \frac{\gamma}{r^2}(\alpha+2\beta) \quad . \tag{5.5}$$

Substituting these expressions in the equation $R_{33} = 0$, we have

$$4D\beta_{12} = -2\beta\alpha_1\gamma_2 + 3(\gamma\alpha_1\beta_2 + \alpha\gamma_2\beta_1) - 4\beta\beta_1\beta_2 + + \frac{1}{r}[-2\alpha\gamma(\beta_1+\beta_2) + \beta(\gamma\alpha_1 + \alpha\gamma_2)] + \beta\frac{D}{r^2} \quad . \tag{5.6}$$

Next we consider the integrability condition of the function α:

$$\alpha_{112} - \alpha_{121} = 0 \; . \tag{5.7}$$

Elimination of the second derivatives which are given in Eqs. (5.2)-(5.6) yields

$$\gamma D\alpha_{11} - 2\beta\gamma\alpha_1\gamma_2 + 3\beta^2\gamma_2\beta_1 + \alpha\gamma\beta\frac{\gamma_2}{r} + 3\gamma^2\alpha_1\beta_2 + 2\,\frac{\gamma}{r}\beta^2\alpha_1 + \frac{\beta}{r}\gamma^2\alpha_1 - - 2\frac{\alpha}{r}\gamma^2\alpha_1 - 4\beta\gamma\beta_1\beta_2 - 2\frac{\alpha}{r}\gamma^2\beta_2 - 2\frac{\gamma}{r}\beta^2\beta_1 + \frac{\alpha\gamma}{r^2}D = 0 \; . \tag{5.8}$$

This relation takes a particularly simple form if we employ here Eqs. (4.19) and (5.1) to get rid of the β derivatives,

$$\beta_1 = \frac{1}{4\alpha}[-\alpha\gamma_2 + 3\beta\alpha_1 - 2\frac{\alpha}{r}(\beta-\gamma)] \tag{5.9}$$

$$\beta_2 = \frac{1}{4\gamma}[-\gamma\alpha_1 + 3\beta\gamma_2 - 2\frac{\gamma}{r}(\beta-\alpha)] \quad . \tag{5.10}$$

Thus we obtain Simon's eikonal equation[5)] from (5.8):

$$4\alpha\alpha_{11} - 3\alpha_1^2 = 0 \qquad . \tag{5.11}$$

Substitution in the remaining field equations yields us trivial identities only. But equation (5.11) can be integrated easily:

$$\alpha = \frac{1}{4\mu^2}(\varepsilon+\lambda)^4 . \qquad (5.12)$$

Here λ and μ are constants of integration. As the imaginary part of the Ernst potential is defined up to an additive constant, we can make λ real by the substitution $\varepsilon \Rightarrow \varepsilon - i\mathrm{Im}\lambda$. The constant μ is complex. We now attempt to integrate the first-order equation (4.19). Multiplying by $\alpha^{-7/4}\,\gamma^{-3/4}r$ and using the analicity of γ in x^2 we get

$$4\left[(\alpha\gamma)^{-3/4} r\beta\right]_{,1} + (\alpha\gamma)^{-3/4}\,(r\gamma_2 - 2\gamma) = 0. \qquad (5.13)$$

This has the integral

$$4b = -\gamma^{-3/4}\int \alpha^{-3/4}(r\gamma_2 - 2\gamma)d\varepsilon + \eta(\bar{\varepsilon}) \qquad (5.14)$$

where $\eta = \eta(\bar{\varepsilon})$ is an analytic function and we have introduced the real function $b = (\alpha\gamma)^{-3/4} r\beta$. Inserting the explicit form of γ from the complex conjugate of Eq. (5.13):

$$b = (\mu\bar{\mu})^{3/2}\,\bar{\mu}^{-2}\,\frac{\varepsilon}{(\varepsilon+\lambda)^2} + \eta(\bar{\varepsilon}) . \qquad (5.15)$$

Observing that b is real on the left hand side, we obtain

$$\eta(\bar{\varepsilon}) = (\mu\bar{\mu})^{3/2}\,\mu^{-2}\,\frac{\bar{\varepsilon}}{(\bar{\varepsilon}+\lambda)^2} + k \qquad (5.16)$$

with k a real constant. Collecting our results we have

$$\beta = \frac{(\varepsilon+\lambda)(\bar{\varepsilon}+\lambda)}{4(\varepsilon+\bar{\varepsilon})}\left\{\bar{\mu}^{-2}\varepsilon(\bar{\varepsilon}+\lambda)^2 + \mu^{-2}\,\bar{\varepsilon}(\varepsilon+\lambda)^2 + \frac{k}{(\mu\bar{\mu})^{3/2}}\,(\varepsilon+\lambda)^2(\bar{\varepsilon}+\lambda)^2\right\}.$$

(5.17)

With the explicit forms (5.12), (5.16) and (4.23) of the functions α,β and ρ, respectively, the process of integration of the field equations in Ernst coordinates is complete. When $\lambda\neq 0$, we can normalize the constants c and λ by rescaling the coordinates such that

$$c = 1 , \qquad \lambda = -1 . \tag{5.17}$$

The metric (4.11) of the 3-space is then given by

$$\alpha = \frac{1}{4\mu^2}(\varepsilon-1)^4, \qquad \gamma = \bar{\alpha} ,$$

$$\beta = \frac{(\varepsilon-1)(\bar{\varepsilon}-1)}{4(\varepsilon+\bar{\varepsilon})\mu\bar{\mu}} \left\{\left[\left(\frac{\bar{\varepsilon}-1}{\bar{\mu}}\right)^2\varepsilon + \left(\frac{\varepsilon-1}{\mu}\right)^2\bar{\varepsilon} + k(\mu\bar{\mu})^{-1}(\varepsilon-1)^2(\bar{\varepsilon}-1)^2\right]\right\}$$

$$\rho^2 = \frac{\alpha\gamma-\beta^2}{4(\alpha\gamma)^{3/2}} (\varepsilon+\bar{\varepsilon})^2 . \tag{5.18}$$

This is the Kerr-NUT metric[13] in Ernst coordinates. We regain the familiar form by performing coordinate transformation

$$\varepsilon = 1 - \frac{m+i\ell}{r+ia\cos\vartheta+i\ell} \tag{5.19}$$

and identifying the parameters

$$\mu = m+i\ell , \quad k = a^2-\ell^2 . \tag{5.20}$$

The metric with $\lambda=0$, together with the Kerr-NUT solution, completes Kinnersley's first class[14] of Petrov type D metrics.

6. CONCLUDING REMARKS

Stationary vacuum space-times with a vanishing Simon tensor have been shown to fall into either of three classes. The first is degenerate in the Petrov classification, and the Simon tensor coincides with the Bach tensor. There are three such metrics, and one of them is the NUT space-time. The second class is type N and consists of Hoffman's plane-fronted standing wave solutions. The third class of space-times contains the Kerr-NUT solution. In showing this, we used Ernst coordinates. The ensuing discussion in Sec. 5 proved somewhat circuitous. Perhaps it would be easier to integrate the SU(2) spin coefficient equations as in Ref.4. The present approach has been carried through with the aid of the R-40 computer of the Central Research Institute for Physics, using Reduce programs written by the author.

APPENDIX: Computation of $C_i^{\ \ell}$

We write the tensor $C_i^{\ \ell}$ in the form

$$C_i^{\ \ell} = \epsilon^{jk\ell}(\phi_{jik} - g_{ij}\, g^{rs}\, \phi_{srk}) \tag{A.1}$$

where [Cf. Eq.(2.4)]

$$\phi_{jik} = G_{j;i}G_k - G_{k;i}G_j \quad . \tag{A.2}$$

The tensor ϕ_{jik} satisfies the antisymmetry relation

$$\phi_{jik} = -\phi_{kij} \ . \tag{A.3}$$

Taking the triad components of Eq. (A.2) we have

$$\begin{aligned}
\phi_{oo+} &= \kappa G_o^2 \\
\phi_{+o-} &= -\kappa G_- G_o \\
\phi_{oo-} &= G_- DG_o - G_o DG_- + \kappa G_-^2 + \bar{\kappa} G_o^2 \\
\phi_{o++} &= \sigma G_o^2 \\
\phi_{++-} &= -\sigma G_o G_-
\end{aligned} \tag{A.4}$$

$$\phi_{o+-} = \delta G_o G_- - \delta G_- G_o + \sigma G_-^2 + \bar{\rho} G_o^2 + \bar{\tau} G_o G_-$$
$$\phi_{o-+} = \rho G_o^2$$
$$\phi_{+--} = -\rho G_o G_- \qquad \text{(A.4)}$$
$$\phi_{o--} = \bar{\delta} G_o G_- - \bar{\delta} G_- G_o + \rho G_-^2 + \bar{\sigma} G_o^2 - \tau G_o G_- \ .$$

The triad components (2.18) of the tensor $C_i^{\ \ell}$ follow easily by transvecting Eq. (A.1) with $z^i_{\underline{m}} m_{\underline{n}\ell}$ and substituting (A.4).

REFERENCES

1. R.P. Kerr, Phys. Rev. Letters 11, 237 (1963)
2. B. Carter, Commun. Math. Phys. 10, 280 (1968)
3. E.T. Newman and A.I. Janis, J. Math. Phys. 6, 915 (1965)
4. Z. Perjès, J. Math. Phys. 11, 3383 (1970)
5. W. Simon, Gen. Rel. Grav. 16, 465 (1984)
6. B. Lukács, Z. Perjès, Á. Sebestyén and G.A.J. Sparling, Gen. Rel. Grav. 15, 511 (1983)
7. B. Lukács, Z. Perjès, Á. Sebestyén and A. Valentini, KFKI-1983-31 preprint, part B.
8. Z. Perjès, INS-Rep. 487 preprint, 1984
9. Greek indices can take the values 0,1,2 and 3 and Roman indices range through 1,2 and 3. Triad components are indexed by the letters $\underline{m}$, $\underline{n}$, $\underline{p}$, ... and may take the values o, + and -. Antisymmetry is indicated by bracketing the indices, and indices excluded from symmetry operations are separated by vertical bars. Covariant derivation referring to the 3-metric g_{ik} is denoted by a semicolòn or by the gradient sign ∇.
10. F.J. Ernst, Phys. Rev. 167, 1175 (1969)
11. W.H. Brinkmann, Proc. Nat. Acad. Sci. USA 9, 1 (1923), J. Ehlers and W. Kundt, in Gravitation: an Introduction to Current Research, Ed. L. Witten (Wiley, 1962), I. Robinson, unpublished (1958)
12. R. Hoffman, J. Math. Phys. 10, 953 (1969)
13. G. Neugebauer and D. Kramer, Commun. Math. Phys. 10, 132 (1968)
14. W. Kinnersley, J. Math. Phys. 10, 1195 (1969)

CASIMIR EFFECT IN SPACE-TIMES WITH NON-EUCLIDEAN TOPOLOGY

S.G.Mamayev, V.M.Mostepanenko
D.I.Mendeleev Metrology Institute,
198005, Leningrad
USSR

ABSTRACT

The non-trivial topology of the space-time leads to the vacuum polarization of quantized fields characterized by non-zero vacuum expectation values of the stress-energy tensor. We present an approach to its computation which allows analytical investigation of the dependence of the effect upon topology and geometrical parameters of the manifold. A new interpretation and computation method for the effective vacuum temperature is also given.

1. INTRODUCTION

In recent years there was a considerable growth of interest to the phenomena known under a generic name of the Casimir effect. The essence of these phenomena is the vacuum polarization of the quantized field which arises in an empty space when either the manifold posesses boundaries or its topology differs from that of the Euclidean space R^n. As a result the vacuum state is characterized by a non-zero expectation value of the stress-energy tensor (SET) of the quantized field which is measurable in principle and leads to various physical effects.

The most widely known example of these phenomena is the original Casimir effect [1]. Two parallel plates of the area S in vacuum are attracted to each other with the force

$$F = \frac{\pi^2}{240} \frac{\hbar c}{a^4} S \tag{1}$$

where a is the separation of the plates, which is assumed to be much smaller than their size ($a \ll \sqrt{S}$). Such an attraction was observed experimentally [2,3]: for $S=1\ cm^2$ and $a=0.5\ \mu m$ the measured force was F=0.2 dyn, in complete accordance with (1).A possible application of this effect in metrology for constructing a standart of small forces based on fundamental constants is discussed in [4].

From the point of view of theoretical physics the interest of the Casimir effect is connected with the modern concepts of possible non-Euclidean topology of space-time. As in the case of material boundaries, topologically non-trivial manifold is characterized by non-zero vacuum expectations. It means that the global structure of the manifold is reflected in the local properties of the physical vacuum. This fact presents an intriguing possibility of reconstruction of the topological structure of the world as a whole on the basis of purely local measurements.

It is well known that in inflationary cosmological scenarios the 3-space is closed (or bounded by a wall of the bubble) and topological components of the vacuum polarization must be taken into account in the early stages of the evolution. For the manifolds with the topology of S^3 and T^3 they were computed in [5-7]. The corresponding energy density for massless fields is $\varepsilon \sim a^{-4}$, a being the geometric dimension of the configuration. Evidently, the Casimir energy becomes comparable to that of the matter fields when $a \sim \ell_{Pl}$, where $\ell_{Pl} = \sqrt{G} =$

$=1.6\cdot 10^{-33}$ cm is the Planck's length (hereafter the units $\hbar = c = 1$ are used).

The role of the Casimir effect may be important in microphysics as well. In the MIT bag model of hadrons [8] the confinement of quarks is ensured by postulating the absence of current through a certain surface (bag) bounding the hadron. In the framework of this model the Casimir energy of quark and gluon fields must be included into the total energy of the bag in calculations of the properties of hadrons [9]. QCD computations of the Casimir effect for the boundaries of various configurations were carried out in [10-13].

In recent years we witness a sharp revival of interest in the unified field theories of the Kaluza-Klein type in the framework of supergravity [14-15]. In modern versions of this theory the real dimension d of the space-time is supposed to be greater than 4 (the most popular value is d=11). However, d-4 spatial dimensions are thought to have undergone what is called a spontaneous compactification, so that the topology of the space-time is $V^d = R^4 \times M^{d-4}$, where M^{d-4} is a compact manifold. The geometric size of the latter must be of the order of ℓ_{Pl}. It means that to construct a realistic self-consistent scheme of the spontaneous compactification it is nesessary to take into account the vacuum polarization of the topological origin [16-18]. Note also that even the macroscopic Casimir effect may prove useful for the elementary particle physics. For example, a precise measurement of the Casimir force between parallel plates allows to obtain the best bounds on the mass of axion and on the scale of the supersymmetry breakdown [19,20].

In the present paper we give field-theoretical foundations for the calculations of the vacuum polarization of topological origin and outline an effective method of computation of the Casimir energy density which allows an analytic study of its dependence on the field mass and geometrical parameters of the manifold. In Sec.2 the methods of regularization of the vacuum expectation values of the SET are considered. Sec.3 contains the results for some 2- and 3-dimensional manifolds. A general approach to the interpretation and calculation of the effective vacuum temperature in space-times with non-trivial topology is presented in Sec.4.

2. METHODS OF REGULARIZATION OF THE STRESS-ENERGY TENSOR

Let us consider the quantization of a free field $\Psi(x)$ (which we shall not specify so far) in a pseudo-Riemannian space-time $V^{n+1} = R^1 \times M^n$. The non-Euclidean geometry and topology of the spatial manifold M^n may be taken into account by using the general covariant forms of the field equations in the case of non-zero curvature and imposing on $\Psi(x)$ boundary conditions. The latter are determined by the topology of M^n and the type of the fibre bundle over M^n which specifies the global structure of the field ("twist" [21]).

Let $\Psi_J^{(\pm)}$ be a complete orthonormal set of positive- and negative-frequency solutions of the field equations obeying the given boundary conditions; J is a cumulative index for quantum numbers. The operator of the quantized field may be expanded as

$$\Psi(x) = \sum_J \left(\Psi_J^{(-)}(x)\, a_J^{(-)} + \Psi_J^{(+)}(x)\, a_J^{(+)} \right) \qquad (2)$$

where the creation and annihilation operators $a_J^{(\pm)}$ obey the standard (anti)commutation relations. The Fock space is constructed according to the usual scheme; the vacuum state $|0\rangle$ is defined by $a_J^{(-)}|0\rangle = 0$ for all J. The natural characteristic of the vacuum $|0\rangle$ is the expectation value of the SET operator of the quantized field. A straightforward calculation yields [22,23]

$$\langle 0|T_{ik}|0\rangle = \sum_{\jmath} T_{ik}\{\overset{*}{\Psi}{}^{(-)}_{\jmath}(x), \Psi^{(+)}_{\jmath}(x)\} \qquad (3)$$

where $T_{ik}\{f, g\}$ is the bilinear form determined by the classical expression for SET.

The main difficulty with the expectations (3) is that they diverge. In the standard field theory in Minkowsky space-time with the vacuum state $|0_M\rangle$ such a divergence is removed by normal ordering of the SET operator in terms of $a^{(\pm)}_{\jmath}$ which is equivalent to the subtraction of the contribution of the zero-point vacuum oscillations. It is obvious that normal ordering in (3) would give zero result. That is why we should use another method of removal of the divergence in (3) in order to obtain the finite terms which would describe the difference between the vacua $|0\rangle$ and $|0_M\rangle$.

First of all note that we study a global effect, depending on the geometrical and topological properties of the manifold M^n as a whole, whereas the ultraviolet divergences in (3) are local, i.e. are determined by the small scale properties of M^n. Since a smooth manifold M^n may be at every point locally approximated by a tangent Euclidean space R^n, the leading divergence must be exactly the same as in the Minkowsky space-time $R^1 \times R^n$. Moreover in the case when M^n is locally flat and differs from R^n only in its topology expectations $\langle 0|T_{ik}|0\rangle$ would contain no other divergences (if M^n has a non-zero curvature, (3) would generally contain weaker divergences).

That is why it is natural to remove the leading divergence in (3) in such a way, that formally it corresponds to the subtraction from (3) of a similar expectation value in the vacuum state $|0_M\rangle$ of a Minkowsky space tangent to our manifold at the given point.

In order to give a sensible interpretation to the difference of two infinite quantities one must introduce an intermediate regularization. Denoting the regularized values by a conventional symbol ε , one may write the finite vacuum expectations as

$$\langle T_{ik}\rangle = \lim_{\varepsilon\to 0}\left[\langle 0|T_{ik}|0\rangle_{\varepsilon} - \langle 0_M|T_{ik}|0_M\rangle_{\varepsilon}\right]. \quad (4)$$

This procedure of the removal of divergences corresponds, as one may easily check, to the renormalization of the cosmological constant Λ in the bare gravitational action. Thus for the scalar field of mass m in (1+3)- dimensional Minkowsky space-time the dimensional regularization gives [24]

$$\langle 0_M|T_{ik}|0_M\rangle_{\varepsilon} = -\frac{m^4}{64\pi^2} g_{ik}\left(\frac{1}{\varepsilon} + \frac{3}{2} - C - \ln\frac{\nu^2}{4\pi}\right) \quad (5)$$

where C=0.577... is the Euler constant, ν =m/M, M being an arbitrary parameter of the dimension of mass. Since (5) is proportional to the metric tensor g_{ik}, the regularization procedure (4) corresponds to the renormalization of Λ.

For the computation of $\langle T_{ik}\rangle$ in specific configurations covariant regularization schemes are not very useful since they are technically ineffective. Usually with this purpose either an explicit cut-off is introduced into the divergent sums (3) [7] or point-splitting in the bilinear form T_{ik} is performed [22]. An alternative approach is based on a formal identity

$$\langle 0|T_{ik}|0\rangle = i\left[\frac{\partial}{\partial x^i}\frac{\partial}{\partial x^k} G(x,x')\right]_{x'=x} \quad (6)$$

where the Green's function $G(x,x')$ is expressed as a sum over various non-homotopic paths connecting the points x and x'. The renormalization (4) of this sum is performed by dropping the term corresponding to an empty Minkowsky space.

An effective method of regularization of vacuum expectation values was proposed in [6] for the case of the scalar field in the space-time with the topology of S^3 and later generalized for various spins and various configurations of M^n in [13,27,28]. Now we shall outline this method for the example of a massive untwisted field on a 1-dimensional manifold S^1.

The S^1 manifold is obtained from the real line R^1 by identification of points with coordinates $x+n\alpha$, $n=0, \pm 1, \ldots$, α being the circumference of S^1. This identification corresponds to the periodicity conditions on $\Psi(t,x)$ and $\partial_x \Psi(t,x)$. A complete set of solutions of the Klein-Fock equation obeying periodicity conditions is

$$\Psi_n^{(\pm)}(t,x) = (2\omega_n \alpha)^{-1/2} \exp[\pm i(\omega_n t - \kappa_n x)], \tag{7}$$
$$\omega_n^2 = \kappa_n^2 + m^2, \quad \kappa_n = 2\pi n/\alpha, \quad n = 0, \pm 1, \ldots .$$

Putting the solutions (7) into (3) with the use of the classical expression for the SET [23] one finds

$$\langle 0|T_{00}|0\rangle = \frac{m}{2\alpha} + \frac{1}{\alpha}\sum_{n=1}^{\infty}\omega_n, \quad \langle 0|T_{11}|0\rangle = \frac{1}{\alpha}\sum_{n=1}^{\infty}\frac{\kappa_n^2}{\omega_n}, \tag{8}$$
$$\langle 0|T_{01}|0\rangle = \langle 0|T_{10}|0\rangle = 0 .$$

As shown in [6,23], the renormalization (4) in this case is equivalent to the following summation formula for the divergent series

$$\mathrm{reg}\sum_{n=1}^{\infty} F(n) = \frac{1}{2}F(0) + i\int_0^{\infty}\frac{F(it) - F(-it)}{\exp(2\pi t) - 1}\,dt . \tag{9}$$

Applying it to the series (8) we obtain

$$\varepsilon \equiv \langle T_{00}\rangle = -\frac{1}{\pi\alpha^2}\Phi^{(0,1)}(m\alpha), \quad P \equiv \langle T_{11}\rangle = -\frac{1}{\pi\alpha^2}\Phi^{(2,-1)}(m\alpha) \tag{10}$$

where the functions $\Phi^{(\alpha,\beta)}(x)$ defined as

$$\Phi^{(\alpha,\beta)}(x) = \int_x^{\infty}\frac{d\xi}{\exp\xi - 1}\,\xi^{\alpha}(\xi^2 - x^2)^{\beta/2} . \tag{11}$$

Expressions (10) describe the vacuum polarization which arises due to the non-trivial topology of the manifold. When $\alpha \to \infty$ the physical distinction between S^1 and R^1 vanishes and ε, $P \to 0$ as one would naturally expect.

As a dimensionless parameter characterizing the difference of S^1 from the Euclidean space R^1 one may take $\lambda = 1/(m\alpha)$. The expressions (10) are non-analytic in λ at the point $\lambda=0$. That is why these results cannot be obtained by means of the perturbation theory.

It is easy to verify that the total vacuum energy $\mathcal{E}=a\varepsilon$ and the pressure P obey the relation $d\mathcal{E}/da=-P$, which holds for the deformation of a string when the entropy is constant. In other words the vacuum $|0\rangle$ behaves as the Maxwell elastic aether with the negative energy density.

For the massless field (10) yields

$$\varepsilon = P = -\frac{1}{\pi}\int_0^\infty \frac{\kappa\, d\kappa}{\exp(a\kappa)-1} = -\frac{\pi}{6a^2}\,. \qquad (12)$$

Note that the spectral density in (12) formally coincides with that of the black body thermal radiation with the temperature $T=1/a$. The interpretation of this fact will be considered in Sec.4.

In the limiting case $ma\gg 1$ the values ε and P are exponentially small

$$\varepsilon = (ma)^{-1}P = -(2\pi)^{-1/2}m^2(ma)^{-3/2}e^{-ma}\,. \qquad (13)$$

For $ma=1$ numerical computations according to (10) give $\varepsilon=-0.219\,m^2$, $P=-0.406\,m^2$.

The mentioned above method based on the representation of the exact Green's function $G(x,x')$ as a sum of contributions of various non-homotopic paths connecting x and x', is equivalent to the generalized summation of divergent series with the help of Poisson's formula

$$\operatorname{reg}\sum_{n=-\infty}^{\infty}\varphi(2\pi n) = \frac{1}{2\pi}\sum_{\substack{n=-\infty\\ n\neq 0}}^{\infty}\mathcal{F}[\varphi](n) \qquad (14)$$

where $\mathcal{F}[\varphi]$ is the Fourier transform of the distribution φ . Regularization using formula (14) always gives the same results as (9), but the answer is expressed in terms of slowly converging series. Thus for the case of S^1 considered above (14) gives

$$\varepsilon = -\frac{m}{\pi a}\sum_{n=1}^{\infty}\frac{1}{n}K_1(man) \qquad (15)$$

where $K_\nu(z)$ is the MacDonald function. This series begins to converge rapidly only from the numbers $n \sim (ma)^{-1}$ and for $ma \ll 1$ it is nesessary to take a large number of terms which decrease only as $1/n^2$.

For higher-dimensional manifolds expressions (3) contain multiple sums and integrosums. They may be regularieżed effectively by consecutive application of formula (9) or similar formula for the summation over semi-integer index

$$\operatorname{reg} \sum_{n=0}^{\infty} F(n+\tfrac{1}{2}) = -i \int_0^{\infty} \frac{F(it)-F(-it)}{\exp(2\pi t)+1}\, dt \qquad (16)$$

which appear in the case of spinor fields as well as for twisted scalar fields.

3. VACUUM POLARIZATION IN TOPOLOGICALLY NON-TRIVIAL STATIC SPACE-TIMES

Now we shall present the results of computation with the method described above of the vacuum polarization in some topologically non-trivial spaces. For the sake of brevity we shall restrict ourselves to the case of untwisted scalar field φ, supposing, however, the coupling constant ξ (in the term $-\xi R\varphi^2/2$ in the Lagrangian) to be arbitrary (the coupling is conformal when $\xi=(n-1)/4n$ for $n+1$ -dimensional space-time).

We shall begin with two-dimensional flat manifolds M^2. There are four types of two-dimensional complete flat manifolds different from R^2 /29/. They may be described by the action of the fundamental group on the universal covering space in Cartesian coordinates (x,y):

a) cylinder

$$(x,y) \longrightarrow (x+ka,\ y), \qquad (17)$$

b) 2-torus

$$(x,y) \longrightarrow (x+ka,\ y+lb), \qquad (18)$$

c) Möbius strip of infinite width

$$(x,y) \longrightarrow (x+ka,\ (-1)^k y), \qquad (19)$$

d) Klein bottle

$$(x,y) \longrightarrow (x+ka,\ (-1)^k y + lb). \qquad (20)$$

Here k , l are integers, a , b - parameters which determine the geometric size of the manifold.

For the cylinder invariance under (17) imposes periodicity conditions in coordinate x for the field $\varphi(t,x,y)$. Regularization of (3) with the help of (9) gives

$$\varepsilon = -P_2 = -\frac{1}{4\pi a^3}\Phi^{(0,2)}(ma), \quad P_1 = -\frac{1}{2\pi a^3}\Phi^{(2,0)}(ma) \tag{21}$$

where $\Phi^{(\alpha,\beta)}(x)$ is determined in (11), $P_{1,2}$ are pressures along the axes x,y respectively. In the case of the massless field (21) reduces to

$$\varepsilon = \frac{1}{2}P_1 = -P_2 = -\frac{\zeta(3)}{2\pi a^3} \tag{22}$$

where $\zeta(z)$ is Riemannian zeta-function, $\zeta(3)\simeq 1.202$.

For the 2-torus (18) imposes periodicity conditions in both coordinates. Here one must apply the formula (9) twice. The result is

$$\varepsilon = -\frac{1}{4\pi a^3}\Phi^{(0,2)}(ma) - \frac{1}{\pi a b^2}\Phi^{(0,1)}(mb) - \frac{2}{ab^2}F^{(0,1)}\left(\frac{b}{a};mb\right) \tag{23}$$

where

$$F^{(\alpha,\beta)}(\lambda;x) = \frac{1}{\pi}\sum_{k=1}^{\infty}\Phi^{(\alpha,\beta)}\left(\sqrt{x^2+(2\pi\lambda k)^2}\right). \tag{24}$$

Pressures P_α may be obtained with the help of the identities

$$\partial\mathcal{E}/\partial a = -bP_1, \quad \partial\mathcal{E}/\partial b = -aP_2 \tag{25}$$

($\mathcal{E} = ab\varepsilon$ is the total energy of vacuum polarization), which follow from the conservation property of the SET. In this case vacuum behaves like an elastic membrane. The apparent lack of symmetry in (23) with respect to interchange $a \rightleftarrows b$ is due to the choice of a definite order of summation over quantum numbers. In fact ε remains invariant under such an interchange while $P_1 \rightleftarrows P_2$. Still when calculating ε according to (23), it is convenient to suppose $b \geqslant a$, since when $\lambda = b/a \gg 1$ the terms (24) are exponentially small. Even for $a = b$ their contribution to (23) does not exceed 1%. For $a = b = m^{-1}$ we have $\mathcal{E} = a^2\varepsilon = -0.349m$. In the massless limit (23) gives

$$\mathcal{E} = -\frac{\zeta(3)}{2\pi a^3} - \frac{\pi}{6ab^2} - \frac{4}{a^2 b}\sum_{k,l=1}^{\infty}\frac{k}{l}K_1\left(2\pi k l\frac{b}{a}\right) \tag{26}$$

For $a = b$ the total energy is $\mathcal{E} = -0.719\,a^{-1}$.

When $b/a \to \infty$ the torus becomes undistinguishable from the cylinder and (23), (26) in this limit reduce respectively to the expressions (21), (22) for $S^1 \times R^1$. Since both cylinder and 2-torus are homogeneous manifolds $\langle T_{ik}\rangle$ here is independent of coordinates; it does not depend upon the coupling constant ξ either.

Another situation takes place in the cases (19) and (20) when there is no homogeneity. First consider the manifold with the topology of the Klein bottle. Identification (20) corresponds to the following boundary conditions

$$\begin{aligned} &\varphi(t,x,0) = \varphi(t,x,b), \quad \partial_y\varphi(t,x,0) = \partial_y\varphi(t,x,b), \\ &\varphi(t,0,y) = \varphi(t,a,b-y), \quad \partial_x\varphi(t,0,y) = \partial_x\varphi(t,0,b-y). \end{aligned} \tag{27}$$

For the sake of brevity we shall restrict ourselves with the massless field. Here the energy density turns out to be dependent upon the y coordinate. It is conveniently presented it in the form

$$\mathcal{E} = \bar{\mathcal{E}} + \Delta\mathcal{E}(y) \tag{28}$$

where $\bar{\mathcal{E}}$ is a constant component of $\mathcal{E}$, while $\Delta\mathcal{E}(y)$ vanishes when averaged over the interval $(0, b)$. With the general method of Sec.2 using formulas (9), (16) we find

$$\Delta\mathcal{E}(y) = -\frac{4\pi}{b^3}\sum_{n=1}^{\infty} n^2 \cos\frac{4\pi n y}{b}\int_1^{\infty}\frac{ds}{\mathrm{sh}(2\pi n s a/b)}\left[\frac{2s^2-1}{\sqrt{s^2-1}} + \frac{1-8\xi}{\sqrt{s^2-1}}\right]. \tag{29}$$

For the conformally coupled field ($\xi = 1/8$ in 2-dimensional space) term in the square brackets with $1-8\xi$ is absent. As to the y-independent part $\bar{\mathcal{E}}$ in (28), it may be presented in two equivalent forms

$$\begin{aligned} \bar{\mathcal{E}} &= -\frac{\zeta(3)}{16\pi a^3} - \frac{\pi}{8a^2 b} - \frac{\pi}{12ab^2} - \frac{1}{a^2 b}\sum_{k,l=1}^{\infty}\frac{k}{l}K_1\left(\pi k l\frac{b}{a}\right) = \\ &= -\frac{\zeta(3)}{2\pi b^3} - \frac{\pi}{6a^2 b} - \frac{2}{ab^2}\sum_{k,l=1}^{\infty}\frac{k}{l}K_1\left(4\pi k l\frac{a}{b}\right). \end{aligned} \tag{30}$$

According to the relation between $2a$ and b one should choose that expression, in which the argument of K_1 is larger. The pressures $P_{1,2}$ may be obtained from (28)-(30) with the help of (25); in this case P_2 is independent of y.

For $a = b$ we get $\mathcal{E} = -0.715\, a^{-1}$. Comparison of (30) with (26) shows that when the manifold is stretched along its a side the difference of its total vacuum energy $ab\bar{\varepsilon}$ from that of the 2-torus of similar configuration becomes exponentially small as one could presume on physical grounds (when comparing (30) with (26) one should interchange $a \rightleftarrows b$ in the latter formula).

For the Möbius strip of infinite width (cf. (19)) we obtain

$$\varepsilon = \bar{\varepsilon} + \Delta\varepsilon(y) = -\frac{\zeta(3)}{16\pi a^3} - \frac{1}{8\pi^2 a^3}\int_0^\infty dq\, q^2 \cos\left(\frac{qy}{a}\right)\int_1^\infty \frac{ds}{\operatorname{sh} qs}\,\frac{s^2-4\zeta}{\sqrt{s^2-1}}. \qquad (31)$$

It is easily checked that (31) is the limiting case of the results (28)-(30) for the Klein bottle when its length $b \to \infty$. In this case $\bar{\varepsilon}$ is four times smaller than the value (22) for the cylinder of circumference a.

In the 3-dimensional case the number of topologically distinct complete flat manifolds is much greater. We shall restrict ourselves with the 3-torus $S^1 \times S^1 \times S^1$ and its limiting cases $S^1 \times S^1 \times R^1$ and $S^1 \times R^2$.

For the 3-torus with the dimensions $a \times b \times c$ vacuum energy density is

$$\varepsilon = -\frac{1}{6\pi^2 a^4}\Phi^{(0,3)}(ma) - \frac{1}{4\pi a b^3}\Phi^{(0,2)}(mb) - \frac{1}{\pi a b c^2}\Phi^{(0,1)}(mc) - \alpha(a,b,c) \qquad (32)$$

where $\alpha(a,b,c)$ denotes positive terms similar to the last term in (23), which are negligibly small when $a < b < c$. The pressures may be found using the relation $P_1 = -(1/bc)\cdot \partial\mathcal{E}/\partial a$ and its two companions obtained by cyclic transposition of a, b and c, $\mathcal{E} = abc\,\varepsilon$ is the total energy. For the fixed volume $V = abc$ the maximum value of $\mathcal{E}$ is achieved when $a = b = c$.

In the case $a = b = c = m^{-1}$ we have $\mathcal{E} = -0.444m$. For the massless field

$$\varepsilon = -\frac{\pi^2}{90a^4} - \frac{\zeta(3)}{2\pi a b^3} - \frac{\pi}{6abc^2} - \alpha \,. \qquad (33)$$

When $a = b = c$ [5] (33) gives $\varepsilon = -0.837a^{-4}$, $P_{1,2,3} = -0.279a^{-4}$.

In the limit $c \to \infty$ we have the manifold with the topology $S^1 \times S^1 \times R^1$; in this case

$$\varepsilon = -\frac{1}{6\pi^2 a^4}\Phi^{(0,3)}(ma) - \frac{1}{4\pi a b^3}\Phi^{(0,2)}(mb) - \lim_{c\to\infty}\alpha(a,b,c). \qquad (34)$$

For m=0 and $a = b$ we obtain $\varepsilon = -0.305\,a^{-4}$ [25].

If both $b, c \to \infty$, the manifold becomes $S^1 \times R^2$ and

$$\varepsilon = -\frac{1}{6\pi^2 a^4}\Phi^{(0,3)}(ma)\,. \qquad (35)$$

For m=0

$$\varepsilon = -\frac{\pi^2}{90a^4} = -0.110\,a^{-4} \qquad (36)$$

in agreement with the result of [22,26].

Now consider the scalar field on a two-dimensional sphere S^2 [28]. The eigenfunctions are

$$\varphi_J^{(+)}(t,\theta,\varphi) = \frac{1}{2\sqrt{\omega_\ell a}}\, e^{i\omega_\ell t}\, Y_{\ell m}(\theta,\varphi)\,, \qquad (37)$$

$\varphi_J^{(-)} = (\varphi_J^{(+)})^*$, $\ell = 0, 1, \ldots$, $m = 0, \pm 1, \ldots, \pm \ell$, $\omega_\ell^2 = m^2 + (\ell + 1/2)^2 a^{-2}$, $Y_{\ell m}$ are the spherical harmonics, a is the radius of the sphere. Putting (37) into (3) and using the addition theorem for the spherical functions and the regularization formula (16), we find

$$\varepsilon = \frac{m^3}{2\pi}\int_0^1 \frac{d\xi\,\xi\sqrt{1-\xi^2}}{\exp(2\pi m a\xi)+1}\,, \quad P = -\frac{m^3}{4\pi}\int_0^1 \frac{d\xi\,\xi^2}{\sqrt{1-\xi^2}\,[\exp(2\pi m a\xi)+1]}. \qquad (38)$$

It is easy to verify the equation $P = -\partial\mathcal{E}/\partial S$, where $\mathcal{E} = S\varepsilon$, $S = 4\pi a^2$ is the surface area of the sphere. In contrast with all the manifolds considered above the vacuum energy here is positive.

For $a \ll m^{-1}$ (38) yields

$$\varepsilon \simeq -P \simeq m^3/(12\pi) \qquad (39)$$

(note that $\varepsilon = 0$ for the massless field). For $a = m^{-1}$ numerical calculation gives $\varepsilon = 2.98 \cdot 10^{-3} m^3$, $P = -3.72 \cdot 10^{-4} m^3$. When $a \gg m^{-1}$ (38) behave asymptotically as

$$\mathcal{E} \simeq \frac{m}{96\pi a^2}\left[1-\frac{7}{40}(ma)^{-2}\right], \quad P \simeq -\frac{7}{3840}\,\frac{1}{ma^4} \quad (40)$$

i.e. $\langle T_{ik} \rangle$ decrease only as powers of $(ma)^{-1}$ (while for 2-torus $\langle T_{ik} \rangle$ in this case are exponentially small). As $a \to \infty$ the total energy approaches a finite value $\mathcal{E}_\infty = m/24$.

For the 3-sphere the results are [6,7]:

$$\varepsilon = \frac{1}{16\pi^6 a^4}\,\Phi^{(2,1)}(ma), \quad P = \frac{1}{48\pi^6 a^4}\,\Phi^{(4,-1)}(ma) \,. \quad (41)$$

Here too, in contrast with the 3-torus, $\varepsilon > 0$. For m=0

$$\varepsilon = 3P = \frac{1}{480\pi^2 a^4} \quad (42)$$

and for $ma \gg 1$

$$\varepsilon \simeq \frac{3}{2\pi}(ma)^{-1} P \simeq \frac{1}{8\pi^3 a^4}(ma)^{5/2} exp(-2\pi ma)\,. \quad (43)$$

If $a = m^{-1}$ we have $\varepsilon = 1.27\cdot 10^{-5} m^4$, $P = 2.24\cdot 10^{-5} m^4$ while the total energy $\mathcal{E} = 2\pi^2 a^3 \varepsilon = 2.51\cdot 10^{-4} m$.

3. EFFECTIVE VACUUM TEMPERATURE

In some cases the spectral density which describes the vacuum polarization in space-times with non-Euclidean topology formally coincides with that of the equilibrium thermal radiation of the temperature T determined by a geometrical parameter. Thus for the manifolds S^1, $S^1 \times R^1$, $S^1 \times R^2$ the corresponding temperature is $T = 1/a$, a being the circumference of S^1, for the 3-sphere S^3 of curvature radius a we have $T = 1/2\pi a$. An effective temperature characterizing the vacuum of quantized fields is known to appear in other situations where the space topology differs from the Euclidean. For example the Hawking radiation has the temperature $T = 1/8\pi GM$ (M is the mass of the black hole), in an accelerated frame there appears a thermal radiation of the temperature $T = w/2\pi$, w being the acceleration (see, e.g., [23,30]).

The procedure for computation of T as well as the treatment of the origin of thermal distribution are generally derived from the specific features of each situation. Here we shall present a unified approach to the computation and interpretation of the effective vacuum temperature applicable to a wide class of problems including all those mentioned above [31,32].

The appearance of the thermal distribution in the vacuum state $|0\rangle$ is obviously a consequence of the global difference between the given manifold and the Minkowsky space-time. The effective temperature, however, is a local quantity and must be determined by the local properties of the state $|0\rangle$. As a characteristic of such local properties one could naturally consider an asymptotic form of the Green's function $G(x,x')$ as $x' \to x$.

The leading singularity of $G(x,x')$ always coincides with the singularity of the Green's function $G^{(0)}(x,x')$ of an empty Minkowsky space; for the massless scalar field

$$G^{(0)} = -i\,[4\pi^2(\tau^2-\rho^2)]^{-1} \tag{44}$$

where $\tau = t - t'$, $\rho = |\vec{\rho}| = |\vec{x}-\vec{x}'|$ (we suppress $i0$ in denominator).

Let us construct a fibre bundle over our manifold adjusting to each point x a tangent Minkowsky space. In this tangent space introduce a thermal Green's function

$$\mathcal{G}_T^{(0)}(x,x') = i\,\mathrm{Tr}\left(e^{-\beta H}\,T\{\varphi(x)\varphi(x')\}\right)/\,\mathrm{Tr}\,e^{-\beta H}. \tag{45}$$

Define the parameter $\beta = 1/T$ in (45) so that the Green's function $G(x,x')$ for our manifold locally coincides with (45):

$$\lim_{x'\to x}\left[G(x,x') - \mathcal{G}_T^{(0)}(x,x')\right] = 0. \tag{46}$$

With the use of the representation

$$\mathcal{G}_T^{(0)}(x,x') = \sum_{n=-\infty}^{\infty} G^{(0)}\left(\tau - \frac{ni}{T}, \vec{\rho}\right) \tag{47}$$

where $G^{(0)}$ is defined in (44) we find

$$\mathcal{G}_T^{(0)}(x,x') = G^{(0)}(x,x') + \frac{iT^2}{12} + O(\sigma), \tag{48}$$

σ being the square of the interval between x and x'. This allows to rewrite the condition (46) in the following way:

$$T = \sqrt{\frac{12}{i}}\,\lim_{x'\to x}\left[G(x,x') - G^{(0)}(x,x')\right]^{1/2}. \tag{49}$$

It means that the effective temperature of the vacuum

$|0\rangle$ is defined as the temperature of such a thermal distribution in the tangent Minkowsky space, whose local properties coincide with the local properties of the state $|0\rangle$ on a given manifold.

As an example consider $S^1 \times R^2$. Scalar Green's function in this case is

$$G(x,x') = -\frac{i}{4\pi^2} \sum_{k=-\infty}^{\infty} \left[\tau^2 - (\vec{\rho} - \vec{e}_1 k a)^2\right]^{-1} \qquad (50)$$

where $\vec{e}_1$ is the unit vector in the x^1 direction. In the limit of $\tau, \rho \to 0$ we obtain

$$\lim_{x' \to x} \left[G(x,x') - G^{(0)}(x,x')\right] = \frac{i}{2\pi^2} \sum_{k=1}^{\infty} (ka)^{-2} = \frac{i}{12a^2}. \qquad (51)$$

Putting (51) into (49) we find $T=1/a$. The same value is obtained from the form of the spectral density.

In [31,32] it was shown how this method works in the cases of accelerated frames and black hole radiation. Finally we must note that not all problems connected with the Casimir effect admit introduction of an effective temperature according to (49). In particular, (49) may give infinite temperature if $G(x,x')$ contains besides (44) weaker singularities. The presence of boundaries may lead to an imaginary effective temperature. In cases of low symmetry the limit (49) may not exist at all (depending on the direction of the splitting of x and x′). However when the limit (49) exists the above construction exhausts completely the physical meaning of the concept of the vacuum effective temperature.

REFERENCES

1. Casimir, H.B.G., Proc.Kon.Ned.Ak.Wet. 51, 793 (1948).
2. Spaarnay, M.J., Physica 24, 751 (1958).
3. Tabor, D., Winterton, R., Proc.Roy.Soc.A312, 435 (1969).
4. Mamayev, S.G. and Mostepanenko, V.M., In: Quantum metrology and fundamental physical constants. L-d, 1982, p.93.
5. Starobinsky, A.A., In: Proc. of the 1-st Marcel Grossman relativity meeting. Amsterdam, 1977, p.499.

6. Mamayev, S.G., Mostepanenko, V.M. and Starobinsky,A.A., ZhETF 70, 1577 (1976).
7. Ford, L.H., Phys. Rev. D11, 3370 (1975).
8. Hasenfratz, P. and Kubi, J., Phys.Rep. C40, 75 (1978).
9. Johnson, K., Acta Phys. Polonica B6, 865 (1975).
10.Brevik,I.,Kolbenstvedt,H., Phys.Rev. D25, 1731 (1982).
11.Baacke,J., Igarashi,Y., Phys. Rev. D27, 460 (1983).
12.Milton, K.A., Phys. Rev. D27, 439 (1983).
13.Mamayev, S.G. and Trunov, N.N.,Teor.Mat.Fiz.38,345(1979).
14.Gremmer,E.,Julia,B.,Scherk,J.,Phys.Lett.B76, 409(1978).
15.Witten, E., Nucl. Phys. B186, 412 (1981).
16.Appelquist,Th.,Chodos,A., Phys.Rev. D28, 772 (1983).
17.Voronov,N.A.,Kagan,Ya.I.,Pisma v ZhETF 38,262 (1983).
18.Candelas,P.,Weinberg,S., Nucl. Phys. B237,397 (1984).
19.Kuzmin, V.A., Tkachev, I.I., Shaposhnikov, M.E., Pisma v ZhETF 36, 49 (1982).
20.Radescu, E.E., Phys. Rev. D27, 1409 (1983).
21.Isham, C.J., Proc.Roy.Soc. A362, 383 (1978).
22.DeWitt, B.S., Phys. Rep. C19, 297 (1975).
23.Grib, A.A., Mamayev, S.G. and Mostepanenko, V.M.,Quantum effects in strong external fields. M., 1980.
24.Mamayev,S.G.,Mostepanenko,V.M.,Preprint ITF-54P,K,1984.
25.Dowker,J.S.,Critchley,R., J.Phys. A9, 535 (1976).
26.DeWitt,B.S.,Hart,C.F.,Isham,C.,Physica A96,187 (1979).
27.Starobinsky,A.A., In: Classical and quantum gravity theory. Minsk, 1976, p.110.
28.Mamayev, S.G. and Trunov, N.N., Izv. vuz.Fiz. N7, 88 (1979); N9, 51 (1979); N7, 9 (1980).
29.Kobayashi, S., Nomizu, K., Foundations of differential geometry. Vol.1. N.Y.-L., 1963.
30.Birrel, N.D. and Davies, P.C.W., Quantum fields in curved space. Cambridge, 1982.
31.Mamayev,S.G.,Trunov,N.N., Yad.Fiz. 34, 1142 (1981).
32.Mamayev,S.G.,Trunov,N.N., Izv.vuz.Fiz. N4, 82 (1982).

AN INERTIAL INTERPRETATION OF ACCELERATION RADIATION

Robert M. Wald

Enrico Fermi Institute and Department of Physics
The University of Chicago
5640 S. Ellis Avenue, Chicago, Illinois
U.S.A.

ABSTRACT

As shown by Unruh in 1976, if a quantum field is in its vacuum state in Minkowski spacetime, an accelerating observer would feel himself immersed in a thermal bath of quanta of the field and a particle detector he carries would get excited. How does an inertial observer interpret this "detection" process, in view of the fact that according to him, no quanta of the field were present initially? Recently, Unruh and I have shown that for a simple model detector the absorption by the detector of a "Rindler particle" corresponds in the inertial viewpoint to emission by the detector of an ordinary Minkowski particle. In this note I shall describe some of our results, focusing particular attention on how a consistent answer is obtained to the question of whether the energy in the quantum field is increased or decreased by the process of detection/emission.

It is a common misconception that the vacuum state, $|0\rangle$, of a quantum field, ϕ , in Minkowski spacetime is interpreted as being "inert", "empty", and devoid of any interesting dynamics. In fact, a great deal of nontrivial dynamics occurs in the so-called "vacuum fluctuations" of the field. Now, an inertial quantum system coupled to the field cannot make transitions upward in energy when the field is in the state $|0\rangle$ because such transitions of the quantum system require a positive frequency perturbation from the field and $|0\rangle$ is annihilated by the positive frequency part of the field. However, the field can induce transitions downward in energy -- a phenomenon misleadingly referred to as "spontaneous emission". More interestingly, if we consider an accelerating quantum system coupled to the field, both upward and downward transitions are possible. In order to make a transition upward in energy in this case, it is still necessary to have a positive frequency perturbation from the field. However, the relevant notion of positive frequency is now defined by Fourier transformation with respect to "accelerating time", i.e., the Rindler time coordinate τ rather than the Minkowski time coordinate t. (The relevant notion of energy here is the one canonically conjugate to τ.) Although $\phi|0\rangle$ has no positive frequency part with respect to t, it does have a nonvanishing positive frequency part with respect to τ. Thus, an accelerating particle detector which is initially in its ground state can become excited as a result of its interaction with the quantum field. Furthermore, since $\phi|0\rangle$ has a nonvanishing positive frequency part with respect to τ, an accelerating observer would naturally say that particles are present in the state $|0\rangle$, and would ascribe the excitation of his detector to its interaction with these particles. (This notion of "particle" -- defined with respect to τ -- will be referred to as "Rindler particle" below to distinguish it from the ordinary notion of "Minkowski particle" which is defined with respect to t.) Indeed, a detailed calculation[1)] shows that the state $|0\rangle$ contains a thermal bath of Rindler particles at temperature $kT = \hbar a / 2\pi c$, where a is the acceleration defining τ. On the other hand, an inertial observer would have to ascribe the excitation of such a detector to its interaction with the vacuum fluctuations of ϕ.

Although the phenomenon of excitation of an accelerating detector was first discussed less than ten years ago, it is not "new physics" in that this prediction is derived in a logically straightforward manner from conventional quantum field theory. It is reasonable to expect, therefore, that in the inertial picture the details of the process can be described in terms of entirely conventional phenomena. Recently, Unruh and I have given such a description in the case of a simple model of a detector coupled to an otherwise free scalar field. We showed that the process described by the accelerating observer as being associated with the absorption of a Rindler particle is described by the inertial observer as being associated with the emission of an ordinary particle. Details of this analysis can be found in Unruh and Wald[2].

However, the interpretation of the detector excitation process as resulting from emission of a particle leads directly to consideration of two rather puzzling issues. The first concerns causality. The Minkowski particle emitted by the detector is partly (in fact, mostly) located in a region noncausally related to the world line of the accelerating detector. In particular, if the detector gets excited, to lowest order, a strictly positive energy density contribution is made to the stress-energy of the field in this noncausally related region. (Conversely, if the detector fails to get excited, the expected field energy in this region decreases.) Thus, it may seem that causality is violated in this process. However, this is not the case. It is indeed true that we have correlations existing over spacelike related regions but -- just as in other examples of the Einstein, Podolsky, Rosen phenomenon -- these correlations cannot be used to communicate information. The existence of correlations in the final state over spacelike related regions should not be surprising in view of the fact that the initial state, $|0\rangle$, of the quantum field already contains such correlations. Interaction of the field with the detector merely transfers some of these correlations from the field to the detector in an entirely causal manner. Further discussion of this causality issue can be found in Unruh and Wald[2] and Wald[3].

The second issue concerns whether energy is transferred into or out of the quantum field in this process and the physical mechanism by which such an energy transfer occurs. Before explaining the puzzling aspects of this issue, it should be emphasized that there are two distinct notions of energy which arise naturally here. The first is the ordinary inertial notion of energy associated with an ordinary Minkowski time translation Killing field t^a. In terms of the stress-energy operator, T_{ab}, of the quantum field, the expected value, E, of this energy is,

$$E = \int_{\Sigma} \langle T_{ab} \rangle t^a d\Sigma^b \qquad (1)$$

where Σ is a Cauchy surface for Minkowski spacetime representing the time at which E is measured. The second is the natural notion of energy that an accelerating observer would use. A uniformly accelerating observer in Minkowski spacetime follows an orbit of the Lorentz boost symmetry. Hence, the vector field τ^a which the inertial observer views as generating this Lorentz boost symmetry would be naturally viewed by the accelerating observer as generating the time translation symmetry he sees in his "Rindler wedge", i.e. in the region bounded by the two null planes to which his world line is asymptotic. The expected value of the energy, $\mathcal{E}$, of the quantum field associated with this notion of time translation symmetry is,

$$\mathcal{E} = \int_{\Sigma_1} \langle T_{ab} \rangle \tau^a d\Sigma_1^b \qquad (2)$$

where Σ_1 is a Cauchy surface for only the "Rindler wedge" of the accelerating observer. We shall refer to $\mathcal{E}$ as the "boost energy" of the field.

Since the vacuum state, $|0\rangle$, is the state which minimizes the expected inertial energy, E, of the quantum field, it is clear that the interaction between the field and detector cannot result in a decrease in E. If the detector is excited, to lowest order, this corresponds to emission of a single Minkowski particle and the integral in eq. (1) is positive everywhere. On the other hand, if the detector is found to be in its ground state at the end of the interaction process, then, to lowest order, this corresponds to a superposition of leaving the field in state $|0\rangle$ and emission of two Minkowski particles. The integrand in eq. (1) is then negative in the region which is noncausally related to the accelerating observer, but the contribution from the Rindler wedge in which the observer is located is sufficiently positive to make the total E positive in this case also. One may ask where the energy that goes into the quantum field comes from. This question is very similar to the question of where the energy radiated into the electromagnetic field by a classical charged particle comes from. The answer in both cases is that the source of energy is the agent responsible for accelerating the radiating object, who must do extra work to counter the radiation reaction forces and keep the radiating object on the desired trajectory.

The puzzling issue concerns the boost energy, $\mathcal{E}$. Suppose that after interaction the detector is found in its excited state. According to the accelerating observer, a particle has been absorbed from the thermal bath, so it would seem clear that he would conclude that $\mathcal{E}$ -- the relevant notion of energy for him -- is decreased. But this contradicts the interpretation of the inertial observer, who says that a Minkowski particle has been emitted. Since $\langle T_{ab}\rangle$ for a one-particle state satisfies the energy conditions of the classical field theory, the integrand of eq. (2) must be positive, so $\mathcal{E}$ must increase.

The resolution of this apparent conflict is that $\mathcal{E}$ does increase when detection occurs in just the manner predicted by the inertial viewpoint, and a more careful analysis from the viewpoint of the accelerating observer yields the identical result. The absorption of a Rindler particle by itself does reduce the field energy. However, that is not the only effect which takes place when the detector gets excited. Although the initial state, $|0\rangle$, of the quantum field is an eigenstate of E, it is <u>not</u> an eigenstate of $\mathcal{E}$. Since the probability of excitation of the detector is proportional to the number of particles present (in modes that interact with the detector), examination of the detector at the end of the interaction performs a partial measurement of the initial boost energy. In particular, if the detector is excited, this provides evidence that more particles (and hence larger $\mathcal{E}$) were present in the initial state than otherwise might have been expected. Thus, although boost energy has been absorbed from the field, the upward revision of $\mathcal{E}$ caused by the partial measurement more than compensates this loss, resulting in an increase in $\mathcal{E}$ in exact agreement with the inertial calculation. Similarly, when the detector is unexcited, the partial measurement effect causes $\mathcal{E}$ to decrease. It is worth noting that, when weighted by the probability of the detector being excited or unexcited, the expected value of $\mathcal{E}$ decreases by exactly the expected increase of detector energy. Thus, conservation of total expected boost energy applies to the full system of detector and field, as should be the case since it does not require any boost energy from an external agent to keep the detector on its uniformly accelerating trajectory.

Thus, although the inertial and accelerating observers agree upon the magnitude and sign of changes which occur in $\mathcal{E}$, they use utterly different language to describe the physical mechanisms by which these changes occur. A similar situation occurs in the process of "mining" of a black hole discussed by Unruh and Wald[4)]. From the viewpoint of stationary observers, this process of extraction of energy from a black hole is described as literally the mining of the thermal bath of particles that surround the black hole. From the viewpoint of a freely falling observer, no such thermal bath is present, and the process is explained in terms of radiation of negative energy into the black hole by the walls of the box used in the "mining".

As we strive toward the formulation of a quantum theory of gravity, it may be useful to keep in mind the subjective nature of many of the notions commonly used in quantum field theory.

Acknowledgement

The research was supported in part by NSF Grant PHY 80 26043 to the University of Chicago.

References

1) Unruh, W. G., Phys. Rev. D14, 870 (1976).

2) Unruh, W. G. and Wald, R. M., Phys. Rev. D29, 1047 (1984).

3) Wald, R. M., "Correlations and Causality in Quantum Field Theory", to appear in proceedings of Oxford Symposium on Quantum Theory, March, 1984.

4) Unruh, W. G. and Wald, R. M., Phys. Rev. D25, 942 (1982).

QUANTUM FIELD THEORY FOR A GENERAL CLASS OF ACCELERATED OBSERVERS

Norma SANCHEZ
Département d'Astrophysique Fondamentale
ER 176, CNRS,
Observatoire de Meudon, 92195 Meudon,
FRANCE.

ABSTRACT

We formulate QFT in a wide class of accelerated coordinates in four dimensions. This generalizes the approach given previously by the author in terms of holomorphic (and / or anti-holomorphic) mappings. We give a characterization of global and asymptotic thermal equilibrium situations, an unicity theorem concerning the Rindler space and a discussion of the quantum detection and the isotropy of thermal radiation.

1. INTRODUCTION

A fundamental problem in the link between General Relativity and Quantum Theory is *the formulation of QFT in general frames of coordinates.* The vacuum and thermal effects characteristic of Hawking emission |1| are not exclusive to black-holes, neither to the curved space-time itself. They are present even in flat space-time when quantum fields are described in terms of accelerated coordinates |2|-|5|. New concepts appear as the *relativity of vacuum* and *even the relativity of the pure or mixed nature of quantum states* (which become a manifestation of the acceleration state of the observers) : what an observer considers as vacuum in a frame of reference may appear as an excited state in another frame. A pure state in a frame of coordinates may look as a mixed state in another one. Most of the litterature discussing this problem is essentially centered around one particular example : the Rindler frame, [6] which describes uniform acceleration and has as vacuum spectrum a purely Panckian formula. However, quantization in a particular frame (inertial or uniformely accelerated) is clearly not enough. In previous papers [7]-[11], the present author has formulated QFT in a wide class of 2-dimensional accelerated frames involving uniform and non-uniform accelerations : one, two or any event horizon ; planckian and non-planckian spectra. The well known Rindler's frame is included here as a particular case. The coordinates in which QFT can be consistently formulated are defined by holomorphic (or anti-holomorphic) mappings in Euclidean space (imaginary time). Each analytic function defines an accelerated frame and a quantum production rate associated with it. Classical, quantum and thermal aspects of the theory are explicitely expressed in terms of the mappings. We have classified the accelerated frames, their associated quantum production rates and temperatures in terms of the nature of the singular and critical points of the mappings. We have interpretated [10] the vacuum spectra predicted by the theory in terms of the measurements carried out by accelerated detectors and generalized Unruh's detector [5] to non-uniformely accelerated motions. Here we generalize all these results to four dimensions and i) we demonstrate an unicity theorem concerning the Rindler space (this space uniquely satisfies the condition of global thermal equilibrium) ii) we discuss the isotropy question of the thermal radiation in accelerated spaces, iii) we include uniform and non-uniform rotation.

2. FOUR DIMENSIONAL APPROACH

We consider a free massive scalar field Ψ in the accelerated coordinates (x', t', y', z') defined by $f(x' \pm t') = x \pm t, y' = y, z' = z$. $(x, t, y, z$ stand for the inertial coordinates) ; f satisfies $f(\pm \infty) = u \pm (u = x - t, v = x + t)$, u+ (u-) can take finite or infinite values corresponding to one, two or zero event horizons. Here we will choose $u_- = 0$, $u_+ = +\infty$, and then $f'(-\infty) = 0$, $f'(+\infty) = +\infty$. Ψ satisfies

$$[\partial^2_{x'} - \partial^2_{t'} + \Lambda(\partial^2_y + \partial^2_z - m^2)]\Psi = 0$$

where $\Lambda = f'(x' + t')\, f'(x' - t')$. The substitution

$$\Psi = \frac{1}{2\pi} e^{i(\lambda_2 y + \lambda_3 z)} \phi(x', t')$$

yields $\{-\partial^2_{t'} + \partial^2_{x'} - \Lambda M'^2\}\phi(x', t') = 0$

with $M'^2 = \lambda_2^2 + \lambda_3^2 + m^2$. The effective mass $\Lambda M'^2$ is zero at $u' = -\infty$ and infinite at $u' = +\infty$ preventing particles escape there. We choose as a complete set of orthogonal solutions, the functions ϕ_λ^{in} satisfying :

$$\lim_{v' \to -\infty} \Phi_\lambda^{in} = \frac{1}{2\sqrt{\pi\lambda_1}} e^{i\lambda_1 u'}$$

$$\lim_{u' \to +\infty} \phi_\lambda^{in} = 0 \qquad , \lambda_1 > 0$$

and given completely by

$$\phi_\lambda^{in} = \frac{i}{2}\sqrt{\frac{\lambda 1}{\pi}} \int_{+\infty}^{u'} d\xi \; e^{i\lambda\xi} J_0 [M'\sqrt{vf(\xi) - u}]$$

ϕ_λ^{in} describes positive frequencies λ with respect to the accelerated time t'. In the formulation of QFT in non-inertial coordinates, the dynamical operators are defined in terms of the accelerated creation and annihilation operators C_λ, C_λ^+ associated to the accelerated modes ϕ_λ^{in}. The vacuum state of the theory ($|0\rangle$) is defined by the inertial operators a_k associated to the inertial modes φ_k ($a_k |0\rangle = 0$). The state $|0'\rangle$ such that $C_\lambda |0'\rangle = 0$ is an excited state with respect to the true vacuum $|0\rangle$. Here

$$\varphi_k \equiv \varphi_{k_1 k_2 k_3} = \frac{1}{4\pi\sqrt{\pi\omega}} e^{i(k_1 x + k_2 y + k_3 z - \omega t)}$$

with $-\infty < k_1, k_2, k_3 < \infty$, $\omega = +\sqrt{k_1^2 + k_2^2 + k_3^2 + m^2} > 0$

C_λ and a_k are related by a Bogoliubov transformation

$$C_\lambda = \int_{-\infty}^{\infty} d^3 k \; [A_\lambda(k) \; a_k + B_\lambda(k) \; a_k^+]$$

where $B_\lambda(k) = \langle \phi_\lambda , \varphi_k^* \rangle$, $A_\lambda(k) = \langle \phi_\lambda , \varphi_k \rangle$

$$\langle \phi, \varphi \rangle = i \int \phi^* \; j_\mu \; d\Sigma^\mu \; , \quad j_\mu = \sqrt{-g}\,\overrightarrow{\partial}_\mu - \overleftarrow{\partial}_\mu \sqrt{-g}$$

We find $B_\lambda(k) = B_{\lambda_1}(k_1, M)\,\delta(k_2 + \lambda_2)\,\delta(k_3 + \lambda_3)$

$$B_{\lambda_1}(k_1, M) = \frac{-(k_1 + E)}{4\pi\sqrt{\lambda_1 E}} \int_{\mu-}^{\mu+} du \; e^{-i_{\lambda_1} F(u) - \frac{i}{2}(k_1 + E)\,u}$$

$$(E = \sqrt{k_1^2 + M^2} \; , \; M^2 = k_2^2 + k_3^2 + m^2 \; , \; F \equiv f^{-1})$$

and $N(\lambda, \lambda') \langle 0 | C_\lambda C'^+_\lambda | 0 \rangle = \int_{-\infty}^{\infty} d^3 k \; B_\lambda^*(k) \, B_{\lambda'}(k) =$

$$= N(\lambda_1, \lambda_1')\,\delta(\lambda_2 - \lambda_2')\,\delta(\lambda_3 - \lambda_3')$$

$R(\lambda, \lambda') = \langle 0 | C_\lambda \; C_\lambda' | 0 \rangle = \int_{-\infty}^{\infty} d^3 k \;\; A_\lambda(k) \, B_\lambda'(k)$

$$= R(\lambda_1, \lambda_1')\,\delta(\lambda_2 + \lambda_2')\,\delta(\lambda_3 + \lambda_3')$$

where
$$N(\lambda_1, \lambda_1') = -\frac{1}{4\pi^2} \frac{1}{\sqrt{\lambda_1 \lambda_1'}} \int_{\mu-}^{\mu+} d\mu \; d\mu_1 \frac{e^{i[\lambda_1 F(\mu + i\varepsilon) - \lambda_1' F(\mu_1 - i\varepsilon)]}}{(\mu - \mu_1 + i\varepsilon)^2} \; , \varepsilon > 0 \quad (1)$$

$$R(\lambda_1, \lambda_1') = +\frac{1}{4\pi^2} \frac{1}{\sqrt{\lambda_1 \lambda_1'}} \int_{\mu-}^{\mu+} d\mu \; d\mu_1 \frac{e^{-i[\lambda_1 F(\mu + i\varepsilon) - \lambda_1' F(\mu_1 - i\varepsilon)]}}{(\mu - \mu_1 + i\varepsilon)^2} \; , \varepsilon > 0 \quad (2)$$

2.1 Unicity Theorem :

Each one of the following statements implies the two others

i) The production function has the form $N_v(\lambda_1, \lambda_1') = N_v(\lambda_1)\,\delta(\lambda_1 - \lambda_1')$

ii) The Bogoliubov transformation can be decomposed as a two-term one

$$C_\lambda = [1 + N_v(\lambda)]^{\frac{1}{2}} \, \tilde{C}_\lambda(+) - [N_v(\lambda)]^{\frac{1}{2}} \, \tilde{C}_\lambda(-)$$

iii) The mapping f is $f(\mu') = e^{2\pi T \mu'}$
where $T = [\lambda N_V(\lambda)]_{\lambda=0}$

The equivalence of statements (i) and (iii) follows from eq. (1). The equivalence of (i) and (ii) follows from the relation

$$A_\lambda(k) = \left[\frac{1 + N_V(\lambda)}{N_V(\lambda)}\right]^{\frac{1}{2}} B_\lambda(k)$$

which is necessary and sufficient condition for the Bogoliubov transformation being decomposable. $N_V(\lambda)$ is the number of created accelerated modes per unit frequency and unit volume, obtained from $N(\lambda,\lambda')$ by introducing wave-packets. The physical counterpart of this theorem is that the vacuum momentum density $\langle 0| T_{0i} |0\rangle$ is zero for all (x',y',z',t'), as well as $\langle 0| T_{ij} |0\rangle$ $(i \neq j \equiv x',y',z')$. At each point of the space-time, particles travelling to the right and to the left are created in equal total numbers. It corresponds to *a global thermal equilibrium situation* over the whole space-time. The mapping (iii) corresponds to Rindler space.

Corollaire : If $N(\lambda,\lambda')$ satisfies statemement (i), then $N_V(\lambda) = \dfrac{1}{(e^{\lambda/T}-1)}$

The converse is not true. There is a large class of accelerated observers for which $N_V(\lambda)$ is planckian but whose $\langle 0| T_{0i} |0\rangle$ are not zero. For these observers the acceleration is uniform only asymptotically for $x'_{\to\pm\infty}$, ie $f(\mu') \underset{\mu'\to\pm\infty}{\sim} e^{2\pi T\mu'}$. This corresponds to *a local (asymptotic) thermal equilibrium situation.*

2.1.1.Asymptotic Temperatures :

The temperature T which characterizes $N_V(\lambda)$ is an asymptotic temperature. In general for $f(u') \underset{u'\to\pm\infty}{\sim} e^{2\pi T_\pm u'}$, $T_\pm$ is given by

$$T_\pm = \frac{1}{2\pi} \lim_{\mu\to\pm\infty} \left[\frac{d}{du'} \ln f(u')\right] \qquad (3)$$

The differential winding number of the mapping,

$$T(x',\tau') = \frac{1}{2\Pi} \frac{\partial}{\partial\tau'} \text{ Im } \log f(x',i\tau') = \frac{1}{2\Pi} \frac{\partial}{\partial X'} \text{ Re } \log f(x'+i\tau')$$

(here, $t=i\tau$) is related to the proper acceleration α_p of the non-inertial observers through

$$\frac{d}{dx'} T(x',\tau') + 2\Pi\,|T(x',\tau')|^2 = \alpha_p\,|f'(u')|\; T(x',\tau')$$

For finite values of (x',τ'), $T(x',\tau')$ can be considered as a local distribution of temperature provided that in a neighbourhood of a point $u'_0 = x'_0 + i\tau_0'$, $T(x',\tau')$ satisfy

$$\left[\frac{1}{T_0^2}\left|\frac{\partial T}{\partial\tau'}(x',\tau')\right]_{(x'_0,\tau_0')}\right| \ll 1 \qquad (4)$$

This condition specifies a situation of thermal equilibrium in the vecinity of the point u'_0. If this condition holds, then the following relation is locally true

$$T(x',\tau') = \frac{1}{2\Pi}\left[f'(u')\right]\alpha_p$$

In particular, it gives

$$T\pm = \frac{1}{2\Pi}\left[|f'(u')|\,\alpha_p\right]_{u'=\pm\infty}$$

That is, the relation between the proper acceleration and the asymptotic temperatures of $N_V(\lambda)$

2.1.2. Rotating and drifting motions :

These correspond to the class of transformations

$$x \pm t = f\,(x' \pm t')$$
$$y = Y\,(x', y', z', t')$$
$$z = Z\,(x', y', z', t')$$

If $f\,(t' \to -\infty) \sim e^{\,2\Pi T(x' \pm t')}$

$Y\,(t' \to -\infty) \sim y' + v\,t'$

$$Z\ (t'_{\to -\infty}) \sim z + \gamma t',$$

then $$N_v(\lambda) = \frac{1}{(e^{(\lambda_1 - v\lambda_2 - \gamma\lambda_3)/T} - 1)}$$

2.2. Interacting Fields :

Thermal features with temperature as given by eq. (3) still survive in the presence of interactions of matter fields. Asymptotically for $u' \to \pm\infty$, the Green functions are periodic in the imaginary accelerated time τ' with period $T_\pm^{-1}$. The spectrum $N_v(\lambda)$ will be a planckian factor plus corrections depending on the coupling constant of the fields. We have not yet performed the calculations for a $g\psi^4$ four dimensional theory but we have found one loop corrections to $R(\lambda,\lambda')$, $N(\lambda,\lambda')$ and $N_v(\lambda)$ for a two dimensional $g\psi^4$ theory. These results are repported in [9]. In the Rindler case we have found

$$N_v(\lambda,g) \underset{\lambda\to\infty}{=} \frac{1}{e^{\lambda/T}-1}\left\{1 - \frac{g}{96\pi^4}\left(\frac{\lambda}{T}\right)^3 e^{-\lambda/2T}\left[\ln(2\ln(\lambda/2\pi T)) + \gamma + 0(1/\ln\lambda)\right] + 0(g^2)\right\},$$

$$\gamma = 0.577\ \ldots$$

2.3 Casimir effect and vacuum effects du to acceleration.

Several authors have stressed the analogies between the Cassimir effect and the Hawking effect but they are different in despite of the fact they give the same $\langle N_v(\lambda)\rangle$ in the Rindler (or in the Black Hole) case. Even in this case, they give differents $\langle\hat{T}_{\mu\nu}\rangle$ and in general they give differents $N(\lambda,\lambda')$ and differents $\langle\hat{T}_{\mu\nu}\rangle$.

The problem of quantum radiation by accelerated mirrors mixes both vacuum effects : those due to acceleration (Unruh's type effect) and those due to boundary conditions on the field (Cassimir effect). Q.F.T. by accelerated observers or in curved space-time can be formulated without to impose additional boundary conditions on the field, i.e. without the introduction of mirrors. In our approach the boundary conditions are imposed on the mapping f which defines the accelerated coordinates, i.e. $u_\pm = f(u')$.

The role of these boundary conditions is clearly illustrated by comparing the mapping $f_1(u') = (\alpha u'+ \beta)/(\gamma u'+\delta)$ and the Rindler one $f_2(u') = \beta\exp(\alpha u')$, ($\alpha,\beta,\gamma,\delta$ being real constants). Both mappings describe uniform acceleration. However, whereas f_2 maps the half axis Re $u > 0$ onto the full real axis u', this is not so for f_1 which does not satisfy. conditions (3). The inverse mapping F_1 takes either the value $+\infty$ or $-\infty$, depending on the sign of $(\alpha\delta - \beta\gamma)$, but not both values. These differences are essential to the description of physical processes, in particular, the formulation of QFT. Conditions (3) guarantee that the accelerated coordinates (x',t') range all values from $-\infty$ to $+\infty$. They mean that for $t'\to\pm\infty$, the world lines x' = const. tend asymptotically to the characteristic lines $x \pm t = u_-$ and $x \pm t = u_+$ where the detector's velocity given by $\mathbf{v} = |f'(v') - f'(u')|/|f'(v')+f'(u')|$ reaches the values plus and minus the speed of light. If these conditions are not satisfied, the self-adjointness of the propagation equations, the completeness and orthogonality of their solutions cease to hold, unless additional assumptions on the wave functions are to be imposed. For instance, Davies and Fulling [14] have considered a motion associated to mapping f_1 to describe quantum radiation by accelerated mirrors. In this case, they impose total reflection boundary conditions, on the field ($\Psi = 0$ on the mirror), which leads to vacuum effects of Casimir type. The time dependent part of the positive frequency modes is the same in both cases, but the spatial part is not.

2.4 Accelerated quantum detectors and anisotropy of the thermal radiation.

Underlying the present day formulation and interpretation of Quantum mechanics, is the use of inertial observers. The eigenvalues of the operators and the mean values computed in the theory are postulated to be the detectable magnitudes. The observer and his measurement device are external classical objects not described by the quantum theory.

In the formulation of Q.F.T. by accelerated observers, the state of motion of the observers appears included in the theory. Hence in such a theory one should provide a description of the observer's detector and of the measurement they carry out.

The quantum theory of detection (as known in quantum optics) for accelerated observers is usually referred to as an accelerated quantum detector (a.q.d) model. It should be pointed out that the response of an a.q.d does not give complete information about the vacuum (and thermal) effects due to acceleration. The detector's response (vacuum frequency spectrum) characterize a class of accelerated motions rather than a particular type. For instance, the a.q.d response does not differenciate between the uniformly accelerated (Rindler) motion and the wide class of non-uniformly accelerated motions which are asymptotically Rindler |10| Moreover, for some questions (for instance the isotropy of the vacuum radiation, such information can be biased by the detection process. Everybody knows that the Planckian radiation in Minkowski inertial space is isotropic. However, the response of a quantum detector with a directional discrimination at rest in a Planckian gas is only isotropic for point-like detectors. The anisotropic behaviour which is found |13,14| when the detector travels with uniform (or asymptotically uniform) acceleration in the vacuum of the Minkowski space will be mixed to an effect of this kind if the detector is extended.

For detectors without directional discrimination, the response of an uniformly accelerated quantum detector (a.q.d.) in the Minkowski vacuum is the same as that of an inertial quantum detector (i.q.d.) in an ordinary thermal gas at given temperature $(2\pi)^{-1}a$. For detectors with directional discrimination, the responses of a.q.d's and i.q.d's at finite temperature are not the same. Point-like and directionally sensitive i.q.d's in an ordinary thermal gas give an isotropic response. For point-like and directionally sensitive a.q.d's travelling with uniform acceleration in the Minkowski vacuum, the response is anisotropic [13,14]. (Rindler's observers have a preferred direction, namely its acceleration). For extended and directionally sensitive i.q.d's in the ordinary thermal gas, the response is anisotropic and for extended and directionally sensitive a.q.d's, the anisotropic response will contain also an effect of this kind.

REFERENCES :

1 - S.W. Hawking, Comm. Math. Phys. 43, 199 (1975) ; G.W. Gibbons and S.W. Hawking, Phys.Rev. D15, 2738 (1977).

2 - S.A. Fulling, Phys.Rev. D7, 2580 (1973).

3 - P.W. Davies, J.Phys. A8, 609-616 (1975).

4 - B.S. DeWitt, Phys.Rep. 19C, 297-357 (1975).

5 - W.G. Unruh, Phys.Rev. D14, 870 (1976).

6 - W. Rindler, Ann.J.Phys. 34, 1174 (1966).

7 - N. Sánchez, Phys.Lett. 87B, 212 (1979).

8 - N. Sánchez, Phys.Rev. 24D, 2100 (1981).

9 - N. Sánchez, Phys.Lett. 81A, 424 (1981).

10 - N. Sánchez, Phys.Lett. 105B, 375 (1981).

11 - N. Sánchez, in "Proceedings of the 2nd Marcel Grossman Meeting", ed. R.Ruffini, North Holland, 501-518, (1982).

12 - P.C.W. Davies and S.A. Fulling, Proc.R.Soc. A348, 393 (1976).

13 - K. Hinton, P.C.W. Davies and J. Pfautsch, Phys.Lett. 120B, 88 (1983).

14 - W. Israel and J.M. Nester, Phys.Lett. 98A, 329 (1983).

ON THE VACUUM DEFINITION IN CURVED SPACE-TIME

Mario Castagnino

Instituto de Física de Rosario
Av. Pellegrini 250, 2000 Rosario
ARGENTINA
and
Instituto de Astronomía y Física del Espacio
Casilla de Correos 67, Sucursal 28,
1428 Buenos Aires
ARGENTINA

ABSTRACT

Two new definitions (Strong and Minimal Vacua) for the vacuum in Robertson-Walker universe are introduced and their physical consequences are given.

1. INTRODUCTION

This work is a new attempt to solve the problem of the vacuum definition in curved space-time. It can be considered as a new version of paper[1]] taking into account the results of papers 2] and 3].

The quantum vacuum state is a well defined and familiar notion if:

i) space-time is flat,

ii) space-time is unbounded and

iii) we use an inertial reference system.

If one or several of these circumstances change we must deal with a new and unconventional vacuum concept,or else we just can't define a reasonable vacuum (e.g. in the case of a non globally hyperbolic manifold).

Thus, there is a set of cases where we can define a vacuum (e.g. when there exists a Killing vector field) and there is another set of cases where we can't define a vacuum -- so it would be very convenient to have a theory that would determinate in which of these two sets, a particular case is, and if it is in the set where a vacuum can be defined would give us the recipe to find it.

We will try to sketch such a theory focusing the problem on spatially flat and unbounded Robertson-Walker universes and observers in the comoving reference system.

Of course the only way we can solve the problem is to obtain a vacuum with as many properties of the normal flat space-time vacuum as possible. But only some vacuum properties can be generalized from flat to curved space-time while other properties can't. We shall choose only two properties to characterize the vacuum, one local and the other global:

a - Local Property. As it is well known, there is a one-to-one correspondence between these three concepts: vacuum, positive and negative basis of solution of the field equation, and the $G_1(x,x')$ Green function. If one defines one of these notions the other two are automatically defined[4]. Thus, the best way to impose a local property to the vacuum is to use $G_1(x,x')$, a geometrical biscalar, i.e. imposing that $G_1(x,x')$ must have a determinated behavior in the coincidence limit $x \to x'$. This behavior is, in fact, dictated by the Equivalence Principle: When $x \to x'$, $G_1(x,x')$ must behave like $G_1^{DS}(x,x')$, the DeWitt-Schwinger Kernel, because this Green Function is the most accurate generalization of the flat space-time $\Delta_1(x,x')$ to curved space-time[5]; and when x is near to x' everything must be, more or less, as in flat space. How to implement this condition is a matter of opinion and we shall give two versions of it, in paragraphs 2 and 3 -- but this property alone is not enough to define a vacuum[6], so we must add also a:

b - Global Property. Normally the vacuum makes some global quantity a minimum, the crucial point is to define what quantity. In the case we are dealing with it turns out to be that a good choice is to take a generalized energy:

$$\langle 0|H|0\rangle = \int_{\Sigma} \langle 0|T^{o}{}_{o}|0\rangle_{ren}\, d\sigma \,, \tag{1}$$

at a Cauchy surface Σ. The vacuum $|0\rangle$ would be the quantum state that minimizes (1), where $\langle 0|T_{oo}|0\rangle_{ren}$ is the renormalized VEV of the

0,0 component of the stress tensor.

We will have a vacuum in all cases where we can find a quantum state with properties a and b. If this is not possible, we can't define a vacuum.

We shall see the main consequences of this idea below.

2. THE STRONG VACUUM

Robertson-Walker metric is:

$$ds^2 = -dt^2 + a^2(t)\delta_{ij}dx^i dx^j \tag{2}$$

$\phi(x)$ is a neutral scalar field the field equation is

$$(\nabla_\mu \nabla^\mu - m^2 - \xi R)\phi = 0 , \tag{3}$$

where we use the usual symbols[3]. $\phi(x)$ can be decomposed as:

$$\phi(x) = \int d^3\underline{k} \frac{e^{i\underline{k}\cdot\underline{x}}}{(2\pi a)^{3/2}} [a_{\underline{k}} f_k(t) + a^+_{-\underline{k}} f^*_k(t)] , \tag{4}$$

and we shall write $f_{\underline{k}}(t)$ as:

$$f_k(t) = [2\Omega_k(t)]^{-1/2} e^{i\int^t \Omega_k(t')dt'} , \tag{5}$$

where $\Omega_k(t)$ must satisfy:

$$\frac{1}{2}\frac{\ddot{\Omega}_k}{\Omega_k} - \frac{3}{4}\left(\frac{\dot{\Omega}_k}{\Omega_k}\right)^2 + \Omega_k^2 = w_k^2 + (\xi - \tfrac{1}{4})R + \tfrac{3}{4}H^2 , \tag{6}$$

where $w_k^2 = m^2 + a^{-2}k^2$ and the dot symbolizes time derivative. Then the VEV of the stress tensor reads:

$$\langle 0|T_{oo}|0\rangle = \int \frac{d^3\underline{k}}{(2\pi a)^3} \frac{1}{4\Omega_k} \left\{\Omega_k^2 + w_k^2 + (\tfrac{9}{4} - 12\xi)H^2 - (6\xi - \tfrac{3}{2})H\frac{\dot{\Omega}}{\Omega} + \frac{1}{4}\left(\frac{\dot{\Omega}}{\Omega}\right)^2\right\}$$

$$\langle 0|T_{ij}|0\rangle = \frac{1}{3}\delta_{ij} \int \frac{d^3\underline{k}}{(2\pi a)^3} \frac{1}{2\Omega_k} \left\{(w_k^2 - m^2) + (6\xi - \tfrac{3}{2})(w_k^2 - \Omega_k^2) + 6\xi(6\xi - 1)R \right.$$

$$\left. + 3(\tfrac{9}{8} - \tfrac{13}{2}\xi)H^2 - 3(4\xi - \tfrac{3}{4})H\frac{\dot{\Omega}_k}{\Omega_k} - \frac{1}{4}(2\xi - \tfrac{1}{2})\left(\frac{\dot{\Omega}_k}{\Omega_k}\right)^2\right\} \tag{7}$$

Let's consider now the problem of the vacuum definition at a time t. We can write the Cauchy data of the negative energy solution related with the vacuum or, which is the same, we can write the Cauchy data of function Ω_k. Then, the Cauchy data that minimize the energy (1) turn out to be[1]:

$$\Omega_k^{ME} = [w_k^2 - 6\xi H^2(6\xi-1)]^{1/2} ,$$
$$\dot{\Omega}_k^{ME} = 3H(4\xi-1)\Omega_k^{ME} . \tag{8}$$

The vacuum obtained from these Cauchy data satisfies condition b.

In this paragraph we shall impose the local property a in its strongest form, i.e. that the $G_1(x,x')$ that define our vacuum should have the same analytical terms that $G_1^{DS}(x,x')$. We shall call a vacuum endowed with this property and property b a "strong vacuum". Let's remember that we only know the analytical expansion of the $G_1^{DS}(x,x')$, thus a strong vacuum has a $G_1(x,x') = G_1^{DS}(x,x')$ + non analytical terms. Now, the Cauchy data of the DS vacuum (up to fourth order in the metric derivatives) read:

$$\Omega_k^{DS} = w_k\Big\{1 + \frac{1}{w_k}[\frac{1}{2}(6\xi-1)\alpha_2] - \frac{1}{w_k^4}[\frac{1}{4}m^2(\alpha_1+\alpha_2) +$$
$$+ \frac{1}{8}(6\xi-1)^2\beta_2 + \frac{1}{4}(6\xi-1)(\beta_2 + \dot{\beta}_3 + \frac{5}{2}\beta_4 + \frac{1}{2}\beta_5)] +$$
$$+ \frac{m^2}{w_k^6}[\frac{5}{8}m^2\alpha_1 + \frac{1}{4}(9\xi\beta_2 - (2-39\xi)\beta_3 +$$
$$+ (15\xi + \frac{1}{4})\beta_4 + \frac{1}{4}\beta_5)] - \frac{1}{4}\frac{m^4}{w_k^8}[\frac{19}{8}(\beta_1+\beta_2) +$$
$$+ \frac{1}{2}(75\xi+39)\beta_3 + \frac{7}{2}\beta_4] + \frac{221}{32}\frac{m^6}{w_k^{10}}(\beta_1+\beta_3) -$$
$$- \frac{1105}{128}\frac{m^8}{w_k^{12}}\beta_1\Big\} + \dots ,$$

$$\dot{\Omega}_k^{DS} = \Big\{-H + \frac{1}{w_k^2}[m^2H + (6\xi-1)(H\alpha_2 + \frac{1}{2}\dot{\alpha}_2)]$$
$$- \frac{m^2}{w_k^4}[(6\xi+\frac{1}{2})H\alpha_2 + \frac{1}{4}\dot{\alpha}_2] + \frac{m^2}{w_k^6}\frac{9}{4}H(\alpha_1+\alpha_2)$$
$$- \frac{15}{4}\frac{m^6}{w_k^8}H\alpha_1\Big\}\Omega_k^{DS} + \dots , \tag{9}$$

where:

$$\alpha_1 = H^2 , \qquad \alpha_2 = R/6 ,$$

$$\beta_1 = \alpha_1^2, \ \beta_2 = \alpha_2^2, \ \beta_3 = \alpha_1\alpha_2, \ \beta_4 = H\dot{\alpha}_2, \ \beta_5 = \ddot{\alpha}_2 . \tag{10}$$

In general, Cauchy data (8) are different than (9), thus we will not have a strong vacuum for every time or for every evolution. The vacuum definition ambiguity is precisely a consecuence of this lack of coincidence. But of course we can easily see that we have a strong vacuum if:

1. a = const, i.e. in Minkowsky space.
2. Also if a = const in the interval $t_1 < t < t_2$, because the universe has a Killing vector field between t_1 and t_2. Also in other cases,[7] every time we have a Killing vector field we have a strong vacuum with its $G_1(x,x')$ invariant under translations along this field.
3. If at time t we have $\dot{a} = 0$, $\ddot{a} = 0$, ... we have a strong vacuum there (in this case we can say that we have a local Killing vector).

All these examples are related with Killing vectors, and a vacuum is obtained because, in each example, we stopped the evolution of the universe. However, there is a completely different example:

4. If $m = 0$, $\xi = 1/6$, we have a strong vacuum, i.e. the conformal vacuum.

Therefore, every respectable vacuum used until now in the literature seems to be a strong vacuum. Besides other criteria have been used to define vacua; at least two of them work all right between two strong vacua:

- The Feynman $G_F(x,x')$ built using two strong vacua is the unique Green Function of Euclidean space translated via a Wick trick.[2,8,9]
- This $G_F(x,x')$ can be obtained also adding a term $i\varepsilon$ to m^2 in the usual way.[2,8,9]

Finally, between two strong vacua the created particle number is obviously finite and the corresponding spectrum is a thermal-like one (in the sense that it vanishes faster than any power of k when $k \to \infty$)

because both vacua have the same analytical terms; being their difference purely non analytical.

All these properties make the strong vacuum concept a reasonable and reliable one

3. THE MINIMAL VACUUM

The strong vacuum is satisfactory enough but it exists only in very special cases, as it is shown by the short list of examples of paragraph 2. Can we, somehow, relax the conditions a and b to obtain a more general vacuum definition? In fact we can relax the version we gave of the local condition a in paragraph 2. What we really need (to implement a backreaction solution of the Einstein equation) is that the renormalized VEV of the stress-tensor (i.e. $\langle 0|T_{\mu\nu}|0\rangle_{ren}$) be finite in the vacuum (or more general quantum state) we choose as state of the universe. This will be our weaker version of the local property a, the vacuum $|0\rangle$ would correspond to a $G_1(x,x')$ with such divergent structure that $\langle 0|T_{\mu\nu}(x)|0\rangle_{ren}$ turns out to be finite for every x. We shall call a "Minimal Vacuum" a vacuum with this property and property b, because we believe this is the minimal set of conditions a vacuum must fulfill: it must be the minimum of some quantity (the generalized energy) and it must yield a finite renormalized stress tensor. Of course every strong vacuum is a minimal vacuum.

Let's now give an important example of minimal vacua (such that they are not strong vacua) to see that the set of minimal vacua is bigger than the set of strong vacua. Let's consider the Robertson-Walker universe in the case $\xi = 1/6$ (and $m \neq 0$). The equations (8) become:

$$\Omega_k^{ME} = w_k, \qquad \dot{\Omega}_k^{ME} = -H\Omega_k^{ME} . \tag{10}$$

Now let's consider a time τ and write the negative frequency solutions that correspond to the vacuum at τ:
$|0\rangle_\tau$ as:

$$f_k^{(\tau)}(t) = \alpha_k(\tau,t)\,\frac{e^{-i\int^t w_k dt'}}{\sqrt{2w_k}} + \beta_k(\tau,t)\,\frac{e^{i\int^t w_k dt'}}{\sqrt{2w_k}} \tag{11}$$

These functions yield solutions of eq. (3) if:

$$\dot{\alpha}_k = \frac{m^2 H}{2w_k^2} \beta_k \, e^{2i\int^t w_k dt'} \quad ; \qquad \dot{\beta}_k = \frac{m^2 H}{2w_k^2} \alpha_k \, e^{-2i\int^t w_k dt'} \tag{12}$$

and are an orthonormal basis if:

$$|\alpha_k|^2 - |\beta_k|^2 = 1 \, . \tag{13}$$

Their $|0\rangle_\tau$ minimize the energy (i.e. they satisfy eq. (10)) if:

$$\alpha_k(\tau,\tau) = 1 \, , \qquad \beta_k(\tau,\tau) = 0 \, . \tag{14}$$

Analogously, the vacuum $|0\rangle_t$, that minimize the energy at t, will have the negative frequency solutions:

$$f_k^{(t)}(t) = \frac{e^{-i\int^t w_k dt'}}{\sqrt{2w_k}} \, . \tag{15}$$

Thus $\alpha(\tau,t)$ and $\beta(\tau,t)$ are the Bogolyubov coefficients between the basis of vacua $|0\rangle_\tau$ and $|0\rangle_t$.

Let's introduce (11) in eq. (7):

$$_\tau\langle 0|T_{oo}(t)|0\rangle_\tau = \int \frac{d^3\underline{k}}{(2\pi a)^3} \frac{1}{2} w_k \left\{1 + 2|\beta_k(\tau,t)|^2\right\} \, ,$$

$$_\tau\langle 0|T_{ij}(t)|0\rangle_\tau = \frac{1}{3} \delta_{ij} \int \frac{d^3\underline{k}}{(2\pi a)^3} \frac{1}{2} w_k \left\{(1 - \frac{m^2}{w_k^2}) \times \right.$$

$$\left. \times \, (1 + 2|\beta_k(\tau,t)|^2) - \frac{2}{H}\frac{d}{dt} (|\beta_k(\tau,t)|^2) \right. \, . \tag{16}$$

To renormalize these infinite quantities we must subtract the $\langle T_{\mu\nu}\rangle^{DS}$(4), i.e. the stress-tensor built with the DeWitt-Schwinger solutions (9) (up to the fourth order in the metric derivatives) that reads:

$$<T_{oo}>^{DS}(4) = \int \frac{d^3\underline{k}}{(2\pi a)^3} \frac{1}{2} w_k - P_{oo} ,$$

$$<T_{ij}>^{DS}(4) = \frac{1}{3} \delta_{ij} \int \frac{d^3\underline{k}}{(2\pi a)^3} \frac{1}{2} w_k \left(1 - \frac{m^2}{w_k^2}\right) - P_{ij} , \tag{17}$$

where

$$2880\pi^2 P_{oo} = -30m^2\alpha_1 + 3\beta_1 - 3\beta_2 + 6\beta_3 + 6\beta_4 ,$$

$$2880\pi^2 P_{ij} = \frac{1}{3} \delta_{ij}[-30m^2(\alpha_1 - 2\alpha_2) + 15\beta_1 - 3\beta_2 - 6\beta_3 - 12\beta_4 - 8\beta_5] . \tag{18}$$

Now the renormalized ${}_\tau<0|T_{\mu\nu}|0>_\tau^{ren}$ is just ${}_\tau<0|T_{\mu\nu}|0>_\tau - <T_{\mu\nu}>^{DS}(4)$, i.e.:

$${}_\tau<0|T_{oo}(t)|0>_\tau^{ren} = P_{oo}(t) + \int \frac{d^3\underline{k}}{(2\pi a)^3} |\beta_k(\tau,t)|^2 w_k(t) ,$$

$${}_\tau<0|T_{ij}(t)|0>_\tau^{ren} = P_{ij}(t) + \frac{1}{3} \delta_{ij} \int \frac{d^3\underline{k}}{(2\pi a)^3} w_k \times \left\{\left(1 - \frac{m^2}{w_k^2}\right)|\beta_k|^2 - \frac{1}{H}\frac{d}{dt} |\beta_k|^2\right\} . \tag{19}$$

As we can see the divergent terms of eqs. (16) and (17) cancel out, thus the renormalized stress tensor turns out to be finite. In fact, the β_k coefficient can be computed using eq. (14) of[10] obtaining $\beta_k \sim k^{-6}$ so also the integrals of (19) are convergent. Thus vacuum $|0>_\tau$ is a minimal one, as all vacua that minimize the energy, at a certain time τ, in a Robertson-Walker universe.

The notion of minimal vacuum allows us to make the following physical interpretation of the two terms of eqs. (19). If we compute the renormalized stress-tensor at time τ using eq. (14) we find:

$${}_\tau<0|T_{\mu\nu}(\tau)|0>_\tau = P_{\mu\nu}(\tau) , \tag{20}$$

i.e. even if the universe is empty at time τ, because we are using precisely $|0>_\tau$ as quantum state, there is a non vanishing stress-

tensor. $P_{\mu\nu}(\tau)$ is, of course, the vacuum polarization tensor, a local object that depends only in the geometry at τ (cfr. eqs. (18)) and these are the first terms of eqs. (19). The second terms are the energy and momentum of the created particles between τ and t because β_k is the Bogolyubov coefficient and thus $|\beta_k|^2$ is the number density of the created particles in mode k. Thus from (19_1) w_k would be the energy of the particles in mode k. Metric (2) is invariant under space-like translations, thus $\underline{p}_k = a^{-1}\underline{k}$ could be considered the particle momentum. So the created particles satisfy the special relativity relation:

$$m^2 = w_k^2 - p_k^2 , \tag{21}$$

and can be considered, in fact, as physical particles.

This canonical decomposition between a polarization term and a created particle term appears also in other examples (e.g. between moving mirrors[11]).

The main difference among strong and minimal vacua is that between two minimal vacua the created particle spectrum could be non thermal. In fact, between two minimal vacua the functions Ω can have an analytical difference. Thus the two types of vacua can be used to deduce directly the characteristics of the spectrum.

4. CONCLUSION

The language of strong and minimal vacua can be used to study also the other problems listed in the introduction, i.e. bounded spaces (cfr. 11]) and non inertial reference systems (cfr. 7]). We believe also that this language can clarify some features of several lines of work, mainly related with the backreaction problem (e.g. paper 12] would have become more clear to us if it had stated that the universe is really in the in vacuum state at $t \to -\infty$, in fact a strong vacuum).

Finally we would like to remark that all our philosophy is based on the energy minimization (i.e. Hamiltonian diagonalization), a method that has been strongly criticized in the literature[13,14,15]. We believe we have overcome the main criticism in paper 2] and also that this method is the basis of several successful papers (e.g. 16-20]). Precisely the principal result of this work is that now we know when we can use the Hamiltonian diagonalization successfully: when the $G_1(x,x')$, related to the vacuum that minimize the energy, has the correct local behavior.

BIBLIOGRAPHY

[1] Castagnino, M., Gen. Rel. Grav., 15, No. 12, 1149 (1983).

[2] Castagnino, M., Mazzitelli, F. D., "Weak and Strong Quantum Vacua in Bianchi Type I Universes", to be published, Phys. Rev. D. (1984).

[3] Castagnino, M., Harari, D., Nuñez, C., "Vacuum Polarization in Curved Background Deduced from Hadamard Kernels", Submitted to Ann. of Phys. (1984).

[4] Castagnino, M., Gen. Rel. Grav., 9, No. 2, 101 (1978).

[5] Castagnino, M., Harari, D., Nuñez, C., "Minimal Hypotheses for Particle Definition in Curved Space-Time", Unification of the Fundamental Particle Interaction II, Ed. Ellis, J., Ferrara, S., Plenum Pub. Co., 455 (1983).

[6] Castagnino, M., Harari, D., Ann. of Phys., 152, No. 1, 85 (1984).

[7] Castagnino, M., Nuñez, C., "On the Vacuum Definition for Non-Inertial Observers", Proceedings Inter. Astron. Congress, Buenos Aires (1982).

[8] Calzetta, E., Castagnino, M., Phys. Rev. D 28, No. 6, 1298 (1983).

[9] Calzetta, E., Castagnino, M., Phys. Rev. D 29, No. 8, 1609 (1984).

[10] Castagnino, M., Harari, D., Chimento, L., Phys. Rev. D 24, No. 2, 290 (1981).

[11] Castagnino, M., Ferraro, R., "A Toy Cosmology: Radiation from Moving Mirrors, The Final Equilibrium State and the Instantaneous Model of Particle", to be published Ann. of Phys. (1984).

[12] Wada, S., Azuma, T., Phys. Lett., 132B, No. 4, 5, 6, 313 (1983).

[13] Parker, L., Phys. Rev., 183, No. 5, 1057 (1969).

[14] Castagnino, M., Verbeure, A., Weder, R., Nuovo Cim., 26B, No. 2, 396 (1975).

[15] Fulling, S., Gen. Rel. Grav., 10, 807 (1979).

[16] Grib, A. A., Mamayev, S. G., Mostepanenko, V. M., Fortsch. der Phys., 28, 173 (1980).

[17] Grib, A. A., Mamayev, S. G., Mostepanenko, V. M., J. Phys. A: Math. Gen., 13, 2057 (1980).

[18] Mamaev, S. G., Mostepanenko, V. M., Sov. Phys. JETP, 51, (1), 9 (1980).

[19] Grig, A. A., Mamayev, S.G., Mostepanenko, V. M., "Self-Consistent Treatment of Vacuum Effects in Isotropic Cosmology", Quantum Gravity, Ed. Markov, M. A., West, P. C., Plenum Pub. Co. (1984).

[20] Grig, A. A., Pendey, S. N., "From Vacuum to Friedman Universe", Proceedings Einstein Found. Int., 1, No. 2, 219 (1983).

SECTION IV

EARLY UNIVERSE

Quantum Fluctuations As The Cause Of Inhomogeneity In The Universe

Jonathan Halliwell

Stephen Hawking

University of Cambridge
Department of Applied Mathematics and Theoretical Physics
Silver Street
Cambridge CB3 9EW
ENGLAND

In this paper we consider the canonical quantization of a cosmological model without any assumptions of homogeneity or isotropy. We treat the two homogeneous and isotropic degrees of freedom of the gravitational and matter fields exactly and the other degrees of freedom to second order in the Hamiltonian. We derive a background Wheeler DeWitt equation for the two homogeneous isotropic degrees of freedom and time dependent Schroedinger equations for the other degrees of freedom which are treated as perturbations on the background. We assume that the universe is in the quantum state defined by a path integral over compact metrics. We justify our treatment of the inhomogeneous modes to second order only by showing that they start out in their ground states.

They remain in the ground state while the background expands exponentially until their wave length becomes greater than the horizon size. The wave function is then "frozen" until the wave length re-enters the horizon in the radiation or matter dominated era. The modes are then in a highly excited state and give rise to a scale free spectrum of density fluctuations that could explain the origin of galaxies and other structures in the universe. The anisotropy in the microwave background would be within the upper limit set by observation if the mass of the scalar field that drives the inflation is less than 10^{14} Gev.

1 Introduction

Observations of the microwave background indicate that the Universe is very close to homogeneity and isotropy on a large scale. Yet we know that the early Universe cannot have been completely homogeneous and isotropic because in that case galaxies and stars would not have formed. In the standard hot big bang model the density perturbations required to produce these structures have to be assumed as initial conditions. However, in the inflationary model of the Universe [1,2,3,4] it was possible to show that the ground state fluctuations of the scalar field that causes the exponential expansion would lead to a spectrum of density perturbations that was almost scale free [5,6,7]. In the simplest GUT inflationary model the amplitude of the density perturbations was too large but an amplitude that was consistent with observation could be obtained in other models with a different potential for the scalar field [8]. Similarly, ground state fluctuations of the gravitational wave modes would lead to a spectrum of long wavelength gravitational waves that would be consistent with observation provided that the Hubble constant H in the inflationary period was not more than about 10^{-4} of the Planck mass [9].

One cannot regard these results as a completely satisfactory explanation of the origin of structure in the Universe because the inflationary model does not make any assumption about the initial or boundary conditions of the Universe. In particular, it does not guarantee that there should be a period of exponential expansion in which the scalar field and the gravitational wave modes would be in the ground state. In the absence of some assumption about the boundary conditions of the

Universe, any present state would be possible: one could pick an arbitrary state for the Universe at the present time and evolve it backwards in time to see what initial conditions it arose from. It has recently been proposed [10,11,12,13] that the boundary conditions of the Universe are that it has no boundary. In other words, the quantum state of the Universe is defined by a path integral over compact 4-metrics without boundary. The quantum state can be described by a wavefunction Ψ which is a function on the infinite dimensional space W called Superspace which consists of all 3-metrics h_{ij} and matter field configurations ϕ_0 on a 3-surface S. Because the wavefunction does not depend on time explicitly, it obeys a system of zero energy Schroedinger equations, one for each choice of the shift N_i and the lapse N on S. The Schroedinger equations can be decomposed into the momentum constraints, which imply that the wavefunction is the same at all points of W that are related by coordinate transformations, and the Wheeler-DeWitt equations, which can be regarded as a system of second order differential equations for Ψ on W. The requirement that the wavefunction be given by a path integral over compact 4-metrics then becomes a set of boundary conditions for the Wheeler-DeWitt equations which determines a unique solution for Ψ.

It is difficult to solve differential equations on an infinite dimensional manifold. Attention has therefore been concentrated on finite dimensional approximations to W, called "Minisuperspaces". In other words, one resticts the number of gravitational and matter degrees of freedom to a finite number and then solves the Wheeler-DeWitt equations on a finite dimensional manifold with boundary conditions that reflect the fact that the wavefunction is given by a path integral over compact 4-metrics. In particular [12,13,14,15] it has been shown that in the case of a homogeneous isotropic closed universe of radius a with a massive scalar field ϕ the wavefunction corresponds in the classical limit to a family of classical solutions which have a long period of exponential or "inflationary" expansion and then go over to a matter dominated expansion, reach a maximum radius and then collapse in a time symmetric manner. This model would be in agreement with observation but, because it is so restricted, the only prediction it can make is that the observed value of the density parameter Ω should be exactly one [15]. The aim of this paper is to extend this Minisuperspace model to the full number of degrees of freedom of the gravitational and scalar fields. We treat the two degrees of freedom of the Minisuperspace model exactly and we expand the other inhomogeneous and anisotropic degrees of freedom to second order in the Hamiltonian. In the region of W in which Ψ oscillates rapidly, one can use the WKB approximation to relate the wavefunction to a family of classical solutions and so introduce a concept of time. As in the Minisuperspace case, the family includes solutions with a long period of exponential expansion. We show that the gravitational wave and density perturbation modes obey decoupled time dependent Schroedinger equations with respect to the time parameter of the classical solution. The boundary conditions imply that these modes start off in the ground state. While they remain within the horizon of the exponentially expanding phase, they can relax adiabatically and so they remain in the ground state. However, when they expand outside the horizon of the inflationary period, they become "frozen" until they re-enter the horizon in the matter dominated era. They then give rise to gravitational waves and a scale free spectrum of density perturbations. These would be consistent with the observations of the microwave background and could be large enough to explain the origins of galaxies if the mass of the scalar field were about 10^{-5} of the Planck mass. Thus the proposal that the quantum state of the Universe is defined by a path integral over compact 4-metrics seems to be able to account for the origin of structure in the Universe: it arises, not from arbitrary initial conditions, but from the ground state fluctuations that have to be present by the Heisenberg Uncertainty Principle.

In section 2 we review the Hamiltonian formalism of classical General Relativity and in section 3, we show how this leads to the canonical treatment of the quantum theory. In section 4 we summarize earlier work [13] on a homogeneous isotropic minisuperspace model with a massive scalar field. We extend this to all the matter and gravitational degrees of freedom in section 5, treating the inhomogeneous modes to second order in the Hamiltonian. In section 6 we decompose the wavefunction into a background term which obeys an equation similar to that of the unperturbed minisuperspace model, and perturbation terms which obey time dependent Schroedinger equations. We use the path integral expression for the wavefunction in section 7 to show that the perturbation wavefunctions start out in their ground states. Their subsequent evolution is described in section 8. In section 9 we calculate the anisotropy that these perturbations would produce in the microwave background and compare with observation. In section 10 we summarize the paper and conclude that the proposed quantum state could account not only for the large scale homogeneity and isotropy but also for the structure on smaller scales.

2 Canonical Formulation of General Relativity

We consider a compact 3-surface S which divides the 4-manifold M into two parts. In a neighbourhood of S one can introduce a coordinate t such that S is the surface $t = 0$ and coordinates x^i, $(i = 1,2,3)$. The metric takes the form

$$ds^2 = -(N^2 - N_i N^i)dt^2 + 2N_i dx^i dt + h_{ij} dx^i dx^j \qquad (2.1)$$

N is called the lapse function. It measures the proper time separation of surfaces of constant t. N_i is called the shift vector. It measures the deviation of the lines of constant x^i from the normal to the surface S. The action is

$$I = \int (L_g + L_m) d^3x dt \qquad (2.2)$$

where

$$L_g = \frac{m_p^2}{16\pi} N(G^{ijkl} K_{ij} K_{kl} + h^{1/2}\, {}^3R) \qquad (2.3)$$

$$K_{ij} = \frac{1}{2N}\left[- \frac{\partial h_{ij}}{\partial t} + 2N_{(i|j)}\right] \tag{2.4}$$

is the second fundamental form of S and

$$G^{ijkl} = ½\, h^{½}(h^{ik}h^{jl} + h^{il}h^{jk} - 2h^{ij}h^{kl}) \tag{2.5}$$

In the case of a massive scalar field Φ

$$L_m = ½Nh^{½}\left[N^{-2}\left[\frac{\partial\Phi}{\partial t}\right]^2 - 2\frac{N^i}{N^2}\frac{\partial\Phi}{\partial t}\frac{\partial\Phi}{\partial x^i}\right. \tag{2.6}$$

$$\left. - \left[h^{ij} - \frac{N^iN^j}{N^2}\right]\frac{\partial\Phi}{\partial x^i}\frac{\partial\Phi}{\partial x^j} - m^2\Phi^2\right]$$

In the Hamiltonian treatment of General Relativity one regards the components h_{ij} of the 3-metric and the field Φ as the canonical coordinates. The canonically conjugate momenta are

$$\pi^{ij} = \frac{\partial L_g}{\partial \dot{h}_{ij}} = - \frac{h^{½}m_p^2}{16\pi}(K^{ij} - h^{ij}K) \tag{2.7}$$

$$\pi_\Phi = \frac{\partial L_m}{\partial \dot{\Phi}} = N^{-1}h^{½}\left[\dot{\Phi} - N^i\frac{\partial\Phi}{\partial x^i}\right] \tag{2.8}$$

The Hamiltonian is

$$H = \int (\pi^{ij}\dot{h}_{ij} + \pi_\Phi \dot{\Phi} - L_g - L_m) d^3x \tag{2.9}$$

$$= \int (NH_0 + N_i H^i) d^3x$$

where

$$H_0 = 16\pi m_p^{-2} G_{ijkl}\pi^{ij}\pi^{kl} - \frac{m_p^2}{16\pi} h^{1/2}\, {}^3R \tag{2.10}$$

$$+ \tfrac{1}{2} h^{1/2} \left[\frac{\pi_\Phi^2}{h} + h^{ij} \frac{\partial\Phi}{\partial x^i}\frac{\partial\Phi}{\partial x^j} + m^2\Phi^2 \right]$$

$$H^i = -2\pi^{ij}{}_{|j} + h^{ij}\frac{\partial\Phi}{\partial x^j}\pi_\Phi \tag{2.11}$$

and

$$G_{ijkl} = \tfrac{1}{2} h^{-1/2} (h_{ik}h_{jl} + h_{il}h_{jk} - h_{ij}h_{kl}) \tag{2.12}$$

The quantities N and N_i are regarded as Lagrange multipliers. Thus the solution obeys the momentum constraint

$$H^i = 0 \tag{2.13}$$

and the Hamiltonian constraint

$$H_0 = 0 \tag{2.14}$$

For given fields N and N^i on S the equations of motion are

$$\dot{h}_{ij} = \frac{\partial H}{\partial \pi^{ij}}$$

$$\dot{\pi}^{ij} = -\frac{\partial H}{\partial h_{ij}}$$

$$\Phi = \frac{\partial H}{\partial \pi_\Phi}$$

$$\dot{\pi}_\Phi = -\frac{\partial H}{\partial \Phi} \tag{2.15}$$

3 Quantization

The quantum state of the Universe can be described by a wavefunction Ψ which is a function on the infinite dimensional manifold W of all 3-metrics h_{ij} and matter fields Φ on S. A tangent vector to W is a pair of fields (γ_{ij}, μ) on S where γ_{ij} can be regarded as a small change of the metric h_{ij} and μ can be regarded as a small change of Φ. For each choice of N > 0 on S there is a natural metric Γ(N) on W [15].

$$ds^2 = \int N^{-1}\left[\frac{m_p^2}{32\pi}G^{ijkl}\gamma_{ij}\gamma_{kl} + \tfrac{1}{2}h^{1/2}\mu^2\right]d^3x \tag{3.1}$$

The wavefunction Ψ does not depend explicitly on the time t because t is just a coordinate which can be given arbitrary values by different choices of the undetermined multipliers N and N_i. This means that Ψ obeys the zero energy Schroedinger equation:

$$H\Psi = 0 \tag{3.2}$$

The Hamiltonian operator H is the classical Hamiltonian with the usual substitutions:

$$\pi^{ij}(x) \rightarrow -i\frac{\delta}{\delta h_{ij}(x)}, \qquad \pi_{\phi}(x) \rightarrow -i\frac{\delta}{\delta\phi(x)} \tag{3.3}$$

Because N and N_i are regarded as independent Lagrange multipliers, the Schroedinger equation can be decomposed into two parts. There is the momentum constraint:

$$\underline{H}\,\Psi \equiv \int N_i H^i d^3x\ \Psi \tag{3.4}$$

$$= \int h^{1/2} N_i \left\{ 2\left[\frac{\delta}{\delta h_{ij}(x)}\right]_{|j} - h^{ij}\frac{\partial\Phi}{\partial x^j}\frac{\delta}{\delta\Phi(x)}\right\} d^3x\ \Psi = 0$$

This implies that Ψ is the same on 3-metrics and matter field configurations that are related by coordinate transformations in S. The other part of the Schroedinger equation, corresponding to $H_{|}\Psi = 0$, where $H_{|} = \int NH_0 d^3x$ is called the Wheeler-DeWitt equation. There is one Wheeler-DeWitt equation for each choice of N on S. One can regard them as a system of second order partial differential equations for Ψ on W. There is some ambiguity in the choice of operator ordering in these equations but this will not affect the results of this paper. We shall assume that H_I has the form [15]

$$(- \tfrac{1}{2}\nabla^2 + \xi \mathbb{R} + V)\Psi = 0 \tag{3.5}$$

where ∇^2 is the Laplacian in the metric $\Gamma(N)$, $\mathbb{R}$ is the curvature scalar of this metric and the potential V is

$$V = \int h^{1/2} N \left[- \frac{m_p^2}{16\pi}\, {}^3R + \epsilon + U \right] d^3x \tag{3.6}$$

where $U = T^{00} - \tfrac{1}{2}\pi_\Phi^2$. The constant ϵ can be regarded as a renormalization of the cosmological constant Λ. We shall assume that the renormalized Λ is zero. We shall also assume that the coefficient ξ of the scalar curvature $\mathbb{R}$ of W is zero.

Any wavefunction Ψ which satisfies the momentum constraint and the Wheeler-DeWitt equation for each choice of N and N_i on S describes a possible quantum state of the Universe. We shall be concerned with the particular solution which represents the quantum state defined by a path integral over compact 4-metrics without boundary. In this case [11,12,13]

$$\Psi = \int d[g_{\mu\nu}]d[\Phi]\exp(- \hat{I}(g_{\mu\nu},\Phi)) \tag{3.7}$$

where $\hat{I}$ is the Euclidean action obtained by setting N negative imaginary and the path integral is taken over all compact 4-metrics $g_{\mu\nu}$ and matter fields Φ which are bounded by S on which the 3-metric is h_{ij} and the matter field is Φ. One can regard (3.7) as a boundary condition on the Wheeler-DeWitt equations. It implies that Ψ tends to a constant, which can be normalized to one, as h_{ij} goes to zero.

4 Unperturbed Friedman Model

References [12,13,14] considered the Minisuperspace model which consisted of a Friedman model with metric

$$ds^2 = \sigma^2(- N^2dt^2 + a^2d\Omega_3^2) \tag{4.1}$$

where $d\Omega_3^2$ is the metric of the unit 3-sphere. The normalization factor $\sigma^2 = \frac{2}{3\pi m_p^2}$ has been included for convenience. The model contains a scalar field $(2^{1/2}\pi\sigma)^{-1}\phi$ with mass $\sigma^{-1}m$ which is constant on surfaces of constant t. One can easily generalize this to the case of a scalar field with a potential $V(\phi)$. Such generalizations include models with higher derivative quantum corrections [16]. The action is

$$I = - \tfrac{1}{2}\int dtNa^3\left[\frac{1}{N^2a^2}\left(\frac{da}{dt}\right)^2 - \frac{1}{a^2} - \frac{1}{N^2}\left(\frac{d\phi}{dt}\right)^2 + m^2\phi^2\right] \tag{4.2}$$

The classical Hamiltonian is

$$H = \tfrac{1}{2}N(- a^{-1}\pi_a^2 + a^{-3}\pi_\phi^2 - a + a^3m^2\phi^2) \tag{4.3}$$

where

$$\pi_a = -\frac{a\,da}{N\,dt} \qquad \pi_\phi = \frac{a^3 d\phi}{N\,dt} \tag{4.4}$$

The classical Hamiltonian constraint is H = 0. The classical field equations are

$$N\frac{d}{dt}\left[\frac{1}{N}\frac{d\phi}{dt}\right] + \frac{3}{a}\frac{da}{dt}\frac{d\phi}{dt} + N^2m^2\phi = 0 \tag{4.5}$$

$$N\frac{d}{dt}\left[\frac{1}{N}\frac{da}{dt}\right] = N^2am^2\phi^2 - 2a\left[\frac{d\phi}{dt}\right]^2 \tag{4.6}$$

The Wheeler-DeWitt equation is

$$\tfrac{1}{2}Ne^{-3\alpha}\left[\frac{\partial^2}{\partial\alpha^2} - \frac{\partial^2}{\partial\phi^2} + 2V\right]\Psi(\alpha,\phi) = 0 \tag{4.7}$$

where

$$V = \tfrac{1}{2}(e^{6\alpha}m^2\phi^2 - e^{4\alpha}) \tag{4.8}$$

and $\alpha = \ell n\ a$. One can regard equation (4.7) as a hyperbolic equation for Ψ in the flat space with coordinates (α,ϕ) with α as the time coordinate. The boundary condition that gives the quantum state defined by a path integral over compact 4-metrics is $\Psi \to 1$ as $\alpha \to -\infty$. If one integrates equation (4.7) with this boundary condition, one finds that the wavefunction starts oscillating in the region $V > 0$, $|\phi| > 1$ (this has been confirmed numerically [14]). One can interpret the oscillatory component of the wavefunction by the WKB approximation:

$$\Psi = \mathrm{Re}\,(\,C\, e^{iS}\,) \tag{4.9}$$

where C is a slowly varying amplitude and S is a rapidly varying phase. One chooses S to satisfy the classical Hamilton-Jacobi equation:

$$H(\pi_\alpha, \pi_\phi, \alpha, \phi) = 0 \tag{4.10}$$

where

$$\pi_\alpha = \frac{\partial S}{\partial \alpha}, \qquad \pi_\phi = \frac{\partial S}{\partial \phi} \tag{4.11}$$

One can write (4.10) in the form

$$\tfrac{1}{2} f^{ab} \frac{\partial S}{\partial q^a}\frac{\partial S}{\partial q^b} + e^{-3\alpha} V = 0 \tag{4.12}$$

where f^{ab} is the inverse to the metric $\Gamma(1)$:

$$f^{ab} = e^{-3\alpha}\mathrm{diag}(\,-1,1) \tag{4.13}$$

The wavefunction (4.9) will then satisfy the Wheeler-DeWitt equation if

$$\nabla^2 C + 2i f^{ab} \frac{\partial C}{\partial q^a} \frac{\partial S}{\partial q^b} + i C \nabla^2 S = 0 \tag{4.14}$$

where ∇^2 is the Laplacian in the metric f_{ab}. One can ignore the first term in equation (4.14) and can integrate the equation along the trajectories of the vector field $X^a = \frac{dq^a}{dt} = f^{ab} \frac{\partial S}{\partial q^b}$ and so determine the amplitude C. These trajectories correspond to classical solutions of the field equations. They are parameterized by the coordinate time t of the classical solutions.

The solutions that correspond to the oscillating part of the wavefunction of the Minisuperspace model start out at $V = 0$, $|\phi| > 1$ with $\frac{d\alpha}{dt} = \frac{d\phi}{dt} = 0$. They expand exponentially with

$$S = -\frac{1}{3} e^{3\alpha} m|\phi|(1 - m^{-2} e^{-2\alpha} \phi^{-2}) \approx -\frac{1}{3} e^{3\alpha} m|\phi| \tag{4.15}$$

$$\frac{d\alpha}{dt} = m|\phi|, \qquad \frac{d|\phi|}{dt} = -\frac{1}{3} m \tag{4.16}$$

After a time of order $3m^{-1}(|\phi_1| - 1)$, where ϕ_1 is the initial value of ϕ, the field ϕ starts to oscillate with frequency m. The solution then becomes matter dominated and expands with e^α proportional to $t^{2/3}$. If there were other fields present, the massive scalar particles would decay into light particles and then the solution would expand with e^α proportional to $t^{1/2}$. Eventually the solution would reach a maximum radius of order $\exp(\frac{9\phi_1^2}{2})$ or $\exp(9\phi_1^2)$ depending on whether it is radiation or matter dominated for most of the expansion. The solution would then recollapse in a similar manner.

5 The Perturbed Friedman Model

We assume that the metric is of the form (2.1) except the right hand side has been multiplied by a normalization factor σ^2. The 3-metric h_{ij} has the form

$$h_{ij} = a^2(\Omega_{ij} + \epsilon_{ij}) \tag{5.1}$$

where Ω_{ij} is the metric on the unit 3-sphere and ϵ_{ij} is a perturbation on this metric and may be expanded in harmonics:

$$\epsilon_{ij} = \sum_{n,\ell,m} \Big[6^{1/2} a_{n\ell m} \tfrac{1}{3}\Omega_{ij} Q^n_{\ell m} + 6^{1/2} b_{n\ell m} (P_{ij})^n_{\ell m} + 2^{1/2} c^o_{n\ell m} (S^o_{ij})^n_{\ell m} + 2^{1/2} c^e_{n\ell m} (S^e_{ij})^n_{\ell m} + 2\, d^o_{n\ell m} (G^o_{ij})^n_{\ell m} + 2\, d^e_{n\ell m} (G^e_{ij})^n_{\ell m} \Big] \tag{5.2}$$

The coefficients $a_{n\ell m}, b_{n\ell m}, c^o_{n\ell m}, c^e_{n\ell m}, d^o_{n\ell m}, d^e_{n\ell m}$ are functions of the time coordinate t but not the three spatial coordinates x^i.

The $Q(x^i)$ are the standard scalar harmonics on the 3-sphere. The $P_{ij}(x^i)$ are given by (suppressing all but the i,j indices)

$$P_{ij} = \frac{1}{(n^2 - 1)} Q_{|ij} + \tfrac{1}{3}\Omega_{ij} Q \tag{5.3}$$

They are traceless, $P_i{}^i = 0$. The S_{ij} are defined by

$$S_{ij} = S_{i|j} + S_{j|i} \tag{5.4}$$

where S_i are the transverse vector harmonics, $S_i{}^{|i} = 0$. The G_{ij} are the transverse traceless tensor harmonics, $G_i{}^i = G_{ij}{}^{|j} = 0$. Further details about the harmonics and their normalization can be found in appendix A.

The lapse, shift and the scalar field $\Phi(x^i,t)$ can be expanded in terms of harmonics:

$$N = N_0\left[1 + 6^{-1/2} \sum_{n,\ell,m} g_{n\ell m}\, Q^n_{\ell m}\right] \tag{5.5}$$

$$N_i = e^{\alpha} \sum_{n,\ell,m} \left[6^{-1/2}\, k_{n\ell m}\, (P_i)^n_{\ell m} + 2^{1/2}\, j_{n\ell m}\, (S_i)^n_{\ell m}\right] \tag{5.6}$$

$$\Phi = \sigma^{-1} \left[\frac{1}{2^{1/2}\pi}\, \phi(t) + \sum_{n,\ell,m} f_{n\ell m}\, Q^n_{\ell m}\right] \tag{5.7}$$

where $P_i = \dfrac{1}{(n^2 - 1)} Q_{|i}$. Hereafter, the labels n,ℓ,m,o and e will be denoted simply by n. One can then expand the action to all orders in terms of the "background" quantities a,ϕ,N_0 but only to second order in the "perturbations" $a_n, b_n, c_n, d_n, f_n, g_n, k_n, j_n$:

$$I = I_0(a,\phi,N_0) + \sum_n I_n \tag{5.8}$$

where I_0 is the action of the unperturbed model (4.2) and I_n is quadratic in the perturbations and is given in appendix B.

One can define conjugate momenta in the usual manner. They are:

$$\pi_\alpha = - N_0^{-1} e^{3\alpha} \dot{\alpha} + \text{quadratic terms} \qquad (5.9)$$

$$\pi_\phi = N_0^{-1} e^{3\alpha} \dot{\phi} + \text{quadratic terms} \qquad (5.10)$$

$$\pi_{a_n} = - N_0^{-1} e^{3\alpha} \left[\dot{a}_n + \dot{\alpha}(a_n - g_n) + \tfrac{1}{3} e^{-\alpha} k_n\right] \qquad (5.11)$$

$$\pi_{b_n} = N_0^{-1} e^{3\alpha} \frac{(n^2 - 4)}{(n^2 - 1)} \left[\dot{b}_n + 4\dot{\alpha} b_n - \tfrac{1}{3} e^{-\alpha} k_n\right] \qquad (5.12)$$

$$\pi_{c_n} = N_0^{-1} e^{3\alpha} (n^2 - 4) \left[\dot{c}_n + 4\dot{\alpha} c_n - e^{-\alpha} j_n\right] \qquad (5.13)$$

$$\pi_{d_n} = N_0^{-1} e^{3\alpha} \left[\dot{d}_n + 4\dot{\alpha} d_n\right] \qquad (5.14)$$

$$\pi_{f_n} = N_0^{-1} e^{3\alpha} \left[\dot{f}_n + \dot{\phi}(3a_n - g_n)\right] \qquad (5.15)$$

The quadratic terms in equations (5.9) and (5.10) are given in appendix B. The Hamiltonian can then be expressed in terms of these momenta and the other quantities:

$$H = N_0 \left[H_{|0} + \sum_n H^n_{|2} + \sum_n g_n H^n_{|1}\right] + \sum_n \left[k_n \, {}^S H^n_{\_1} + j_n \, {}^V H^n_{\_1}\right] \qquad (5.16)$$

The subscripts 0,1,2 on the $H_|$ and H_- denote the orders of the quantities in the perturbations and S and V denote the scalar and vector parts of the shift part of the Hamiltonian. $H_{|0}$ is the Hamiltonian of the unperturbed model with N = 1 :

$$H_{|0} = \tfrac{1}{2}e^{-3\alpha}\left[- \pi_\alpha^2 + \pi_\phi^2 + e^{6\alpha}m^2\phi^2 - e^{4\alpha}\right] \qquad (5.17)$$

The second order Hamiltonian is given by $H_{|2} = \sum_n H^n_{|2} = \sum_n ({}^S H^n_{|2} + {}^V H^n_{|2} + {}^T H^n_{|2})$ where

$${}^S H^n_{|2} = \tfrac{1}{2}e^{-3\alpha}\Bigg\{\left[\tfrac{1}{2}a_n^2 + \frac{10(n^2-4)}{(n^2-1)} b_n^2\right]\pi_\alpha^2 + \left[\frac{15}{2}a_n^2 + \frac{6(n^2-4)}{(n^2-1)} b_n^2\right]\pi_\phi^2$$

$$- \pi_{a_n}^2 + \frac{(n^2-1)}{(n^2-4)} \pi_{b_n}^2 + \pi_{f_n}^2 + 2a_n\pi_{a_n}\pi_\alpha + 8b_n\pi_{b_n}\pi_\alpha - 6a_n\pi_{f_n}\pi_\phi$$

$$- e^{4\alpha}\left[\frac{1}{3}(n^2-\frac{5}{2})a_n^2 + \frac{(n^2-7)}{3}\frac{(n^2-4)}{(n^2-1)} b_n^2 + \frac{2}{3}(n^2-4)\, a_n b_n - (n^2-1)f_n^2\right]$$

$$+ e^{6\alpha}m^2\left[f_n^2 + 6a_n f_n\phi\right] + e^{6\alpha}m^2\phi^2\left[\frac{3}{2}a_n^2 - \frac{6(n^2-4)}{(n^2-1)}b_n^2\right]\Bigg\} \qquad (5.18)$$

$${}^V H^n_{|2} = \tfrac{1}{2}e^{-3\alpha}\Bigg\{(n^2-4)c_n^2\left[10\pi_\alpha^2 + 6\pi_\phi^2\right] + \frac{1}{(n^2-4)}\pi_{c_n}^2 + 8c_n\pi_{c_n}\pi_\alpha +$$

$$(n^2-4)c_n^2\left[2e^{4\alpha} - 6e^{6\alpha}m^2\phi^2\right]\Bigg\} \qquad (5.19)$$

$${}^T H^n_{|2} = \tfrac{1}{2}e^{-3\alpha}\Bigg\{d_n^2\left[10\pi_\alpha^2 + 6\pi_\phi^2\right] + \pi_{d_n}^2 + 8d_n\pi_{d_n}\pi_\alpha +$$

$$d_n^2\left[(n^2+1)e^{4\alpha} - 6e^{6\alpha}m^2\phi^2\right]\Big] \tag{5.20}$$

The first order Hamiltonians are

$$H^n_{|1} = \tfrac{1}{2}e^{-3\alpha}\left[- a_n\left[\pi_\alpha^2 + 3\pi_\phi^2\right] + 2\left[\pi_\phi\pi_{f_n} - \pi_\alpha\pi_{a_n}\right]\right.$$

$$\left. + m^2e^{6\alpha}\left[2f_n\phi + 3a_n\phi^2\right] - \frac{2}{3}e^{4\alpha}\left[(n^2-4)b_n + (n^2+\tfrac{1}{2})a_n\right]\right] \tag{5.21}$$

The shift parts of the Hamiltonian are

$${}^S H^n_{\_1} = \frac{1}{3}e^{-3\alpha}\left[- \pi_{a_n} + \pi_{b_n} + \left[a_n + \frac{4(n^2-4)}{(n^2-1)}b_n\right] \pi_\alpha + 3f_n\pi_\phi\right] \tag{5.22}$$

$${}^V H^n_{\_1} = e^{-\alpha}\left[\pi_{c_n} + 4(n^2-4)\ c_n\pi_\alpha\right] \tag{5.23}$$

The classical field equations are given in appendix B.

Because the Lagrange multipliers N_0, g_n, k_n, j_n are independent, the zero energy Schroedinger equation

$$H\Psi = 0 \tag{5.24}$$

can be decomposed as before into momentum constraints and Wheeler-DeWitt equations. As the momentum constraints are linear in the momenta, there is no ambiguity in the operator ordering. One therefore has

$$
{}^{S}H^{n}_{\_1}\Psi = -\frac{1}{3}e^{-3\alpha}\left[\frac{\partial}{\partial a_n} - \left[a_n + \frac{4(n^2-4)}{(n^2-1)}\,b_n\right]\frac{\partial}{\partial \alpha}\right.
$$

$$
\left. - \frac{\partial}{\partial b_n} - 3f_n\frac{\partial}{\partial \phi}\right]\Psi = 0 \tag{5.25}
$$

$$
{}^{V}H^{n}_{\_1}\Psi = e^{-\alpha}\left[\frac{\partial}{\partial c_n} + 4(n^2-4)\,c_n\frac{\partial}{\partial \alpha}\right]\Psi = 0 \tag{5.26}
$$

The first order Hamiltonians $H^{n}_{|1}$ give a series of finite dimensional second order differential equations, one for each n. In the order of approximation that we are using, the ambiguity in the operator ordering will consist of the possible addition of terms linear in $\frac{\partial}{\partial \alpha}$. The effect of such terms can be compensated for by multiplying the wavefunction by powers of e^{α}. This will not affect the relative probabilities of different observations at a given value of α. We shall therefore ignore such ambiguities and terms.

$$
\tfrac{1}{2}e^{-3\alpha}\left[a_n\left[\frac{\partial^2}{\partial\alpha^2} + 3\frac{\partial^2}{\partial\phi^2}\right] - 2\left[\frac{\partial}{\partial f_n}\frac{\partial}{\partial\phi} - \frac{\partial}{\partial a_n}\frac{\partial}{\partial\alpha}\right]\right.
$$

$$
\left. + m^2e^{6\alpha}\left[2\phi f_n + 3a_n\phi^2\right] - \frac{2}{3}e^{4\alpha}\left[(n^2-4)b_n + (n^2+\tfrac{1}{2})a_n\right]\right]\Psi = 0 \tag{5.27}
$$

Finally, one has an infinite dimensional second order differential equation

$$
\left[H_{|0} + \sum_n({}^{S}H^{n}_{|2} + {}^{V}H^{n}_{|2} + {}^{T}H^{n}_{|2})\right]\Psi = 0 \tag{5.28}
$$

where $H_{|0}$ is the operator in the Wheeler-DeWitt equation of the unperturbed Friedman Minisuperspace model:

$$H_{|0} = \tfrac{1}{2}e^{-3\alpha}\left[\frac{\partial^2}{\partial\alpha^2} - \frac{\partial^2}{\partial\phi^2} + e^{6\alpha}m^2\phi^2 - e^{4\alpha}\right] \qquad (5.29)$$

and

$$S_{H^n_{|2}} = \tfrac{1}{2}e^{-3\alpha}\left[- \left[\tfrac{1}{2}a_n^2 + \frac{10(n^2-4)}{(n^2-1)} b_n^2\right]\frac{\partial^2}{\partial\alpha^2} - \left[\frac{15}{2}a_n^2 + \frac{6(n^2-4)}{(n^2-1)} b_n^2\right]\frac{\partial^2}{\partial\phi^2}\right.$$

$$+ \frac{\partial^2}{\partial a_n{}^2} - \frac{(n^2-1)}{(n^2-4)}\frac{\partial^2}{\partial b_n{}^2} - \frac{\partial^2}{\partial f_n{}^2} - 2a_n\frac{\partial}{\partial a_n}\frac{\partial}{\partial\alpha} - 8b_n\frac{\partial}{\partial b_n}\frac{\partial}{\partial\alpha} + 6a_n\frac{\partial}{\partial f_n}\frac{\partial}{\partial\phi}$$

$$- e^{4\alpha}\left[\frac{1}{3}(n^2-\frac{5}{2})a_n^2 + \frac{(n^2-7)}{3}\frac{(n^2-4)}{(n^2-1)} b_n^2 + \frac{2}{3}(n^2-4)\, a_n b_n - (n^2-1)f_n^2\right]$$

$$\left. + e^{6\alpha}m^2\left[f_n^2 + 6a_n f_n\phi\right] + e^{6\alpha}m^2\phi^2\left[\frac{3}{2}a_n^2 - \frac{6(n^2-4)}{(n^2-1)}b_n^2\right]\right] \qquad (5.30)$$

$$V_{H^n_{|2}} = \tfrac{1}{2}e^{-3\alpha}\left[- (n^2-4)c_n^2\left[10\frac{\partial^2}{\partial\alpha^2} + 6\frac{\partial^2}{\partial\phi^2}\right] - \frac{1}{(n^2-4)}\frac{\partial^2}{\partial c_n{}^2} - 8c_n\frac{\partial}{\partial c_n}\frac{\partial}{\partial\alpha} + \right.$$

$$\left.(n^2-4)c_n^2\left[2e^{4\alpha} - 6e^{6\alpha}m^2\phi^2\right]\right] \qquad (5.31)$$

$$T_{H^n_{|2}} = \tfrac{1}{2}e^{-3\alpha}\left[- d_n^2\left[10\frac{\partial^2}{\partial\alpha^2} + 6\frac{\partial^2}{\partial\phi^2}\right] - \frac{\partial^2}{\partial d_n{}^2} - 8d_n\frac{\partial}{\partial d_n}\frac{\partial}{\partial\alpha} + \right.$$

$$d_n^2\left[(n^2+1)e^{4\alpha} - 6e^{6\alpha}m^2\phi^2\right]\Bigg] \tag{5.32}$$

We shall call equation (5.28) the master equation. It is not hyperbolic because, as well as the positive second derivatives $\frac{\partial^2}{\partial\alpha^2}$ in $H_{|0}$, there are the positive second derivatives $\frac{\partial^2}{\partial a_n^2}$ in each ${}^{S}H^{n}_{|2}$. However, one can use the momentum constraint (5.25) to substitute for the partial derivatives with respect to a_n and then solve the resultant differential equation on $a_n = 0$. Similarly, one can use the momentum constraint (5.26) to substitute for the partial derivatives with respect to c_n and then solve on $c_n = 0$. One thus obtains a modified equation which is hyperbolic for small f_n. If one knows the wavefunction on $a_n = 0 = c_n$, one can use the momentum constraints to calculate the wavefunction at other values of a_n and c_n.

6 The Wavefunction

Because the perturbation modes are not coupled to each other, the wavefunction can be expressed as a sum of terms of the form

$$\Psi = \mathrm{Re}\ (\ \Psi_0(\alpha,\phi)\prod_n \Psi^{(n)}(\alpha,\phi,a_n,b_n,c_n,d_n,f_n)) \tag{6.1}$$

$$= \mathrm{Re}\ (\ Ce^{iS})$$

where S is a rapidly varying function of α and ϕ and C is a slowly varying function of all the variables. If one substitutes (6.1) into the master equation and divides by Ψ, one obtains

$$-\frac{\nabla_2^2\Psi_0}{2\Psi_0} - \sum_n \frac{\nabla_2^2\Psi^{(n)}}{2\Psi^{(n)}} - \sum_{n\leq m}\frac{(\nabla_2\Psi^{(n)}).(\nabla_2\Psi^{(m)})}{2\Psi^{(n)}\Psi^{(m)}} - \frac{(\nabla_2\Psi_0)}{\Psi_0}.\left[\sum_n \frac{\nabla_2\Psi^{(n)}}{\Psi^{(n)}}\right] \tag{6.2}$$

$$+ \sum_n \frac{H^n_{|2}\Psi}{\Psi} + e^{-3\alpha}\ V(\alpha,\phi) = 0$$

where ∇_2^2 is the Laplacian in the Minisuperspace metric $f_{ab} = e^{3\alpha}\mathrm{diag}(-1,1)$ and the dot product is with respect to this metric.

An individual perturbation mode does not contribute a significant fraction of the sums in the third and fourth terms in equation (6.2). Thus these terms can be replaced by

$$- \frac{(\nabla_2\Psi)}{\Psi} \cdot \sum_n \frac{(\nabla_2\Psi^{(n)})}{\Psi^{(n)}} + \frac{1}{2}\left[\sum_n \frac{\nabla_2\Psi^{(n)}}{\Psi^{(n)}}\right]^2 \qquad (6.3)$$

$$\approx \; - i(\nabla_2 S) \cdot \sum_n \frac{(\nabla_2\Psi^{(n)})}{\Psi^{(n)}} + \frac{1}{2}\left[\sum_n \frac{\nabla_2\Psi^{(n)}}{\Psi^{(n)}}\right]^2$$

In order that the ansatz (6.1) be valid, the terms in (6.2) that depend on a_n, b_n, c_n, d_n, f_n have to cancel out. This implies

$$\frac{(\nabla_2\Psi)}{\Psi} \cdot (\nabla_2\Psi^{(n)}) + \frac{1}{2}\nabla_2^2\Psi^{(n)} = \frac{H^n_{|2}\Psi}{\Psi}\Psi^{(n)} \qquad (6.4)$$

$$(- \frac{1}{2}\nabla_2^2 + e^{-3\alpha} V + \frac{1}{2} J.J)\Psi_0 = 0 \qquad (6.5)$$

where $J = \sum_n \frac{\nabla_2\Psi^{(n)}}{\Psi^{(n)}}$

In regions in which the phase S is a rapidly varying function of α and ϕ, one can neglect the second term in (6.4) in comparison with the first term. One can also replace the π_α and π_ϕ which appear in $H^n_{|2}$ by $\frac{\partial S}{\partial \alpha}$ and $\frac{\partial S}{\partial \phi}$ respectively. The vector $X^a = f^{ab}\frac{\partial S}{\partial q^b}$ obtained by raising the covector $\nabla_2 S$ by the inverse minisuperspace metric f^{ab} can be regarded as $\frac{\partial}{\partial t}$ where t is the time parameter of the classical Friedman metric that corresponds to Ψ by the WKB approximation. One then obtains a time dependent Schroedinger equation for each mode along a trajectory of the vector field X^a:

$$i\frac{\partial\Psi^{(n)}}{\partial t} = H^{n}_{|2}\Psi^{(n)} \qquad (6.6)$$

Equation (6.5) can be interpreted as the Wheeler–DeWitt equation for a two dimensional minisuperspace model with an extra term ½J.J arising from the perturbations. In order to make J finite, one will have to make subtractions. Subtracting out the ground state energies of the $H^{n}_{|2}$ corresponds to a renormalization of the cosmological constant Λ. There is a second subtraction which corresponds to a renormalization of the Planck mass m_p and a third one which corresponds to a curvature squared counterterm. The effect of such higher derivative terms in the action has been considered elsewhere [16].

One can write Ψ^n as

$$\Psi_n = {}^{S}\Psi^{(n)}(\alpha,\phi,a_n,b_n,f_n)\;{}^{V}\Psi^{(n)}(\alpha,\phi,c_n)\;{}^{T}\Psi^{(n)}(\alpha,\phi,d_n) \qquad (6.7)$$

where ${}^{S}\Psi^{(n)}$, ${}^{V}\Psi^{(n)}$ and ${}^{T}\Psi^{(n)}$ obey independent Schroedinger equations with ${}^{S}H^{n}_{|2}$, ${}^{V}H^{n}_{|2}$ and ${}^{T}H^{n}_{|2}$ respectively.

7 The Boundary Conditions

We want to find the solution of the master equation that corresponds to

$$\Psi[h_{ij},\Phi] = \int d[g_{\mu\nu}]d[\Phi]\, \exp(-\hat{I}) \qquad (7.1)$$

where the integral is taken over all compact 4-metrics and matter fields which are bounded by the 3-surface S. If one takes the scale parameter α to be very negative but keeps the other parameters fixed, the Euclidean action $\hat{I}$ tends to zero like $e^{2\alpha}$. Thus one would expect Ψ to tend to one as α tends to minus infinity.

One can estimate the form of the scalar, vector and tensor parts $S_\Psi(n)$, $V_\Psi(n)$, $T_\Psi(n)$ of the perturbation $\Psi^{(n)}$ from the path integral (7.1). One takes the 4-metric $g_{\mu\nu}$ and the scalar field Φ to be of the background form

$$ds^2 = \sigma^2(- N^2dt^2 + e^{2\alpha(t)}d\Omega_3^2) \tag{7.2}$$

and $\phi(t)$ respectively plus a small perturbation described by the variables (a_n, b_n, f_n), c_n and d_n as functions of t. In order for the background 4-metric to be compact, it has to be Euclidean when $\alpha = -\infty$ ie N has to be purely negative imaginary at $\alpha = -\infty$, which we shall take to be $t = 0$. In regions in which the metric is Lorentzian, N will be real and positive. In order to allow a smooth transition from Euclidean to Lorentzian, we shall take N to be of the form $-ie^{i\mu}$ where $\mu = 0$ at $t = 0$. In order that the 4-metric and the scalar field be regular at $t = 0$, a_n, b_n, c_n, d_n, f_n have to vanish there.

The tensor perturbations d_n have the Euclidean action

$${}^T\hat{I}_n = {}^{\prime}\tfrac{1}{2}\int dt\ d_n\ {}^TD\ d_n + \text{boundary term} \tag{7.3}$$

where

$$ {}^{T}D = \left[-\frac{d}{dt}\left(\frac{e^{3\alpha}}{iN_0}\frac{d}{dt}\right) + iN_0 e^{\alpha}(n^2-1)\right] \tag{7.4} $$

$$ + 4iN_0 e^{3\alpha}\left[+ \tfrac{1}{2}e^{-2\alpha} - \tfrac{3}{2}m^2\phi^2 - \frac{3\dot{\phi}^2}{2(iN_0)^2} - \frac{3\dot{\alpha}^2}{2(iN_0)^2} - \frac{1}{iN_0}\frac{d}{dt}\left(\frac{\dot{\alpha}}{iN_0}\right)\right]\Bigg] $$

The last term in (7.4) vanishes if the background metric satisfies the background field equations. The action is extremized when d_n satisfies the equation

$$ {}^{T}D\, d_n = 0 \tag{7.5} $$

For a d_n that satisfies (7.5), the action is just the boundary term

$$ {}^{T}\hat{I}^{cl}_n = \frac{1}{2iN_0}e^{3\alpha}\left[d_n\dot{d}_n + 4\dot{\alpha}d_n^2\right] \tag{7.6} $$

The path integral over d_n will be

$$ \int d[d_n]\, \exp(-\,{}^{T}\hat{I}_n) = (\det {}^{T}D)^{-1/2}\, \exp(-\,{}^{T}\hat{I}^{cl}_n) \tag{7.7} $$

One now has to integrate (7.7) over different background metrics to obtain the wavefunction ${}^{T}\Psi(n)$. One expects the dominant contribution to come from background metrics that are near a solution of the classical background field equations. For such metrics one can employ the adiabatic approximation in which one regards α to be a slowly varying function of t. Then the solution of (7.5) which obeys the boundary condition $d_n = 0$ at $t = 0$ is

$$d_n = A(e^{\nu\tau} - e^{-\nu\tau}) \tag{7.8}$$

where $\nu = e^{-\alpha}(n^2-1)^{1/2}$ and $\tau = \int iN_0\, dt$. This approximation will be valid for background fields which are near a solution of the background field equations and for which

$$\left|\frac{\dot{\alpha}}{N_0}\right| \ll ne^{-\alpha} \tag{7.9}$$

For a regular Euclidean metric, $\left|\frac{\dot{\alpha}}{N_0}\right| = e^{-\alpha}$ near $t = 0$. If the metric is a Euclidean solution of the background field equations, then $\left|\frac{\dot{\alpha}}{N_0}\right| < e^{-\alpha}$. Thus the adiabatic approximation should hold for large values of n into the region in which the solution of the background field equations becomes Lorentzian and the WKB approximation can be used. The wavefunction $T_\Psi(n)$ will then be

$$T_\Psi(n) = B \exp\left[- \left[\tfrac{1}{2} ne^{2\alpha}\coth(\nu\tau) + \frac{2}{iN_0}\dot{\alpha}e^{3\alpha}\right] d_n^2 \right] \tag{7.10}$$

In the Euclidean region, τ will be real and positive. For large values of n, $\coth(\nu\tau) \approx 1$. In the Lorentzian region where the WKB approximation applies, τ will be complex but it will still have a positive real part and $\coth(\nu\tau)$ will still be approximately 1 for large n. Thus

$$T_\Psi(n) = B \exp\left\{ - 2i \frac{\partial S}{\partial \alpha} d_n^2 - \tfrac{1}{2} n e^{2\alpha} d_n^2 \right\} \qquad (7.11)$$

The normalization constant B can be chosen to be 1. Thus, apart from a phase factor, the gravitational wave modes enter the WKB region in their ground state.

We now consider the vector part $V_\Psi(n)$ of the wavefunction. This is pure gauge as the quantities c_n can be given any values by gauge transformations parameterized by the j_n. The freedom to make gauge transformations is reflected quantum mechanically in the constraint

$$e^{-\alpha}\left[\frac{\partial}{\partial c_n} + 4(n^2-4)c_n \frac{\partial}{\partial \alpha}\right]\Psi = 0 \qquad (7.12)$$

One can integrate (7.12) to give

$$\Psi(\alpha, \{c_n\}) = \Psi(\alpha - 2 \sum_n (n^2-4)c_n^2, 0) \qquad (7.13)$$

where the dependence on the other variables has been suppressed. One can also replace $\frac{\partial \Psi}{\partial \alpha}$ by $i\frac{\partial S}{\partial \alpha}\Psi$. One can then solve for $V_\Psi(n)$:

$$V_\Psi(n) = \exp\left\{2i\, (n^2-4)c_n^2 \frac{\partial S}{\partial \alpha}\right\} \qquad (7.14)$$

The scalar perturbation modes a_n, b_n and f_n involve a combination of the behaviour of the tensor and vector perturbations. The scalar part of the action is given in appendix B. The action is extremized by solutions of the classical equations

$$N_0 \frac{d}{dt}\left(e^{3\alpha}\frac{\dot{a}_n}{N_0}\right) + \frac{1}{3}(n^2 - 4)N_0^2 e^{\alpha}(a_n + b_n) + 3e^{3\alpha}(\dot{\phi}\dot{f}_n - N_0^2 m^2 \phi f_n) =$$

$$N_0^2\left[3e^{3\alpha}m^2\phi^2 - \frac{1}{3}(n^2+2)e^{\alpha}\right]g_n + e^{3\alpha}\dot{\alpha}\dot{g}_n - \frac{1}{3}N_0 \frac{d}{dt}\left[e^{2\alpha}\frac{k_n}{N_0}\right] \quad (7.15)$$

$$N_0 \frac{d}{dt}\left(e^{3\alpha}\frac{\dot{b}_n}{N_0}\right) - \frac{1}{3}(n^2 - 1)N_0^2 e^{\alpha}(a_n + b_n) = \frac{1}{3}(n^2 - 1)N_0^2 e^{\alpha} g_n$$

$$+ \frac{1}{3}N_0 \frac{d}{dt}\left[e^{2\alpha}\frac{k_n}{N_0}\right] \quad (7.16)$$

$$N_0 \frac{d}{dt}\left(e^{3\alpha}\frac{f_n}{N_0}\right) + 3e^{3\alpha}\dot{\phi}\dot{a}_n + N_0^2\left[m^2 e^{3\alpha} + (n^2-1)e^{\alpha}\right]f_n =$$

$$e^{3\alpha}\left[- 2N_0^2 m^2 \phi g_n + \phi \dot{g}_n - e^{-\alpha}\dot{\phi}k_n\right] \quad (7.17)$$

There is a three parameter family of solutions to (7.15) to (7.17) which obey the boundary condition $a_n = b_n = f_n = 0$ at $t = 0$. There are however, two constraint equations:

$$\dot{a}_n + \frac{(n^2-4)}{(n^2-1)}\dot{b}_n + 3f_n\dot{\phi} = \dot{\alpha}g_n - \frac{e^{-\alpha}}{(n^2-1)} k_n \qquad (7.18)$$

$$3a_n (- \dot{\alpha}^2 + \dot{\phi}^2) + 2(\dot{\phi}\dot{f}_n - \dot{\alpha}\dot{a}_n)$$

$$+ N_0^2 m^2(2f_n\phi + 3a_n\phi^2) - \tfrac{2}{3}N_0^2 e^{-2\alpha}\left[(n^2-4)b_n + (n^2 + \tfrac{1}{2})a_n\right]$$

$$= \tfrac{2}{3}\dot{\alpha}e^{-\alpha}k_n + 2g_n(- \dot{\alpha}^2 + \dot{\phi}^2) \qquad (7.19)$$

These correspond to the two gauge degrees of freedom parameterized by k_n and g_n respectively. The Euclidean action for a solution to equations (7.15) to (7.19) is

$$\hat{S}^{cl}_{I_n} = \frac{1}{2iN_0}e^{3\alpha}\left[- a_n\dot{a}_n + \frac{(n^2-4)}{(n^2-1)}b_n\dot{b}_n + f_n\dot{f}_n\right.$$

$$+ \dot{\alpha}\left[-a_n^2 + \frac{4(n^2-4)}{(n^2-1)}b_n^2\right] + 3\dot{\phi}a_n f_n$$

$$\left. + g_n\left[\dot{\alpha}a_n - \dot{\phi}f_n\right] - \tfrac{1}{3}e^{-\alpha}k_n\left[a_n + \frac{(n^2-4)}{(n^2-1)}b_n\right]\right] \qquad (7.20)$$

where the background field equations have been used.

In many ways the simplest gauge to work in is that with $g_n = k_n = 0$. However, this gauge does not allow one to find a compact 4-metric which is bounded by a 3-surface with arbitrary values of a_n, b_n and f_n and which is a solution of the equations (7.15) to

(7.17) and the constraint equations. Instead, we shall use the gauge $a_n = b_n = 0$ and shall solve the constraint equations (7.18) and (7.19) to find g_n and k_n:

$$g_n = 3\frac{\left[(n^2-1)\dot{\alpha}\dot{\phi}f_n + \ddot{\phi}f_n + N_0^2m^2\phi f_n\right]}{\left[(n^2-4)\dot{\alpha}^2 + 3\dot{\phi}^2\right]} \tag{7.21}$$

$$k_n = 3(n^2-1)e^{\alpha}\frac{\left[\dot{\alpha}\ddot{\phi}f_n + N_0^2m^2\phi f_n\dot{\alpha} - 3f_n\dot{\phi}(-\dot{\alpha}^2 + \dot{\phi}^2)\right]}{\left[(n^2-4)\dot{\alpha}^2 + 3\dot{\phi}^2\right]} \tag{7.22}$$

With these substituted, (7.17) becomes a second order equation for f_n

$$N_0\frac{d}{dt}\left[e^{3\alpha}\frac{\dot{f}_n}{N_0}\right] + N_0^2\left[m^2e^{3\alpha} + (n^2-1)e^{\alpha}\right]f_n =$$

$$e^{3\alpha}\left[-2N_0^2m^2\phi g_n + \dot{\phi}\dot{g}_n - e^{-\alpha}\dot{\phi}k_n\right] \tag{7.23}$$

For large n we can again use the adiabatic approximation to estimate the solution of (7.23) when $|\phi| > 1$:

$$f_n = A\sinh(\nu\tau) \tag{7.24}$$

where $\nu^2 = e^{-2\alpha}(n^2-1)$. Thus for these modes

$$S_{\Psi}^{(n)}(\alpha,\phi,\ 0\ ,\ 0\ ,f_n) \approx \exp\left[-\ ½ n e^{2\alpha} f_n^2 - ½ i\frac{\partial S}{\partial \phi} g_n f_n\right] \quad (7.25)$$

This is of the ground state form apart from a small phase factor. The value of $S_{\Psi}^{(n)}$ at non-zero values of a_n and b_n can be found by integrating the constraint equations (5.25) and (5.27).

The tensor and scalar modes start off in their ground states, apart possibly from the modes at low n. The vector modes are pure gauge and can be neglected. Thus the total energy $E = \sum_n \frac{H_{|2}^{(n)}\Psi^{(n)}}{\Psi^{(n)}}$ of the perturbations will be small when the ground state energies are subtracted. But $E = i(\nabla_2 S).J$ where $J = \sum_n \frac{\nabla_2 \Psi^{(n)}}{\Psi^{(n)}}$. Thus J is small. This means that the wavefunction Ψ_0 will obey the Wheeler-DeWitt equation of the unperturbed minisuperspace model and the phase factor S will be approximately $-\ i\ln\Psi_0$. However the homogeneous scalar field mode ϕ will not start out in its ground state. There are two reasons for this: first, regularity at $t = 0$ requires $a_n = b_n = c_n = d_n = f_n = 0$, but does not require $\phi = 0$. Second, the classical field equation for ϕ is of the form a damped harmonic oscillator with a constant frequency m rather than a decreasing frequency $e^{-\alpha}n$. This means that the adiabatic approximation is not valid at small t and that the solution of the classical field equation is ϕ approximately constant. The action of such solutions is small, so large values of $|\phi|$ are not damped as they are for the other variables.

Thus the WKB trajectories which start out from large values of $|\phi|$ have high probability. They will correspond to classical solutions which have a long inflationary period and then go over to, a matter dominated expansion. In a realistic model which included other fields of low rest mass, the matter energy in the oscillations of the massive scalar field would decay into light particles with a thermal spectrum. The model would then expand as a radiation dominated universe.

8 Growth of Perturbations

The tensor modes will obey the Schroedinger equation

$$i\frac{\partial T_\Psi(n)}{\partial t} = T_{H}{}^{n}_{|2}\, T_\Psi(n) \tag{8.1}$$

$$= \tfrac{1}{2}e^{-3\alpha}\left[+ d_n^2\left[10\left(\frac{\partial S}{\partial\alpha}\right)^2 + 6\left(\frac{\partial S}{\partial\phi}\right)^2\right] - \frac{\partial^2}{\partial d_n{}^2} - 8d_n i\frac{\partial S}{\partial\alpha}\frac{\partial}{\partial d_n} + \right.$$

$$\left. d_n^2\left[(n^2+1)e^{4\alpha} - 6e^{6\alpha}m^2\phi^2\right]\right] \tag{8.2}$$

One can write

$$T_\Psi(n) = \exp(-2\alpha)\,\exp\left[-2i\,\frac{\partial S}{\partial\alpha}\,d_n^2\right]\,T_{\Psi_0}(n) \tag{8.3}$$

then

$$i\frac{\partial T_{\Psi_0}(n)}{\partial t} = \tfrac{1}{2}e^{-3\alpha}\left[-\frac{\partial^2}{\partial d_n^2} + d_n^2(n^2-1)e^{4\alpha}\right]T_{\Psi_0}(n) \tag{8.4}$$

The WKB approximation to the background Wheeler-DeWitt equation has been used in deriving (8.4). Then (8.4) has the form of the Schroedinger equation for an oscillator with a time dependent frequency $\nu = (n^2-1)^{1/2}e^{-\alpha}$. Initially the wavefunction ${}^{T}\Psi_0(n)$ will be in the ground state (apart from a normalization factor) and the frequency ν will be large compared to $\dot{\alpha}$. In this case one can use the adiabatic approximation to show that ${}^{T}\Psi_0(n)$ remains in the ground state

$$ {}^{T}\Psi_0(n) \approx \exp\left[- \tfrac{1}{2} n e^{2\alpha} d_n^2 \right] \qquad (8.5) $$

The adiabatic approximation will break down when $\nu \approx \dot{\alpha}$ ie the wavelength of the gravitational mode becomes equal to the horizon scale in the inflationary period. The wavefunction ${}^{T}\Psi_0(n)$ will then " freeze "

$$ {}^{T}\Psi_0(n) \approx \exp\left[- \tfrac{1}{2} n e^{2\alpha_*} d_n^2 \right] \qquad (8.6) $$

where α_* is the value of α at which the mode goes outside the horizon. The wavefunction ${}^{T}\Psi_0(n)$ will remain of the form (8.6) until the mode re-enters the horizon in the matter or radiation dominated era at the much greater value α_e of α. One can then apply the adiabatic approximation again to (8.4) but ${}^{T}\Psi_0(n)$ will no longer be in the ground state: it will be a superposition of a number of highly excited states. This is the phenomenon of the amplification of the ground state fluctuations in the gravitational wave modes that was discussed in references [9,17,18].

The behaviour of the scalar modes is rather similar but their description is more complicated because of the gauge degrees of freedom. In the previous section we evaluated the wavefunction $S_{\Psi}(n)$ on $a_n = b_n = 0$ by the path integral prescription. The ground state form (in f_n) that we found will be valid until the adiabatic approximation breaks down ie until the wavelength of the mode excedes the horizon distance during the inflationary period. In order to discuss the subsequent behaviour of the wavefunction, it is convenient to use the first order Hamiltonian constraint (5.27) to evaluate $S_{\Psi}(n)$ on $a_n \neq 0, b_n = f_n = 0$. One finds that

$$S_{\Psi}(n)(\alpha,\phi,a_n, 0 , 0) = B \exp\left[iCa_n^2\right] S_{\Psi_0}(n)(\alpha,\phi,a_n) \quad (8.7)$$

The normalization and phase factors B and C depend on α and ϕ but not a_n.

$$C = \frac{1}{2}\left[\frac{\partial S}{\partial\alpha}\right]^{-1}\left[\left[\frac{\partial S}{\partial\alpha}\right]^2 - \frac{1}{3}(n^2-4)e^{4\alpha}\right] \quad (8.8)$$

At the time the wavelength of the mode equals the horizon distance during the inflationary period, the wavefunction $S_{\Psi_0}(n)$ has the form

$$S_{\Psi_0}(n) = \exp\left[- ½\, n\, y_*^{-2}\, e^{2\alpha_*}\, a_n^2\right] \quad (8.9)$$

where y_* is the value of $y = \frac{\partial S}{\partial\alpha}\left[\frac{\partial S}{\partial\phi}\right]^{-1}$ when the mode leaves the horizon, $y_* = 3\phi_*$. More generally, in the case of a scalar field with a potential $V(\phi)$, $y = 6V\left(\frac{\partial V}{\partial \phi}\right)^{-1}$.

One can obtain a Schroedinger equation for ${}^{S}\Psi_{0}^{(n)}$ by putting $b_n = f_n = 0$ in the scalar Hamiltonian ${}^{S}H^{n}_{|2}$ and substituting for $\frac{\partial}{\partial b_n}$ and $\frac{\partial}{\partial f_n}$ from the momentum constraint (5.25) and the first order Hamiltonian constraint (5.27) respectively. This gives

$$i\frac{\partial\, {}^{S}\Psi_{0}^{(n)}}{\partial t} = \tfrac{1}{2}e^{-3\alpha}\left\{ - y^2 \frac{\partial^2}{\partial a_n^2} + e^{4\alpha}(n^2-4)\left[\frac{1}{y^2} - \frac{1}{3}e^{4\alpha}\left(\frac{\partial S}{\partial \alpha}\right)^{-2}\right] a_n^2 \right\} {}^{S}\Psi_{0}^{(n)} \qquad (8.10)$$

where terms of order $\frac{1}{n^2}$ have been neglected. The term $e^{4\alpha}\left(\frac{\partial S}{\partial \alpha}\right)^{-2}$ will be small compared to $\frac{1}{y^2}$ except near the time of maximum radius of the background solution. The Schroedinger equation for ${}^{S}\Psi_{0}^{(n)}(a_n)$ is very similar to the equation for ${}^{T}\Psi_{0}^{(n)}(d_n)$, (8.4), except that the kinetic term is multiplied by a factor y^2 and the potential term is divided by a factor y^2. One would therefore expect that for wavelengths within the horizon, ${}^{S}\Psi_{0}^{(n)}$ would have the ground state form $\exp(-\tfrac{1}{2}ny^{-2}e^{2\alpha}a_n^2)$ and this is bourne out by (8.9). On the other hand, when the wavelength becomes larger than the horizon, the Schroedinger equation (8.10) indicates that ${}^{T}\Psi_{0}^{(n)}$ will freeze in the form (8.9) until the mode re-enters the horizon in the matter dominated era. Even if the equation of state of the Universe changes to radiation dominated during the period that the wavelength of the mode is greater than the horizon size, it will still be true that ${}^{S}\Psi_0^{(n)}$ is frozen in the form (8.9). The ground state fluctuations in the scalar modes will therefore be amplified in a similar manner to the tensor modes. At the time of re-entry of the horizon the rms fluctuation in the scalar modes, in the gauge in which $b_n = f_n = 0$, will be greater by the factor y_* than the rms fluctuation in the tensor modes of the same wavelength.

9 Comparison with Observation

From a knowledge of ${}^{T}\Psi_0^{(n)}$ and ${}^{S}\Psi_0^{(n)}$ one can calculate the relative probabilties of observing different values of d_n and a_n at a given point on a trajectory of the vector field X^i ie at a given value of α and ϕ in a background metric which is a solution of the classical field equations. In fact, the dependence on ϕ will be unimportant and we shall neglect it. One can then calculate the probabilities of observing different amounts of anisotropy in the microwave background and can compare these predictions with the upper limits set by observation.

The tensor and scalar perturbation modes will be in highly excited states at large values of α. This means that we can treat their development as an ensemble evolving according to the classical equations of motion with initial distributions in d_n and a_n proportional to $|{}^{T}\Psi_0^{(n)}|^2$ and $|{}^{S}\Psi_0^{(n)}|^2$ respectively. The initial distributions in $\dot{d}_n$ and $\dot{a}_n$ will be proportional to $|{}^{T}\Psi_0^{(n)}\pi_{d_n}$ TPno$|$ and $|{}^{S}\Psi_0^{(n)}\pi_{a_n}{}^{S}\Psi_0^{(n)}|$ respectively. In fact, at the time that the modes re-enter the horizon, the distributions will be concentrated at $\dot{d}_n = \dot{a}_n = 0$.

The surfaces with $b_n = f_n = 0$ will be surfaces of constant energy density in the classical solution during the inflationary period. By local conservation of energy, they will remain surfaces of constant energy density in the era after the inflationary period when the energy is dominated by the coherent oscillations of the homogeneous background scalar field ϕ. If the scalar particles decay into light particles and heat up the universe, the surfaces with $b_n = f_n = 0$ will be surfaces of constant temperature. The surface of last scattering of the microwave background will be such a surface with temperature T_s. The microwave radiation can be considered to have propagated freely to us from this surface. Thus the observed temperature will be

$$T_o = \frac{T_s}{1 + z} \tag{9.1}$$

where z is the redshift of the surface of last scattering. Variations in the observed temperature will arise from variations in z in different directions of observation. These are given by

$$1 + z = \ell^{\mu} n_{\mu} \tag{9.2}$$

evaluated at the surface of last scattering where n_{μ} is the unit normal to the surfaces of constant t in the gauge $g_n = k_n = j_n = 0$ and $b_n = f_n = 0$ on the surface of last scattering and ℓ^{μ} is the parallelly propagated tangent vector to the null geodesic from the observer normalized by $\ell^{\mu} n_{\mu} = 1$ at the present time. One can calculate the evolution of $\ell^{\mu} n_{\mu}$ down the past light cone of the observer:

$$\frac{d}{d\lambda}\left[\ell^{\mu}n_{\mu}\right] = n_{\mu;\nu}\ell^{\mu}\ell^{\nu} \tag{9.3}$$

where λ is the affine parameter on the null geodesic. The only non-zero components of $n_{\mu;\nu}$ are

$$n_{i;j} = e^{2\alpha}\left[\dot{\alpha}\Omega_{ij} + \sum_{n}(\dot{a}_n + \dot{\alpha}a_n)\tfrac{1}{3}\Omega_{ij}Q + \sum_{n}(\dot{b}_n + \dot{\alpha}b_n)P_{ij}\right.$$

$$\left. + \sum_{n}(\dot{d}_n + \dot{\alpha}d_n)G_{ij}\right] \tag{9.4}$$

In the gauge that we are using, the dominant anisotropic terms in (9.4) on the scale of the horizon, will be those involving $\dot{\alpha}a_n$ and $\dot{\alpha}d_n$. These will give temperature anisotropies of the form

$$\langle(\Delta T/T)^2\rangle \approx \langle a_n^2\rangle \quad \text{or} \quad \approx \langle d_n^2\rangle \tag{9.5}$$

The number of modes that contribute to anisotropies on the scale of the horizon is of the order of n^3. From the results of the last section

$$\langle a_n^2\rangle = y_*^2 n^{-1} e^{-2\alpha_*} \tag{9.6}$$

$$\langle d_n^2\rangle = n^{-1} e^{-2\alpha_*} \tag{9.7}$$

The dominant contribution comes from the scalar modes which give

$$\langle (\Delta T/T)^2 \rangle \approx y_*^2 n^2 e^{-2\alpha_*} \tag{9.8}$$

But $ne^{-\alpha_*} \approx \dot{\alpha}_*$, the value of the Hubble constant at the time that the present horizon size left the horizon during the inflationary period. The observational upper limit of about 10^{-8} on $\langle (\Delta T/T)^2 \rangle$ restricts this Hubble constant to be less than about $5.10^{-5} m_p$ [8] which in turn restricts the mass of the scalar field to be less than 10^{14} GeV .

10 Conclusion snd Summary

We started from the proposal that the quantum state of the Universe is defined by a path integral over compact 4-metrics. This can be regarded as a boundary condition for the Wheeler-DeWitt equation for the wavefunction of the Universe on the infinite dimensional manifold, superspace, the space of all 3-metrics and matter field configurations on a 3-surface S. Previous papers had considered finite dimensional approximations to superspace and had shown that the boundary condition led to a wavefunction which could be interpreted as corresponding to a family of classical solutions which were homogeneous and isotropic and which had a period of exponential or inflationary expansion. In the present paper we extended this work to the full superspace without restrictions. We treated the two basic homogeneous and isotropic degrees of freedom exactly and the other degrees of freedom to second order. We justified this approximation by showing that the inhomogeneous or anisotropic modes started out in their ground states.

We derived time dependent Schroedinger equations for each mode. We showed that they remained in the ground state until their wavelength exceded the horizon size during the inflationary period. In the subsequent expansion the ground state fluctuations got frozen until the wavelength re-entered the horizon during the radiation or matter dominated era. This part of the calculation is similar to earlier work on the development of gravitational waves [9] and density perturbations [5,6] in the inflationary universe but it has the advantage that the assumptions of a period of exponential expansion and of an initial ground state for the perturbations are justified. The perturbations would be compatible with the upper limits set by observations of the microwave background if the scalar field that drives the inflation has a mass of 10^{14} GeV or less.

In section 8 we calculated the scalar perturbations in a gauge in which the surfaces of constant time are surfaces of constant density. There are thus no density fluctuations in this gauge. However, one can make a transformation to a gauge in which $a_n = b_n = 0$. In this gauge the density fluctuation at the time that the wavelength comes within the horizon is

$$\langle(\Delta\rho/\rho)^2\rangle \approx y^2 \frac{\dot{\rho}_e^2}{\dot{\alpha}_e^2 \rho_e^2} \dot{\alpha}_*^2 \qquad (10.1)$$

Because y and $\dot{\alpha}_*$ depend only logarithmically on the wavelength of the perturbations, this gives an almost scale free spectrum of density fluctuations. These fluctuations can evolve according to the classical field equations to give rise to the formation of galaxies and all the other structure that we observe in the Universe. Thus all the complexities of the present state of the Universe have their origin in the ground state fluctuations in the inhomogeneous modes and so arise from the Heisenberg Uncertainty Principle.

Appendix A: Harmonics on the 3-sphere

In this appendix we describe the properties of the scalar, vector and tensor harmonics on the 3-sphere S^3. The metric on S^3 is Ω_{ij} and so the line element is

$$d\ell^2 = \Omega_{ij}dx^i dx^j = d\chi^2 + \sin^2\chi(d\theta^2 + \sin^2\theta d\phi^2) \qquad (A1)$$

A vertical stroke will denote covariant differentiation with respect to the metric Ω_{ij}. Indices i,j,k are raised and lowered using Ω_{ij}.

Scalar Harmonics

The scalar spherical harmonics $Q^n_{\ell m}(\chi,\theta,\phi)$ are scalar eigenfunctions of the Laplacian operator on S^3. Thus, they satisfy the eigenvalue equation

$$Q^{(n)}{}_{|k}{}^{|k} = -(n^2-1)\,Q^{(n)} \qquad n = 1,2,3... \qquad (A2)$$

The most general solution to (A2), for given n, is a sum of solutions

$$Q^{(n)}(\chi,\theta,\phi) = \sum_{\ell=0}^{n-1}\sum_{m=-\ell}^{\ell} A^n_{\ell m}\, Q^n_{\ell m}(\chi,\theta,\phi) \qquad (A3)$$

where $A^n_{\ell m}$ are a set of arbitrary constants. The $Q^n_{\ell m}$ are given explicitly by

$$Q^n_{\ell m}(\chi,\theta,\phi) = \Pi^n_\ell(\chi) Y_{\ell m}(\theta,\phi) \qquad \text{(A4)}$$

where $Y_{\ell m}(\theta,\phi)$ are the usual harmonics on the 2-sphere, S^2, and $\Pi^n_\ell(\chi)$ are the Fock harmonics [19,20]. The spherical harmonics $Q^n_{\ell m}$ constitute a complete orthogonal set for the expansion of any scalar field on S^3.

Vector Harmonics

The transverse vector harmonics $(S_i)^n_{\ell m}(\chi,\theta,\phi)$ are vector eigenfunctions of the Laplacian operator on S^3 which are transverse. That is, they satisfy the eigenvalue equation

$$S^{(n)}_{i\ \ |k}{}^{|k} = -(n^2 - 2)\, S^{(n)}_i \qquad n = 2,3,4\ldots \qquad \text{(A5)}$$

and the transverse condition

$$S^{(n)|i}_i = 0 \qquad \text{(A6)}$$

The most general solution to (A5) and (A6) is a sum of solutions

$$S^{(n)}_i(\chi,\theta,\phi) = \sum_{\ell=1}^{n-1} \sum_{m=-\ell}^{\ell} B^n_{\ell m}\, (S_i)^n_{\ell m}(\chi,\theta,\phi) \qquad \text{(A7)}$$

where $B^n_{\ell m}$ are a set of arbitrary constants. Explicit expressions for the $(S_i)^n_{\ell m}$ are given in reference [20] where it is also explained how they are classified as odd (o) or even (e) using a parity transformation. We thus have two linearly independent transverse vector harmonics S^o_i and S^e_i (n, ℓ, m suppressed).

Using the scalar harmonics $Q^n_{\ell m}$ we may construct a third vector harmonic $(P_i)^n_{\ell m}$, defined by (n, ℓ, m suppressed)

$$P_i = \frac{1}{(n^2 - 1)} Q_{|i} \quad n = 2,3,4... \tag{A8}$$

It may be shown to satisfy

$$P_{i|k}{}^{|k} = -(n^2 - 3) P_i \quad \text{and} \quad P_i{}^{|i} = -Q \tag{A9}$$

The three vector harmonics S^o_i, S^e_i and P_i constitute a complete orthogonal set for the expansion of any vector field on S^3.

Tensor Harmonics

The transverse traceless tensor harmonics $(G_{ij})^n_{\ell m}(\chi,\theta,\phi)$ are tensor eigenfunctions of the Laplacian operator on S^3 which are transverse and traceless. That is, they satisfy the eigenvalue equation

$$G^{(n)}_{ij\,|k}{}^{|k} = -(n^2 - 3)\, G^{(n)}_{ij} \quad n = 3,4,5... \tag{A10}$$

and the transverse and traceless conditions

$$G^{(n)|i}_{ij} = 0 \quad , \quad G^{(n)i}_{i} = 0 \tag{A11}$$

The most general solution to (A11) and (A12) is a sum of solutions

$$G^{(n)}_{ij}(\chi,\theta,\phi) = \sum_{\ell=2}^{n-1} \sum_{m=-\ell}^{\ell} C^{n}_{\ell m} \, (G_{ij})^{n}_{\ell m}(\chi,\theta,\phi) \tag{A12}$$

where $C^{n}_{\ell m}$ are a set of arbitrary constants. As in the vector case they may be classified as odd or even. Explicit expressions for $(G^{o}_{ij})^{n}_{\ell m}$ and $(G^{e}_{ij})^{n}_{\ell m}$ are given in reference [20].

Using the transverse vector harmonics $(S^{o}_{i})^{n}_{\ell m}$ and $(S^{e}_{i})^{n}_{\ell m}$, we may construct traceless tensor harmonics $(S^{o}_{ij})^{n}_{\ell m}$ and $(S^{e}_{ij})^{n}_{\ell m}$ defined, both for odd and even, by (n, ℓ, m suppressed)

$$S_{ij} = S_{i|j} + S_{j|i} \tag{A13}$$

and thus $S_{i}{}^{i} = 0$ since S_{i} is transverse. In addition, the S_{ij} may be shown to satisfy

$$S_{ij}{}^{|j} = -\,(n^2 - 4)\, S_{i} \tag{A14}$$

$$S_{ij}{}^{|ij} = 0 \tag{A15}$$

$$S_{ij|k}{}^{|k} = -\,(n^2 - 6)\, S_{ij} \tag{A16}$$

Using the scalar harmonics $Q^n_{\ell m}$, we may construct two tensors $(Q_{ij})^n_{\ell m}$ and $(P_{ij})^n_{\ell m}$ defined by $(n, \ell, m$ suppressed)

$$Q_{ij} = \tfrac{1}{3}\Omega_{ij}Q \qquad n = 1,2,3 \tag{A17}$$

and
$$P_{ij} = \frac{1}{(n^2 - 1)} Q_{|ij} + \tfrac{1}{3}\Omega_{ij}Q \qquad n = 2,3,4 \tag{A18}$$

The P_{ij} are traceless $P_i^{\ i} = 0$, and in addition, may be shown to satisfy

$$P_{ij}^{\ \ |j} = -\tfrac{2}{3}(n^2 - 4)\, P_i \tag{A19}$$

$$P_{ij|k}^{\ \ \ \ |k} = -(n^2 - 7)\, P_{ij} \tag{A20}$$

$$P_{ij}^{\ \ |ij} = \tfrac{2}{3}(n^2 - 4)\, Q \tag{A21}$$

The six tensor harmonics $Q_{ij}, P_{ij}, S^o_{ij}, S^e_{ij}, G^o_{ij}$ and G^e_{ij} constitute a complete orthogonal set for the expansion of any symmetric second rank tensor field on S^3.

Orthogonality and Normalization

The normalization of the scalar, vector and tensor harmonics is fixed by the orthogonality relations. We denote the integration measure on S^3 by $d\mu$. Thus

$$d\mu = d^3x \ (\det\Omega_{ij})^{1/2} = \sin^2\chi \ \sin\theta \ d\chi d\theta d\phi \tag{A22}$$

The $Q^n_{\ell m}$ are normalized so that

$$\int d\mu \ Q^n_{\ell m} \ Q^{n'}_{\ell' m'} = \delta^{nn'} \ \delta_{\ell\ell'} \ \delta_{mm'} \tag{A23}$$

This implies

$$\int d\mu \ (P_i)^n_{\ell m} \ (P^i)^{n'}_{\ell' m'} = \frac{1}{(n^2 - 1)} \ \delta^{nn'} \ \delta_{\ell\ell'} \ \delta_{mm'} \tag{A24}$$

and

$$\int d\mu \ (P_{ij})^n_{\ell m} \ (P^{ij})^{n'}_{\ell' m'} = \frac{2(n^2 - 4)}{3(n^2 - 1)} \ \delta^{nn'} \ \delta_{\ell\ell'} \ \delta_{mm'} \tag{A25}$$

The $(S_i)^n_{\ell m}$, both odd and even, are normalized so that

$$\int d\mu \ (S_i)^n_{\ell m} \ (S^i)^{n'}_{\ell' m'} = \delta^{nn'} \ \delta_{\ell\ell'} \ \delta_{mm'} \tag{A26}$$

This implies

$$\int d\mu \ (S_{ij})^n_{\ell m} \ (S^{ij})^{n'}_{\ell' m'} = 2(n^2 - 4) \ \delta^{nn'} \ \delta_{\ell\ell'} \ \delta_{mm'} \tag{A27}$$

Finally, the $(G_{ij})^n_{\ell m}$, both odd and even, are normalized so that

$$\int d\mu \ (G_{ij})^n_{\ell m} \ (G^{ij})^{n'}_{\ell' m'} = \delta^{nn'} \ \delta_{\ell\ell'} \ \delta_{mm'} \tag{A28}$$

The information given in this appendix about the spherical harmonics is all that is needed to perform the derivations presented in the main text. Further details may be found in references [19,20].

Appendix B

The action (5.8) is

$$I = I_0(\alpha, \phi, N_0) + \sum_n I_n \qquad (B1)$$

where I_0 is the action of the unperturbed model (4.2):

$$I_0 = -\tfrac{1}{2}\int dt\ N_0 e^{3\alpha}\left[\frac{\dot{\alpha}^2}{N_0^2} - e^{-2\alpha} - \frac{\dot{\phi}^2}{N_0^2} + m^2\phi^2\right] \qquad (B2)$$

I_n is quadratic in the perturbations and may be written

$$I_n = \int dt(L_g^n + L_m^n) \qquad (B3)$$

where

$$L_g^n = \tfrac{1}{2}e^{\alpha}N_0\left\{\tfrac{1}{3}(n^2-\tfrac{5}{2})a_n^2 + \frac{(n^2-7)}{3}\frac{(n^2-4)}{(n^2-1)}b_n^2 - 2(n^2-4)c_n^2 - (n^2+1)d_n^2 + \tfrac{2}{3}(n^2-4)a_n b_n\right.$$

$$\left. g_n\left[\tfrac{2}{3}(n^2-4)b_n + \tfrac{2}{3}(n^2+\tfrac{1}{2})a_n\right] + \frac{1}{N_0^2}\left[-\frac{1}{3(n^2-1)}k_n^2 + (n^2-4)j_n^2\right]\right\}$$

$$+ \tfrac{1}{2}\frac{e^{3\alpha}}{N_0}\Bigg[- \dot{a}_n^2 + \frac{(n^2-4)}{(n^2-1)}\dot{b}_n^2 + (n^2 - 4)\dot{c}_n^2 + \dot{d}_n^2$$

$$+ \dot{\alpha}\left[- 2a_n\dot{a}_n + 8\frac{(n^2-4)}{(n^2-1)}b_n\dot{b}_n + 8(n^2 - 4)c_n\dot{c}_n + 8d_n\dot{d}_n\right]$$

$$+ \dot{\alpha}^2\left[- \frac{3}{2}a_n^2 + 6\frac{(n^2-4)}{(n^2-1)}b_n^2 + 6(n^2 - 4)c_n^2 + 6d_n^2\right]$$

$$+ g_n\left[2\dot{\alpha}\dot{a}_n + \dot{\alpha}^2(3a_n - g_n)\right]$$

$$+ e^{-\alpha}\left[k_n\left[- \frac{2}{3}\dot{a}_n - \frac{2(n^2-4)}{3(n^2-1)}\dot{b}_n + \frac{2}{3}\dot{\alpha}g_n\right] - 2(n^2-4)\dot{c}_n j_n\right]\Bigg] \quad (B4)$$

and

$$L_m^n = \tfrac{1}{2}N_0 e^{3\alpha}\Bigg[\frac{1}{N_0^2}\left[\dot{f}_n^2 + 6a_n\dot{f}_n\dot{\phi}\right] - m^2\left[f_n^2 + 6a_n f_n\phi\right] - e^{-2\alpha}(n^2-1)f_n^2$$

$$+ \frac{3}{2}\left[\frac{\dot{\phi}^2}{N_0^2} - m^2\phi^2\right]\left[a_n^2 - \frac{4(n^2-4)}{(n^2-1)}b_n^2 - 4(n^2-4)c_n^2 - 4d_n^2\right] + \frac{\dot{\phi}^2}{N_0^2}g_n^2$$

$$- g_n\left[2m^2 f_n\phi + 3m^2 a_n\phi^2 + 2\frac{\dot{f}_n\dot{\phi}}{N_0^2} + 3\frac{a_n\dot{\phi}^2}{N_0^2}\right] - 2\frac{e^{-\alpha}}{N_0^2}k_n f_n\dot{\phi}\Bigg] \quad (B5)$$

The full expressions for π_α and π_ϕ are

$$\pi_\alpha = \frac{e^{3\alpha}}{N_0}\left[-\dot{\alpha} + \sum_n \left[-a_n\dot{a}_n + \frac{4(n^2-4)}{(n^2-1)}b_n\dot{b}_n + 4(n^2-4)c_n\dot{c}_n + 4d_n\dot{d}_n\right]\right.$$

$$\dot{\alpha}\sum_n \left[-\frac{3}{2}a_n^2 + \frac{6(n^2-4)}{(n^2-1)}b_n^2 + 6(n^2-4)c_n^2 + 6d_n^2\right]$$

$$\left.\sum_n g_n\left[\dot{a}_n + \dot{\alpha}(3a_n - g_n) + \frac{1}{3}e^{-\alpha}k_n\right]\right] \qquad (B7)$$

$$\pi_\phi = \frac{e^{3\alpha}}{N_0}\left[\dot{\phi} + \sum_n \left[3a_n\dot{f}_n + \dot{\phi}\left[\frac{3}{2}a_n^2 - \frac{4(n^2-4)}{(n^2-1)}b_n^2 - 4(n^2-4)c_n^2 - 4d_n^2\right]\right] +\right.$$

$$\left.\sum_n \left[\dot{\phi}g_n^2 - g_n(\dot{f}_n + 3a_n\dot{\phi}) - e^{-\alpha}k_n f_n\right]\right] \qquad (B7)$$

The classical field equations may be obtained from the action (B1) by varying with respect to each of the fields in turn. Variation with respect to α and ϕ gives two field equations, similar to those obtained in section 4, but modified by terms quadratic in the perturbations:

$$N_0\frac{d}{dt}\left[\frac{1}{N_0}\frac{d\phi}{dt}\right] + 3\frac{d\alpha}{dt}\frac{d\phi}{dt} + N_0^2m^2\phi = \text{quadratic terms} \qquad (B8)$$

$$N_0\frac{d}{dt}\left[\frac{\dot{\alpha}}{N_0}\right] + 3\dot{\phi}^2 - N_0^2e^{-2\alpha}$$

$$-\frac{3}{2}\left[-\dot{\alpha}^2 + \dot{\phi}^2 - N_0^2e^{-2\alpha} + N_0^2m^2\phi^2\right] = \text{quadratic terms} \qquad (B9)$$

Variation with respect to the perturbations a_n, b_n, c_n, d_n and f_n leads to five field equations:

$$N_0 \frac{d}{dt}\left(e^{3\alpha}\frac{\dot{a}_n}{N_0}\right) + \frac{1}{3}(n^2 - 4)N_0^2 e^{\alpha}(a_n + b_n) + 3e^{3\alpha}(\dot{\phi}\dot{f}_n - N_0^2 m^2 \phi f_n) =$$

$$N_0^2\left[3e^{3\alpha}m^2\phi^2 - \frac{1}{3}(n^2+2)e^{\alpha}\right]g_n + e^{3\alpha}\dot{\alpha}\dot{g}_n - \frac{1}{3}N_0 \frac{d}{dt}\left[e^{2\alpha}\frac{k_n}{N_0}\right] \quad (B10)$$

$$N_0 \frac{d}{dt}\left(e^{3\alpha}\frac{\dot{b}_n}{N_0}\right) - \frac{1}{3}(n^2 - 1)N_0^2 e^{\alpha}(a_n + b_n) = \frac{1}{3}(n^2 - 1)N_0^2 e^{\alpha} g_n$$

$$+ \frac{1}{3}N_0 \frac{d}{dt}\left[e^{2\alpha}\frac{k_n}{N_0}\right] \quad (B11)$$

$$\frac{d}{dt}\left(e^{3\alpha}\frac{\dot{c}_n}{N_0}\right) = \frac{d}{dt}\left[e^{2\alpha}\frac{j_n}{N_0}\right] \quad (B12)$$

$$N_0 \frac{d}{dt}\left(e^{3\alpha}\frac{\dot{d}_n}{N_0}\right) + (n^2 - 1)N_0^2 e^{\alpha} d_n = 0 \quad (B13)$$

$$N_0 \frac{d}{dt}\left(e^{3\alpha}\frac{\dot{f}_n}{N_0}\right) + 3e^{3\alpha}\dot{\phi}\dot{a}_n + N_0^2\left[m^2 e^{3\alpha} + (n^2-1)e^{\alpha}\right]f_n =$$

$$e^{3\alpha}\left[- 2N_0^2 m^2 \phi g_n + \dot{\phi}\dot{g}_n - e^{-\alpha}\dot{\phi}k_n\right] \quad (B14)$$

In obtaining (B10) – (B14), the field equations (B8) and (B9) has been used and terms cubic in the perturbations have beon droppped.

Variation with respect to the Lagrange multipliers k_n, j_n, g_n and N_0 leads to a set of constraints. Variation with respect to k_n and j_n leads to the momentum constraints:

$$\dot{a}_n + \frac{(n^2-4)}{(n^2-1)}\dot{b}_n + 3f_n\dot{\phi} = \dot{\alpha}g_n - \frac{e^{-\alpha}}{(n^2-1)}k_n \tag{B15}$$

$$\dot{c}_n = e^{-\alpha}j_n \tag{B16}$$

Variation with respect to g_n gives the linear Hamiltonian constraint:

$$3a_n(-\dot{\alpha}^2 + \dot{\phi}^2) + 2(\dot{\phi}\dot{f}_n - \dot{\alpha}\dot{a}_n)$$

$$+ N_0^2 m^2(2f_n\phi + 3a_n\phi^2) - \tfrac{2}{3}N_0{}^2 e^{-2\alpha}\left[(n^2-4)b_n + (n^2 + \tfrac{1}{2})a_n\right]$$

$$= \tfrac{2}{3}\dot{\alpha}e^{-\alpha}k_n + 2g_n(-\dot{\alpha}^2 + \dot{\phi}^2) \tag{B17}$$

Finally, variation with respect to N_0 yields the Hamiltonian constraint, which we write as

$$\tfrac{1}{2}e^{3\alpha}\left[-\frac{\dot{\alpha}^2}{N_0^2} + \frac{\dot{\phi}^2}{N_0^2} - e^{-2\alpha} + m^2\phi^2\right] = \text{quadratic terms} \tag{B18}$$

References

1 A. H. Guth, Phys. Rev. D23 347 (1981)

2 A. D. Linde, Phys. Lett. 108B 389 (1982)

3 S. W. Hawking & I. G. Moss, Phys. Lett. 110B 35 (1982)

4 A. Albrecht & P. J. Steinhardt, Phys. Rev. Lett. 48 120 (1982)

5 S. W. Hawking, Phys. Lett. B115 295 (1982)

6 A. H. Guth & S. Y. Pi, Phys. Rev. Lett. 49 1110 (1982)

7 J. M. Bardeen, P. J. Steinhardt & M. S. Turner, Phys. Rev. D28 679 (1983)

8 S. W. Hawking "Limits on Inflationary Models of the Universe", Phys. Lett. B (to be published)

9 V. A. Rubakov, M. V. Sazhin & A. V. Veryaskin, Phys. Lett. 115B 189 (1982)

10 S. W. Hawking in: Astrophysical Cosmology, Pontificiae Academiae Scientiarum Varia 48, 563 (1982)

11 J. B. Hartle & S. W. Hawking, Phys. Rev. D28, 2960 (1983)

12 S. W. Hawking in: "Relativity, Groups and Topology II", Les Houches Session XL (1984), North-Holland.

13 S. W. Hawking, Nucl. Phys. B239 257 (1984).

14 S. W. Hawking & Z. C. Wu, "Numerical Calculations of Minisuperspace Models", Phys. Lett. B (to be published)

15 S. W. Hawking & D. N. Page "Operator Ordering and the Probabilty Measure in Quantum Cosmology", D. A. M. T. P. preprint (1984).

16 S. W. Hawking & J. C. Luttrell, Nucl. Phys. B247 250 (1984)

17 L. P. Grishchuk, Zh. Eksp. Teor. Fiz. 67 825 (1974); Ann. New York Acad. Sci. 302 439 (1977)

18 A. A. Starobinsky, Pis'ma Zh. Eksp. Teor. Fiz. 30 719 (1979)

19 E. M. Lifschitz & I. M. Khalatnikov, Adv. Phys. 12 185 (1963)

20 U. H. Gerlach & U. K. Sengupta, Phys. Rev. D18 1773 (1978)

INFLATIONARY STAGES IN COSMOLOGICAL MODELS WITH A SCALAR FIELD

V.A.Belinsky[*)],L.P.Grishchuk,[+)] Ya.B.Zeldovich['),
I.M.Khalatnikov[*)]

[*)]L.D.Landau Institute for Theoretical Physics, 117940 MoscowV-334, USSR

[+)]Sternberg Astronomical Institute, 119899 Moscow V-234, USSR

[')]Institute for Physical Problems, 117940 Moscow V-334

INTRODUCTION

This work provides an analysis of all possible solutions to the equations for a homogeneous isotropic Universe with a scalar field. This study, though having an independent interest, is largely motivated by the ideas of the so-called inflationary cosmological models. Although a possibility of the inflationary (De Sitter) stage in the evolution of the early Universe was mentioned long ago (see, e.g., references in the review [1]), its role in the solution of cosmological problems has been explicitly clarified in [2]. Yet, so far it has been found out that not all versions of inflationary models can stand the comparison with observational data. A viable version of the theory seems to be the one which assumes that in the early Universe there is a scalar field φ with the values exceeding m_p, where $m_p = 1{,}22 \cdot 10^{19}$ GeV is the Planck mass ["].

["] In the article we are using a system of units where the velocity of light, Planck constant and Boltzmann constant equal unity. Latin indices take the values o,1,2,3, Greek indices - 1,2,3. The metric is written

It has been pointed out [1] that if the initial value of $\dot{\varphi}$ is sufficiently small, the scale factor of the homogeneous isotropic Universe $a(t)$ grows according to the law close to the exponential law, i.e., the quasi-De Sitter stage of the expansion is realized. However, the degree of generality of solutions with the inflationary stage, possibility of occurence of these stages at large initial values of $\dot{\varphi}$, quantitative characteristics of 2favourable" and "unfavourable" cases, modifications in the set of possible solutions due to the nonzero spatial curvature and other important issues have not been investigated. As will be shown below, these problems can be solved by applying the methods of the qualitative theory of dynamical systems.

We shall study the simplest case of massive (with the mass m) minimally coupled scalar field described by the Lagrangian

$$L = -\frac{1}{2}\varphi_{;k}\varphi^{;k} - \frac{1}{2}m^2\varphi^2 \qquad (1.1)$$

in the homogeneous isotropic Universe with the metric:

$$-ds^2 = -dt^2 + \frac{a^2(t)\left(dx_1^2 + dx_2^2 + dx_3^2\right)}{\left[1 + k\left(x_1^2 + x_2^2 + x_3^2\right)/4\right]} \qquad (1.2)$$

where $k = 1$, $k = -1$ and $k = 0$ correspond to a closed, open and flat models, respectively. In this case only the diagonal components of the energy-momentum tensor of the scalar field are nonzero; these components have the form identical to the one for the perfect fluid

---

") in the form $-ds^2 = g_{ik}dx^i dx^k$ where the signature is $(-+++)$. The time is denoted by $x^0 = t$. Time differentiation is denoted by a dot.

with certain effective energy density ε and pressure p. We have $T_0^0 = -\varepsilon$, $T_\alpha^\beta = p\delta_\alpha^\beta$ where

$$\varepsilon = \frac{1}{2}\dot{\varphi}^2 + \frac{1}{2}m^2\varphi^2 , \quad p = \frac{1}{2}\dot{\varphi}^2 - \frac{1}{2}m^2\varphi^2 \qquad (1.3)$$

It is clear already from these formulas that at $\dot{\varphi}^2 \ll m^2\varphi^2$ the effective equation of state is $p = -\varepsilon$, which makes an appearance of the quasi-De Sitter stage possible. In the opposite case $\dot{\varphi}^2 \gg m^2\varphi^2$ one has $p = \varepsilon$, i.e., the stiff equation of state. Cosmological solutions for this case have been studied in [3]. This equation of state is proposed previously in [4]. Finally in the regime of oscillating φ the averaged p is zero which imitates a dust-like medium. This dependence of the effective state of state on the regime of the scalar field has been worked out in [5].

We shall consider solutions of the classical equations for $\varphi(t)$ and $a(t)$ up to the singularities which are typical of the cosmological models. Yet, from the physical point of view the solutions to the classical equations are not applicable at densities and curvatures reaching and exceeding the Planck values. The values of the fields at which $\varepsilon = \frac{1}{2}\dot{\varphi}^2 + \frac{1}{2}m^2\varphi^2 \approx m_p^4$ can be regarded as the boundary of applicability for classical solutions. The initial data for classical stages of the evolution should be given on this boundary. This or that set of initial data can be represented as a consequence of the solution corresponding to quantum gravity problem in the region $\varepsilon \gtrsim m_p^4$. We shall stick to the ideas of [6] according to which the cosmological singularity should be somehow replaced by an act of spontaneous creation of the Universe. In [6] it has been shown that the inflationary stage of the expansion is necessary so that the Universe created with characteristic Planck sizes could within a certain period of time grow up to macroscopic sizes and finally up to the sizes of the Universe we observe now. Leaving for a while problems of quantum-mechanical analysis and discussion of probability of these or those initial data, it is important to find out the fate of solutions characterized by arbitrary initial values of φ and $\dot{\varphi}$ in the classical regime. Of a particular interest are closed models ($K=1$) since for them the concept of "creation" is more sensible.*) The conclusions of this work testify to the fact that the inflationary stage is a fairly general property of the solutions under study and thus this concept of "quantum creation" of the Universe with the subsequent inflationary stage does not require that any particularly special conditions should be imposed.

2. BASIC EQUATIONS

The Lagrangian (1.1) and the metric (1.2) lead to the following joint system of equations:

$$\ddot{\varphi} + 3H\dot{\varphi} + m^2\varphi = 0 \tag{2.1}$$

$$\dot{H} + H^2 = \frac{1}{6}\varkappa m^2\varphi^2 - \frac{1}{3}\varkappa\dot{\varphi}^2 \tag{2.2}$$

$$H^2 + K a^{-2} = \frac{1}{6}\varkappa\left(m^2\varphi^2 + \dot{\varphi}^2\right) \tag{2.3}$$

where $\varkappa = 8\pi G = 8\pi m_p^{-2}$ and

$$H = (\ln a)^{\cdot} \tag{2.4}$$

*) Some literature has already been devoted to the problem of quantum creation of the Universe (see [7] and references therein).

Note that Eqs. (2.1) and (2.2) form a three-dimensional system in the phase space $\varphi, \dot{\varphi}, H$. It is however more convenient to use these variables and time t alongside with the dimensionless quantities x, y, z and η:

$$\varphi = \left(\frac{3}{4\pi}\right)^{1/2} m_p x, \quad \dot{\varphi} = \left(\frac{3}{4\pi}\right)^{1/2} m m_p y, \quad H = mz, \quad t = m^{-1}\eta \tag{2.5}$$

Then Eqs. (2.1) and (2.2) become

$$\begin{aligned} x_\eta &= y \\ y_\eta &= -x - 3zy \\ z_\eta &= x^2 - 2y^2 - z^2 \end{aligned} \tag{2.6}$$

and the relation (2.3) is written as

$$x^2 + y^2 - z^2 = k m^{-2} a^{-2} \tag{2.7}$$

Here and below the index η denotes a derivative with respect to this variable. Formula (2.4) for the Hubble parameter acquires the form

$$z = (\ln a)_\eta \tag{2.8}$$

Henceforth it will be convenient to give results both in terms of x, y, z, η and in terms of $\varphi, \dot{\varphi}, H, t$. The use of one set of variables alongside with the other should not cause any difficulties.

Eqs. (2.6) do not contain the parameter k and thus describe all the three models in the phase space

x, y, z . Eq.(2.7) indicates regions of the phase space where trajectories of different models lie. It is clear that the surface of the cone $x^2+y^2-z^2=0$ corresponding to the flat model $k=0$ separates the regions containing trajectories of the open model $k=-1$ (the interior of the cone which includes the axis z) from the region containing trajectories of the closed model $k=1$ (the exterior of the cone comrising the plane x, y). Trajectories of the flat model lie on the cone and, consequently, form a two-dimensional invariant phase space.

Note that the trajectories in the upper half of the phase space ($z>0$) correspond to the expansion of the model ($\dot{a}>0$) whereas those in the lower half - to the contraction ($z<0, \dot{a}<0$). The trajectories of the open and flat models cannot be continued from the upper half into the lower one. They are separated by a singular point - the origin $(x, y, z) = (0, 0, 0)$ which is a vertex of the upper and lower parts of the cone. Physically this point (if to approach it from the region $z>0$) corresponds to the final stages of boundless expansion ($a\to\infty$), which, as is known, terminate the evolution of the models with $k=-1$ and $k=0$. For $k=1$ the trajectories can cross the plane $z=0$ corresponding to the extrema moments of the scale factor ($\dot{a}=0$).

Thus, expansion and contraction for the models with $k=-1$ and $k=0$ should be investigated separately. Then all the trajectories describing contraction in these models can be obtained from the trajectories associated with the expansion via two symmetry transformations which the system of Eqs. (2.6-2.8) possesses.

$$\begin{aligned} \eta &\to -\eta & \eta &\to -\eta \\ z &\to -z & z &\to -z \\ x &\to -x \quad , & y &\to -y \end{aligned} \tag{2.9}$$

Let us study the simplest case $k=0$.

3. FLAT MODEL

For $k=0$ the system of Eqs.(2.6-2.7) is simplified. We shall deal only with an expanding model for which $z \geqslant 0$. Inserting $z=+\sqrt{x^2+y^2}$ into the second equation of (2.6) we get a two-dimensional dynamical system in terms of the variables x, y :

$$\begin{aligned} x_\eta &= y \\ y_\eta &= -x - 3y\sqrt{x^2+y^2} \end{aligned} \tag{3.1}$$

The plane x, y will be treated as the phase space of this system; yet it should be borne in mind that the genuine trajectories of the system (3.1) lie on the cone $z=+\sqrt{x^2+y^2}$ and the phase portrait in the plane x, y is an orthogonal projection of the genuine diagram upon the horizonthal plane. It is clear that in the finite region of variation of x, y the system (3.1) has only one singular point, i.e., the origin $(x, y)=(0,0)$. A simple analysis reveals that this point is a stable focus and the asymptotics of the solution in its vicinity is of the form:

$$x=\frac{2}{3\eta}\sin(\eta-\eta_0),\ y=\frac{2}{3\eta}\cos(\eta-\eta_0),\ (\eta_0=\text{const}),\ \eta\to+\infty \tag{3.2}$$

Then for the variable z we have:

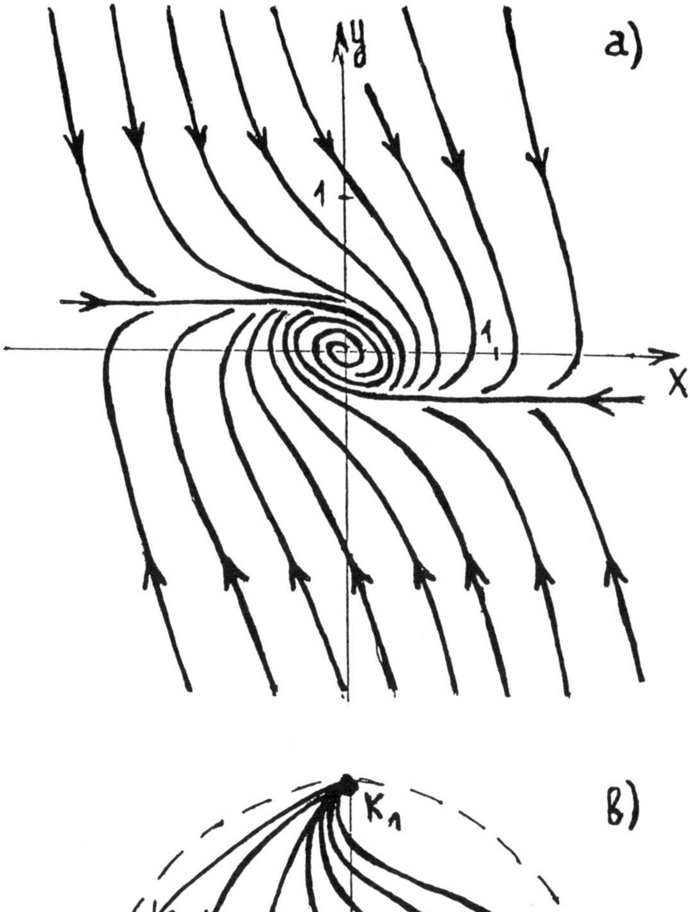

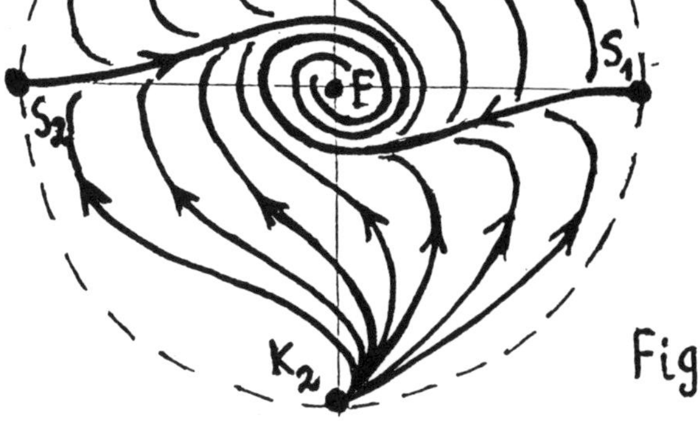

Fig. 1

$$z = \frac{2}{3\eta} \tag{3.3}$$

Thus the point $(x, y) = (0, 0)$ corresponds, as has already been mentioned, to the final stages of boundless expansion at which the field φ oscillates with damping and the scale factor tends to infinity according to the law

$$a \sim \eta^{2/3} \tag{3.4}$$

All the remaining singular points of the system (3.1) lie on the infinity and to find out their properties it is possible to employ a standard method of mapping the plane x, y (imagining it in the horizonthal position) onto the lower half-sphere of a unit radius, tangential to the plane at the origin (the so-called Poincare sphere). The mapping of the points of the plane onto the sphere is performed by means of the central projection (from the centre of the sphere). Then infinitely remote points of the plane are projected upon the equator of the sphere. If to project now the whole lower half of the sphere upon the horizonthal plane but by means of a vertical projection we shall finally get a mapping of the plane x, y onto the interior of the unit radius circle. The study of the character of the singular points on the boundary of this circle provides the necessary data for describing the infinity of the phase space. If to carry out the necessary analysis associated with the described procedure we shall come to the results illustrated by Fig. 1. Fig. 1a shows the behaviour of trajectories in the x, y - plane and Fig. 1b shows the mapping of this phase diagram onto the unit radius circle. On the boundary of this circle ($x^2 + y^2 = \infty$) the system under study has 4 singular points, 2 repulsive knots K_1 and K_2 and

2 saddle points S_1 and S_2 . In the centre of the circle (i.e., in the origin $x = 0$, $y = 0$) there is the above-mentioned stable focus with the asymptotics (3.2). All trajectories go out from the 4 infinitely remote singular points and then are woven around this central focus.

Write out the asymptotics of the solutions of the system (2.1-2.3) in the vicinity of the singular points. Near K_1 we have

$$\varphi = (12\pi)^{-1/2} m_p \ln(t/t_0), \quad H = 1/3t, \quad t \to +0 \quad (3.5)$$

where $t_0 > 0$ is an arbitrary constant. In the vicinity of this point the effective equation of state is $p = \varepsilon$. Near the saddle point S_1 , the asymptotics of the solution corresponding to the outgoing separatrix S_1F is

$$\varphi = -(12\pi)^{-1/2} m m_p t, \quad H = -\frac{1}{3} m^2 t, \quad t \to -\infty \quad (3.6)$$

Near this separatrix in the region of sufficiently large φ the effective equation of state is $p = -\varepsilon$.

Note that the asymptotical value of $\dot{\varphi}$ in (3.6) is

$$\dot{\varphi} = (\dot{\varphi})_\infty, \quad (\dot{\varphi})_\infty = -(12\pi)^{-1/2} m m_p \quad (3.7)$$

It means that in the plane φ, $\dot{\varphi}$ the separatrix S_1F has a horizontal asymptote (3.7) to which S_1F tends at $\varphi \to \infty$.

Asymptotics near the other 2 singular points K_2 and S_2 and properties of the separatrix S_2F are similar and entail from self-evident symmetry properties.

Since the effective equation of state $p = -\varepsilon$ is realized near the separatrix S_1F at sufficiently large φ (specified below), it is in this region that we should expect appearance of inflationary stages.

Let us pass over to analytical construction of solutions near this separatrix. As is evident from (1.3), the equation of state $p = -\varepsilon$ is realized for positive φ in the region

$$|\dot{\varphi}| \ll m\varphi \tag{3.8}$$

This is the region where the separatrix S_1F lies.

As follows from Eqs.(2.2-2.3) at $k=0$ the condition (3.8) is equivalent to the requirement

$$|\dot{H}| \ll H^2 \tag{3.9}$$

With (3.8) from (2.3) at $K=0$ we get

$$H \approx (4\pi/3)^{1/2} m_p^{-1} m\varphi \tag{3.10}$$

Inserting this expression for H into (2.1) we obtain the following approximate equation for phase trajectories

$$[\dot{\varphi} - (\dot{\varphi})_\infty] \exp[\dot{\varphi}/(\dot{\varphi})_\infty] = C \exp(6\pi m_p^{-2} \varphi^2) \tag{3.11}$$

where C is an arbitrary constant. From the same equations (2.1) and (3.10) we can also get a ratio of the scale factor in a certain initial moment t_i to the scale factor in a certain finite moment t_f. Since

$$a(t_f)/a(t_i) = \exp \int_{t_i}^{t_f} H(t)\,dt$$

from the above-mentioned equations it follows that

$$\frac{a(t_f)}{a(t_i)} = \left| \frac{\dot{\varphi}(t_i) - (\dot{\varphi})_\infty}{\dot{\varphi}(t_f) - (\dot{\varphi})_\infty} \right|^{1/3} \tag{3.12}$$

Hence it is clear that a considerable growth of the scale factor is possible only for those trajectories which approach by the moment t_f the separatrix S_1F (see (3.6-3.7)) sufficiently closely. It is well known that for cosmological applications the ratio $a(t_f)/a(t_i)$ on the inflationary stage should be of the order of 10^{30}. It is clear that the ratio of this order can be obtained only on those parts of phase trajectories which start at $|\dot{\varphi}-(\dot{\varphi})_\infty| \sim m m_p$ and terminate at $|\dot{\varphi}-(\dot{\varphi})_\infty| \sim 10^{-90} m m_p$. As for the parts of the trajectories where $|\dot{\varphi}|$ vaies from maximally large values $\sim m\varphi$ up to values corresponding to $|\dot{\varphi}-(\dot{\varphi})_\infty| \sim m m_p$, in these regions the ratio $a(t_f)/a(t_i)$ does not exceed the value $(m_p/m)^{1/3}$. Actually this follows from Formula (3.12), if one takes for $|\dot{\varphi}(t_i)|$ a maximally possible value $\sim m\varphi(t_i) \sim m_p^2$. Since additional cosmological constraints imposed to the growth of small perturbations and constraints imposed by the isotropy of the background radiation require $m/m_p \sim 10^{-5}-10^{-6}$, then the growth of the scale factor by $(m_p/m)^{1/3}$ times is actually negligibly small in comparison with the necessary. It is also possible to show that beyond the applicability region of the approximate solution (3.11-3.12), i.e., on the parts of the trajectories where $\dot{\varphi}^2 \gg m^2\varphi^2$, the growth of the scale factor is also very small and the ratio $a(t_f)/a(t_i)$ is determined by the same value $(m_p/m)^{1/3}$.

The ratio $a(t_f)/a(t_i)$ can be expressed in terms of the initial and finite values of the field φ. Formula (3.12) with (3.11) and the approximate equality

$$\exp[\dot{\varphi}/(\dot{\varphi})_\infty] \approx 1$$

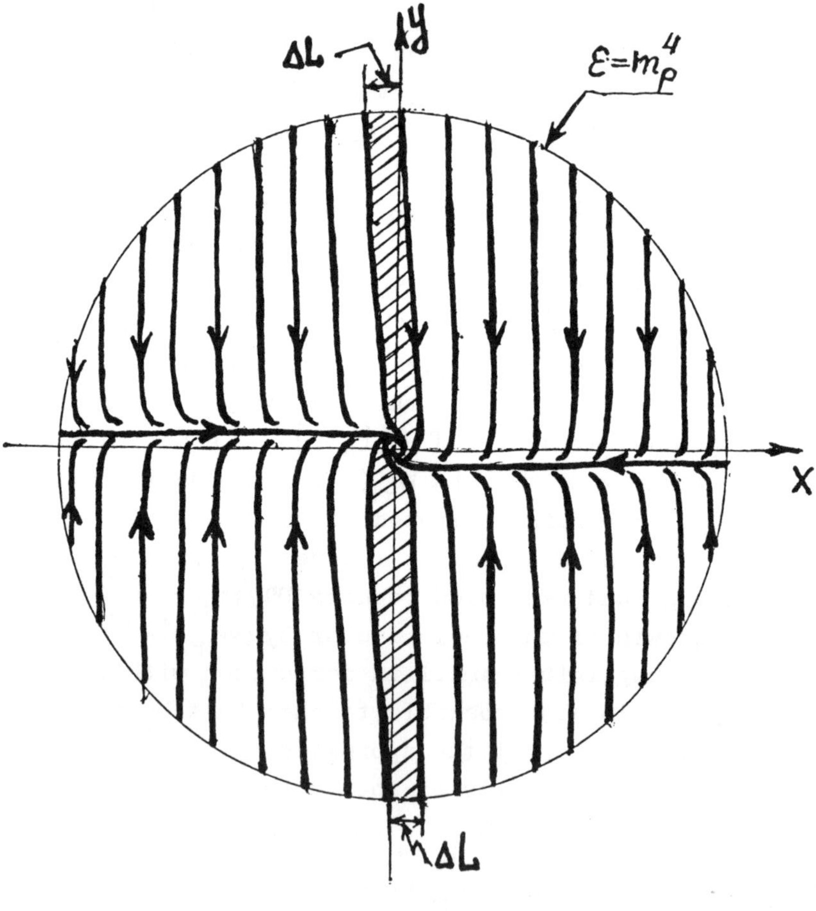

Fig. 2

which is valid in the $\dot{\varphi}$ -variation region (from $|\dot{\varphi}-(\dot{\varphi})_\infty| \sim m m_p$ up to $|\dot{\varphi}-(\dot{\varphi})_\infty| \sim 10^{-90} m m_p$ we are interested in) can be rewritten as

$$a(t_f)/a(t_i) = \exp\left[2\pi m_p^{-2}(\varphi_i^2 - \varphi_f^2)\right] \quad (3.13)$$

If φ_f is equal to the values of $\sim m_p$ in the region where the final expansion stage with damping oscillations of φ begins, then for the inflationary stage to be realized, suffice it to take $\varphi_i \sim (3 \div 4) m_p$. If the initial values are $\varphi_i \gg m_p$, the duration of the nacessary inflationary stage in terms of $\Delta\varphi = \varphi_i - \varphi_f$ is determined by the condition

$$\Delta\varphi \approx 6 m_p^2/\varphi_i$$

In particular, if φ_i is taken on the boundary of the quantum region, then $\Delta\varphi \sim m \ll m_p$.

The overall analysis still holds (with the accuracy up to the substitution $\varphi \to -\varphi$, $\dot{\varphi} \to -\dot{\varphi}$) for the region in the vicinity of the separatrix $S_2 F$.

Let us now discuss the problem to what extent the inflationary stage is a property characteristic of these solutions. It is most simple to solve this problem by means of a precisely drawn phase diagram of the system (3.1) in the rgion inside the quantum boundary where we specify the initial data. The phase diagram drawn in large scales (up to the quantum boundary) is qualitatively represented in Fig. 2. Here it is shown that phase trajectories are practically vertical beyond the central region with a radius $\sim m_p$. The trajectories in the region $|\varphi| > m_p$ sharply turn near the separatrices and go along them as far as the central region where they are woven around the focus $X = 0$, $y = 0$.

As has already been pointed out, the favourable trajectories (i.e., possessing the required inflationary stage) are those for which $\varphi_i > (3\div 4) m_p$. These trajectories start in the quantum boundary points positioned everywhere on the circumference except 2 of its sections near the axis y (see Fig.2). A number of unfavourable trajectories (i.e., starting from the points of these sections) is small, since the length of each of these sections $\Delta L \sim (6\div 8) m_p$ whereas the length of the whole quantum circumference $L \sim 2\pi m_p^2/m$. Thus the ratio $\Delta L/L$ determining a measure of unfavourable trajectories is

$$\Delta L/L \approx m/m_p \ll 1$$

Thus the inflationary stage, occuring sooner or later, is inherent in almost all trajectories, starting on the quantum boundary. This conslusion seems to be very important since it testifies to a great generality of inflationary regions in the models with $k = 0$.

Now discuss a model with $k \neq 0$.

4. BEHAVIOUR OF TRAJECTORIES ON THE INFINITY OF THE PHASE SPACE

First of all let us supplement the description of the three-dimensional phase space of the system (2.6), initiated in Sec.2. From (2.6) it is evident that for the finite values of x, y, z this system has only one equilibrium state - the origin $(x, y, z) = (0, 0, 0)$. The remaining singular points may lie only on the infinity. To investigate the character of these points and their number it is convenient to make compact the phase space and supply it with an infinitely remote boundary $x^2 + y^2 + z^2 = \infty$. It can be done by transforming the Cartesian coordinates x, y, z to spherical $x = r\sin\theta\cos\psi$, $y = r\sin\theta\sin\psi$, $z = r\cos\theta$ with the subsequent transformation of the radius, i.e.,

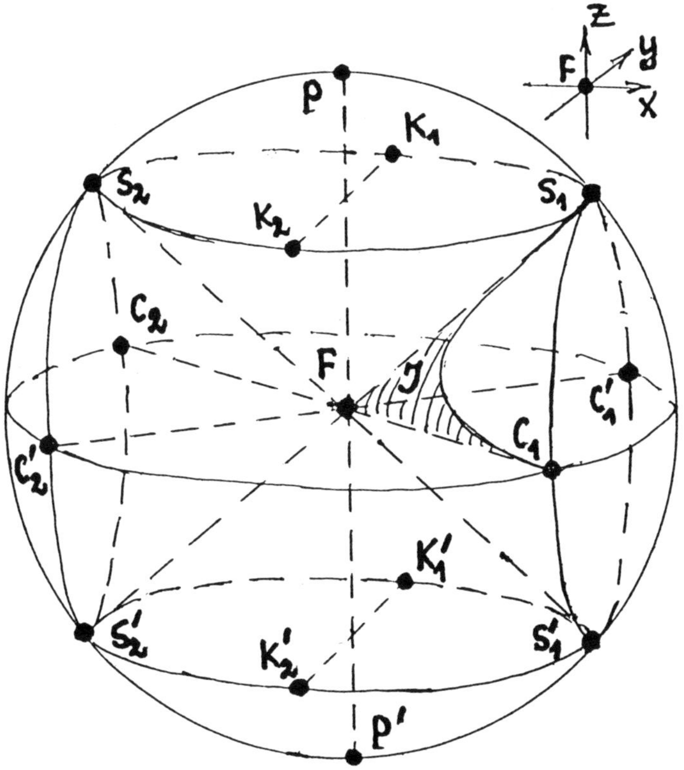

Fig. 3

e.g., according to the law $r = r'(1 - r')$ where $0 \leq r' \leq 1$. In terms of the variables r', θ, ψ the phase space is compact and can be represented as a sphere of a unit radius positioned in the same phase space x, y, z with the center in the origin. Each point of the original space is mapped along its radial direction onto a certain point of the interior of the sphere and the points on the infinity $x^2 + y^2 + z^2 = \infty$ are mapped onto the surface of the sphere $r' = 1$. The study reveals that the system (2.6) has 14 singular points on this surface. All of them together with the ones on the interior of the sphere are shown in Fig.3. Only 4 of them are essentially different: one point of each group P, K, S and C. Properties of all remaining points are obtained from those chosen 4 points by an appropriate combination of symmetry transformations (2.9).

The saddle point P is on the northern pole of the sphere: a two-dimensional bunch of trajectories (a two-dimensional separatrix) goes out from this point inside the phase space. These solutions in the vicinity of P have the asymptotics

$$\varphi = \varphi_0 - \frac{1}{8} m^2 \varphi_0 t^2 , \quad H = 1/t , \quad t \to +0 \qquad (4.1)$$

where φ_c is an arbitrary constant. The scale factor near P behaves according to the law

$$a \sim t \qquad (4.2)$$

Among these solutions there is a purely vacuum solution $\varphi \equiv 0$ (at $\varphi_c = 0$) whose trajectory is represented as the pole axis PF. It corresponds to the metric of the flat (in terms of 4 dimensions) Universe (the so-called Milne model).

The parallel $\theta = \pi/4$ is an intersection of

the surface of the sphere with the upper half of the cone $x^2+y^2-z^2=0$, i.e., is the boundary of the two-dimensional phase space of the flat model discussed in Sec.3. The saddle points S_1 and S_2 from which only one trajectory goes out inside the sphere, lie on the parallel. These two trajectories lie on the cone and are the separatrices S_1F and S_2F of the flat model considered in Sec.3. The points K_1 and K_2 are repulsive knots and they eject trajectories both into the space of the flat model and into the phase space of the open and closed models. These knots correspond to the initial cosmological singularities and the asymptotics in their vicinity are irrespective of the type of the model. For K_1 this is the same asymptotics (3.5), whereas for K_2 - the same formulas with the substitution $\varphi \to -\varphi$.

Additional 4 singular points of the saddle character are located on the equator of the surface. The points C_1 and C_2 are infinitely remote ends of a straight line lying on the intersection of the planes $z=1/\sqrt{2}$ and $x=-\sqrt{2}\,y$. This line is an asymptote for two-dimensional bunches of trajectories outgoing from C_1 and C_2 into the phase space of the closed model. The first terms of the asymptotics of these solutions in the vicinity of C_1 and C_2 are

$$\varphi = \text{const}\cdot e^{-mt/\sqrt{2}},\ H=m/\sqrt{2},\ t\to-\infty \quad (4.3)$$

where the constant prior to the exponent is positive for C_1 and negative for C_2). The scale factor here grows from zero exponentially fast:

$$a \sim e^{mt/\sqrt{2}} \quad (4.4)$$

yet it does not imply appearance of the inflationary

stage in this region. In fact. it follows from (1.3) that near C_1 and C_2 the effective equation of state is $\varepsilon + 3p = 0$ and

$$\varepsilon \approx \frac{3}{4} m^2 \varphi^2 \sim e^{-\sqrt{2} m t} \sim a^{-2}$$

Thus the energy density falls with increasing t too rapidly, which does not satisfy our conventional requirements to the notion of inflationary stages.

The saddle points C_1' and C_2' (ends of the line being the intersection of the planes $z=-1/\sqrt{2}$ and $x=\sqrt{2}y$) correspond to the lower half of the phase space (contraction) and attract the respective two-dimensional bunches.

The singular points P', K', S', symmetric with respect to P, K, S, are also positioned on the lower half of the surface of the sphere. The southern pole attracts a two-dimensional bunch of trajectories including the vacuum trajectory FP'. The knots K_1' and K_2' also attract trajectories and correspond to finite cosmological singularities. The asymptotics near these points of the collapse is irrespective of the type of the model also. Only one trajectory comes into the saddle points S_1' and S_2' : these are the separatrices FS_1' and FS_2' of the contracting flat model.

5. OPEN MODEL

Let us briefly discuss solutions for the expanding open model. Their compact phase diagram is restricted by the surface of the cone $z = +\sqrt{x^2+y^2}$ and a section of the spherical surface tangent to the northern pole P (see Fig.3).

The quantum boundary now is the surface of a cylindre $x^2+y^2 = 8\pi m_p^2/3m^2$ (i.e., $\varepsilon = m_p^4$)

intersecting the surface of the cone by the circumference which is the quantum boundary of the flat models discussed previously (see Fig.2). For non-flat models the boundary surface is two-dimensional since in comparison with the flat case it is necessary to specify another independent parameter - the initial value of H (or a). It is evident that the trajectories starting on the surface of the cylindre but close to its intersection with the cone will have qualitatively the same properties as the flat model trajectories, i.e., most of them approach and then go along the separatrices S_1F and S_2F on the surface of the cone and are subjected to prolonged inflation. These trajectories are such that the spatial curvature unimportant already in the region where the initial data are set, in the course of time will decrease on the inflationary stage. As for the other trajectories crossing the quantum boundary far from the mentioned circumference, their fate is less clear. However, it is expected that part of them will approach the cone in the regions of the separatrices S_1F and S_2F inside the classical region and will experience an inflationary regime. Others will reach the regions of oscillations near the focus F without experiencing inflation.

It is noteworthy that in all solutions even in those where the spatial curvature considerably decreased on the inflationary stage, it will finalyy become dynamically important at final stages of boundless expansion in the region of oscillations near the focus F (the effective equation of state is $p=0$). Hence it follows that all the trajectories approaching the focus F are closer and closer woven around the vertical vacuum line PF.

Thus the inflationary stage is an inevitable intermediate stage of most solutions in the open model although it is difficult to give a numerical estimate of their measure.

6. CLOSED MODEL

As has been said, the entire phase space for the case $K=1$ is the interior of the whole sphere minus the interior regions of the cone $x^2+y^2-z^2=0$ where trajectories of the open model lie. In comparison with the cases $K=0$ and $K=-1$ a new feature arises: it is a possibility for the trajectories to cross the plane $H=0$ ($z=0$), i.e., there emerge points of regular maxima or minima of the scale factor $a(t)$. These points can be strictly separated. To describe them we shall first find a surface in the phase space where $\dot{H}=0$ ($z_\eta=0$). As is clear from the last relation of (2.6) this surface is a cone with the equation $x^2-2y^2-z^2=0$. As is shown in Fig.3, the lobes of this cone are oriented horizonthally (they contain the axis x) and cross the infinitely remote boundary of the phase space along the curves $S_1C_1S_1'C_1'$ and $S_2C_2S_2'C_2'$. These curves cross the equator of the surface of the sphere in the singular points C and the whole cone intersects the horizonthal plane $z=0$ along the two straight lines $C_1'FC_2'$ and C_1FC_2 having the equations $x=\sqrt{2}\,y$ and $x=-\sqrt{2}\,y$, respectively. The cone under study lies in the region of the closed model space and has two common lines with the phase space of the flat model (i.e., with the cone $x^2+y^2-z^2=0$). They are two generatrices $z=\pm x$. The lines $x=\pm\sqrt{2}\,y$ on the plane $z=0$ are those which separate the points where the scale factor reaches its maximum or minimum. Any trajectory crossing the plane $z=0$ in the region $x^2>2y^2$ is a solution with a regular minimum $a(t)$. The trajectories intersecting the plane $z=0$ in the points of the region $x^2<2y^2$

have the maximum of $a(t)$ in these points. On any trajectory inside the cone $\dot{H}=0$ the Hubble parameter H increases with time, i.e., the motion in this region is oriented only upwards. Outside the cone H decreases and the motion along the trajectories is oriented downwards.

To find out the degree of generality of inflationary stages in the closed model it is important to find the points of the cone $\dot{H}=0$ where trajectories can leave. To determine a curve on the surface of the cone where $\ddot{H}=0$ it is possible to show that in the expansion phase ($H>0$) and at positive values of φ an exit from the cone is possible only through a narrow region on its surface denoted by $\mathcal{J}$ and shaded in Fig.3. At negative φ a similar region exists in the left-hand side of the cone in the triangle S_2C_2F. At large values of $|\varphi|$ which we are mainly interested in, the outgoing trajectories come into the vicinity of the separatrices S_1F and S_2F and the respective solutions experience a prolonged inflationary regime. Since the region of the exit from the cone is close to the separatrices S_1F and S_2F it means that almost all trajectories leaving it come into the inflationary regime. Such are all the trajectories appearing in the expansion phase inside the cone $\dot{H}=0$, i.e., from the points of the regular minimum of the scale factor and from the singular points C_1 and C_2. Part of the trajectories, starting from the singularities K come to the inflationary regime in the same way, i.e., first entering the inside region of the cone $\dot{H}=0$. The other part of the trajectories, starting in K, approach the separatrices S_1F and S_2F without entering inside the cone $\dot{H}=0$. These are the trajectories near flat model along the routes $K_1 \rightarrow S_2F$

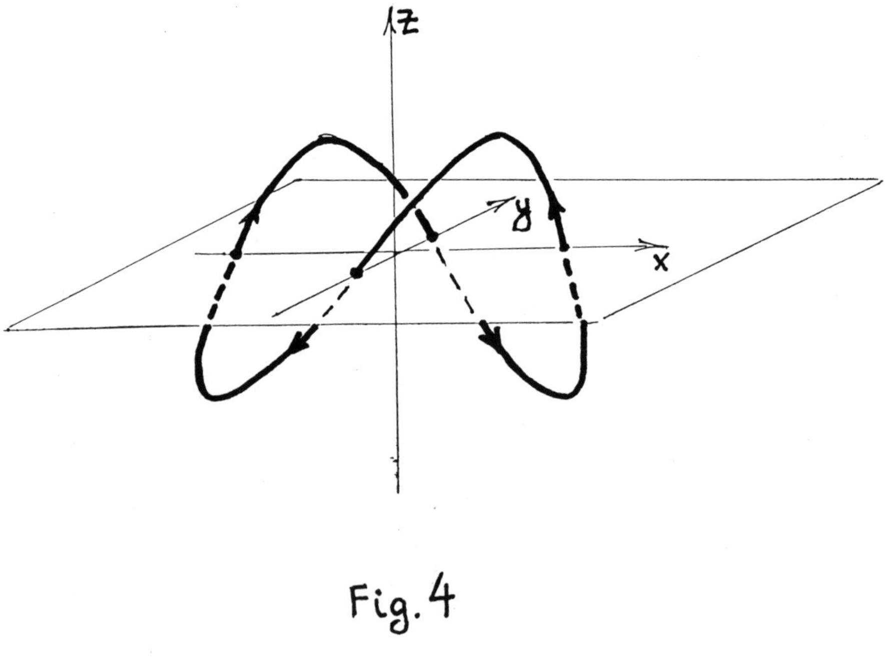

Fig. 4

and $K_2 \to S_1F$. It is difficult to estimate numerically a relative number of solutions with the required inflationary stage, yet, the above considerations show that this nimber is rather large.

Alongside with the studied solutions there are numerous trajectories along which it is possible to pass directly (avoiding the region near the separatrices S_1F and S_2F) from the initial singularity K to the collapse K' either through one expansion maximum or by experiencing a finite number of oscillations of the scale factor between regular minima and maxima.

A closed chain of trajectories on the boundary of the phase space of the system (e.g., $S_1FS_2'S_2FS_1'S_1$) implies a possibility of the existence of periodic solutions represented by closed trajectories in the vicinity of this chain. Realization of these solutions requires a special choice of initial data since on the section FS_2' , for example, going towards the saddle point S_2' , a trajectory should for a long time go along the unstable solution FS_2' .

The existence of the t-symmetric solutions with respect to moments of regular maxima or minima of $a(t)$ is noteworthy. It is possible to show that these trajectories cross the plane $z = 0$ in the points of the axis y or x . The existence of the periodic solutions, t-symmetric with respect to the both moments, when a_{max} is reached and the moments corresponding to a_{min} (see Fig.4) is also possible. Probable existence of such solutions has been mentioned in [8] and the problem of existence of infinitely oscillating non-periodic solutions has been discussed in [9]. As has been asserted in [8], periodic solutions represent a discrete set. Under such conditions the existence of trajectories one end of which starts or terminates in the singularities K, K' and the other is woven around one of the periodic solutions, also becomes possible.

Besides, the existence of trajectories performing a finite number of rotations around the periodic solutions and passing successively from one to another is not excluded.

Note that the existence of the classes of trajectories with the above-described properties leads to stochastization of the regime of its behaviour.

REFERENCES

1. Linde A.D. Rep.Prog.Phys.47,925,(1984).
2. Guth A. Phys.Rev. D23,347,(1981).
3. Belinsky V.A.,Khalatnikov I.M. ZhETF,63,1121,(1972).
4. Zeldovich Ya.B. ZhETF,41,1609,(1961).
5. Zeldovich Ya.B. Uspekhi Fiz.Nauk,1985 (in press).
6. Grishchuk L.P.,Zeldovich Ya.B. In "Quantum Structure of Space and Time", Eds. M.Duff,C.Isham,CUP,1982.
7. Linde A.D. ZhETF,87,369,(1984).
8. Hawking S.W., Preprint DAMTP, July (1983).
9. Page D. Preprint Phys.Dept.Pennsylvania, (1984).

THE PRESENT STATUS OF THE INFLATIONARY UNIVERSE SCENARIO x)

A.D.Linde

Lebedev Physical Institute, Moscow 117924, USSR

Different versions of the inflationary universe scenario (IUS) are considered and it is argued that the most natural realization of the idea of inflation can be achieved in the context of the chaotic inflation scenario. The possibility of generation of isothermal density perturbations and a new mechanism of baryosynthesis in the inflationary universe are suggested.

There is a large interest now in the inflationary universe scenario (IUS). According to this scenario the universe at the very early stages of its evolution expanded exponentially, like de Sitter space. This scenario in its present form [1] makes it possible to solve simultaneously about ten different problems related to the elementary particle theory and cosmology. The main idea of the IUS is very simple and attractive. Therefore during the last three years about two hundred papers on this scenario have been written. Unfortunately, the simplicity of this scenario is deceptive. The inflationary universe scenario is based on an investigation of effects related to the elementary particle theory, to the theory of superdense matter, to cosmology and even to quantum gravity. The investigation of these effects is very complicated, and therefore it was impossible to suggest a complete scenario which would take into account all these effects simultaneously. Usually some simplified versions of this scenario have been suggested, then they were improved (or disproved), and, as a result, at present there exist about four different versions of the IUS. This complexity of the situation combined with the apparent simplicity of first versions of the IUS has led to some confusion in the current literature. For example, only two simplest versions of the IUS are widely known and are extensively studied at present. These versions are the "old" scenario suggested by Guth [2] and the "new" inflationary universe scenario [3] (see also [4]). Meanwhile, some other, more elaborated versions of the IUS, which have a much better chance to provide us with a correct description of the

x) This is an extended version of the paper submitted to Comments on Astrophysics.

very early stages of the universe evolution, remain almost unknown.

In the present paper I would like to express my point of view on the current status of the IUS and on some controversal questions related to it.

The possibility that the universe at the very early stages of its evolution was exponentially expanding in the vacuum - like state was first suggested by Gliner [5] about twenty years ago and was discussed later by many authors [6,7]. In particular, Gliner and Dymnikova [6] have first estimated the duration of the inflationary (de Sitter) stage to be $O(50)\,H^{-1}$, where H is the Hubble "constant", $H=\dot{a}/a$, $a(t)$ is the scale factor of the universe. However, the origin of the vacuum - like state investigated by these authors remained obscure.

Later it became clear, that the constant classical scalar field φ which necessarily appears in all unified theories of elementary particles, looks just like the vacuum state with the energy density $V(\varphi)$, where $V(\varphi)$ is its effective potential [8-10]. It was shown, that this energy density is temperature dependent [8], and that during the phase transitions with the change of the field φ the energy of this field transforms into heat [11]. If the phase transition proceeds from a strongly supercooled vacuum-like state φ, the total entropy of the universe after the phase transition considerably grows [12] (the so-called Grand Bang [13]).

On the other hand, it was noted by Starobinsky [14], that the exponentially expanding de Sitter space is an unstable solution of the Einstein equations with radiative corrections [15], and this space after its decay transforms into the hot Friedmann universe [14].

The actual significance of all these facts became clear in the middle of 1980 after the remarkable paper by Guth [2], who suggested using the de Sitter (inflationary) stage of the universe expansion in a supercooled vacuum state in order to solve three longstanding cosmological problems: the flatness problem, the horizon problem and the primordial monopole problem. This suggestion was very important, the main idea of the scenario proposed by Guth was very clear and attractive, and many scientists worked enthusiastically in order to obtain some realisation of this scenario. However, as was pointed out by Guth himself, the universe after the phase transition in his scenario becomes extremely, inhomogeneous [2]. This problem has been thoroughly investigated by many authors, and the main conclusion was that the difficulties of this scenario were insurmountable [16].

Fortunately, in October 1981 the "new" inflationary universe scenario (NIUS) was suggested [3], which was free of the main difficulties of the "old" scenario. This scenario was widely discussed in the end of 1981 [17,18], and it became especially popular in the beginning of 1982, when Albrecht and Steinhardt have written a paper suggesting essentially the same scenario [4].

However, as it was noted in [3], this first version of the new inflationary universe scenario was oversimplified since it had not taken into account the effects connected with the rapid expansion of the universe during the phase transition from the supercooled vacuum state $\varphi = 0$. The improvement of this scenario has been accomplished in a series of papers in the beginning of 1982 [18-23], and has been discussed at the Nuffield Workshop in Cambridge [24]. It was, shown in particular that, the high-temperature effects are irrelevant for the description of the tunneling of the field φ from the supercooled state $\varphi = 0$ during inflation [19,20]. It was shown also that this scenario can be realized even without any <u>tunneling</u> from $\varphi = 0$, but just due to the growth of the long-range quantum fluctuations $\langle\varphi^2\rangle$ of the field φ [19, 21, 23]. This modification of the NIUS is very important. However, the papers [19-24] are much more complicated and much less populatizable (and, consequently, they are much less known) than the simplified versions of the new inflationary universe scenario [3,4]. One should bear this in mind when discussing the present status of the NIUS and the reliability of the corresponding investigations.

Resently Mazenko <u>et al</u> [25] have claimed that in many models which have been considered in the new inflationary universe scenario no inflation can actually exist, and that the standard picture of inflation is wrong. Since their paper is widely discussed now, and one of the authors of the second version of the NIUS [4] has admitted [26] a possible validity of the arguments of Mazenko <u>et al</u>, it is necessary to make some comments on this issue.

Mazenko <u>et al</u> have discussed the simplified version of the new inflationary universe scenario only [3,4]. Their main argument against it is that during the phase transition from the state $\varphi = 0$ to the minimum of $V(\varphi)$ at $\varphi = \varphi_0$ the dispersion $\tilde{\varphi}(T_c) = \sqrt{\langle\varphi^2\rangle}$ of the field φ may be of the same order as φ_0. If this is the case, one may argue that the phase transition proceeds by formation of domains of the field $\varphi \sim \varphi_0$ rather than by tunneling or rolling of the field φ down to φ_0, which was necessary for inflation.

However, Mazenko *et al* have not supported their expectations by a quantitative investigation of the phase transitions in gauge theories. Fortunately, the corresponding investigation has been made by many different authors several years ago [11,12,24,28]. This investigation shows that for the simplest theory of one scalar field φ with $V(\varphi) = -\frac{m^2\varphi^2}{2} + \frac{\lambda}{4}\varphi^4$ the argument of Mazenko *et al* [11,12,28] is actually correct, $\tilde{\varphi}(T_c) = \frac{T_c}{2\sqrt{3}} = \frac{\varphi_0}{\sqrt{3}}$, where T_c is the critical temperature of the phase transition in this model. However, nobody have ever expected that the NIUS can be realized in this simple theory. On the other hand, the same argument is completely wrong for gauge theories with $\lambda \ll g^2 \ll 1$, where g^2 is the gauge coupling constant [11,12]. For example, in the SU(5) Coleman-Weinberg theory with $\lambda \sim g^4$, $g^2 \sim 0.3$, which was used as a basis for the new inflationary universe scenario, $\tilde{\varphi}(T_c) \approx \frac{5g}{8\pi\sqrt{3}}\varphi_0 \approx \frac{\varphi_0}{15}$, which means that the formation of domains with $\varphi \sim \varphi_0$ in this theory is exponentially suppressed.*) This result is not unexpected, since the investigation of the domain formation via thermodynamical fluctuations of the field φ is equivalent to the investigation of the high-temperature "tunneling" with the bubble production [29], and the results of this investigation also confirm the validity of the new inflationary universe scenario. [24]

We will not give here a list of other statements of Mazenko *et al* which are not valid for the Coleman-Weinberg theory, since they have admitted that their criticism could be inapplicable to the Coleman-Weinberg model with $g^2 \ll 1$ [27]. (A detailed numerical investigation of this problem shows that $g^2 \sim 0.3$ is quite sufficient for inflation to occur [3,19-24]). Recently they also admitted [30] that their criticism do not apply to the primordial inflation and to the chaotic inflation scenario (see below). However in that case it becomes unclear to which models of inflation their criticism could be applied.

Recently the arguments by Mazenko *et al* have been discussed also by Albrecht and Brandenberger [26]. They have admitted, without investigation, that due to the interaction of the field φ with the gauge field A_μ in the expanding universe this field actually can form domains of the field $\varphi \sim \varphi_0$. They have succeeded in ob-

*) It is suppressed by a factor $\sim \exp(-\frac{\varphi_0^2}{\tilde{\varphi}^2}) \sim \exp(-200)$.

taining constraints on g^2 and λ at which the interaction terms can be neglected. However, the investigation of the improved version of the new inflationary universe scenario shows [19,21, 23], that inflation occurs even with an account taken of these interaction terms, and the constraints on g^2 and λ obtained in [26] are unnecessary [24].

Nevertheless, it is actually extremely difficult to obtain a completely consistent cosmological theory in the context of the new inflationary universe scenario. The corresponding difficulty has been revealed more than a year ago [1, 21, 32] **and has nothing in common with the** comments made by Mazenko et al. The point is that in order to obtain small density perturbations after inflation, $\delta\rho/\rho \sim 10^{-4}$, which are necessary for the galaxy formation, one must consider a theory of an extremely weakly interacting field φ [23,33]. For example, the constant λ in the theory $\lambda\varphi^4/4$ should be of the order 10^{-12} [23]. The theories of such a weakly interacting field φ do actually exist. One may consider, for example, the theory of the gauge singlet field φ coupled to the $SU(5)$ Coleman-Weinberg theory [34], or the theory of the chiral superfield z coupled to $N=1$ supergravity (the so called primordial inflation scenario [35]). However, in such theories the high-temperature corrections to the effective potential $V(\varphi)$ are negligibly small (even if the field φ was in the thermodynamically equilibrium state in the early universe). Therefore the high-temperature effects in such theories could not drive the field φ into the metastable state $\varphi = 0$ during the small time interval at which the temperature in the universe was suffciently high [1,32]. In order to obtain both inflation and small density perturbations $\delta\rho/\rho \sim 10^{-4}$ in the NIUS one must develop such theories in which radiative corrections to the effective coupling constant $\lambda \sim 10^{-12}$ are vanishingly small (some cancellation?), whereas the high-temperature corrections to $V(\varphi)$ are sufficiently large. This is a considerable complication [1,32], which, together with many unnatural constraints on the shape of the effective potential in this scenario, makes the realization of the new inflationary universe scenario extremely difficult, if possible at all. Despite many efforts, no models of new inflation which satisfy all the above-mentioned conditions have been suggested so far.

Fortunately, these difficulties, which preclude a successful realization of the NIUS, are absent in the chaotic inflation scenario [31]. This scenario differs both from the old and from the new inflationary universe scenario since it is not based on the theory of high-temperature phase transitions in the early universe. The main idea of the chaotic inflation scenario is that if the universe initially was singular, then near the Planck time $t_P \sim M_P$ the energy density ε could be determined only with the accuracy $\Delta\varepsilon \sim M_P^4$. Therefore, one may argue that the chaotically distributed field φ in the universe at $t \sim t_P$ may take any value for which $(\partial_0\varphi)^2 \lesssim M_P^4$, $(\partial_i\varphi)^2 \lesssim M_P^4$, $V(\varphi) \lesssim M_P^4$ with approximately the same probability. For the theory $\lambda\varphi^4/4$ with $\lambda \sim 10^{-12}$ this would mean that the most natural condition for the field φ at $t \sim t_P$ is $\varphi = \varphi_0 \sim M_P \lambda^{-1/4} \sim 10^3 M_P$. In this case it can be shown, that any domain of the universe of a size exceeding $H^{-1} \sim M_P/\sqrt{V(\varphi)} \sim M_P^{-1}$, in which initially $(\partial_\mu\varphi)^2$ is several times smaller than $V(\varphi) \sim M_P^4$, expands quasi-exponentially and grows $\sim \exp(\pi\varphi_0^2/M_P^2)$ times before the end of inflation, which stops when the field φ becomes smaller than $M_P/3$. Note, that in any "classical " space-time (i.e. in space-time, in which quantum fluctuations of metric are not too large) the energy density should be smaller than M_P^4, which means in particular that $(\partial_\mu\varphi)^2 \lesssim M_P^4$. In that case an existence of a domain of a size $\Delta\ell \gtrsim M_P^{-1}$ in which $V(\varphi) \sim M_P^4$ and $(\partial_\mu\varphi)^2 < V(\varphi)$ seems quite probable, or, at least, we do not know any reasons why the number of domains with $(\partial_\mu\varphi)^2 < V(\varphi)$ should be strongly suppressed as compared with the number of domains with $(\partial_\mu\varphi)^2 > V(\varphi)$. Therefore the conditions which are necessary for the realization of the chaotic inflation scenario seem to be quite natural. In any case, in the open (infinite) universe there should exist infinitely many domains of size $\ell \gtrsim M_P^{-1}$ with $V(\varphi) \sim M_P^4 > (\partial_\mu\varphi)^2$, and just these domains expand exponentially and presumably contain the main part of the physical volume of the universe after inflation. A typical magnitude of inflation in this scenario is extremely large. For $\lambda \sim 10^{-12}$ (which is necessary to have adiabatic perturbations $\delta\rho/\rho \sim 10^{-4}$) the universe typically expands $\exp(1/\sqrt{\lambda}) \sim \exp(10^6)$ times before the end of inflation [1,31].

In our opinion, the chaotic inflation scenario is much superior to all other versions of the inflationary universe scenario considered so far. One of the most important features of this scenario is that inflation in this scenario may occur in a wide class of theories including e.g. all theories in which $V(\varphi) \sim \varphi^n$ at $\varphi \gtrsim M_P$, and there is no need for $V(\varphi)$ to satisfy numerous constraints imposed on the shape of the effective potential in the new inflationary universe scenario. This scenario can be easily implemented either in the context of N=1 supergravity [32, 36] or in the context of the $SU(5)$ theory coupled to a weakly interacting gauge singlet field φ [37, 38]. It is very easy to incorporate into this scenario an extended version of the Starobinsky model [39] and similar models based on the higher dimensional gravity [40]. A general study of this scenario has been accomplished with the initial conditions for the field φ fixed either at $t \sim t_P$ [31] or at $t \to 0$ [39,41]. In the context of this scenario inflation seems to be a rather general property of a wide class of cosmological models, and it is even surprising that this scenario has not been discovered many years ago.

Recently a new version of the chaotic inflation scenario has been suggested [42, 43]. This version is based on the idea that the initial conditions in the universe are determined by quantum gravity effects at the moment at which our "classical" universe emerges out of the primary space-time-field foam. [44] The first attempts to estimate the probability P of such an event by means of the euclidean approach to quantum gravity gave the result $P \sim \exp(3M_P^4/8V(\varphi))$ [45,42]. This result was rather surprising, since it implied that the universe most probably was created in the state with $V(\varphi) = 0$, in which quantum gravity effects were negligible (and no inflation could occur). A possible resolution of this paradox has been suggested in [43], where it was noted that the usual euclidean methods are not directly applicable to the quantization of the universe. The reason is that our universe can be described as a system of elementary

---

*) One should note, that if one wishes to solve the hierarchy problem in gauge theories with the help of light weakly interacting Polonyi fields z, which appear in supergravity [51], one encounters the so-called Polonyi problem, or the entropy crisis in cosmology [52,53]. Only two ways to solve this problem of models based on supergravity are known to us at present. One way is to consider a complicated hidden sector with the O' Raifeartaigh-type superpotential [54]. This possibility does not seem very natural [53]. Another way is to consider such a superpotential for the Polonyi field z, that the effective potential $V(z)$ at a sufficiently large distance from the minimum of $V(z)$ at z_0 is dominated not by the quadratic term $\sim m^2|z-z_0|^2$, but by the quartic term $\sim|z-z_0|^4$ or by some other term which grows at large $|z-z_0|$ more rapidly than $|z-z_0|^4$. In such a case it can be shown that the entropy crisis can also be circumvented since large energy density cannot be stored in a theory with a very steep $V(z)$.

particles and classical fields with positive energy and of the scale factor a with negative energy (the total energy of a closed universe is zero). This theory has an unstable vacuum state, which is crucially important for the very possibility of our existence (there is an exponentially rapid pumping of energy from the scale factor a to the scalar field φ during inflation). There are no general euclidean or Hamiltonian prescriptions for quantization of such unstable systems (compare with the situation in the nonequilibrium quantum statistics). Fortunately, at present quantum fluctuations of the scale factor a can be neglected, and one can use the standdard euclidean methods for quantization of the matter fields with positive energy. On the other hand, the evolution of the scalar field φ at the moment of the universe creation can be neglected [43], and one can use another version of euclidean methods, which is applicable for quantization of the scale factor a with negative energy. An improved expression for the probability of the universe creation is $P \sim \exp\left(-\frac{3M_P^4}{8V(\varphi)}\right)$ [43]. The same result has been obtained later by a different method by Starobinsky [46], Rubakov [47] and Vilenkin [48]. This semiclassical result may be somewhat modified with an account taken of particle production during the creation of the universe [47]. But in any case it is clear that the quantum creation of the universe becomes possible only if the energy density inside the created universe is of the order of M_P^4. This means (for a sufficiently homogeneous field φ, $(\partial_i\varphi)^2 < V(\varphi)$) that $V(\varphi) \gtrsim M_P^4$, which is quite sufficient for the realization of the chaotic inflation scenario after the universe creation [43].

In the last part of this paper I would like to discuss a more practical problem related to the inflationary universe scenario. The flatness of the universe, $\Omega = \rho/\rho_c = 1$, and the flatness of spectrum of adiabatic perturbations, $\delta\rho(l)/\rho$ = const, are considered now as two important predictions of the inflationary universe scenario which can be experimentally tested. One may worry, however, that these two predictions, combined together, may form something like the Procrustean bed for the theory of formation of the large-scale structure of the universe. Let us assume for a moment that these predictions will be in a contradiction either with the theory of galaxy formation, or with the observed isotropy of the microwave background ratiation. Would it mean that we must abandon the idea of inflation?

In order to answer this question let us first

recall, that there is a functional freedom in the choice of the effective potential $V(\varphi)$ in supergravity [36]. By a proper choice of $V(\varphi)$ one can obtain $\Omega \neq 1$ in the chaotic inflation scenario, though this does not seem very natural. Much easier is to change the shape of the spectrum of adiabatic perturbations by a proper change of $V(\varphi)$. However, the first thing to be investigated is the possibility of generation of other types of perturbations in the IUS.

Long-wave density perturbations $\delta\rho$ in the IUS usually are proportional to the perturbations of the scalar field φ generated during inflation. After the reheating of the universe the perturbations of the scalar field energy density $\delta\rho_\varphi$ give rise to adiabatic perturbations of hot matter, $\delta\rho_\varphi \sim \delta\rho_{tot} \sim T^3\delta T$. There exist many other scalar fields Φ, different from the "inflaton" field φ which governs inflation. Inflation induces perturbations of all such fields, but initially $\rho_\varphi \gg \rho_\Phi$, and perturbations $\delta\rho_\Phi$ have a little effect on $\delta\rho_{tot}$. However, if some of the Φ-particles cannot decay into light ultrarelativistic particles, the energy density of the Φ-particles in the expanding universe decreases more slowly than the energy of the ultrarelativistic hot matter. Such particles may give the leading contribution to the present density of the universe, $\rho_\Phi \sim \rho_{tot}$. Perturbations of ρ_Φ appear even if perturbations of temperature $\delta T \sim T^{-3}\delta\rho_\varphi$ are vanishingly small. In this sense perturbations $\delta\rho_\Phi$ are isothermal. An investigation of $\delta\rho_\Phi$ shows that, contrary to the usual belief [49], isothermal perturbations in some models may be greater than the usual adiabatic perturbations, and their spectrum in some cases differs from the flat spectrum. [50] For example, in the axion-like models with the effective potential $V(\Phi) \sim 1 - \cos\Phi/\Phi_0$ there is a cut-off of the flat spectrum of isothermal perturbations in the long-wave region at $\ell \simeq H^{-1}\exp(4\pi^4\Phi_0^2H^{-2})$ [50]. This observation provides a possibility of a more flexible approach to the theory of formation of the large-scale structure of the universe in the inflationary universe scenario.

Generation of long-wave perturbations of light scalar fields during inflation, which is necessary for galaxy formation, simultaneously solves many other difficult problems. The same mechanism solves the problem of symmetry breaking in SUSY GUTs [55]. A similar mechanism proves to be extremely useful for the generation of the baryon asymmetry of the universe. The baryon charge of the universe can be carried not only by quarks, but by their superpartners - squarks [56]. Squarks are relatively light bosons, and therefore classical squark-slepton fields, just as all other classical scalar fields with $m \ll H$, are generated via long-wave quantum fluctuations of these fields during inflation 57. Later these almost homogeneous classical fields decay into quarks and leptons, thus producing the baryon asymmetry of fermions in our universe. This mechanism of baryosynthesis proves to be very efficient and can easily produce the baryon asymmetry of the univerce even greater than $n_B/n_\gamma \sim 10^{-9}$. An important feature of this mechanism is that the decay of the classical squark-slepton field occurs at a temperature $T \sim m_W \sim$ ~ 100 GeV 57. This removes many constraints usually imposed on the inflationary universe scenario , connected with the reheating after inflation. The same fact makes it possible to avoid many difficulties associated with the gravitino problem in N = I supergravity since this problem arises only if the reheating temperature after inflation should be very large. Moreover, during the decay of the squark-slepton classical fields (and during the decay of other particles, which may be even desirable now in order to reduce the large baryon asymmetry produced by this mechanism) the total entropy of the universe grows considerably. This makes the relative abundance of gravitinos, monopoles etc. extremely small and simplifies the solution of the Polonyi problem 57.

In conclusion I would like to note that one of the differences between man and other types of animals is a comparatively large number of stages of evolution of a man from an embryo to a grown-up person. Similarly, there were many stages of development of the inflationary universe scenario from its embryonal period 5-7 to its birth 2,I4, childhood 3,4 and to the present stage of its evolution I,3I,42,43. Presumably, the inflationary universe scenario now is still far from its maturity, but the continuous progress of this scenario makes us rather optimistic about its future.

REFERENCES

1. A.D.Linde, Rep.Prog. Phys. 47,925 (1984).
A.D.Linde, a plenary talk at the XXII International Conference on High Energy Physics, Leipzig 1984 (to be published in the Proceedings).

2. A.H.Guth, Phys. Rev. D23,347 (1981).

3. A.D.Linde, in Quantum Gravity, proceedings of the second seminar on quantum gravity, Moscow, October 1981, edited by M.A.Markov and P.C.West (Plenum Press, New York 1983).
A.D.Linde, Phys. Lett. 108B,389 (1982).

4. A.Albrecht and P.J.Steinhardt, Phys. Rev. Lett. 48, 1220 (1982).

5. E.B.Gliner, Sov. Phys. - JETP 22, 378 (1965).
E.B.Gliner, Dokl. Akad. Nauk SSSR 192, 771 (1970).

6. E.B.Gliner and I.G.Dymnikova, Pis. Astron. Zh. 1, No 5, 7 (1975).

7. A.D.Sakharov, Sov. Phys. - JETP 22, 241 (1965).
B.L.Altshuler, in Proc. 3rd Soviet Gravitational Conf., Erevan, p.6 (1972).
L.E.Gurevich, Astrophys. Space Sci. 38, 67 (1975).

8. A.D.Linde, JETP Lett. 19,183 (1974).

9. M.Veltman, Rockefeller Univ. preprint (1974).
M.Veltman, Phys. Rev. Lett. 34, 77 (1975).

10. J.Dreitlein, Phys. Rev. Lett. 33, 1243 (1974).

11. D.A.Kirzhnits and A.D.Linde, Ann. Phys. (N.Y.) 101, 195 (1976).

12. A.D.Linde, Rep. Prog. Phys. 42, 389 (1979).

13. A.D.Linde, Phys. Lett. 99B, 391 (1981).

14. A.A.Starobinsky, JETP Lett. 30, 682 (1979).
A.A.Starobinsky, Phys. Lett. 91B,99 (1980).

15. J.S.Dowker and R.Critchley, Phys. Rev. D13, 3224(1976).

16. S.W.Hawking, I.G.Moss and J.M.Stewart, Phys. Rev. D26, 2681 (1982).
A.H.Guth and E.Weinberg,Nucl.Phys.B212, 321 (1983).

17. S.W.Hawking, a talk at the Pennsylvania university, November 1981.

18. S.W.Hawking and I.G.Moss, Phys. Lett. 110B, 35 (1982).

19. A.D.Linde, Phys. Lett. 114B, 431 (1982).

20. A.D.Linde, Phys. Lett. 116B, 340 (1982).

21. A.D.Linde, Phys. Lett. 116B,335 (1982).

22. A.D.Dolgov and A.D.Linde, Phys. Lett. 116B,339 (1982).

23. A.A.Starobinsky, Phys. Lett. 117B, 175 (1982).

24. A.D.Linde, in The Very Early Universe, edited by G.W. Gibbons, S.W.Hawking and S.Siklos (Cambridge Univ. YPress 1983).

25. G.F.Mazenko, W.G.Unruh and R.M.Wald, Chicago Univ. preprint (1984).

26. A.Albrecht and R.Brandenberger, Santa Barbara preprint NSF-ITP-84-146 (1984).

27. G.F.Mazenko, W.G.Unruh and R.M.Wald, a revised version of the Chicago Univ. preprint, to be published in Phys. Rev. D.

28. S.Weinberg, Phys. Rev. D9, 3357 (1974). L.Dolan and R.Jackiw, Phys. Rev. D9, 3320 (1974). D.A.Kirzhnits and A.D.Linde, ZhETF, 1263 (1974).

29. A.D.Linde , Nucl. Phys. B216,421 (1983).

30. W.G.Unruh, a talk at the 3rd seminar on quantum gravity, Moscow, October 1984.

31. A.D.Linde, JETP Lett. 38, 176 (1983). A.D.Linde, Phys. Lett. 129B,177 (1983). A.D.Linde, in Proc. Shelter Island Conf. II (Cambridge, Mass: MIT Press 1984).

32. A.D.Linde, Phys. Lett. 132B,317 (1983).

33. S.W.Hawking, Phys. Lett. 115B,295 (1982). A.H.Guth and S.-Y.Pi, Phys. Rev. Lett. 49, 1110 (1982). J. Bardeen, P.J.Steinbardt and M.Turner, Phys. Rev. D28,679 (1983).

34. Q.Shafi and A.Vilenkin, Phys. Rev. Lett. 52,691(1984).

35. J.Ellis, D.V.Nanopoulos, K.A.Olive and K.Tamvakis, Nucl. Phys. B221,524 (1983). D.V.Nanopoulos, K.A.Olive, M.Srednicki and K.Tamvakis, Phys. Lett. 123B, 41 (1983).

36. A.S.Goncharov and A.D.Linde, Phys. Lett. 139B, 27 (1984).
A.S.Goncharov and A.D.Linde, Class. Quant. Gravity, 1, L 75 (I984).

37. B.L.Spokoiny, Pisma ZhETF 40, 354 (1984).

38. A.S.Goncharov and A.D.Linde, in preparation.

39. L.A.Kofman, A.D.Linde and A.A.Starobinsky, submitted to Phys. Lett.

40. Q.Shafi and C.Wetterich, preprint BA-84-36 (1984).

41. V.A.Belinsky, L.P.Grishchuk, Ya.B.Zeldovich and I.M.Khalatnikov to be published.

42. S.W.Hawking, Nucl. Phys. B239, 257 (1984).

43. A.D.Linde, ZhETF 87, 369 (1984). A.D.Linde, Lett. Nuovo Cim. 39, 401 (1984).

44. P.I.Fomin, Preprint Kiev ITF-73-137 P (1973). E.P.Tryon, Nature 246,396 (1973). P.I.Fomin, Dokl. Akad. Nauk USSR, 9A,831 (1975).

45. A.Vilenkin, Phys. Lett. 117B, 25 (1982). J.B.Hartle and S.W.Hawking, Phys. Rev. D28,2960 (1983).

46. A.A.Starobinsky, unpublished. Ya.B.Zeldovich and A.A.Starobinsky, Pis. Astron. Zh. 10, 323 (I984).

47. V.A.Rubakov, Pisma ZhETF 39, 89 (1984).

48. A.Vilenkin, Phys. Rev. D30, 509 (1984).

49. M.Axenides, R.Brandenberger and M.Turner, Phys. Lett. 126B,178 (1983).

50. A.D.Linde, JETP Lett. 40, 496 (1984).

51. H.P.Nilles, Phys. Rep., to be published.

52. G.D.Coughlan, W.Fischler, E.W.Kolb, S.Raby and G.G.Ross, Phys. Lett. 131B,59 (1983).

53. A.S.Goncharov, A.D.Linde and M.I.Vysotsky, Phys. Lett., I47B, 279 (I984).

54. M.Dine, W.Fischler and D.Nemeschansky, Phys. Lett. 136B,169 (1984). G.D.Coughlan, R.Holman, P.Ramond and G.G.Ross, Phys. Lett. 140B, 44 (1984).

55. A.D.Linde, Phys.Lett., I3IB, 330 (I983).

56. I.Affleck and M.Dine, Princeton Univ. preprint (I984).

57. A.D.Linde, The new mechanism for baryogenesis and the inflationary universe, to be published.

DYNAMICS OF INFLATING BUBBLES IN THE EARLY UNIVERSE

V.A.Berezin, V.A.Kuzmin, I.I.Tkachev

Institute for Nuclear Research of the Academy of Sciences of the USSR, 60-th October Anniversary Prospect, 7a, Moscow II73I2, USSR

We show that (under reasonable assumptions) the inflating fluctuation in the early universe can be equivalent only to a wormhole (such a possibility can take place both in the open universe and in the closed one) or the field fluctuation occupies more than a half of a closed universe. Other possibilities are highly exotic.

A great attention was paid quite recently to the study of such a new and important subject as bubbles arising in the course of cosmological phase transitions[I-9]. These are both new phase bubbles in interiors of an old phase[I-3,5-9] and old phase remnants surrounded by the new phase[4,5,9]. It has become clear that an account of general relativity effects is necessary in investigations of these objects[3-6,9]. It is clear also that though in the proposed inflationary universe scenarios[I0-I2] all visible part of the universe is inside a single bubble[II-I2] the latter is nevertheless nothing but a bubble, so one should study it as such, using the formalism of thin shells in general relativity[I3] elaborated in ref[5] for the case of vacuum bubbles.

The main idea of the recent scenarios of the inflationary universe may be formulated in the following way. There exist mechanisms of the development of a region occupied by the classical scalar field φ possessing the correct symmetry but non-equilibrium magnitude that the whole observable part of the universe is inside this (initially small) region.

If one supposes that in the case of the 'new inflationary' scenario[11] the phase transition proceeds by tunneling of the field φ through the potential barrier then both regions inside the new phase bubble and outer regions filled by the false phase will inflate. The energy density inside the bubble is smaller (the pressure correspondingly larger) than that outside the bubble. This bubble is a normal one in all respects (though it may become relatively large at the later stages of its evolution), so its dynamics is clear in general[5]. However, the another assumption about the initial state of the system seems to be also realistic[14]. It may happen[14] that the percolation through the new phase takes place before conditions for the new phase bubble inflation are satisfied so only potentially inflating regions are the domains occupied by the false phase in the sea of the correct phase with the vanishing Λ-term. Formally, the same situation takes place in the chaotic universe scenario. Here one needs not any phase transition to occur. It is belived that a region occupied by the strong enough classical scalar field should inflate after the scalar field energy density becomes larger than the particle thermal energy density. In both latter cases the energy density inside the region occupied by the false phase or by the field fluctuation is larger (and the pressure, correspondingly, smaller) than that in the outer region. If gravity effects are not very significant then it is clear that such a bubble should collapse. It does not matter in fact that the scale factor in the inner region behaves as $a \sim \exp(Ht)$ because this is nothing but a coordinate effect. It is important to know whether the physical volume of the bubble is increasing, $V_{phys} = V_{coord}(t) \exp(3Ht)$, V_c being the coordinate volume of the bubble. Since V_c may vanish, V_{phys} may vanish as well despite the 'inflating' exponent. This just takes place in the case of old phase remnants collapsing into black holes, as it was shown in ref[5].

On the other hand it is clear by intuition that if the size of the region occupied by the field fluctuation exceeds the inner Hubble radius I/H then the shell moving with the velocity smaller than the velocity of light can not collapse in time necessary for the realization of the inflation $\tau \sim 60/H$. However, in this case the fluctuation region can not be got into the (arbitrary) a priori prepared universe (or should change it cardinally). In such a situation gravity effects in the transient layer are rather significant and a detailed analysis is necessary.

So let us consider the transient layer which can be represented as a 3-dimensional hypersurface separating space-time on two parts. It is clear that the form of this hypersurface (or the motion of the fluctuation boundary) should obey the general equtions[13,5]

$$\left(\left[K_i^{\ j}\right] - \delta_i^{\ j} \left[K_l^{\ l}\right] \right) = 8\pi æ S_i^{\ j} \qquad (I)$$

where $\left[K_i^{\ j}\right]$ is the discontinuity of the $K_i^{\ j}$ component of the shell outer curvature tensor, $S_i^{\ j}$ is the surface density of energy-momentum tensor $T_\mu^{\ \nu}$ on the shell

$$S_i^{\ j} = \lim_{\delta \to 0} \int_{-\delta}^{\delta} dn \ T_i^{\ j} \qquad (2)$$

and n is the coordinate measured in the direction of the outer normal to the hypersurface. We know the metric of the inner part by the requirement that it should correspond to that of the de Sitter world while the metric and the matter equation of state of the outer region are unknown. We restrict ourselves to the consideration of spherically symmetric case only (though the factor of such high symmetry may prove in principle to be rather essential*)).

---

*) It is clear that small deviations from spherical symmetry are not essential. One may wonder, however, what happens if the fluctuation region at the fixed moment is equivalent not to a sphere but, say, to a torus.

Let us wright the metric of the outer region as follows

$$ds^2 = e^{\nu} dt^2 - e^{\lambda} dq^2 - r^2(q,t) d\Omega^2 , \quad (3)$$

where $r^2 \, d\Omega^2$ is a surface element of 2-dimensional sphere. We choose the coordinate q raising from the center of a bubble on the outside. The metric (3) still possesses a coordinate freedom t'(t,q), q'(t,q), so one can put on them one arbitrary condition (say, $\nu = 0$, or $\lambda = 0$, or $q = r$).

In the curved space-time the normal vector to the surface r = const may be space-like as well as time-like. In the first case

$$\Delta \equiv g^{\alpha\beta} r_{,\alpha} r_{,\beta} < 0 \quad (4)$$

and the corresponding region is called a R-region (in the flat space R-region occupies the whole space). In the second case

$$\Delta > 0. \quad (5)$$

Such a region is called a T-region. In a chosen coordinate frame (3) we have

$$\Delta = e^{-\nu} \dot{r}^2 - e^{-\lambda} r'^2 . \quad (6)$$

As far as $\Delta > 0$ in a T-region it is impossible to fulfil the condition $\dot{r} = 0$ there. That is in a T-region either $\dot{r} > 0$ or $\dot{r} < 0$ remains true under any continuose coordinate change. The region where $\dot{r} > 0$ we shall call the T-region of expansion (or T_+ -region), while the region where $\dot{r} < 0$ we shall call the T-region of contraction (or T_-).

Similarly the sign of $r' = dr/dq$ does not depend upon the coordinate choosing in R - region (remember that the q is raising from the center of a bubble). We shall call the region where $r' > 0$ the R_+ -region while that one with $r' < 0$ - R_- -region.

It is convenient for our purposes to use the Gaussian normal coordinates

$$ds^2 = -dn^2 + e^{\nu} d\tau^2 - r^2(\tau, n) d^2\Omega \, . \tag{7}$$

These coordinates are chosen in such a way that $n = 0$ is the equation of the shell and $\nu(\tau, n = 0) = 0$, so that at $n=0$ the coordinate τ coinsides with the proper time on the shell.

The Eq.(I) for the outer metrics (3) and the inner de Sitter metric takes the form[5)]

$$\sigma_{in}\sqrt{\dot{\rho}^2 + I - \rho^2/a_{in}^2} - \sigma_{out}\sqrt{\dot{\rho}^2 - \Delta} = 4\pi æ \rho S_0^0 \tag{8}$$

where $\rho \equiv r(\tau, n=0)$, $\sigma \equiv \text{sing}\, \partial r/\partial n$, $a_{in}^{-2} = 8\varepsilon_{in} r^2 \pi æ/3$. In R-regions we have also $\sigma = \text{sign}\partial r/\partial q$ for any space-like coordinate q raising from the center of a bubble.

First of all let us prove the following

<u>Statement I</u>.

The shell may inflate only in T_+ -region of infinite expansion or in R_- -region of outer metric.

Indeed, the shell can not inflate in T_- -region since here the physical volume of the fluctuation region decreases.

Further, the inflating fluctuation at some moment has the size exceeding the inner Hubble radius

$$-\Delta_{in} = 1 - \frac{\rho^2}{a_{in}^2} < 0 \qquad (9)$$

which means that the shell crosses the T-region of the inner metric. Let us suppose that at the same time the shell crosses R_+- region, or $\sigma_{out} = +1$. It follows from Eq. (8) that in this case $\sigma_{in} = +1$ (we restrict ourselves by the consideration of shells with $S_0^0 > 0$ only) and it then immediately follows from Eqs. (8) and (9) that $\Delta_{out} > 0$. This means that the shell crosses the T-region in the outer metric too. In other words the shell can not cross the R_+-region of the outer metric and be simultaneously in the T-region of the inner metric. One arrives therefore at the following

Statement 2.

If the inflating shell has once crossed the R_+-region of the outer metric it should first enter the T_+-region of the outer metric and only then the T-region of the inner metric.

Thus, the inflation of the shell is only possible in R_--region and in T_+-region of the outer metric.

If the fluctuation moves in the R_--region this means that either the whole confuguration is equivalent to a wormhole (such a possibility may take place both in the open Universe and in the closed one) or the field fluctuation occupies more than a half of the closed mother Universe.

Let us find the conditions at which there may exist a T_+-region of infinite expansion in the outer metric.

In the case of spherical symmetry the Einstein equations can always be written as follows

$$\left[r(1+\Delta)\right]_{,\mu} = 8\pi æ r^2 \left[T r_{,\mu} - T_\mu^{\nu} r_{,\nu}\right], \quad (10)$$

where $T = T_0^{\,0} + T_1^{\,1}$. Let us restrict ourselves to the case when T_μ^{ν} is the energy-momentum tensor of perfect fluid

$$T_\mu^{\nu} = (\varepsilon + p) u_\mu u^\nu - p\,\delta_\mu^{\nu}, \quad (11)$$

where u_μ is a medium 4-velocity, $u_\mu u^\mu = 1$. Then the Eq.(10) takes the form

$$\left[r(1+\Delta)\right]_{,\mu} = 8\pi æ r^2 \left[\varepsilon r_{,\mu} - (\varepsilon + p) u_\mu u^\nu r_{,\nu}\right] \quad (12)$$

For one of the Eqs.(12) we now have in coordinates (7)

$$r\dot{\Delta} = -\dot{r}(1+\Delta) - 8\pi æ r^2 \left[p\dot{r} + (\varepsilon + p)(u^1)^2 \dot{r} + (\varepsilon + p) u_0 u^1 r'\right], \quad (13)$$

where $\dot{r} = dr/d\tau$, $r' = dr/dn$. Let us substitute the equation of the shell $n=0$. For expanding shells ($\dot{r} > 0$) all terms in the right-hand side of Eq.(13), except for one, are negatively defined when $p > 0$ in the T-region. The only sign indefinite term is as follows

$$A \equiv -8\pi æ r^2 r'(\varepsilon + p)u_0 u^1 = -8\pi æ r^2 r' T_0^1 \quad . \tag{14}$$

If $A < 0$ then the total right-hand side of Eq.(13) is negatively defined, hence $\dot{\Delta} < 0$ and therefore during the shell expansion Δ becomes negative at some moment*) and the shell trajectory enters the R-region of the outer metric. In view of the above statements this may be the R_- region only. Let us find therefore at what conditions A may be positively defined (at these conditions only the shell could expand in T-region at $p_{out} > 0$). Since σ = sign r' we have to consider two cases.

I. $\sigma_{out} = +1$. In this case $A > 0$ only if $u^1 < 0$. This means that the outer medium moves towards the shell. Let us consider how this medium should be prepared in order to maintain this regime. From Eq.(8) the condition follows

*) This moment can not take place at $r \to \infty$ since the solution of Eq.(13) is

$$\Delta = -1 + 2æ m(n=0)/r - (8\pi æ \int r^2 B dt)/r, \tag{15}$$

where B is the quantity in the square brackets in Eq.(13), m(n) is an arbitrary integral of Eq.(13). The value of the radius where Δ vanishes may tend to infinity only if $m(n=0) \to \infty$ which means that the region occupied by the fluctuation just from the beginning has the Schwarzschild mass equal to the mass of the whole universe.

$$1 + \Delta > 8\pi æ r^2 \mathcal{E}_{in}/3 \quad . \tag{I6}$$

It is useful to wright Eq.(I2) in the matter comoving coordinates $u^1 = 0$. In this frame the derivative of Δ in the direction L^μ of the shell motion ($L^0=I$, $L^1 = \dot{Q}$, $q = Q(t)$ is the shell trajectory in this coordinates) is

$$r\Delta_{,\mu}L^\mu = \left[8\pi æ r^2 \mathcal{E}_{out} - (I+\Delta)\right] r_{,\mu}L^\mu - \\ - 8\pi æ r^2(\varepsilon + p)(\partial r/\partial t) \quad . \tag{I7}$$

In the expanding universe ($\partial r/\partial t \geq 0$) the right-hand side of Eq (17) may be positive only if along the way of the shell motion the condition is fulfilled permanently

$$\mathcal{E}_{out}(t,q(t)) > \mathcal{E}_{in}/3 \quad . \tag{I8}$$

This requirement can be satisfied only if the matter in the outer universe is distributed in an extremely peculiar way, namely, the matter density should increase rapidly from the center of the fluctuation region to the infinity.

2. $\sigma_{out} = -I$. In this case $A > 0$ if $u^1 > 0$, that means that the outer medium velocity is directed outside the shell, furthermore this condition should be satisfied yet at the shell in the shell frame system. It is clear that in this case the point is not in the preparation of the outer medium but the shell itself has to create such a medium. In any case there is a vacuum inside the shell $(\varepsilon + p)_{in} = 0$ at $u^1_{out} > 0$. Let us use the following junct-

ion equation on the shell[7)]

$$dS_o^o/d\tau + 2\dot{\rho}(S_o^o - S_2^2)/\rho = (T_o^1)_{in} - (T_o^1)_{out} = \\ = -(T_o^1)_{out} \qquad (I9)$$

where $\rho \equiv (\tau, n=0)$. Substituting now the value of $(T_o^1)_{out}$ from this expression into Eq.(I4) one obtaines

$$A = 8\pi\varkappa r^2 r'(dS_o^o/d\tau + 2\dot{\rho}(S_o^o + S_2^2)/\rho). \qquad (20)$$

We see that $A > 0$ if either $\dot{S}_o^o < 0$ that means that the shell emits outside the matter at the expense of the energy contained in S_o^o or $S_2^2 > S_o^o$ that implies the vacuum burning[7)].

However, as one can show, the existence of such shells with the magnitude of A sufficient for maintaining the inflation (so that the shell does not leave the T-region) requires also the whole universe to be specially prepared - it must be either closed (with fluctuation occupying more than a half of it) or it must be a wormhole (see Fig. I,3), or infinite in spatial coordinates without R-region anywhere outside the shell (see Fig.2). We shall not present here the rigorious proof of this statement but give some physical arguments[I5)].

The existence of a shell under consideration would imply that there is a mechanism of conversion into particles of the vacuum energy of the inflating bubble not inside the bubble after the field fluctuation rolling down to the potential minimum, but outside the bubble yet during the process of the inflation. If the inflation of a fluctuation were possible in any arbitrarily prepared universe then one has to conclude that it is possible also in our world. Let us imagine now that one has somehow created (say, at superhigh energy collisions) a region occupied by the non-equilibrium field φ of the magnitude sufficient for the inflation ($\varphi > M_{Pl}$). Let the dimensions of the region be larger than $r_o \gtrsim M_{Pl}/(\lambda\varphi^4)^{1/2}$ (one has to spend for that the energy $E > M_{Pl}^3/(\lambda\varphi^4)^{1/2}$). Then one might argue that all conditions for the inflation of this region to take place would be satisfied, so one would be unable to stop the catastrophe. Note that in our space the region boundary expands of course with the velocity not exceeding the velocity of light. The inflation duration $\tau \gtrsim 60/H$ is small for the inner observer while for the outer observer the corresponding time duration is $t \gtrsim \exp(H\tau)/H$, i.e. the known twins effect takes place here (this would be valid also for the fluctuation which gave rise to the creation of our universe). Thus, having once spent the energy portion $E \gtrsim M_{Pl}^3/(\lambda\varphi^4)^{1/2}$ one could then release into the outer space arbitrarily large energy. This seems to be somewhat like a perpetuum mobile.

Thus, we see that at reasonable assumptions the T-region of infinite expansion may persist in the outer metric only if $p_{out} < 0$. This means that outside the fluctuation region there is the nonvanishing positive Λ-term too. The whole universe in the global sense is the de Sitter one, so there inflates the universe as a whole and not the fluctuation region only. The outer metric is of the Schwarzschild-de-Sitter type now. In this metric there exist both the R_- -region and the $T_\pm$-region of infinite expansion. However, in this metric, according to our Statement 2, the shell may cross the R_+ -region and enter the $T_\pm$-region of expansion only if the Λ -term in the outer metric exceeds that in the inner one. Such a configuration can not obviously be considered as the inflating 'fluctuation'. Therefore, in the Schwarzschild- de Sitter metric having R-regions the inflating fluctuation ($\Lambda_{in} > \Lambda_{out}$) should necessarily be wormholes again (see Fig. I,3). The Schwarzschild-de Sitter metric has necessarily R-regions if $m^2 < M_{Pl}^4 \, a_{in}^2 /27$. In the opposite case this metric does not contain either R-regions or horizons and it describes one all-round T-region. The shell with such an outer metric will undoubtedly inflate(see Fig.2).

One may wonder , however, could this trick please anybody?

Thus, considering the global geometry of the Universe, we have shown that "fluctuation" in the chaotic inflationary scenario may inflate only if it occupies more than a half of the whole closed Universe, or if the global geometry of the Universe corresponds to one of the cases presented in Figs I -3. In Fig. I,3 the fluctuation forms a wormhole being seen both in the outer (closed or open) and in the inner universes as a black hole. Such a black hole evaporates thus splitting these universes and leaving us with the closed inner world.

The wormholes shown in Figs I,3 and geometry shown in Fig.2 require special boundary conditions to be fixed just in the singularity. Such a situation does not seem attractive. On the contrary, the quastion seems to be quite reasonable on whether a wormhole can be created in the course of the universe evolution,

The most attractive scenario, however, seems to be as follows. The inflating universe is the closed one filled from the very beginning by the classical scalar field. Such initial conditions are possibly not unnatural. We have in mind at the moment one or another mechanism of the quantum creation of the universe. It has been argued in ref[I6) for example, that in the case of the tunneling universe the system prefers, due to a kind of a balance of (negative) gravitational energy and (positive) matter excitation energy, to come out being relatively hot. Having this in mind, we may conjecture that the creation of an universe just in an excited state could be a rather general property of the system when one deals with the quantum gravity. If the universe at the creation prefers to be filled with the smooth scalar field throughout, then it will inflate.

We are grateful to V.A.Matveev,V.A.Rubakov,A.N.Tavkhelidze for the interest in the work and discussions.

REFERENCES

I. M.B.Voloshin,I.Yu.Kobzarev and L.B.Okun', Yad.Fiz. 20, I229 (I974) / Sov.J.Nucl.Phys.20 (I975) 644

2. S.Coleman, Phys.Rev. DI5, 2922 (I977).

3. Coleman S., DeLuccia F.,Phys.Rev. D2I,3305 (I980)

4. Sato K.,Sasaki M.,Kodama M.,Maeda K.,Prog.Theor. Phys.,65,3 (I98I)

5. Berezin V.A.,Kuzmin V.A.,Tkachev I.I.,Phys.Lett., I20B,9I (I983)

6. Berezin V.A.,Kuzmin V.A.,Tkachev I.I.,Phys.Lett. I30B,23 (I983)

7. Berezin V.A.,Kuzmin V.A.,Tkachev I.I., Phys.Lett. I24B, 479 (I983)

Zh.Eksp.Theor.Fiz. 86,785 (I984) /Sov.Phys. JETF 59(3),459 (I984)

8. DeGrand T, Kajantie K., Phys.Lett.,I47B, 273 (I984)

9. Aurilia A.,Denardo G.,Legovini F.,Spallucci E., Phys.Lett. I47B, 258 (I984)

IO. Guth A.H., Phys.Rev. D23, 347 (I98I)

II. Linde A.D., Phys.Lett., I08B, 382 (I982)

Albrecht A.,Steinhardt P.J., Phys.Rev.Lett.,48, I220 (I982)

I2. Linde A.D., Phys.Lett., I29B, I77 (I983)

I3. Israel W., Nuovo Cim., 44B, I (I966),48B,463 (I967)

I4. Masenko G.F., Unruh W.G., Wald R.M., Phys.Rev., D3I, 273 (I985)

I5. Berezin V.A., Kuzmin V.A., Tkachev I.I., to be published

I6. Rubakov V.A., Phys.Lett., I48B, 280 (I984)

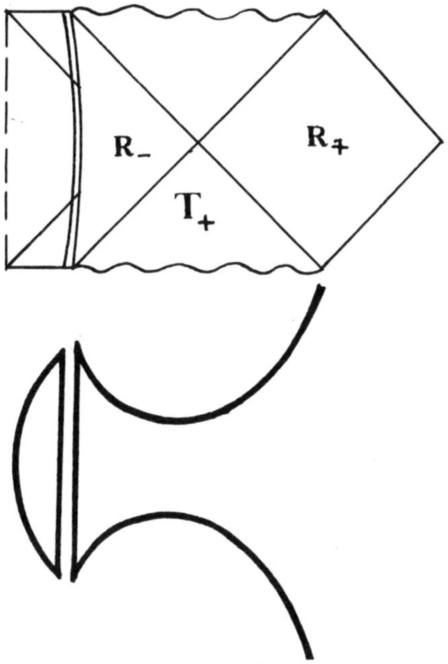

Fig.I.

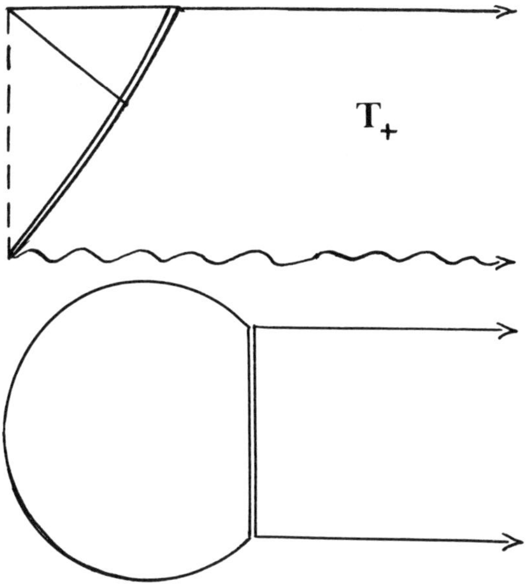

Fig.2.

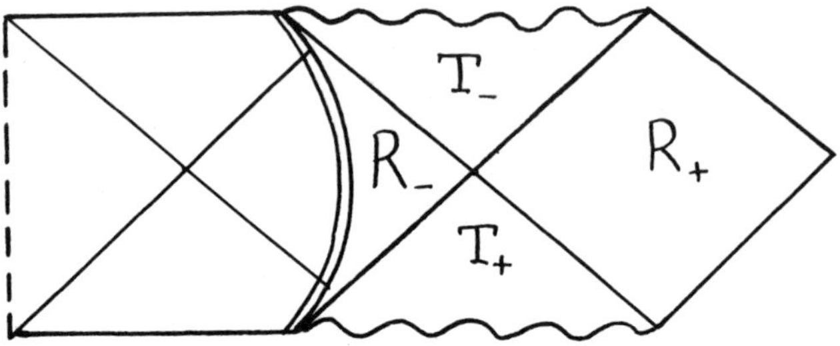

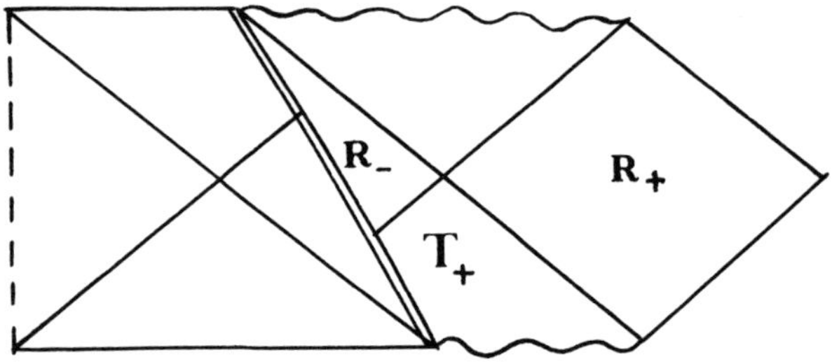

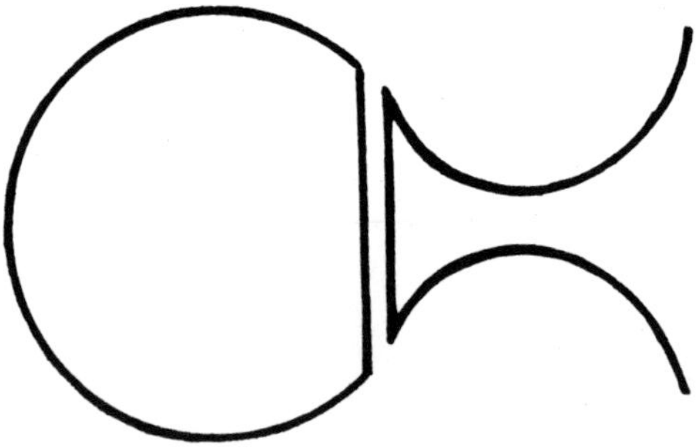

Fig. 3.

Figs. I-3. Global geometries of the space-time with the inflating region. Below each of the Penrose diagrams we present schematically the corresponding spatial section at some moment t = const. For the sake of simplicity we restrict Penrose diagrams to the nearest R_+ region to the right of the shell (which is denoted by the double line). More R -regions, for example in the case of Schwarzschild de Sitter geometry, are possible.

INFLATION?

W.G. Unruh

Department of Physics
University of British Columbia
Vancouver, B.C., Canada V6T 2A6

ABSTRACT

Questions are raised about the use of the effective potential in determining whether or not inflation occurs in the early universe due to a supercooled phase transition.

The issue I want to raise in this talk is whether or not the scenario, which leads to the so called new inflation in the early universe, is valid. Although I will not be able to definitively answer the question of validity, I do want to raise the possibility that the phase transition responsible for the inflation may take place in a different way than in the standard picture, a way which does not lead to any period of inflation. In particular, I wish to point out that the question is a dynamical one, for which the usual equilibrium calculations may be very misleading. These comments will be essentially those which have been made in the paper by G. Mosenko, R. Wald and myself.[1] For the details of the argument, and for references, I will refer the reader to that paper and simply outline the arguments here.

The first point I would like to raise is the use of the "effective potential" techniques and a misinterpretation which seems to be placed on the results of these calculations. The effective potential is defined by the following sequence of operations for a field theory containing a scalar field ϕ.

Define the function

$$F(J)= -\langle \ell n\ e^{-\beta (H-J\phi)}\rangle /\beta v \qquad (1)$$

where J is a constant, H is the hamiltonian for the field ϕ, β is the inverse temperature and v is the volume of the region under consideration (which is assumed to go to infinity).

We now have

$$\langle \int \phi d^3x \rangle_J / v = \frac{\partial}{\delta J} F(J) \tag{2}$$

This average field we will designate by ϕ_0. We can invert this relation to find J as a function of ϕ_0, which I will designate by $J(\phi_0)$.

Now, let us define

$$W(\phi_0) = F\big(J(\phi_0)\big) - J(\phi_0)\, \phi_0 / \beta V \tag{3}$$

$W(\phi_0)$ is the so called effective potential. What it corresponds to is the free energy (E-TS) of the system with the spatial average of the field constrained to be ϕ_0. Unfortunately, $W(\phi_0)$ is in general incalculable. What is used instead is the so called one loop effective potential.

To be definite, I will assume that we are dealing with a pure scalar field theory, with H given by

$$H = \frac{1}{2}\left(\pi^2 + (\nabla\phi)^2\right) + V(\phi) \tag{4}$$

The one loop effective potential is derived by rewriting the true effective potential as a path integral

$$W(\phi_0) = \frac{1}{\beta v} \ln \int \delta\phi \, \exp - \left\{ \iint_{vo}^{\beta} \left[\frac{1}{2}\left((\tfrac{d\phi}{d\tau})^2 + (\nabla\phi)^2 \right) + V(\phi) - J\phi \right] d\tau\, d^3x \right\} - J\phi_0 \tag{5}$$

where J is chosen so that $\partial W/\partial J = 0$. Now, rewrite $\phi = \phi_0 + \tilde{\phi}$, and retain terms in the exponent only to quadratic order in $\tilde{\phi}$, dropping the linear term in $\tilde{\phi}$ (i.e. use $J = V'(\phi_0)$). A similar quadraticisation of the exponential is used also for more complex theories.

The general features of the effective potential used in most models are that at high temperatures, the effective potential has a sharp minimum at $\phi_0 = 0$ with a large curvature, $W''(\phi_0 = 0)$, of order $1/\beta^2$ there. At low temperatures ($\beta \sim \infty$) the effective potential develops two equal minima at $\phi_0 = \pm\, \phi_c$. (See figure 1.)

FIG. 1

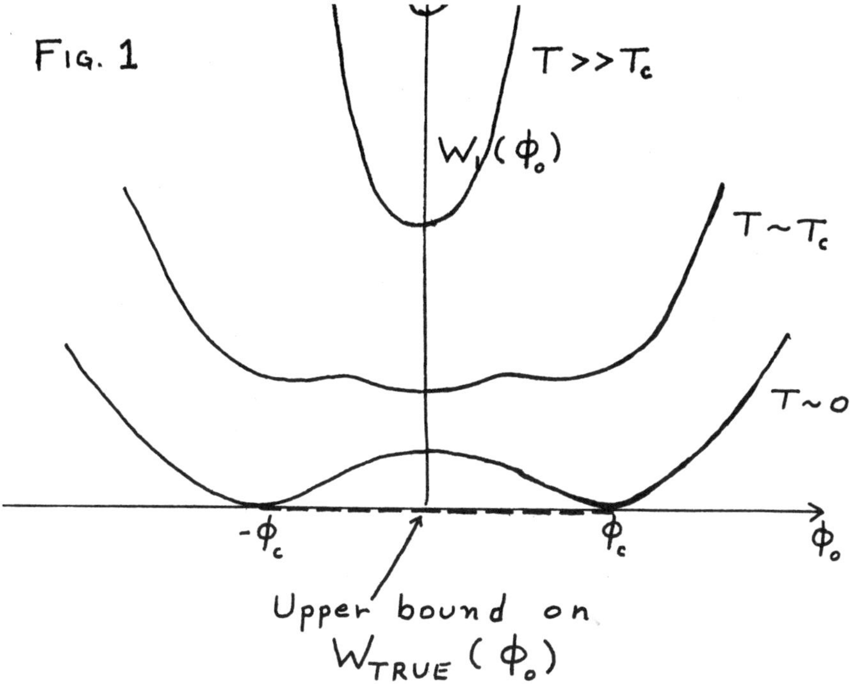

The scenario is now that at high temperatures the field is constrained to lie at 0 by the sharp minimum in the effective potential. As the universe rapidly cools due to expansion, the field is depcsited at the top of the T = 0 ($\beta = \infty$) potential with small fluctuations. The energy momentum tensor

$$T_{\mu\nu} = \frac{1}{2}\left[\partial_\mu\phi\partial_\nu\phi - \frac{1}{2}\, g_{\mu\nu}\left(\partial_\mu\phi\partial_\nu\phi - V(\phi)\right)\right] \tag{6}$$

is now potential dominated since $\partial\phi/\partial t \approx 0$ because of the very shallow slope of the effective potential at $\phi_0 = 0$, $\partial\phi/\partial x$ is roughly zero because the expansion has made all wavelengths very long, and thus

$$\langle T_{\mu\nu}\rangle \approx \langle \frac{1}{4}\, V\ (\phi = 0) g_{\mu\nu}\rangle \tag{7}$$

which produces an exponential expansion of the universe.

One point to note is that the above discussion is rather heuristic. Both the derivative terms, which have been neglected, and the potential are actually divergent and must be renormalised. The assumption usually made is that $\langle V(0)\rangle$ may be replaced by W(0) in eq. 7. But, even at T = 0, W(0) is the energy of the state in question. One must separately determine the pressure of the field in that state. It is not a priori obvious that the pressure is just $-W(\phi)$ as is required for $T_{\mu\nu}$ to be equal to a multiple of $g_{\mu\nu}$.

However, if the pressure in the one loop approximation can be determined to be $-W(\phi)$, the argument has deeper difficulties. The first is that the double humped one loop effective potential for low temperatures cannot be a good approximation to the true effective potential. The true effective potential is the free energy for the field with the average value of the field constrained to be ϕ_0. Let us assume that the density matrices for the state of the field at $\phi_0 = \pm\phi_c$, the minima of the low temperature effective potential, are given by $\rho\pm$. Then the density matrix $\rho = \lambda\rho_- + (1-\lambda)\rho_+$ has a mean value of the field of

$$\langle\phi\rangle (= \langle\int \phi d^3x/v\rangle \text{ by homogeneity})$$

$$= \mathrm{tr}(\phi\rho)$$

$$= \lambda \ \mathrm{tr}\ \phi\rho_- + (1-\lambda)\ \mathrm{tr}\ \phi\rho_+$$

$$= -\lambda\phi_c + (1-\lambda)\ \phi_c = (\phi_c - 2\lambda\ \phi_c)$$

As λ can be chosen to lie between 0 and 1, the mean value of the field can be chosen anywhere between $-\phi_c$ and ϕ_c. But we also have that the free energy of the state ρ

$$F(\rho) = \langle E\rangle - T\left(\mathrm{tr}(\rho\ \ell n\ \rho)\right)$$

is bounded by

$$F(\rho) \leqslant \lambda\ F(\rho) + (1-\lambda)\ F(\rho-)$$

because of the additivity of the energies of two density matrices and the convexity of the entropy function.[2] I.e., the true free energy must always be less than the free energy along any straight line joining two points on the free energy curve.

Although this certainly throws doubt on the one loop approximation as a valid approximation to the true equilibrium free energy, it may still be useful in calculating the dynamical behaviour of the field. The one loop approximation essentially assumes that the density matrix is Gaussian in the field strength. If the dynamical state of the field is such a Gaussian density matrix, then the one loop approximation will give a better approximation to the free energy of the field in that state than does the true equilibrium free energy. One must, however, show that the state of the field, after the expansion of the universe has dropped the temperature to near zero, is actually given by that Gaussian distribution which the one loop approximation assumes.

This brings me to the heart of the difficulty with the usual scenarios. The question of which state (by which I mean density matrix) the field is left in after it has cooled through the phase transition temperature is a dynamical question, a question concerned with non-equilibrium statistical mechanics. The equilibrium free energy (or the one loop approximation thereto) does not tell one much about the way in which the field actually behaves in such a non-equilibrium situations. In particular, for the problem under consideration, one can think of an alternative way in which the field can approach equilibrium. This way is that which actually occurs for an Ising model. At high temperatures the field ϕ is a rapidly fluctuating one, taking on a large values (much larger than ϕ_c). (This is not in contradiction with the sharp minimum in the effective potential $W(\phi_0)$. ϕ_0 is the spatial average of the field and the sharp minimum simply states that the average value of the field over large volumes must be zero. Although this is true if the field everywhere is near zero, it is also true if the field has large fluctuations which, because of small correlation lengths, average out to zero.)

As the field cools, the field ϕ becomes trapped in the potential minima at $\pm\phi_c$ with however short correlation lengths. One has now innumerable regions in which the field is randomly oriented near $\pm\phi_c$. The mean spatial field is still zero. Gradually however, the dynamic motion of the walls of these regions causes one or the other to dominate, and eventually one region grows to include all of spacetime. At this point the phase transition has been completed.

The important point is that in this scenario, the energy momentum tensor is never "potential dominated". For regions where $\phi \sim \phi_c$, the potential is, by assumption, zero (in order that the universe not expand exponentially in its current zero temperature phase). In the boundary regions where two areas of $\phi = \pm\phi_c$ meet, the field will be highly dynamic as these boundaries move around under the field dynamics. Neither $\partial\phi/\partial t$ nor $\partial\phi/\partial x^i$ are expected to be zero in these areas, and the size of these boundary regions are probably far too small to allow them to inflate.

The crucial point is not that this latter, non inflationary, scenario is the most likely one for a realistic model. The point is that the equilibrium, or quasi (one loop) equilibrium calculations do not give any information about the actual dynamical behaviour of the fields as the universe cools through the phase transition temperature. The key assumption in the inflationary scenario is that the universe is left in a state which is far from equilibrium. In order to determine whether or not the universe goes through a period of exponential expansion, one must know what that state is, a knowledge which cannot be obtained by looking at equilibrium, approximately equilibrium configurations, or one loop approximations to equilibrium configurations.

REFERENCES

[1] Mazenko, G.F., Unruh, W.G., Wald, R.M. Phys. Rev. D. 31, 273 (1985).

[2] Wehrl, A., Rev. Mod. Phys. 50, 221 (1978).

TUNNELING TRANSITIONS WITH GRAVITATION: BREAKING OF THE QUASICLASSICAL APPROXIMATION

G.V.Lavrelashvili, V.A.Rubakov, P.G.Tinyakov
Institute for Nuclear Research of the Academy of Sciences of the USSR, 117312, Moscow, 60th October Anniversary prospect 7 a, USSR

ABSTRACT

We discuss the problem of the validity of the quasiclassical approximation for the description of tunneling transitions with gravitational effects taken into account. We present two model examples in which quasiclassics is not applicable. The first example is the closed homogeneous and isotropic Universe with positive cosmological constant undergoing the transition from the Friedman regime to DeSitter-like one, This example is relevant to the problem of creation of the Universe from "nothing". The second example is the transition leading to the materialization of the bubble of a new phase in the metastable vacuum in a model with a certain relation between parameters. We discuss the physical reasons for breaking of the quasiclassical description of tunneling in these models and speculate on possible alternative processes. We point out that the quasiclassical approximation might be inapplicable for other tunneling transitions with gravity.

1. INTRODUCTION

Presently, much attention is paid to studying various tunneling process which could occur in the very early Universe. Two kinds of them are of particular interest, namely the processes associated with the first order phase transitions [1] and those related to the possibility of the creation of the Universe from "nothing" [2,3]. The former transitions are widely accepted to proceed through the materialization of the bubbles of the new phase in the metastable vacuum [4-8], while the latter could occur homogeneously in the entire Universe[3]. In both cases, the existing attempts of the quantitative description of tunneling are based on the quasiclassical approximation, much in common to ordinary quantum mechanics. Presently, we see no reason to doubt the validity of this approximation in the situation where the gravitational effects are negligible. However, the gravitational effects being included, the quasiclassical description of tunneling becomes less obvious.

The most evident reason is the non-renormalizability of quantum gravity. We have nothing to say on this problem here, so we assume, as most authors do, that there exists a (natural?) cut-off at momenta of order of Planck mass, M_{pl} . Another reason is that the Euclidean gravitational action, unlike the actions for matter fields, is not positive-definite. The latter property can make the classical euclidean trajectories unstable, which would be a clear signal for breaking of quasiclassics. The main purpose of this paper is to illustrate this possibility by two toy examples, one of which is related to the problem of the creation of the Universe from "nothing", another is the well-known process of the materialization of the bubble.

Our strategy will be as follows. In both examples, we begin by trying to apply the quasiclassical approximation and consider the fluctuations around the euclidean classical trajectories. We find that this trajectories are unstable in the following sence: there exists a large number of negative modes among the fluctuations (this number tends to infinity in the limit infinite ultraviolet cut-off), so that quasiclassics is, in fact, not applicable. Of course we cannot consider in this way all possible classical trajectories, so we restrict ourselves to the most symmetric trajectories (homogeneous and isotopic Universe in the first example and O(4)-invariant bounce in the second one), which are widely accepted to be the most important ones. However, we think that these instabilities are not peculiar to the particular trajectories: we think that there exist physical reasons for the inapplicability of quasiclassics. It is worth noting that breaking of quasiclassics has nothing to do with the Planck scales; indeed, we will choose the parameters of our models in such a way that the tunneling transitions will be characterized by the energy densities much less than Planck ones and length scales much larger than Planck ones.

This paper is organized as follows. In sect.2 we discuss our first model, namely, the homogeneous and isotropic Universe with positive cosmological constant which undergoes the tunneling transition from Friedmann regime to De Sitter-like one. To the zeroth order in quasiclassics this transition was considered in refs.[9,10] and the related transition from "nothing" to De Sitter-like Universe was discussed, also to the zeroth order, in refs.[3,11-13]. We recapitulate the classical regimes and zeroth order results in subsects. 2.1 and 2.2 respectively.

Our main purpose in Sect.2 is to study the fluctuations around the classical euclidean trajectory*).

---

*) The main results of this study were briefly reported earlier [11,14].

This is done in subsect.2.3. For the sake of definiteness, we consider only fluctuations of massive scalar field conformally coupled to gravity; other matter fields, as well as gravitons, can be treated in the same way with essentially the same results. We find it useful to apply the formalism based on the Dirac-Wheeler-DeWitt equation of the canonical quantum gravity[15,16], in which the first correction to the zeroth-order quasiclassics is naturally interpreted as particle creation during the tunneling transition. Our technique is in several respects analogus to that used to describe particle creation during the non-gravitational tunneling[17,18]. Our main observation is that the particle creation in the classicaly forbidden region is very rapid and the larger the particle momenta, the larger the creation rate. We give the physical explanation of this phenomenon. The number of particles in each mode becomes (formally) infinite at some point in the classicaly forbidden region; for high momentum modes these points correspond to the very beginning of tunneling. Clearly, this means that the first correction to quasiclassics is very large (formaly, infinite) everywhere in the forbidden region, so that the quasiclassical approximation breaks down.

We then discuss the same phenomenon in the functional integral language. We point out that in this model the zeroth order quasiclassics can be obtained from the functional integral only if the usual measure e^{-S} is replaced by e^{+S}. This prescription was also suggested from slightly different reasoning in ref.[12]. According to this prescription, all modes of matter fields (as well as gravitons) are negative, in the sence that they give positive contribution to the expotential. Therefore, there exists an infinite number of negative modes, and we again come to the conclusion that the quasiclassical approximation is broken.

In subsect.2.4 we speculate on an alternative process

which is likely to occur instead of the quasiclassical tunneling in this model.

In sect.3 we consider the fluctuations around the bounce solution[7] describing the formation of the bubble of a new phase in the metastable vacuum (in this case the functional measure is the standard e^{-S}). In subsect.3.1 we recapitulate some features of the bounce, and in subsect.3.2 we find that, for some range of parameters, there exists an infinite number of negative modes, so that quasiclassics is again broken. These modes are O(4)-invariant; in subsect 3.3 we check, within the thin wall approximation, that the number of negative modes, non-trivially transforming under O(4) is finite (if any).

In this model, the physical explanation of the breaking of quasiclassics is less straightforward. However, we think we have some understanding of this phenomenon. We discuss this point in subsect.3.4, where we also speculate on a possible altrnative to the quasiclassical formation of a bubble.

Sect.4 is devoted to concluding remarks. The main conclusion is negative of course: the quasiclassical approximation should sometimes be replaced by some other technique, but presently we have no idea on what this new technique is. Furthermore, it is even unclear whether the non-quasiclassical process are essential or not in models with stable classical euclidean solutions. We think that the present understanding of tunneling transitions with gravity effects is rather incomplete.

2. THE HOMOGENEOUS TRANSITION FROM THE FRIEDMANN REGIME TO DE SITTER-LIKE ONE

As the first example, we consider the closed Friedmann-Robertson-Walker Universe with positive cosmological constant Λ and free massive scalar field $\tilde{\varphi}$ and free massless scalar field $\tilde{\chi}$ both conformally coupled to gravity. To introduce notations, we begin with the brief description of classical motions.

2.1 Classical Motions.

Let $N(t)$ be the lapse function and $R(t)$ be the radius of the Universe, so that the most general closed FRW metrics is

$$ds^2 = N^2 dt^2 + R^2 d\Omega^2$$

where $d\Omega^2$ is the metrics of the unit 3-sphere. Let $\varphi = R\tilde{\varphi}$, $\chi = R\tilde{\chi}$ be conformal variables. At the classical level one can consistently set $\varphi = 0$ and take the field χ to be homogeneous, $\chi = (2\pi^2)^{-1/2}\chi_0(t)$. Then the classical field equations, including the Einstein equations can be derived from the following action functional,

$$S_1 = S_G + S_{\chi_0} \tag{2.1}$$

where

$$S_G = \frac{3\pi}{4} M_{pe}^2 \int \left(-\frac{R\dot{R}^2}{N} + NR - \frac{\Lambda}{3} NR^3 \right) dt \tag{2.2}$$

is the standard Einstein-Hilbert action (plus cosmological constant term) with all variables except N and R frozen out,

$$S_{\chi_0} = \int \left(\frac{R}{2N} \dot{\chi}_0^2 - \frac{N}{2R} \chi_0^2 \right) dt$$

is the action for massless conformal scalar field. We assume $0 < \Lambda \ll M_{pe}^2$ in what follows.

For our purposes, it is convenient to introduce the canonical momenta

$$\pi_R = -\frac{3\pi}{2} M_{pe}^2 \frac{R\dot{R}}{N} , \quad \pi_{\chi_0} = \frac{R}{N} \dot{\chi}_0$$

and rewrite the action in the hamiltonial form,

$$S_1 = \int (\pi_R \cdot \dot{R} + \pi_{\chi_0} \cdot \dot{\chi}_0 - N H_1) dt$$

where

$$H_1 = \frac{1}{R} \left[- \frac{1}{3\pi M_{pe}^2} \pi_R^2 - \mathcal{U}(R) + \mathcal{E}_0 \right]$$

$$\mathcal{U}(R) = \frac{3\pi}{4} M_{pe}^2 R^2 - \frac{\pi}{4} \Lambda M_{pe}^2 R^4$$

$$\mathcal{E}_0 = \frac{1}{2} \pi_{\chi_0}^2 + \frac{1}{2} \chi_0^2$$

One can see that $\mathcal{E}_0$ is conserved, so that the Einstein equations are reduced to the only equation

$$H_1 = 0 \tag{2.3}$$

Note that this equation can be formally interpreted as the requirement that the total energy of the Universe is zero. Indeed, $\mathcal{E}_0$ can be viewed as the energy (defined with respect to conformal time, for which $N = R$) of the field χ_0 , so that the quantity

$$H_G^c \equiv R H_G \equiv - \frac{1}{3\pi M_{pe}^2} \pi_R^2 - \mathcal{U}(R) \tag{2.4}$$

can be interpreted as the (conformal) energy of the gravitational field. Although this interpretation is purely formal, we shall adopt this language in what follows.

Eq.(2.3) can be viewed also as describing the motion of a"particle" with "energy" $\mathcal{E}_0$ in the "potential" $\mathcal{U}(R)$ (of course, this analogy is also purely formal), which is shown in fig.1. For $1 \ll \mathcal{E}_0 < \mathcal{U}_{max}$ there exist two classically allowed regions, I and II, which correspond to the Friedmann regime and De Sitter-like one respectively. For. $\mathcal{E}_0 \lesssim 1$, the classical turning point R_1 corresponds to the Universe of the Planck size, so that the classical considerations in the region I are not reliable while the classical motion right to the barrier is still possible. To avoid uncertaincies connected with the dynamics at the

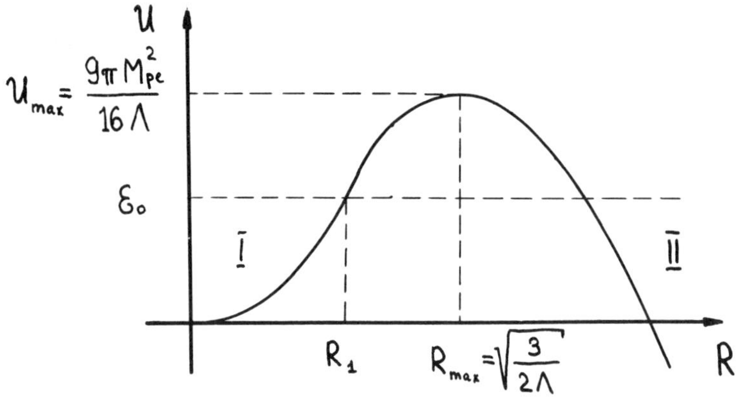

Fig. 1

Planck scales, we concentrate on the case $1 \ll \mathcal{E}_0 < \mathcal{U}_{max}$ in what follows (unless explicitly indicated).

2.2 The Zeroth Order Approximation.

In the quantum theory, the transition from the region I to the region II is allowed. At first sight, it seems possible to describe this transition within the semiclassical approximation[9,10]. The main purpose of sect.2 is to show that this is not true, since the corrections to the zeroth order quasiclassics are extremely large (formally, infinite). Before turning to these corrections, let us briefly discuss the zeroth order quasiclassics itself[9,10].

To the zeroth order, one can still neglect the non-homogeneous and anisotropic fluctuations of the gravitational field (gravitons) as well as of the field φ. We also recall that the field χ decouples to this order; the only trace of it is the constant $\mathcal{E}_0$. Keeping in mind these observations, we write the basic equation determining the wave function

$$H_1\Psi(R) \equiv \frac{1}{R}\left(-\frac{1}{3\pi M_{pe}^2}\tilde{\pi}_R^2 - \mathcal{U}(R) + \mathcal{E}_0\right)\Psi(R) = 0 \qquad (2.5)$$

which is nothing but Dirac-Wheeler-DeWitt equation[15,16] with all degrees of freedom except R frozen out. Here $\hat{\pi}_R = -i\,\partial/\partial R$; the choice of the ordering of the non-commuting operators R and $\hat{\pi}_R$ is not essential for our purposes (in connection with this see, e.g.[19]). We note in passing that equations of this kind are widely discussed in "quantum cosmology" (for a review see, e.g.[20]).

The semiclassical wave function in the forbidden region is found from eq. (2.5) to be

$$\Psi^{(0)}(R) = e^{-A(R)} \tag{2.6}$$

where

$$A(R) = \int_{R_1}^{R} dR \sqrt{3\pi M_{pe}^2 (\mathcal{U} - \mathcal{E}_0)} \tag{2.7}$$

Note that $\Psi^{(0)}(R)$ exponentially decreases in the forbidden region, as it should. Note also that $A(R)$ can be understood as the value of the "euclidean action"

$$A(R) = S_E(R(\tau)), \tag{2.8}$$

$$S_E = \int \left[\frac{3\pi}{4} M_{pe}^2 \left(\frac{R R'^2}{N_E} + N_E R - \frac{\Lambda}{3} N_E R^3\right) - \frac{N_E}{R}\,\mathcal{E}_0\right] d\tau \tag{2.9}$$

(where τ is the "euclidean time", N_E is an arbitrary "euclidean lapse function" and prime denotes $\partial/\partial\tau$) evaluated at the"euclidean trajectory" $R(\tau)$ connecting R_1 and R and obeying the euclidean Einstein equations. The latter equations are in fact reduced to the only equation

$$\begin{aligned}\frac{dR(\tau)}{d\tau} &= \frac{2}{3\pi M_{pe}^2}\cdot\frac{N_E}{R}\sqrt{3\pi M_{pe}^2(\mathcal{U}-\mathcal{E}_0)}\,\Bigg|_{R=R(\tau)} \\ &\equiv \frac{2}{3\pi M_{pe}^2}\cdot\frac{N_E}{R}\cdot\frac{dA}{dR}\,\Bigg|_{R=R(\tau)}\end{aligned} \tag{2.10}$$

For future use, we present the explicit solution to eq. (2.10) near the turning point R_1, with the conformal choice of the "euclidean time":

$$N_E = R$$

$$R(\tau) = R_1 + \frac{R_1}{2}\left(1 - \frac{2\Lambda}{3} R_1^2\right)\tau^2 \qquad (2.11)$$

It is worth pointing out that the sign of the "euclidean action" (2.9) is unusual: this sign could formally be obtained if one rotated the time variable as follows: $t \to +i\tau$ while the standard prescription is $t \to -i\tau$. We shall return to this point in the next subsection.

We note finally that for $\varepsilon_0 \lesssim 1$ the tunneling transition just described can be considered as a toy model for the creation of the Universe from "nothing"[11-13]. Indeed, in this case the expanding DeSitter Universe appears as a result of the transition from the purely quantum state with all parameters of order of Planck ones; the latter state can be identified with "quantum puff"[21] or "mini-Universe"[19] or "nothing"[11-13].

2.3 First Order Corrections: Particle Creation in the Tunneling Universe.

We now turn to the discussion of the first order corrections to the quasiclassical wave function (2.6). To this end, we have to introduce new terms to the action (2.1) which describe the fluctuations of the gravitational field (gravitons) as well as the fields φ and χ. For the sake of simplicity, we discuss here only the fluctuations of the field φ: the field χ is massless and its effects are solely due to the conformal anomaly which we are not going to discuss here, while the consideration of gravitons parallels that of φ-particles (with technical complications due to the tensor indices and gauge invariance, cf.[22]).

Turning on fluctuations changes the form of the Dirac-Wheeler-DeWitt equation: instead of eq.(2.5) we now have

$$(H_1+H_\varphi)|\Psi\rangle \equiv \left\{ \frac{1}{R}\left[-\frac{1}{3\pi M_{pe}^2}\left(-i\frac{\partial}{\partial R}\right)^2 - \mathcal{U}(R)+\mathcal{E}_0\right] + H_\varphi \right\}|\Psi\rangle = 0 \qquad (2.12)$$

where we disregard the fluctuations of the field χ and gravitons. The state vector $|\Psi\rangle$ is now a function of R taking values in the matter Hilbert space where the operators $\varphi(\vec{x})$ and $\pi_\varphi(\vec{x})$ act; for instance, $|\Psi(R)\rangle$ can be viewed as the Fock column with R -dependent coefficients. H_φ is, roughly speaking, the hamiltonian for the field φ ; its explicit form will be given later (of course, H_φ contains all partial waves of the field φ).

To derive the equation governing the behaviour of fluctuations, we extract the semiclassical part from $|\Psi(R)\rangle$ and write (cf.[23,17,18])

$$|\Psi(R)\rangle = e^{-A(R)-A_1(R)}\,|\Phi(R)\rangle \qquad (2.13)$$

where $A(R)$ is defined by eq.(2.7) and $A_1 = \frac{1}{2}\ln\sqrt{\mathcal{U}-\mathcal{E}}$. Inserting (2.12) into eq. (2.12) and neglecting the second derivatives of $|\Phi\rangle$ with respect to R (they are second order in $\hbar$) we obtain

$$-\frac{2}{3\pi M_{pe}^2}\cdot\frac{1}{R}\frac{dA}{dR}\frac{\partial|\Phi\rangle}{\partial R} + H_\varphi|\Phi\rangle = 0$$

We now introduce the euclidean "time" τ performing the change of variables from R to τ according to eq.(2.10). Considering $|\Phi\rangle$ as a function of τ, $|\Phi(\tau)\rangle \equiv |\Phi(R(\tau))\rangle$, we find

$$\frac{\partial|\Phi\rangle}{\partial\tau} = N_E H_\varphi|\Phi\rangle \qquad (2.14)$$

which is the desired equation. In what follows it will be convenient to use the conformal choice of the euclidean time, $N_E = R$; with this choice eq.(2.14) reads

$$\frac{\partial|\Phi\rangle}{\partial\tau} = H_\varphi^{(c)}\,|\Phi\rangle\,; \quad N = R \tag{2.15}$$

where $H_\varphi^{(c)} = R\,H_\varphi$ is the conformal hamiltonian of the field φ .

Let us stop at this point to discuss the main peculiarity of this equation, namely, the unusual sign on the right hand side. Indeed, the sign in the corresponding equation in the case of non-gravitational tunneling is opposite (negative)[17,18]. The formal reason for this difference is the negative sign of the gravitational "energy" (2.4). The underlying physics is quite different. In the non-gravitational case, raising the energy of fluctuations lowers the (positive) energy of the tunneling subsystem thus lowering the probability of the barrier penetration [17,18]. In our case, raising the energy of fluctuations (i.e., excitations of the field φ) also leads to lowering the energy of the tunneling subsystem (the tunneling subsystem is now characterized by the only variable R); however, in our case the tunneling subsystem has negative energy, so that the probability of penetration becomes <u>larger</u>. In other words, raising the total energy of matter filling the Universe (by, e.g., creation of φ -particles) corresponds, roughly speaking, to raising the value of $\mathcal{E}_0$ which clearly makes easier the barrier penetration (see fig.1). This is reflected by eq.(2.14): for positive matter energy $N_\varepsilon H_\varphi$, the wave function $|\Psi(\tau)\rangle$ exponentially grows in the forbidden region (it is worth noting that in any case the total wave function $|\Psi(\tau)\rangle$, eq. (2.13), should decrease in the forbidden region, otherwise it would sease to describe the tunneling from the region I to the region II). These qualitative remarks explain the physical reason for very rapid particle creation during the tunneling transition: particle creation exponentially raises the probability of the barrier penetration, therefore, it is a favourable process.

To proceed quantitatively, we first note that the field $\varphi(\vec{x})$ and its conjugate momentum $\pi_\varphi(\vec{x})$ are conveniently decomposed over the harmonics on a unit 3-sphere,

$$\begin{pmatrix} \varphi(\vec{x}) \\ \pi_\varphi(\vec{x}) \end{pmatrix} = \sum_{k,\ell,m} \begin{pmatrix} \varphi_{k\ell m} \\ \pi_{k\ell m} \end{pmatrix} Y_{k\ell m}\left(\frac{\vec{x}}{r}\right)$$

$$k = 1, 2, \ldots \; ; \; \ell = 0, 1, \ldots k-1 \; ; \; m = -\ell, \ldots, \ell .$$

Here $\varphi_{k\ell m}$ and $\pi_{k\ell m}$ are quantum mechanical variables obeying the standard commutation relation,

$$[\pi_{k\ell m}, \varphi_{k'\ell' m'}] = -i\, \delta_{kk'}\, \delta_{\ell\ell'}\, \delta_{mm'}$$

In terms of these variables, the conformal hamiltonian reads

$$H^{(c)}_\varphi = \sum_{k,\ell,m} H^{(c)}_{k\ell m} \tag{2.16}$$

where

$$H^{(c)}_{k\ell m} = \frac{1}{2}\left(\pi^2_{k\ell m} + \omega^2_k(\tau)\, \varphi^2_{k\ell m}\right)$$
$$\omega^2_k = k^2 + \mu^2 R^2 \tag{2.17}$$

μ being mass of the field φ. Note that k and ω_k are conformal momentum and conformal energy of a particle in the mode (k, ℓ, m); the physical momentum and energy are

$$p = \frac{k}{R}, \quad \varepsilon_p = \frac{\omega_k}{R} \tag{2.18}$$

We now have to solve eq.(2.15). Since modes with diferent quantum numbers decouple, we can consider them separately, i.e. instead of eq. (2.15) we can solve the following equation

$$\frac{\partial |\Phi_{k\ell m}\rangle}{\partial \tau} = H^{(c)}_{k\ell m} |\Phi_{k\ell m}\rangle \tag{2.19}$$

where $H^{c}_{k\ell m}$ explicitly depends on τ through $R(\tau)$. The wave function $|\Phi\rangle$ is a product of all $|\Phi_{k\ell m}\rangle$. One way to solve eq.(2.19) is provided by the non-unitary Bogoliubov transformation technique[18]. However, we prefer to use here a simpler technique based on the notion of the n-particle amplitudes[17].

We first note that, for each τ, the conformal hamiltonian (2.17) is nothing but the hamiltonian for a harmonic oscillator. Let $|n\rangle_\tau$ be the ("time"-dependent) eigenstates of $H^{(c)}$ (we omit subscripts (k, ℓ, m) where-ever possible), the eigenvalues are ($n + 1/2$) ω_k. The state $|n\rangle_\tau$ can be identified with the state with n particles in a mode (k, ℓ, m) at "time" τ (i.e. when the radius of the Universe is equal to $R(\tau)$). This interpretation is unambiguous for slowly varying hamiltonians; the harmonic oscillator hamiltonian (2.17) is indeed slowly varying if

$$\omega_k^{-2}\, \partial_\tau \omega_k \ll 1 \tag{2.20}$$

which is correct at least for high momentum modes. For our purposes, it is sufficient to discuss the modes for which (2.20) is valid. At each τ, we decompose $|\Phi(\tau)\rangle$ over the set $|n\rangle_\tau$:

$$|\Phi(\tau)\rangle = \exp\left(\int_0^\tau \frac{\omega_k}{2}\, d\tau \right) \sum_n C_n(\tau)\, |n\rangle_\tau$$

where the common expotential factor is introduced for convenience, and $C_n(\tau)$ are n-particle amplitudes. It is straightforward to see that eq.(2.19) leads to the following set of equations

$$\frac{\partial C_n}{\partial \tau} = n\,\omega_k C_n + \sum_{n' \neq n} \frac{C_n}{\omega_k (n' - n)} \langle n | \frac{\partial H^{(c)}}{\partial \tau} | n' \rangle \tag{2.21}$$

We also have $\partial_\tau H^{(c)} = \frac{1}{2}\, \partial_\tau\, \omega_k^2\, \varphi^2$, so the sum on the right hand side of eq.(2.21) consists only of two terms with $n' = n - 2$ (two particle creation term) and

$n' = n + 2$ (two particle annihilation one). We get finally

$$\frac{\partial C_n}{\partial \tau} = n\omega_k C_n + \frac{1}{\omega_k}\frac{\partial \omega_k}{\partial \tau}\sqrt{n(n-1)}\, C_{n-2} + \frac{1}{\omega_k}\frac{\partial \omega_k}{\partial \tau}\sqrt{(n+1)(n+2)}\, C_{n+2} \quad (2.22)$$

To see the effect of particle creation we choose the state of the field φ at the classical turning point R_1 to be the vacuum state $|\Phi(\tau=0)\rangle = |0\rangle_{\tau=0}$, i.e. $C_n(\tau=0) = \delta_{n0}$. Then the last term on the right hand side of eq.(2.22) (two particle annihilation) is small compared to the first term essentially everywhere (recall the inequality (2.20)). Neglecting this term we obtain the solution of eq.(2.22):

$$C_{2n+1} = 0_\tau$$

$$C_{2n}(\tau) = \frac{\sqrt{(2n)!}}{n!}\left[\int_0^\tau d\tau'\, e^{2\int_{\tau'}^{\tau}\omega_k d\tau'} \frac{1}{4\,\omega_k(\tau')} \cdot \frac{\partial \omega_k(\tau')}{\partial \tau'}\right]^n \quad (2.23)$$

This formula can be interpreted as follows. The factor $\frac{1}{\omega_k(\tau)} \cdot \frac{\partial \omega_k(\tau')}{\partial \tau'}$ can be understood as the amplitude of creation of a pair of φ-particles at "time" τ (up to some Bose factors, see eq.(2.22)), while the exponential factor corresponds to the growth of the wave function with two extra particles during the "time" $(\tau - \tau')$ (this, as discussed above, corresponds to the larger probability of the barrier penetration).

Eq. (2.23) shows that the n-particle amplitudes grow with τ, and the number of created particles in the mode (k, ℓ, m) becomes infinite when the integral in the square brackets in eq. (2.23) becomes equal to $1/2$ (indeed, at this "time" $C_n \propto n^{-1/2}$ for large n; note that the annihilation term in eq. (2.22) is still negligible at this "time"). For sufficiently large momenta, this integral is easily calculated since one can neglect the "time"-dependence of all quantities except $\partial_\tau \omega_k$ and use eq. (2.11) for the explicit form of $R(\tau)$. We find that the number of created particles in the mode (k, ℓ, m) becomes infinite at"time"

$$\tau_{n_p=\infty} = \frac{1}{2\varepsilon_p R_1} \ln \left[\frac{2\varepsilon_p^4 R_1^2}{\mu^2 (1 - \frac{2\Lambda}{3} R_1^2)} \right] \qquad (2\text{-}24)$$

where we use the physical quantities (2.18). The radius of the Universe at this "time" can be read off from eq.(2.11). Eq. (2.24) shows that $\tau_{n_p=\infty}$ tends to zero as $p \to \infty$; this means that the number of created particles with high energy becomes large when the radius of the Universe is very close to the classical turning point R_1 (but still in the forbidden region). We conclude that high energy particles are very intensively created at the very beginning of tunneling, so that their total number becomes formally infinite everywhere in the forbidden region. The first correction $|\Phi(R)\rangle$ is divergent and quasiclassics is not applicable.

What is the functional integral signal of the breaking of quasiclassics in our model? To answer this question we first recall that the standard functional integral description of tunneling begins with the rotation of the real time integral $\int e^{iS}$ to the euclidean domain, $t = -i\tau$ and the euclidean integral, $\int e^{-S}$, emerges. The extremum of the euclidean action is the path of least resistance [24,5] and it determines the tunneling amplitude. However, the rotation in the same direction in the theory with the action (2.2) would lead to the integral $\int e^{+S_E}$, where S_E is given by eq.(2.9), so that one could find that the lowest order tunneling amplitude was equal to e^{+A} (where A is defined by eq.(2.8)) which is obviously wrong. This is of course due to the unusual sign of the right hand side of eq.(2.2). To obtain the correct lowest order result one has to perform the rotation in the opposite direction, $t = +i\tau$. However, after this rotation, the contribution to the exponential due to matter fields (as well as gravitons) is positive, i.e. the functional integral over these fields is divergent. This divergency just signalizes the breaking of the quasiclassical approximation.

To conclude this subsection, we would like to note that although we have considered explicitly only the particular field (massive scalar field conformally coupled to gravity), our main conclusions are expected to be the same for other fields (including gravitons themselves) which quanta can be created by the gravitational field of the non-stationary closed FRW Universe. We note finally that the quasiclassical approximation seems to break down also in the case of the decay of the "quantum puff" into the DeSitter-like expanding Universe in our toy model; indeed, the functional integral arguments are directly applicable to this process. So, we think that the quantitative description of the creation of the Universe from "nothing" is to be based on a technique different from the standard quasiclassics.

2.4. A Possible Alternative Process.

Let us now speculate on a possible alternative process which presumably occurs instead of the quasiclassical tunnelig in our toy model. We have seen that the Universe tends to create particles during the tunneling process thus raising the probability of the barrier penetration. It seems reasonable that at each R in the forbidden region and left to R_{max}, the most probable state corresponds to the Universe filled with particles (and gravitons) having the total conformal energy roughly equal to $V(R) - \mathcal{E}_0$. Indeed, for smaller conformal energy the wave function $|\Psi(R)\rangle$ is exponentially suppressed, while higher energy states correspond to classical motions, not to the classically forbidden ones. So, we expect that the most probable state right to the barrier is the De Sitter-like expanding Universe which has enough particles to start its expansion from the top of the potential of fig.1. Instead of penetrating through the barrier, the Universe seems to prefer to jump on the top of it!

The results of subsect.2.3, in particular eq.(2.24), indicate that this expanding Universe is initially filled with high energy (and most heavy) particles since these are created most rapidly in the forbidden region.

We also expect that the transition probability, instead of being suppressed by $\exp(-\mathrm{const}/\Lambda)$, is suppressed only by some power of Λ^{-1}. Thus, both qualitative and quantative aspects of the tunneling process are likely to be much altered by the particle creation phenomenon discussed in this section.

3. FORMATION OF A BUBBLE OF A NEW PHASE IN THE METASTABLE VACUUM.

As the second example we consider the tunneling transition leading to the materialization of a bubble of a new phase in the metastable vacuum. Our main observation is that for some range of parameters, the gravitational effects make the quasiclassical description of this process inapplicable because of the existence of an infinite number of negative modes around the bounce solution. In what follows we shall often use the thin wall bounce[7] as an illustration although our main results are not restricted ti this particular case. Before turning to the discussion of the fluctuations around the bounce solution, let us briefly recapitulate some features of this solution itself[7].

3.1 The Bounce Solution.

The (real-time) action defining the model to be considered in the section is

$$S = \int d^4x \sqrt{-g} \left[\frac{1}{2} \partial_\mu \varphi \partial^\mu \varphi - V(\varphi) - \frac{1}{2\varkappa} R \right] \tag{3.1}$$

where R is the curvative scalar, φ is real scalar field and $V(\varphi)$ has the form shown in fig.2. To describe

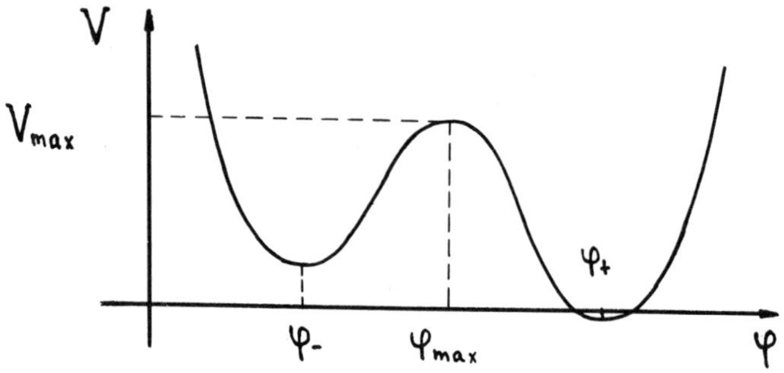

Fig 2

the transition from the false vacuum, $\langle \varphi \rangle = \varphi_-$, to the true one, $\langle \varphi \rangle = \varphi_+$, one usually begins with performing the Wick rotation to the euclidean space-time, so that instead of the action (3.1) one considers its euclidean counterpart,

$$S_E = \int d^4x \sqrt{g} \left[\frac{1}{2} \partial_\mu \varphi \partial^\mu \varphi + V(\varphi) - \frac{1}{2\varkappa} R \right] \tag{3.2}$$

The main contribution to the functional integral determining the amplitude of the bubble formation is assumed to come from neighbourhood of the bounce solution. In what follows we discuss the O(4) invariant bounce which is widely accepted to be the most important one. The most general O(4) invariant metrics reads

$$ds^2 = N(\xi)^2 d\xi^2 + \rho(\xi)^2 d\Omega^2$$

where $d\Omega^2$ is the metrics on the unit 3-sphere. With this ansatz (and $\varphi = \varphi(\xi)$) one rewrites the action (3.2) in the following form,

$$S_E = 2\pi^2 \int d\xi \left[\frac{\rho^3}{2N} \varphi'^2 - \frac{3\rho}{\varkappa N} \rho'^2 + N\rho^3 V(\varphi) - \frac{3N\rho}{\varkappa} \right] \tag{3.3}$$

where prime denotes $d/d\xi$. This action leads to the field equations,

$$\varphi'' + \left(3\frac{\rho'}{\rho} - \frac{N'}{N}\right)\varphi' = N^2 \frac{dV}{d\varphi} \tag{3.4a}$$

$$\rho^2 \varphi'^2 - \frac{6}{\varkappa}\rho'^2 = 2N^2\left(\rho^2 V - \frac{3}{\varkappa}\right) \tag{3.4b}$$

which coincide with the Coleman-DeLuccia equations for $N = 1$ (variation of (3.3) with respect to ρ does not lead to any new equation).

The bounce configuration $\rho_c(\xi), \varphi_c(\xi), N_c = 1$ which is the solution to eq. (3.4), has the most simple form in the thin wall case. This case is realized when the difference between the energy densities in the false and true vacua, $\varepsilon \equiv V(\varphi_-) - V(\varphi_+)$, is small[7]. In this case the bounce solution has $\varphi \approx \varphi_+$ inside the sphere of radius $\bar{\rho}$, $\varphi \approx \varphi_-$ outside this sphere and a sharp (though continuous) transition between these values at $\rho \approx \bar{\rho}$. It is essential for our purposes that the radius of the wall, $\bar{\rho}$ is essestially a free parameter of the model and can be arbitrarly large. For instance, for $V(\varphi_-)=0$ one has[7].

$$\bar{\rho} = \frac{\bar{\rho}_0}{1 - \frac{\varkappa}{12}\bar{\rho}_0 \varepsilon} \tag{3.5}$$

where $\bar{\rho}_0 = \frac{3S_1}{\varepsilon}$; $S_1 = \int_{\varphi_+}^{\varphi_-} \sqrt{2(V(\varphi) - V(\varphi_-))}\, d\varphi$

3.2 O(4) - Invariant Fluctuations.

We now turn to the discussion of the fluctuations around the bounce solution. In this section we consider the O(4)-invariant fluctuations. These fluctuations are described by the only independent variable, which we choose to be $\hat{\varphi}(\xi) = \varphi(\xi) - \varphi_c(\xi)$; the gauge condition being imposed, the O(4) invariant variables are expressed through $\hat{\varphi}(\xi)$ with the use of "Gauss' law" (3.4b).

We find it convenient to proceed in the hamiltonian-like formalism and first rewrite the action (3.3) in the hamiltonian-like form (c-f.[25]).

$$S_E = 2\pi^2 \int d\xi \left[\Pi_\varphi \varphi' + \Pi_\rho \cdot \rho' - NH\right]$$

where $\Pi_\varphi = \dfrac{\varphi' \rho^3}{N}$, $\Pi_\rho = -\dfrac{6\rho\rho'}{\varkappa N}$

To discuss the fluctuations, we write

$$\begin{aligned} N &= 1 + \hat{N} \\ \rho &= \rho_c + \hat{\rho} \ , \quad \Pi_\rho = \Pi_{\rho,c} + \hat{\Pi}_\rho \\ \varphi &= \varphi_c + \hat{\varphi} \ , \quad \Pi_\varphi = \Pi_{\varphi,c} + \hat{\Pi}_\varphi \end{aligned} \tag{3.6}$$

where the subscript c denotes the quantities evaluated at the bounce solution while hat denotes the fluctuations. The quadratic part of the action reads

$$S_E^{(2)} = 2\pi^2 \int d\xi \left(\hat{\Pi}_\rho \hat{\rho}' + \hat{\Pi}_\varphi \hat{\varphi}' - NH^{(1)} - H^{(2)} \right) \tag{3.7}$$

where

$$H^{(1)} = \varphi_c' \hat{\Pi}_\varphi + \rho_c' \hat{\Pi}_\rho - \rho_c^3 \cdot \frac{dV(\varphi_c)}{d\varphi_c} \cdot \hat{\varphi}$$

$$H^{(2)} = \frac{\hat{\Pi}_\varphi^2}{2\rho_c^3} - \frac{\varkappa}{12} \hat{\Pi}_\rho^2 - \frac{1}{2} \rho_c^2 \frac{d^2 V}{d\varphi_c^2} \hat{\varphi}^2$$

are linear and quadratic parts of H . Eq. (3.7) implies that there is a constraint (linear in fluctuations part of Gauss' law),

$$H^{(1)} = 0 \tag{3.8}$$

According to the standard procedure[26], one has to eliminate the gauge degrees of freedom by imposing a gauge condition and solving eq.(3.8). The simplest choice of gauge is

$$\hat{\rho} = 0 \tag{3.9}$$

With this choice we obtain from eq. (3.8)

$$\hat{\pi}_\rho = \frac{1}{\rho_c'}\left[-\varphi_c'\hat{\pi}_\varphi + \rho_c^3\frac{dV(\varphi_c)}{d\varphi_c}\cdot\hat{\varphi}\right] \tag{3.10}$$

Inserting these $\hat{\rho}$ and $\hat{\pi}_\rho$ into eq. (3.7) we find the quadratic part of the action in terms of the unconstrained variables $\hat{\varphi}, \hat{\pi}_\varphi$:

$$S_E^{(2)}[\hat{\varphi},\hat{\pi}_\varphi] = 2\pi^2\int d\xi\,(\hat{\pi}_\varphi\hat{\varphi}' - H_{\hat{\varphi}}^{(2)}) \tag{3.11}$$

where

$$H_{\hat{\varphi}}^{(2)} = \frac{1}{2\rho_c^3\rho'^2}\left(\rho_c'^2 - \frac{\varkappa}{6}\rho_c^2\varphi_c'^2\right)\hat{\pi}_\varphi^2 + \frac{\varkappa}{6\rho_c'^2}\rho_c^2\frac{dV(\varphi_c)}{d\varphi_c}\varphi_c'\hat{\varphi}\hat{\pi}_\varphi + \frac{\rho_c^3}{2\rho_c'^2}\left[\frac{\varkappa}{6}\rho_c^2\left(\frac{dV}{d\varphi_c}\right)^2 + \rho_c'^2\frac{d^2V}{d\varphi_c^2}\right]\hat{\varphi}^2 \tag{3.12}$$

Eqs. (3.11) and (3.12) imply that there exists an infinite number of negative modes provided that the parameters of the model are chosen so that

$$\rho_c'^2 - \frac{\varkappa}{6}\rho_c^2\varphi_c'^2 < 0 \tag{3.13}$$

in some region of the euclidean space-time. Indeed, in this case the first term on the right hand side of eq. (3.12) is negative. Performing the gaussian integration over $d\hat{\pi}_\varphi$ in the functional integral over fluctuations,

$$I \equiv \int d\hat{\pi}_\varphi\, d\hat{\varphi}\, \exp\left(-S_E(\hat{\varphi},\hat{\pi}_\varphi)\right), \tag{3.14}$$

we obtain

$$I = \int d\hat{\varphi}\, e^{-S_E(\hat{\varphi})} \tag{3.15}$$

where

$$S_E^{(2)}(\hat{\varphi}) = 2\pi^2 \int d\xi \left[\frac{1}{2} \frac{\rho_c^3 \rho_c'^2}{\rho_c'^2 - \frac{\varkappa}{6} \rho_c^2 \varphi_c'^2} (\hat{\varphi}')^2 + W(\xi) \hat{\varphi}^2 \right] \tag{3.16}$$

Here $W(\xi)$ is expressed through ρ_c and φ_c ; its explicit form is not essential for our purposes. In the region where (3.13) holds, the first term in the integrand of eq.(3.16) is negative, so that high momentum perturbations concentrated in this region give negative contributions to the action; they are negative modes. The functional integral over these fluctuations is divergent, which means that the quasiclassical approximation breakes down. Clearly, the number of negative modes is infinite in the limit of infinite ultraviolet cut-off.

A remaining question is whether there exist sets of parameters of the model for which (3.13) holds in some region while the characteristics of the bounce are far from being Planck ones. To see that the answer is positive, we consider the thin wall case. We first note that the equivalent form of (3.13) is

$$Q \equiv 1 - \frac{\varkappa}{3} \rho_c^2 V(\varphi_c) < 0 \tag{3.17}$$

(since ρ_c and φ_c obey eq.(3.4 b) with $N = 1$). For (3.17) to be valid in some region inside the wall, it is sufficient to require

$$\frac{\varkappa}{3} \bar{\rho}^2 V_{max} > 1 \tag{3.18}$$

If the left hand side of (3.18) is of order of one (but still larger than one) then $\bar{\rho} \approx \bar{\rho}_0$ (see eq.(3.5); recall that $\varepsilon \ll V_{max}$ in the thin wall case). For sufficiently small ε we can have $V_{max} \ll M_{pe}^4$, $S_L \ll M_{pe}^3$ and still

$$\bar{\rho}_0^2 > \frac{3}{8\pi} \cdot \frac{M_{pe}^2}{V_{max}}$$

which essentially coincides with (3.18). We conclude that at least for thin wall bounces, there exist sets of parameters of the model for which the quasiclassical approximation is broken far from the Planck scales.

3.3 O(4) Non-Invariant Fluctuations.

Let us now discuss the fluctuations which are not invariant under O(4). The main purpose of this subsection is to show that the number of negative modes among them is finite, if any. We evaluate the quadratic part of the action in terms of the unconstrained variables and show that it is always positive for rapidly varying O(4) non-invariant fields, as opposed to the case discussed in subsect. 3.2. Namely, we find that the terms in the unconstrained action which contain two derivatives are always positive. This means that the existence of an infinite number of negative modes is peculiar to O(4) invariant perturbations.

As in subsect.3.2. we proceed in the hamiltonian-like formalism and write the euclidean metrics as follows

$$ds^2 = (N^2 + N_i N^i)\, d\xi^2 + 2 N_i\, d\xi\, dx^i + h_{ij}\, dx^i dx^j$$

where x^i $(i = 1, 2, 3)$ are coordinates on hypersurfaces $\xi = \text{const}$ (3-spheres in the absence of perturbations) and latin indices are raised by h^{ij} $(h^{ij} h_{jk} = \delta^j_k$ by difenition). In terms of N, N_i, canonical coordinates h_{ij}, φ and their conjugate momenta

$$\pi^{ij} = \frac{\sqrt{h}\, N}{2\varkappa} \Gamma^0_{mn} (h^{ij} h^{mn} - h^{im} h^{jn})$$

$$\pi_\varphi = \frac{\sqrt{h}}{N} (\varphi' - N^i \partial_i \varphi)$$

(where $h \equiv \det h_{ij}$) the euclidean action can be written as follows (cf.[25])

$$S_E = \int d^4x \, (\pi^{ij} h'_{ij} + \pi_\varphi \varphi' - NH - N_i H^i) \qquad (3.18)$$

where

$$H = 2\varkappa\, \pi^{ij}\pi^{mn} G_{ijmn} + \frac{1}{2\sqrt{h}}\pi_\varphi^2 + \frac{\sqrt{h}}{2\varkappa}\, {}^3R$$
$$- \frac{\sqrt{h}}{2} h^{ij}\partial_i\varphi\partial_j\varphi - \sqrt{h}\cdot V(\varphi) \tag{3.19a}$$

$$H^i = -2\, {}^3\nabla_j \pi^{ij} + \pi_\varphi h^{ij}\partial_j\varphi \tag{3.19b}$$

$$G_{ijmn} = \frac{1}{2\sqrt{h}}\left(h_{im}h_{jn} + h_{in}h_{jm} - h_{ij}h_{mn}\right) \tag{3.19c}$$

${}^3\nabla_j$ and ${}^3R$ are covariant derivative and curvature scalar evaluated with respect to metrics h_{ij}.

To discuss perturbations, we again define

$$N = 1 + \hat{N}, \quad N_i = \hat{N}_i$$
$$h_{ij} = h_{ij,c} + \hat{h}_{ij}, \quad \pi^{ij} = \pi^{ij}_c + \hat{\pi}^{ij}$$
$$\varphi = \varphi_c + \hat{\varphi}, \quad \pi_\varphi = \pi_{\varphi,c} + \hat{\pi}_\varphi .$$

where notations are the same as in eq. (3.6).

The quadratic part of the action (3.18) has the following form

$$S_E^{(2)} = \int d^4x \left(\hat{\pi}^{ij}\hat{h}'_{ij} + \hat{\pi}_\varphi\hat{\varphi}' - NH^{(1)} - N_i H^{i(1)} - H^{(2)}\right) \tag{3.20}$$

where $H^{(1)}$ and $H^{i(1)}$ are linear parts of (3.19a) and (3.19b) respectively, while $H^{(2)}$ is a quadratic part of (3.19a). Eq. (3.20) implies that there are four constraints

$$H^{(1)} = 0 \tag{3.21a}$$
$$H^{i(1)} = 0 \tag{3.21b}$$

which, as usual, are to be solved by imposing the gauge conditions.

The further analysis is simplified by the observation that, since we are interested in high momentum perturbations, we can neglect the curvative of a background sphere $\xi = const$, i.e. we can insert $h_{cij} = \rho_c^2(\xi)\,\delta_{ij}$ into eqs.(3.20) and (3.21) For this (flat) background metrics we can

Fourier transform the perturbations with respect to x^i and decompose each Fourier component into the spin components (the case $\vec{k} = 0$ corresponds to spherically symmetric perturbations and should be considered separately),

$$h_{ij}(\xi,\vec{k}) = e_{ij} + (k_i f_j + k_j f_i) + \frac{k_i k_j}{\bar{k}^2} P + \left(\delta_{ij} - \frac{k_i k_j}{\bar{k}^2}\right) S$$

$$\pi^{ij}(\xi,\vec{k}) = \lambda^{ij} - \frac{1}{2\vec{k}^2}(k^i \chi^j + k^j \chi^i) + \frac{k^i k^j}{\bar{k}^2}\theta + \frac{1}{2}\left(\delta_{ij} - \frac{k^i k^j}{k^2}\right)\zeta$$

where f_i and χ^i are transverse and e_{ij} and λ^{ij} are transverse and traceless.

The convenient choice of gauge is

$$P = 0\ ,\ S = 0\ ,\ f_i = 0 \tag{3.22}$$

With this choice, eqs.(3.21) can be solved in terms of canonical momenta conjugate to P, S and f_i

$$\chi^i = 0\ ,\ \theta = \frac{1}{2}\rho_c \varphi_c' \cdot \varphi$$
$$\zeta = -\frac{\varphi_c'}{2\rho_c \rho_c'}\hat{\pi}_\varphi + \frac{1}{2\rho_c'}\left[\rho_c^2 \frac{dV(\varphi_c)}{d\varphi_c} - \rho_c \rho_c' \varphi_c'\right] \tag{3.23}$$

where $\hat{\varphi}$ and $\hat{\pi}_\varphi$ are now the 3-dimensional Fourier transforms of the scalar field perturbations. Inserting eqs. (3.22) and (3.23) into eq.(3.18), we find the unconstrained action

$$S_E^{(u)} = \int d\xi\, d^3k \left(\lambda^{*ij} e_{ij}' + \hat{\pi}_\varphi^* \hat{\varphi}' - H^{(2),(u)}\right)$$

where

$$\begin{aligned} H^{(2),(u)} = {} & 2\varkappa G^{(0)}_{ijmn} \lambda^{*ij}\lambda^{mn} - \varkappa \rho_c \theta^*(\theta - 3\zeta) \\ & - \frac{2\rho_c'}{\rho_c}\lambda^{*ij} e_{ij} + \frac{1}{2\rho_c^3}|\hat{\pi}_\varphi|^2 \\ & + \left[-\frac{\bar{k}^2}{8\varkappa \rho_c^3} + \frac{5}{4}\frac{\rho_c'}{\varkappa^2 \rho_c^3} + \frac{\varphi_c'^2}{8\rho_c} + \frac{V(\varphi_c)}{2\rho_c}\right] e_{ij}^* e_{ij} \\ & - \left[\frac{\rho_c \bar{k}^2}{2} + \frac{\rho_c^3}{2}\frac{d^2V(\varphi_c)}{d\varphi_c^2}\right]|\hat{\varphi}|^2 \end{aligned} \tag{3.24}$$

$$G^{(0)}_{ijmn} = \frac{\rho_c}{2}\left(\delta_{im}\delta_{jn} + \delta_{in}\delta_{jm} - \delta_{ij}\delta_{mn}\right)$$

Here θ and ζ are assumed to be expressed trough the unconstrained variables according to eqs.(3.23).

Eq.(3.24) implies that the high momentum perturbations have positive definite action. Indeed, integrating out the canonical momenta λ^{ij} and $\hat{\pi}_\varphi$, we obtain the unconstrained action in the lagrangian form

$$S_E^{(2)} = \int d\xi\, d^3k \left(\frac{1}{8\varkappa\rho_c}|e'_{ij}|^2 + \frac{\rho_c^3}{2}|\hat{\varphi}'|^2 + \frac{\vec{k}^2}{8\varkappa\rho_c^3}|e_{ij}|^2 + \frac{\rho_c}{2}\vec{k}^2|\hat{\varphi}|^2 + \dots\right)$$

where we omitted terms independent of $\vec{k}$ or derivatives with respect to ξ. This action is clearly positive definite for large $\vec{k}$ and/or $|e'_{ij}|$, $|\hat{\varphi}'|$, which is the desired result.

The case $\vec{k} = 0$ is to be considered separately. This case corresponds to the perturbations independent of the coordinates on the 3-spheres $\xi =$ const. The traceless part of $\hat{h}_{ij}$ can be studied in the same way as before and it can be shown that it does not contain negative modes with large $\hat{h}'_{ij}$. On the other hand, perturbations of the form $\hat{h}_{ij} = \hat{\rho}^2\delta_{ij}$ correspond to spherically symmetric fluctuations discussed in subsect.3.2. We have checked that the consideration of these fluctuations along the lines of this subsection leads just to the results of subsect. 3.2: the unconstrained action has precisely the form of eq.(3.16) with the expression for $W(\xi)$ being slightly changed (the latter change, which is of no importance for our purposes, is due to the fact that in this subsection we neglect the curvative of the 3-spheres $\xi =$ const). This result can be considered as a consistency check for the approach utilized in this subsection.

3.4. Discussion

In sucsect.3.2. we have shown that there exists an infinite (in the limit of infinite ultraviolet cut-off) number of negative modes among the $O(4)$ -invariant perturbations around the bounce provided that the latter has negative $Q(\xi)$ somewhere in euclidean space-time (for definition of Q see eq.(3.17)). At first sight this fact seems to indicate that besides the Coleman-De Luccia bounce there could exist another classical euclidean solution describing the formation of a bubble and having smaller action. However, we think that this is not the case.

The argument goes as follows. First, in subsect.3.3. we have shown that the number of $O(4)$ non-invariant negative modes is finite, if any, so that the above property is peculiar to O(4)- invariant perturbations. This implies that the new classical solution would be O(4)-invariant. Now, the arguments presented in ref.[5] show that, in the non-gravitational case, the O(4)-invariant bounce solution is unique. It seems unlikely that turning on gravitation would lead to breaking this property of classical equations (recall that we are discussing the bounce with the parameters which are far from being Planck ones). To check the latter point, we have studied numerically the solutions to the Coleman-DeLuccia equations for a particular potential

$$V = \lambda(\varphi^2 - a^2)^2 + \gamma(\varphi + a)$$

$$a = 3.3 \cdot 10^{-2} \varkappa^{-1/2} \; ; \quad \gamma = -1.8 \cdot 10^{-5} \varkappa^{-3/2} \; ; \quad \lambda = 9$$

This potential corresponds to the thin wall bounce with $\bar{\rho} = 5.8 \cdot 10^2 \varkappa^{1/2}$, $Q_{min} = -0.2$. We have obtained numerically that, indeed, no solutions with positive-definite Q and required asymptotic properties ($\varphi(\xi = +\infty) = \varphi_-$) exist, while the Coleman-DeLuccia bounce is correctly reproduced.

Of course, this argument is not complete (we have no

proof of the absence of relevant O(4) non-invariant solutions). However, we think it strongly suggests that the real meaning of our observation of the existence of infinite number of negative modes is the breaking of the quasiclassical approximation.

To discuss the reason for the breaking of quasiclassics, and to speculate on the possible alternative to the quasiclassical tunneling, we repeat the arguments of subsect. 3.2 in slightly different terms. For the configurations obeying the constraint (3.4b) ("Gauss' law") the action (3.3) can be rewritten as follows,

$$S_E = -\frac{12\pi^2}{\varkappa}\int d\xi\, N\rho\left(1-\frac{\varkappa}{3}\rho^2 V\right) \tag{3.25}$$

To derive the unconstrained action for perturbations we impose the gauge condition (3.9) and express N through φ with the use of eq.(3.4b) (we do not use the hamiltonian-like formalism here)

$$N = \left[\frac{1}{2}\cdot\frac{1}{\rho_c^2 V - \frac{3}{\varkappa}}\cdot\left(\rho_c^2\varphi'^2 - \frac{6}{\varkappa}\rho_c'^2\right)\right]^{1/2} \tag{3.26}$$

Now, we consider rapidly varying fluctuations of the field φ and derive the part of the unconstrained action, which is quadratic in $\hat{\varphi}'$.

It is clear that the only source of the quadratic in $\hat{\varphi}'$ term in the action (3.25) is dependence of N on $\hat{\varphi}'$, i.e.

$$S_E^{(2)}(\textit{quadratic in } \hat{\varphi}') = -\frac{12\pi^2}{\varkappa}\int d\xi\, \hat{N}_{(2)}\rho_c Q \tag{3.27}$$

where $\hat{N}_{(2)}$ is quadratic in $\hat{\varphi}'$ part of N and $Q(\xi)$ is defined by eq.(3.17). Now $\hat{N}_{(2)}$ is evaluated from eq. (3.26) to be

$$\hat{N}_{(2)} = -\frac{\varkappa}{12}\,\frac{\rho_c^2\rho_c'^2}{Q^2}\,(\hat{\varphi}')^2 \tag{3.28}$$

Therefore, the relevant part of the action is

$$S_E^{(2)}(\text{quadratic in } \hat{\varphi}') = \pi^2 \int d\xi \, \frac{\rho_c^3 \, \rho_c'^2}{Q} \, \hat{\varphi}'^2$$

This derivation makes clear the source of the negativeness of the action for rapidly varying perturbations in the region of negative Q. Indeed, for these perturbations one can substitute the classical bounce expressions for all terms in the integrand of eq.(3.25) except N. Eq.(3.28) shows that the perturbations always lead to decrease in N. In the region of negative Q this decrease is favourable in the sence that it leads to decrease in action (see also eq.(3.27)). Geometrically, $N(\xi)d\xi$ is the element of the radial distance; we conclude that it is favourable for the system to shrink the radial size of the region where $Q < 0$ by developing rapidly varying O(4)-invariant fluctuations of the field φ.

If this tendency persists not only for small perturbations, we come to the following speculation concerning the non-quasiclassical euclidean process of the bubble formation. In the region where $Q < 0$ the field φ is expected to be rapidly varying around the bounce value φ_c, so that $N(\xi)$ be zero in this region. Thus the radial size of this region is likely to be zero while $\rho(\xi)$ still has a finite change. Recall that $\rho(\xi)$ has direct physical significance since it determines the area of the sphere $\xi = \text{const}$. If we plot ρ as a function of the radial distance, $r = \int N d\xi$, then the bounce configuration has the form shown in fig.3 where (ρ_1, r_1) and (ρ_2, r_2) are the boundaries of the region where $Q < 0$. According to the above discussion, it seems likely that the fluctuations of the field φ shrink the region (r_1, r_2) to a point, so that the actual euclidean metrics seem to have the form shown in fig.4.

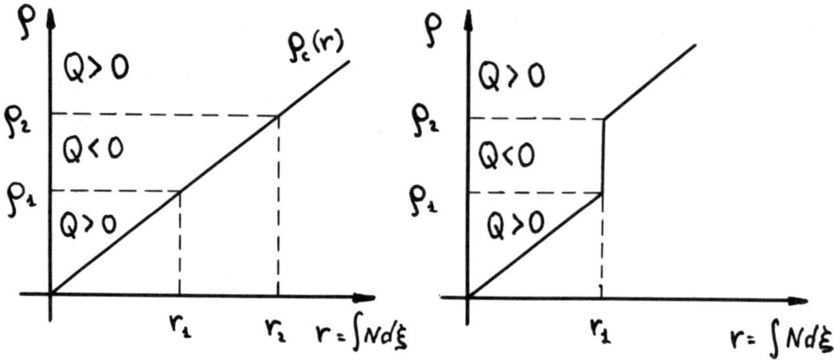

Fig 3,4

The latter configuration corresponds to a singular geometry with discontinuous metrics. We conclude that the materialized bubble (i.e. the configuration at $x^4 = 0$) is likely to be described by a singular metrics and have many high energy φ -particles near the singularity sphere. Of course, this picture can be destroyed by the quantum gravity effects, so that the region of the Planck radial size could emerge at $r \approx r_1$ instead of singularity. In any case, the materialized bubble is expected to be quite different from that described by the Coleman-DeLuccia bounce; the tunneling probability is likely to be much larger than that predicted by quasiclassics.

4. CONCLUSION.

In this paper we have discussed the problem of the applicability of the quasiclassical approximation to the description of the tunneling process with gravity. We have seen that gravitational effects sometimes break quasiclassics even if the parameters of models are far from being Planck ones. Clearly, this means that one has to develop some other technique for describing tunneling; at the moment, we have no idea on what this new technique is.

Although we have speculated on the possible alternative processes, these speculations have themselves been based upon quasiclassics; furthermore, we have been unable to discuss these alternative processes quantitatively even in a speculative way. Thus, we think that the present understanding of the tunneling transitions with non-trivial gravitational effects is far from being satisfactory, at least in the two models discussed in this paper.

Although this paper is devoted to studying only two model examples, we think that the problem of breaking of quasiclassics might be relevant to other models. In fact, our results imply that there exist non-quasiclassical mechanisms of tunneling, so even in cases where the classical euclidean solutions are stable (i.e. there is only one negative mode per a solution as it should be[6]), one might wonder whether these non-quasiclassical mechanisms are more operative than quasiclassical ones. We think that one cannot be sure that tunneling is correctly described within quasiclassics (even if the classical solutions are stable) unless the non-quasiclassical tunneling is understood.

The authors are deeply indebted to V.A.Matveev and A.N.Tavkhelidze for their interest and support and to V.A.Beresin, M.E.Shaposhnikov and I.I.Tkachev for many helpful discussions during various stages of this work.

REFERENCES.

1. Kirzhnits, D.A. and Linde, A.D., ZhETF 67, 1263 (1974); Dolan, L. and Jackiw, R., Phys.Rev. D9, 3320 (1974); Weinberg, S. Phys. Rev. D9, 3357 (1974).
2. Fomin, P.I. Inst.Theor.Phys. Kiev preprint (1973).
3. Zel'dovich, Ya.B. Pis'ma Astron.Zh. 7, 579 (1981). Grishchuk, L.P. and Zel'dovich, Ya.B. in Proc. 2nd Int. Seminar on Quantum Gravity, eds. Markov, M.A., Berezin, V.A. and Frolov V.P., INR Acad.Sci. USSR, Moscow, 1982;

Vilenkin, A. Phys.Lett. 117B, 25, 1982.
4. Voloshin, M.B., Kobzarev, I.Yu and Okun, L.B. Yad.Fiz. 20, 1229 (1974).
5. Coleman, S. Phys.Rev. D15, 2929 (1977).
6. Callan, C.G. and Coleman, S. Phys.Rev. D16, 1762 (1977).
7. Coleman, S. and De Luccia, F. Phys.Rev. D21, 3305(1980).
8. Berezin, V.A., Kuzmin, V.A. and Tkachev I.I. Phys. Lett. 120B, 91(1983); 124B, 497(1983); 130B, 23(1983); ZhETF 86, 785 (1984).
9. Kalinin, M.I. and Mel'nikov, V.N. Proc. VNIIFTRI, 16 [46], 43, State Standard Committee, Moscow, 1972.
10. Hartle, J. and Hawking, S.W. Phys. Rev. D28, 2960 (1983).
11. Rubakov, V.A. Pis'ma ZhETF 39, 89 (1984),
12. Linde, A.D: Lett. Nuovo Cimento, 39, 401 (1984).
13. Starobinsky, A.A. Talk given at the Seminar of P.K.Sternberg State Astronomical Inst., (November, 1983).
14. Rubakov,V.A. "Quantum mechanics in the tunneling Universe", to appear in Physics Letters B.
15. DeWitt, B.S. Phys.Rev. 162, 1113 (1967).
16. Ashtekar, A. and Geroch, R. Rep.Prog.Phys. 37,1211(1974); Isham, C.J. in "Quantum Gravity. An Oxford Symposium", eds. Isham, C.J. Penrose, R. and Sciama, D.W., Clarendon Press, Oxford (1975).
17. Rubakov, V.A. INR Acad.Sci. preprint P-0340, Moscow (1984).
18. Rubakov, V.A. Nucl. Phys. B245, 481 (1984).
19. Mal'tsev, V.K. and Markov, M.A. INR Acad.Sci. preprint P-0160 (1980).
20. McCallum, M.A.H. in "Quantum Gravity. An Oxford Symposium", eds. Isham, C.J., Penrose, R. and Sciama, D.W., Clarendon Press, Oxford (1975).
21. Misner, C.W. in "Magic without Magic", ed. Klauder, J., Freeman, S.Francisco (1972).
22. Lifshits, E.M. ZhETF 16, 587 (1946);
Grishchuk, L.P. ZhETF 67, 825 (1974).

23. Lapchinsky, V.G. and Rubakov, V.A. Acta Phys. Polonica B10, 1041 (1979).
24. Banks, T., Bender, C. and Wu, T.T. Phys.Rev. D8, 3346 (1973);
Banks, T. and Bender, C. Phys.Rev. D8, 3366 (1973).
25. Arnowitt, R. Deser, S. and Misner, C.W. in "Gravitation: an Introduction to Current Research", McGrow-Hill (1962).
26. Faddeev, L.D. Teor.Mat.Fiz. 1, 3(1969).

KALUZA-KLEIN COSMOLOGY AND THE ACTUAL STATE OF THE UNIVERSE

Ulrich Bleyer

Dierck-Ekkehard Liebscher

Zentralinstitut für Astrophysik der
Akademie der Wissenschaften der DDR
DDR - 1502 Potsdam-Babelsberg,
Rosa-Luxemburg-Str. 17 A

ABSTRACT

We consider generalized Friedmann-Robertson-Walker cosmologies in 4+D dimensions. By use of a canonical decomposition of matter distributions we give a systematic analysis of the possibilities to compensate the large curvature of the internal space necessary today to make the additional dimensions undetectable. This leads to general statements about the existence of solutions with static internal space. Our scheme covers the special solutions derived from models of spontaneous compactification. An outlook is given on solutions with changing internal space scale.

1. INTRODUCTION

The recent revival of the Kaluza-Klein ideas of unification using higher dimensional space-time was induced by the development of supergravity, which looks like a good candidate for a finite theory of quantum gravity. The way seems to be open to understand the unification of gauge interactions by deducing all gauge interactions from pure gravity in a higher dimensional space-time /1-3/. The modern versions of Kaluza-Klein theories do not exclude the internal dimensions by periodicity or symmetry conditions /4,5/, but take the internal space as real. Compact internal spaces are found as solutions of the field equations during spontaneous compactification /6,7/. Additional dimensions are undetectable at present if their linear dimensions are of the Planck scale. Therefore, we have to follow the development of the internal space during the history of the universe in order to construct scenarios compatible with the actual state of the universe. Quite different methods of realising spontaneous compactification lead to the common result that only solutions with static internal space are able to describe the actual state of the universe. We obtain this result by a model-independent phenomenological consideration using a canonical decomposition of matter and linear equations of state for each matter component. This phenomenological picture is able to give general conditions for the existence of static internal space solutions and covers a number of popular models of spontaneous compactification.

2. EINSTEIN's FIELD EQUATIONS IN N DIMENSIONS

2.1 The Generalized Friedmann-Robertson-Walker Ansatz

Let us start the description of the cosmological model by choosing the metric, which the field equations will be deduced from. According to the Kaluza-Klein philosophy described in the introduction, we start from a N-dimensional space-time splitted into a product space

$$M^N = M^{1+d_0} \times \prod_{i=1}^{\alpha} M^{d_i}, \quad 1 + \sum_{i=0}^{\alpha} d_i = N. \qquad (2.1)$$

The index i denumerates the internal factor spaces, which are assumed to be compact manifolds as a result of spontaneous compactification. The dimension of the ordinary (outer) space is taken to be d in order to cover as much as possible results known from the literature /8/. Of course, special attention will be given to the case d=3 and i=1, d =D.

Following the work of Freund /9/ and the recent paper of Taylor and coworkers /10/ we assume that the compact manifolds are Einstein spaces (i.e. that the Ricci tensor of M^{d_i} is proportional to its metric). In that case there is a single scale associated with each compact manifold. This is required in the case of one internal space by the mechanism for compactifying the extra spatial dimensions by the help of the expectation value of a fourth rank antisymmetric tensor field strength /11/. Compactification by quantum fluctuations of matter fields does not require this assumption /12/, but it seems to be very convenient.

In the notation of Weinberg /13/ we get for the internal spaces

$$R^{(i)}_{m_i n_i} = -(d_i - 1) K_i \, g^{(i)}_{m_i n_i} , \tag{2.2}$$

where K_i is a positive constant produced by the curvature in each factor space. These assumptions let the generalized Friedmann-Robertson-Walker (FRW) model start with the metric

$$\begin{aligned} ds^2 &= g_{KL} \, dx^K dx^L \\ &= -dt^2 + \sum_{i=0}^{\alpha} R_i^2(t) \, g^{(i)}_{k_i l_i} \, dx^{k_i} dx^{l_i} , \end{aligned} \tag{2.3}$$

where K, L are running from 0 to N-1, and k , l from 1 to d . denotes the different factor spaces, i=0 for the ordinary space (d_0 = 3 in conservative interpretations). The metric $g^{(0)}_{m_0 n_0} = g_{mn}$ is tthe maximally symmetric metric of ordinary space, and $g^{(i)}_{m_i n_i}$ the metric tensors of the d -dimensional internal factor spaces, which are assumed to be Einstein spaces.

A model of special interest is given by /14/:

$$M^{4+D} = R^1 \times S^3 \times S^D . \tag{2.4}$$

In this case $g^{(1)}_{m_1 n_1}$ and $g^{(0)}_{m_0 n_0}$ are the metrics of the D-dimensional and the 3-dimensional spheres respectively. Instead of the sphere S one can choose an open space of constant curvature and instead of the sphere S one can choose a compact quotient space G/H of D dimensions. From (2.2) and (2.1) we derive the Ricci tensor

$$R^0{}_0 = \sum_{i=0}^{\alpha} d_i \frac{\ddot{R}_i}{R_i} , \tag{2.5}$$

$$R^{m_i}{}_{n_i} = \delta^{m_i}_{n_i} \left(\frac{d}{dt}\left(\frac{\dot{R}_i}{R_i}\right) + \frac{d_i - 1}{R_i^2} k_i + \frac{\dot{R}_i}{R_i} \sum_{j=0}^{\alpha} d_j \frac{\dot{R}_j}{R_j} \right) , \tag{2.6}$$

where (2.6) is to be required for each factor space, including the ordinary space.

2.2 The Generalized Perfect Fluid

We shall consider the Einstein equation with matter

sources in N dimensions

$$E^K_{\ L}=R^K_{\ L}-\frac{1}{2}\delta^K_{\ L}R=-\kappa T^K_{\ L}\ ;\ R^K_{\ L}=-\kappa\left(T^K_{\ L}-\delta^K_{\ L}\frac{T}{N-2}\right). \qquad (2.7)$$

A non-vanishing cosmological constant will be taken into consideration too. We treat the energy-momentum tensor as phenomenological equivalent without bothering about its origin. Some examples of its derivation via spontaneous compactification will be reviewed in section 4.

The metric chosen in eq. (2.3) admits a lot of isometries. For the special example, eq. (2.4), the symmetry group of spatial sections is $O(4)\times O(D+1)$. In order to preserve homogeneity and isotropy of the FRW ansatz, eq. (2.3), we take the matter tensor in the generalized form of a perfect fluid with different pressure in each factor space, the non-vanishing components being:

$$T_0^{\ 0}=\varrho,\ T^{m_i}_{\ \ n_i}=-p_i\,\delta^{m_i}_{\ \ n_i}\,. \qquad (2.8)$$

We now substitute this form into eqs. (2.7) and write instead of the equation for $R^0_{\ 0}$ (2.5) the equation for $E^0_{\ 0}$:

$$\left(\sum_{i=0}^{\alpha} d_i\frac{\dot R_i}{R_i}\right)^2-\sum_{i=0}^{\alpha} d_i\left(\frac{\dot R_i^2}{R_i^2}-\frac{(d_i-1)}{R_i^2}k_i\right)=2\kappa\varrho, \qquad (2.9)$$

$$\frac{d}{dt}\left(\frac{\dot R_i}{R_i}\right)+\frac{\dot R_i}{R_i}\sum_j d_j\frac{\dot R_j}{R_j}+\frac{d_i-1}{R_i^2}k_i=\kappa\left(p_i+\frac{\varrho-\sum_j d_j p_j}{N-2}\right). \qquad (2.10)$$

For the special case of $\alpha=1$, $d_0=3$, $d_1=D$, we use the notation $R_0=R$, $R_1=S$, $p_0=p$, $p_1=p^*$, and get

$$\left(3\frac{\dot R}{R}+D\frac{\dot S}{S}\right)^2-3\frac{\dot R^2}{R^2}-D\frac{\dot S^2}{S^2}=2\kappa\varrho-\frac{6K_0}{R^2}-\frac{D(D-1)}{S^2}K_1, \qquad (2.11)$$

$$\left(\frac{\dot R}{R}\right)^{\cdot}+\frac{\dot R}{R}\left(3\frac{\dot R}{R}+D\frac{\dot S}{S}\right)=\kappa\left(p+\frac{\varrho-3p-Dp^*}{D+2}\right)-2\frac{K_0}{R^2}, \qquad (2.12)$$

$$\left(\frac{\dot S}{S}\right)^{\cdot}+\frac{\dot S}{S}\left(3\frac{\dot R}{R}+D\frac{\dot S}{S}\right)=\kappa\left(p^*+\frac{\varrho-3p-Dp^*}{D+2}\right)-(D-1)\frac{K_1}{S^2}. \qquad (2.13)$$

3. DYNAMICAL EQUATIONS AND CANONICAL DECOMPOSITION OF MATTER

As in 4-dimensional GRT the Einstein equations (2.9), (2.10) are not independent. They are connected by the Bianchi identity

$$E^{K}{}_{L;K} = 0 . \tag{3.1}$$

This may be seen immediately for the system of equations (2.9), (2.10) itself. In this case the continuity equation implied by eq. (3.1) is simply

$$\frac{d\varrho}{dt} + \sum_{j=0}^{\kappa} d_j \frac{\dot{R}_j}{R_j} (\varrho + p_j) = 0 . \tag{3.2}$$

This is an obvious generalization of four dimensional FRW cosmology with perfect fluid source where the continuity equation reads

$$\frac{d\varrho}{dt} + 3\frac{\dot{R}}{R}(\varrho + p) = 0 . \tag{3.3}$$

For simplicity, and in analogy to the analysis of the 4-dimensional cosmology we use linear equations of state. It is well known that equation (3.3) may be integrated for the equation of state

$$p = \left(\frac{m}{3} - 1\right)\varrho \tag{3.4}$$

to give the solution

$$\varrho = M_m R^{-m} \tag{3.5}$$

with the integration constant M_m . Various matter components are included in this scheme. For instance, dust matter yields

$$p = 0 , \quad m = 3 , \quad \varrho = M_3 R^{-3} , \tag{3.6}$$

for radiation we have

$$p = \frac{1}{3}\varrho , \quad m = 4 , \quad \varrho = M_4 R^{-4} , \tag{3.7}$$

cold nucleon gas (Zeldovich matter) implies

$$p = \varrho \,, \quad m = 6 \,, \quad \varrho = M_6 R^{-6} \,. \tag{3.8}$$

We may, therefore, in usual FRW cosmology, write the Friedmann equation with all matter contributions collected on the r.h.s. in the form

$$\varrho = \sum_m M_m R^{-m} \,. \tag{3.9}$$

This formal expansion covers the cosmological constant by

$$\Lambda = \kappa M_0 \,, \quad m = 0 \,, \quad p = -\varrho \,, \tag{3.10}$$

and the curvature term with

$$-2K = \kappa M_2 \,, \quad m = 2 \,, \tag{3.11}$$

too. A general discussion of the Friedmann equation with the canonical decomposition, eq.(3.9), allows a simple phenomenological description of the phase transitions in FRW cosmology /15/.

The canonical decomposition may be extended immediately to Kaluza-Klein cosmologies. For every matter component a linear equation of state

$$p_i = \left(\frac{m_i}{d_i} - 1 \right) \varrho \tag{3.12}$$

leads by the continuity equation to

$$\varrho = \sum_{m_0, \dots, m_\alpha} M_{m_0, \dots, m_\alpha} R_0^{-m_0} \dots R_\alpha^{-m_\alpha} \,. \tag{3.13}$$

From this equation we find the components $\Lambda = \kappa M_{0,\dots,0}$ corresponding to the cosmological constant, and the curvature terms

$$-(d_i - 1) k_i = \kappa M_{0,\dots,2,\dots,0} \quad \left(\begin{matrix} \text{index 2 at} \\ \text{the place } i \end{matrix} \right), \tag{3.14}$$

where we have to note that positive curvature enters the r.h.s. as negative density component.

By use of those conventions we get the following form of the Einstein field equations in the case $M^N = M^4 \times M^D$:

$$\left(3\frac{\dot{R}}{R}+D\frac{\dot{S}}{S}\right)^2-3\frac{\dot{R}^2}{R^2}-D\frac{\dot{S}^2}{S^2}=2\sum_{mn}M_{mn}R^{-m}S^{-n}, \qquad (3.15)$$

$$\left(\frac{\dot{R}}{R}\right)^{\cdot}+\frac{\dot{R}}{R}\left(3\frac{\dot{R}}{R}+D\frac{\dot{S}}{S}\right)=\sum_{mn}\frac{6+(D-1)m-3n}{3(D+2)}M_{mn}R^{-m}S^{-n}, \qquad (3.16)$$

$$\left(\frac{\dot{S}}{S}\right)^{\cdot}+\frac{\dot{S}}{S}\left(3\frac{\dot{R}}{R}+D\frac{\dot{S}}{S}\right)=\sum_{mn}\frac{2D+2n-Dm}{D(D+2)}M_{mn}R^{-m}S^{-n}. \qquad (3.17)$$

We consider cosmological models which are solutions of these equations.

4. MODELS OF SPONTANEOUS COMPACTIFICATION AND THE TRACE CONDITION OF THE MATTER TENSOR

Quantum derivations of phenomenological matter components show that the large variety of matter components in the canonical expansion (3.13) may be restricted by a further condition: We demand the trace of the energy-momentum tensor to vanish. This assumption is suggested by interpreting the unification idea of Klein--Kaluza theories phenomenologically. In unified models matter fields (gauge fields) are introduced as massless excitations above the ground state (s. e.g./16/). For instance, we have only massless fields in 11-dimensional supergravity, where mass terms are introduced by spontaneous symmetry breaking from a super-Higgs effect.

By the trace condition we consider all components as generalized massless in the N-dimensional space-time. All components are protoradiation. The partition in the characteristics m, n of the equations of state allows us to understand also ordinary massive dust as massless in this sense: Dust particles are interpreted as relativistic, their velocity components being essential in the internal dimensions only. We show that popular models of spontaneous compactification are covered by this procedure. The assumption

$$T^K{}_K=0 \qquad (4.1)$$

leads with (3.13) to the condition

$$-\varrho+\sum_{i=0}^{\alpha}p_i=0\,. \qquad (4.2)$$

If we insert the equations (3.12) we get

$$\sum_{i=0}^{\alpha} m_i = \sum_{i=0}^{\alpha} d_i + 1 . \qquad (4.3)$$

Let us consider some examples for the case $d_0=3$, i=1, $d_1=D$, which will be used in what follows. p=0 (m=3) describes dust matter in ordinary space. From the trace condition, eq. (4.3), we get n = D + 1 , and $p^*= \varrho/D$, which means radiation-like behaviour of this component in the internal space. The opposite is given by the radiation in the ordinary space, eq. (3.7). This leads to n = D, $p^*=0$, and we have dust-like behaviour in the internal space: The essential velocity components all lie in the ordinary space directions.

Now we show how special models of spontaneous compactification fit into this scheme of canonical decomposition of matter. At first we consider the source terms arising by compactification derived from the expectation value of an antisymmetric tensor field of fourth rank /11/, and the compactification from quantum fluctuations of matter fields (Candelas and Weinberg /12/), which are reviewed in /17/ and /18/. The model of Freund and Rubin as referred to in /17/ leads to

$$\varrho \sim S^{-2D} , \qquad (4.4)$$

which means in our notation m=0 and $n=2D$. The pressure components are

$$p=-\varrho , \quad p^*=\varrho. \qquad (4.5)$$

The point is that ϱ is independent of R and that for this model the trace condition is not fulfilled. In the static case of Candelas and Weinberg we have /18/:

$$\varrho \sim S^{-N} . \qquad (4.6)$$

The pressures are with m=0 and n=D+4

$$\varrho = -p , \quad p^* = \frac{4}{D}\varrho . \qquad (4.7)$$

Therefore, is independent of R and the trace condition is fulfilled. The same equation of state (4.7) is used by Taylor and coworkers /10/, who start with Einstein-Maxwell theory in 4+D dimensions. The models reviewed by Moss /18/ contain in addition the case m=N, n=0, which also fulfils the trace condition. Sahdev /8/ (s. also /19/) uses in a (d+D+1)-dimensional space-time the equation of state

$$p = p^* = \frac{\varrho}{d+D} . \qquad (4.8)$$

This leads to the characteristics

$$m = \frac{d}{d+D} + d \;, \; n = \frac{D}{d+D} + D \;, \qquad (4.9)$$

and the trace condition (4.4) is fulfilled again.

Special attention we draw to the model considered by Randjbar-Daemi, Salam and Strathdee /14/. They assume that the density and pressures are due to a gas of non-interacting, spinless particles in thermal equilibrium. The free energy of such a system can be expressed by a one-loop integral, which is carried out in detail. As a result they get two terms of our canonical decomposition of matter:

$$\rho = M_{0,N} S^{-N} + M_{4,D} R^{-4} S^{-D} . \qquad (4.10)$$

The latter is the thermal part, which enters the potential energy of the system in the form

$$U = \sim \frac{J^{4/3}}{R} \qquad (4.11)$$

in order to give

$$T = \frac{\partial U}{\partial J} \sim \frac{J^{\frac{1}{3}}}{R} \qquad (4.12)$$

(J being the entropy of the system). We see that both contributions fulfil the trace condition, eq. (4.3). A discussion of the generalized FRW cosmology in 4+D dimensions using the canonical decomposition of matter covers these models.

5. STATIC INTERNAL SPACE SOLUTIONS

Let us now consider the FRW cosmological model in 4+D dimensions, which is described by the field equations (3.15-17). We rescale the M_{mn} in an appropriate way to get convenient initial values for R, $\dot{R}$, and S. The actual state of the universe is characterized by the values of the Hubble constant, the deceleration parameter, and a vanishing cosmological constant. If we put to-day

$$\dot{R}_0 / R_0 = 1 \;, \qquad (5.1)$$

we choose the linear age of the universe

$$t = H_0^{-1} \qquad (5.2)$$

for the time scale. The additional choice

$$R_0 = 1 \qquad (5.3)$$

fixes the actual expansion parameter as length unit in the ordinary space. Again, the value of S today is chosen to be the lenght unit in the internal dimensions:

$$S_0 = 1 \qquad (5.4)$$

Both choices destroy the usual gauge ($k=\pm 1$, or 0) of the curvature constants, k_i , because this gauge fixes the length unit to be the curvature radius. Large values of curvature are now connected with large values of k_i . The internal dimensions shall be hidden by the effect of a large curvature in this factor space, most probably of the order of the Planck value. At the same time the curvature of the real space-time is negligibly small. The problem of the construction of a realistic cosmological model in Kaluza-Klein theory consists in the need of the compensation of the large contribution of the internal curvature term.

It is well known that a variation of the scale factor S of the internal space leads to variations in the four-dimensional gravitational and cosmological constants. Using our canonical decomposition of matter we derive the variation of the mean density of the universe to be

$$-\frac{d\rho}{dt} = \sum M_{mn} R^{-m} S^{-n} \left(m\frac{\dot R}{R} + n\frac{\dot S}{S} \right). \qquad (5.5)$$

The term proportional m is the conservative one, the term proportional n describes a variation in time of the effective FRW quantities of the matter components. Because these variations seem to be very small at present, we have to consider two possibilities: The case n=0 and the case S=0. If we use our trace condition in the first case the only matter constribution allowed is M_{N0}. This case represents the type I solution of Moss /18/ given as a N-dimensional de Sitter space.

Now we intend to give a detailed study of the second case, where the internal space is a static one. If we put S=1, S=0 into the equation (3.17) we get the condition

$$\sum_n (2D - Dm + 2n) M_{mn} = 0, \qquad (5.6)$$

which has to be fulfilled for every m. Eq. (5.6) is a condition on the constituents of the effective Friedmann equation

$$3\frac{\dot{R}^2}{R^2} = \sum_m \frac{M_m}{R^m}\ , \quad M_m = \sum_n M_{mn}\ . \tag{5.7}$$

The components of matter in usual FRW cosmology are characterized by the exponent m. We have generalized this concept and got an additional exponent n for the dependence on S. By eq.(5.7), we have the choice, formally, to take as dust $M_{3,0}$, or $M_{3,3}$ or any $M_{3,n}$. This is to do analogously for every m. Condition (5.6) tells us, that for every m only one generalization leads to a component, which does not need compensation for the static internal space to be possible. These components have the characteristics

$$m,\ n = \frac{D}{2}(m-2). \tag{5.8}$$

This is the case, for instance, for $M_{2,0}$, $M_{4,D}$, $M_{6,2D}$. The physical interpretation in other cases seems to be difficult.

In any case, we have to compensate the large curvature term M_{02}. We may do this by M_{00}, or $M_{0,N}$, and exclude other possibilities by the trace condition. These are the contributions derived by Candelas and Weinberg, as by Taylor and others. The contribution $M_{0,2D}$ by Freund and Rubin may also do. Taken all together, we get an effective cosmological constant in the ordinary spacetime:

$$\Lambda = \kappa M_0 = \kappa \sum_n M_{0n} S^{-n} = \Lambda(S). \tag{5.9}$$

The requirement of a vanishing constant today yields

$$\Lambda(1) = 0. \tag{5.10}$$

This condition is implied qualitatively by eq.(3.15). In order to get an acceptable deceleration parameter, the corresponding contribution to eq.(3.16) has to vanish too (eq.(5.7)):

$$\Lambda(S) - S\frac{d\Lambda}{dS} = 0 \quad \text{for } S = 1. \tag{5.11}$$

The three conditions, eqs. (5.6), (5.10) and (5.11), are not independent, because they are derived from equations connected by the Bianchi identities. Therefore, we get the following statements:

1) For static internal space solutions the high internal curvature M_{02} may be compensated by a combination of M_{00}

and $M_{0,N}$. One of these terms alone will not do.

2) The solutions of eq. (5.8), $M_{2,0}$, $M_{4,D}$, $M_{6,2D}$ do not compensate the internal curvature. They contribute to the effective FRW development.

The conditions, eqs. (5.10) and (5.11), are found in /14/ already. We derived them in a model-independent way from the canonical decomposition of matter as a purely phenomenological result.

We found general conditions, required to get at present the usual FRW cosmology. Some additional remarks are necessary. If we drop the trace condition further contributions with m=0 may be included. The conditions for coefficients and exponents remain unchanged. We see that a component with the equation of state used by Sahdev cannot compensate the internal curvature to give a static internal solution. Therefore, a possible Sahdev phase has to be connected by a phase transition to a static internal space solution. The matter component $M_{N,0}$ described by Moss /10/ leads to a static internal solution at most approximately for very small S and very large R. This makes the instability of the solution IIa in /18/ including $M_{N,0}$ quite evident.

6. CONCLUSIONS

Using our purely phenomenological decomposition of matter we derived the common result of the special models reviewed in section 4 that only static internal space solutions can be fitted to the actual values of the characteristics of the universe. But this does not exclude the possibility of periods with time-depending internal space scale at early times. Such solutions have to describe the contraction of the internal spaces to the Planck scale. The usual scenario provides a phase transition to pass to the solution with static internal space describing the actual development of our universe. In the language of canonical decomposition of matter the following condition has to be fulfilled during the phase transition:

$$\sum_{mn}\left(\underset{2}{M}_{mn} - \underset{1}{M}_{mn}\right) R^{-m} S^{-n} = 0. \qquad (6.1)$$

This is the fitting condition at discontinuities of the sources of the field equations (3.15-17), derived from the continuity equation (3.2). It rules the discontinuous changes in the composition of the source, i.e. of equations of state. It ensures formally a smooth change of the matter density. Before the phase transition occurs, the matter contributions involved are model dependend (s./8/, /18/, and section 4). We only mention that in classical cosmology the curvature terms M_{20} and M_{02} as well as M_{00} are of geometrical origin and cannot be changed by a phase tran-sition. Their use as matter components is allowed by the quantum interpretation of M_{00} , and the quantum jumps in curvature only.

7. REFERENCES:

/1/ Duff, M., in: Quantum Gravity, Inst.Nucl.Research, Moscow (1983), p. 248.

/2/ Mecklenburg, W., Fortschr.Phys. 32 (1984), 207.

/3/ Appelquist, T., and Chodos, A., Phys.Rev. D28 (1983), 772.

/4/ de Sabbata, V., and Schmutzer, E. (eds.): "Unified Field Theories of More Than 4 Dimensions Including Exact Solutions", World Scientific, Singapore 1983.

/5/ Ne'eman, Y., and Sternberg, S., in "Gauge Theories: Fundamental Interactions and Rigorous Results", Birkhäuser, Boston 1982, p. 103.

/6/ Cremmer, E., in: Supergravity '81, Cambridge UP, Cambridge 1982, p. 313.

/7/ Duff, M., and Pope, C.N., in Supersymmetry and Supergravity '82, World Scientific, Singapore 1983, p. 183.

/8/ Sahdev, D., Phys.Lett. 137 B (1984), 155.

/9/ Freund, P.G.O., Nucl.Phys. B 209 (1982), 146.

/10/ Gleiser, M., Rajpoot, S., and Taylor, J.G., "Higher Dimensional Cosmologies", King's College preprint 1984.

/11/ Freund, P.G.O., and Rubin, M., Phys.Lett. 97 B (1980), 233.

/12/ Candelas, P., and Weinberg, S., Texas preprint UTT6-8-83 (1983).

/13/ Weinberg, S., "Gravitation and Cosmology", J.Wiley & Sons, New York 1972.

/14/ Randjbar-Daemi, S., Salam, A., and Strathdee, J., Phys.Lett. 135 B (1984), 388.

/15/ Liebscher, D.-E., Die Sterne 60 (1984), 153.

/16/ Galli, D., Martellini, M., Phys.Lett. 141 B (1984), 359.

/17/ Bailin, D., Love, A., and Veronakis, C.E., Phys.Lett. 142 B (1984), 344.

/18/ Moss, I.G., Phys.Lett. 140 B (1984), 29.

/19/ Alvarez, E., and Cavela, M.B., Phys.Rev.Lett. 51 (1983), 931.

ANISOTROPY OF THE RELIC RADIATION AS A TEST OF THE EARLY UNIVERSE THEORIES

V.N.Lukash[*], P.D.Naselskij[**], I.D.Novikov[*]

[*]Space Research Institute Academy of Sciences of the USSR, Moscow 117810, USSR
[**]Rostov State University, Rostov on Don 344090, USSR

ABSTRACT

The correlation and multipole characteristics of anisotropy in the intensity distribution $\Delta T/T$ of the microwave background are calculated for a number of cosmological models with $\Omega \leq 1$. Different spectra of primordial perturbations and different assumptions as to the nature of the missing mass are considered. We emphasize the importance of measurements of the sky $\Delta T/T$ - pattern on large angular scales ($\alpha > 5^{\circ}$) and compare the results of calculations with the available observational data.

1. THE NECESSITY TO MEASURE THE $\Delta T/T$ - ANISOTROPY

The investigation of possible anisotropy of the microwave relic background is very important for cosmology. Such investigations are important due to the following.

First, observations of the background microwave anisotropy are a clue to the Early Universe physics. They provide direct information about the primordial homogeneity perturbations in the Universe and, consequently, about the processes which caused them.

Second, the features of angular correlations in the background temperature fluctuations over large scales tell us about the overall space curvature and, thus, about the total matter density in the Universe including the missing mass. To gain this sort of information in any other way is quite difficult.

The background anisotropy explorations are stimulated nowdays by the Early Universe theories. New theories predict both the spectrum of the primordial perturbations of the matter and the overall matter density in the Universe. So, calculations of the anisotropy of the relic radiation and their confrontation with the observational data provide a unique opportunity to test new theories of the very early stages of the Universe expansion.

By no means belittling the importance of searching for small-scale $\Delta T/T$ - fluctuations which correspond to linear dimensions encompassing the masses typical of galaxy clusters, we have always tried to draw special attention to the measurements of the $\Delta T/T$ - anisotropy of large angular scales (from $\alpha \sim 5^{\circ}$ and up to the dipole component). Our arguments are as follows.

(i) In many models the large-scale fluctuations are expected to be of greater amplitudes than the small-scale ones.

(ii) They bear direct information as to the primordial metric fluctuations and are not coupled to the details of recombination dynamics.

(iii) They cannot be smoothed out by possible secondary heating of cosmic gas.

(iv) Detection of large-scale $\Delta T/T$ - variations (say, a quadrupole moment) - or even setting a reliable upper limit for them - would enable most important conclusions on the spectrum of primordial perturbations and, as a consequence, on the underlying theoretical hypotheses.

(v) The "spottiness effect" in the large-scale $\Delta T/T$ - correlations makes it possible to evaluate the total matter density $\Omega = \rho/\rho_{cr}$ (including the invisible mass). Combined measurements of $\Delta T/T$ on both small and large angular scales would enable one to determine the spectrum of primordial perturbations as well as the total matter density Ω and, separately, the barion density Ω_b ; one could also place restrictions on possible rest masses of relic particles comprising the invisible mass.

Here we present the results of our calculations of the $\Delta T/T$ - anisotropy in the whole range of angular scales for different cosmological models [1).]

2. THEORETICAL GROUNDS FOR $\Delta T/T$ - CALCULATIONS AND CONFRONTATION WITH OBSERVATIONS

How to confront the observed and theoretically predicted anisotropies?

Theory predicts the existence in the Early Universe of the primordial cosmological perturbations which lately developed into galaxies and formed the large-scale structure of the Universe. These homogeneity perturbations gave rise to background temperature fluctuations. So the theory allows one to derive the amplitude of temperature fluctuations

$$\frac{\Delta T}{T}(\vec{e}) = f\left(q\,,\ \begin{matrix}\text{Early Universe}\\ \text{dynamics}\end{matrix}\,,\Omega\right) \quad (1)$$

which is a function of: the field of primordial cosmological perturbations $q = q(\vec{x})$; the dynamics of the Early

Universe that relates the primordial perturbations with the temperature fluctuations and includes missing-mass-parameters, the hydrogen recombination dynamics, reheating and other factors; the parameter Ω that is the total matter density of the present Universe in the critical density units.

Observations give the map of the sky-readiation temperature $\frac{\Delta T}{T}(\vec{e})$, where $\vec{e}$ is the unit vector along the line of sight.

To compare these two functions $\frac{\Delta T}{T}(\vec{e})$, we employ correlation analysis. This possibility is based on the hypothesis about the statistical independence (randomness) of the primordial perturbation amplitudes on different scales. It means that the random function $q(\vec{x})$ is characterized by a single one-parameter family of q_k, $k \geqslant 0$, which we call the spectrum:

$$\overline{q_{\vec{k}}\, q_{\vec{k}'}} = 2\pi^2 k^{-3} q_k^2 \,\delta(\vec{k} - \vec{k}') \; ; \qquad (2)$$

where the bar denotes the average over the state of q-field; $q_{\vec{k}}$ is the Fourier transformation of $q(\vec{x})$, $k = |\vec{k}|$. By definition $\overline{q^2} = \int q_k^2 \, d\ln k$.

Eq. (2) shows that the observed sky $\Delta T/T$ - pattern is the result of a random superposition of the independent perturbation amplitudes. In particular, we can calculate the correlation function

$$K(\alpha) \equiv \overline{\frac{\Delta T}{T}(\vec{e}_1)\,\frac{\Delta T}{T}(\vec{e}_2)}, \qquad (3)$$

that obviously depends on the angle between the observation directions, $\cos\alpha = \vec{e}_1 \vec{e}_2$. Ergodicity theorem allows identifying eq. (3) with the observed correlation function

$$K(\alpha) = \langle \frac{\Delta T}{T} (\vec{e}_1) \frac{\Delta T}{T} (\vec{e}_2) \rangle , \qquad (4)$$

where the brackets mean the average over all directions on the celestial sphere with the fixed angle α. Eq.(4) holds for $\alpha > \alpha_o$, the beamwidth of the antenna to be used. Here we present the results of our calculations for the root-mean-square temperature fluctuation that is simply related to the correlation function

$$\frac{\Delta T}{T}(\alpha) = \langle (\frac{T(\vec{e}_1)-T(\vec{e}_2)}{T})^2 \rangle^{1/2} = \sqrt{2} \, (K(0) - K(\alpha))^{1/2} \qquad (5)$$

This function is directly detected in the small-scale experiments ($\alpha \sim 4' \div 50'$, see [2),3),4)]). It can also be derived from the sky $\Delta T/T$ - pattern ($\alpha > 5^o$) obtained in the large-scale experiments [5),6),7)] . (For confrontation, one should substitute $K(\alpha_o)$ instead of $K(0)$ in eq. (5). Observers give also the amplitudes of dipole, quadrupole and other low-multipole moments of $\Delta T/T$. Therefore, in addition to $K(\alpha)$, we construct the expected amplitudes $\langle a_{lm}^2 \rangle^{1/2}$ of multipole components of the large-scale anisotropy for the spherical-harmonic expansion:

$$\frac{\Delta T}{T} (\vec{e}) = \sum_{l,m} a_{lm} \Psi_{lm} (\vec{e}) \quad . \qquad (6)$$

The confidence level for $\langle a_{lm}^2 \rangle^{1/2}$ to coincide with the observed amplitude is calculated for the Gaussian distribution of the primordial amplitudes.

So, the investigation of the cosmic temperature fluctuations is based on eq. (2) . Fortunately, statistical independence of primordial perturbation amplitudes is predicted by almost all theories of the Early Universe. In particular, it is so for cosmological perturbations, produced by the parametric effect [8),9),10)] or by vacuum phase transitions in the Grand Unification era [11),12),13)].

3. THE COSMOLOGICAL MODELS

Let us now come to the investigated models.

To separate different effects that form the sky $\Delta T/T$- pattern, we consider first the flat model and then show how to determine the total density Ω from the $\Delta T/T$ - distribution if the Universe is open. In both cases we have investigated three variants of missing mass: (a) barions, (b) massive thermodynamic neutrinos ($m_\nu \simeq 6 \div 30$ eV) and (c) very heavy relic particles ($m_R > 100$ eV) such as axions, primordial black holes, monopoles, gravitino, etc. The physical nature of the superheavy particles is not important for $\Delta T/T$ - calculations. It matters only that they become non-relativistic rather early, before the recombination epoch. We have also considered three types of the primordial spectra:

$$q_k \sim k^n , \qquad n = \begin{cases} 0 \; , & \text{(flat spectrum);} \\ -1 \; , & \text{(white noise);} \\ \Theta(\lambda - \lambda_0), & (20 < \lambda_0 < 600 \,[\mathrm{Mpc}]\,) \; . \end{cases} \tag{7}$$

The third spectrum with the changing slope $(0 \longrightarrow 1)$ appears in the parametric amplification theory [10)]. We used the following normalization of the primordial matter density perturbations:

$$\overline{\left(\frac{\delta\rho}{\rho}\right)^2}\Bigg|_{z=z_s} = 1 \tag{8}$$

where the redshift z_s corresponds to the moment when the first structure of the Universe emerges. The following results are given for $z_s = 3$.

4. THE PHYSICAL CAUSES OF $\Delta T/T$ - FORMATION

Before concentrating on the conclusions we briefly recall what processes cause fluctuations of the background temperature at different angular scales.

When we observe the relic radiation we see the wall of the last scattering of the photons by the ionized matter. It corresponds to the moment of the hydrogen recombination in the past, when the optical depth by Thompson scattering is unity. There were primordial perturbations of matter and, consequently, perturbations of gravitational field as well. They led to the formation of fluctuations in the relic radiation.

For the largest angular scales $\alpha \gtrsim 5^{\circ}$, corresponding to the horizon linear scale as it is seen on the recombination sphere, the fluctuations of $\Delta T/T$ are caused by the perturbations of the gravitational field [14] in combination with the overall space curvature [15]-[19].

On scales $\alpha \lesssim 11'$, corresponding to the angular resolution of the recombination width taken by the cosmic plasma to transform from the opaque to transparent state, the dominant effect is the Doppler shift experienced by the relic photons having been scattered off the moving condensations [20],[21],[22],[23]. At $11' < \alpha < 5^{\circ}$ this process is augmented by the density perturbation effect accounting for the fact that just before the recombination plasma density fluctuations were accompanied by radiation temperature variations [24]. On smaller scales this effect vanishes because the radiation has time to escape density enhancements, and the dominant contribution becomes that due to the Doppler shift since translucent condensation boundaries move freely.

For scales $\alpha \lesssim 11'$ one should take into account the mutual compensation effect when light rays pass successively through many condensations and rarefactions along the recombination width. Here we use the numerical simulation results [25].

For scales $\alpha > 11'$ we can treat the recombination as an intantaneous process and explicitly calculate the background anisotropy:

$$\frac{\Delta T}{T}(\vec{e}) = -\frac{1}{2} e^{\alpha} e^{\beta} \int_{\eta_{rec}}^{\eta_0} h'_{\alpha\beta}\, d\eta + (v' + e^{\alpha} v_{,\alpha})\Big|_{\eta=\eta_{rec}}, \quad (9)$$

where $h_{\alpha\beta}$, $\delta_b = 3v'$ and $u_b^i = \{1, -v_{,\alpha}/a\}$ are metric, barion density and 4-velocity perturbations in the synchronous co-moving to relic particles reference system, η is the conformal time, $(') = \partial/\partial\eta$, a is the scale factor, functions $h_{\alpha\beta}$ and v are taken on the light-cone. Three terms on the right-hand-side of eq. (9) are obviously responsible for the corresponding effects mentioned above.

5. THE RESULTS FOR $\Omega = 1$

The results for the root-mean-square temperature fluctuation $\frac{\Delta T}{T}(\alpha)$ (see eq. (5)) in the entire angular range $(0 < \alpha < 180^{\circ})$ are shown in the Figure. Arrows fix the up-to-date observational limits for the small [2] and large [7] scale anisotropies. We chose here the flat spectrum of primordial perturbations (see eq. (7)) which is predicted by a standard model of the inflationary universe. Solid lines demonstrate the models with the flat 3-space. Curves (a) and (b) show that the models with usual

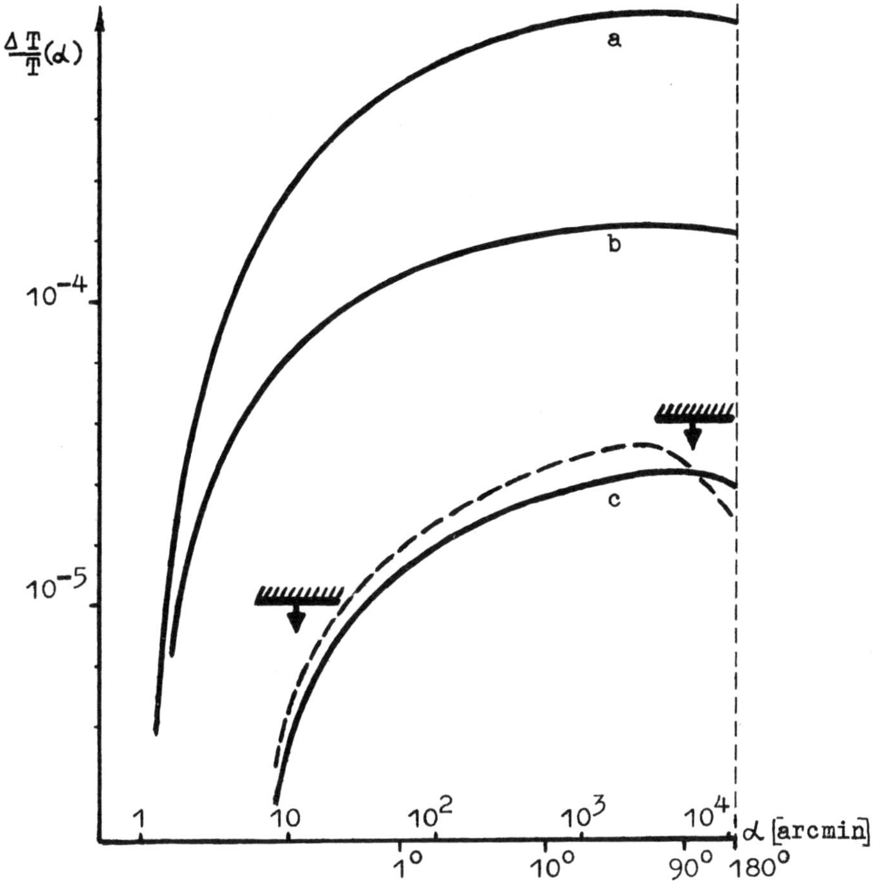

FIGURE. The anisotropy of the relic radiation $\frac{\Delta T}{T}(\alpha)$
solid lines: $\Omega = 1$, $n = 0$, missing mass: (a) barions, (b) massive neutrinos, (c) superheavy relic particles;
dashed line: variant (c) for $\Omega = 0.5$;
(for the curve (c), objects with masses $\sim 10^{14} M_{\odot}$ form at $z = 3$).
Dipole moment is subtracted. Arrows - observational data. Data for $\alpha = 90^{\circ}$ correspond to the upper limit for the quadrupole anisotropy. Data for $\alpha \sim 10'$ correspond to $\frac{\Delta T}{T}(\alpha)$.

barions and thermodynamic neutrinos *) as the missing mass, are in contradiction with both observational limits. The model with superheavy free particles is in agreement with the existing observations. It is still on the margin of the observational limits if one chooses normalization of the primordial density perturbation spectrum corresponding to the curve (c).

The expected amplitudes of dipole and quadrupole moments (see eq. (6)) for all considered models are summarized in the Table. Two different normalizations of the

TABLE. Dipole $\langle a_{1m}^2 \rangle^{1/2}$ and quadrupole $\langle a_{2m}^2 \rangle^{1/2}$ anisotropies of the relic radiation in the massive neutrino (b) and superheavy particle (c) dominated universes for three variants of the primordial perturbation spectra (see eq. (7))

	l	n = 0	n = -1	n = 1
b	1	$1.5\ 10^{-2}$	$2.2\ 10^{-2}$	$1.5\ 10^{-2}$
	2	$3\ 10^{-5}$	$9\ 10^{-4}$	$< 2\ 10^{-5}$
c	1	$1.8\ 10^{-3}$	$2.8\ 10^{-2}$	$1.8\ 10^{-3}$
	2	$2\ 10^{-5}$; $3\ 10^{-6}$	$6\ 10^{-5}$	$< 10^{-6}$

primordial perturbations in c-model - the one corresponding to the curve (c) in the Figure and the other taken by Peebles [26] - account for two values of the quadrupole anisotropy. The latter corresponds to the formati-

*) We deal here with one sort of massive neutrino (and two other massless). Increase of massive neutrino sorts (eq. $\Omega = 1$ holds) will only raise the $\Delta T/T$-anisotropy level.

on at $z = 3$ of objects with masses $\sim 10^{12} M_{\odot}$, in this case our result for the quadrupole $\Delta T/T$ - amplitude confirms the earlier calculations [26),27)] $(\langle a^2_{2m} \rangle^{1/2} \simeq 3 \cdot 10^{-6})$.

The dipole moments are calculated for the rather long waves that evolve linearly up to now. For confrontation with the observations, one will have to take into account the kinematics of the local group of galaxy clusters in order to subtract the contribution of the non-linear modes to the observed dipole anisotropy.

6. HOW TO DETERMINE Ω FROM THE SKY $\Delta T/T$ - PATTERN

Let us now turn to the structure of the relic radiation in the open Universe ($\Omega < 1$).

We consider only those states of the primordial perturbation field $q(x)$ that we call as homogeneous (on the average) ones [16)]: they are uniformly limited in the Lobachevski space (3-space of the open Universe) and equally excited at every point of the space. Homogeneous states can be represented as random superpositions of the full sets of wavefunctions with constant amplitudes over the space. These functions were found in ref. [16)]:

$$f^{\pm}_{k,\vec{e}} = \left(\sqrt{1+(h\vec{x})^2} \pm h\vec{e}\,\vec{x}\right)^{\pm i\frac{k}{h}} \qquad (10)$$

and are plane-wave analogs in the Lobachevski space. Here $\vec{x}$ labels the locally Cartesian coordinates, h^{-1} is the space curvature radius. For $h \longrightarrow 0$ functions (10) turn to usual plane-waves in the Euclidean space ($\vec{k} = k\vec{e}$). Being homogeneous on the average waves (10) become spatially homogeneous in the infinite-wavelength limit mit [16),17),18)] for $k \longrightarrow 0$ they acquire the group of motions of Bianchi type V models. In general, every uniformly limited function $|q(\vec{x})| <$ const in the Lobachevski space can be Fourier-expanded over the plane-waves (10) [16),18)].

Let us now clarify the "spotty structure" of the large scale $\Delta T/T$ - distribution in the open Universe [15)-19)]. A single density perturbation plane-wave in the flat Friedmann model produces the sky $\Delta T/T$ - pattern with the large-scale shape presented as a quadrupole dependence on the angle θ between the wavevector and the line of sight ($\Delta T/T \sim \sin^2\theta$). The large-scale $\Delta T/T$ - distribution produced by a single plane-wave in the open Friedmann model has the form of a "spot":

$$\frac{\Delta T}{T} \sim \left(\frac{2\varkappa \tan \theta/2}{1 + (\varkappa \tan \theta/2)^2} \right)^2 , \qquad (11)$$

where $\varkappa = e^{-h\chi_{rec}}$, $\chi_{rec} = \eta_0 - \eta_{rec}$ is the geodetic 3-distance of the observer (η_0) from the recombination sphere (η_{rec}). The spot shape (11) is independent of the wavevector modulus k and identical with that of Bianchi V model [15)]. The $\Delta T/T$ - anisotropy fully develops inside the annulus of the angular radius $\theta_0 = 2\tan^{-1}\varkappa$ and the width $\sim \theta_0$. For $\Omega = 1$ ($h = 0$), the spot (11) reduces to a quadrupole dependence. For small Ω, $\varkappa = \Omega/4$ and the angular size of the spot is equal to Ω.

We make the following prediction about large-scale $\Delta T/T$ - distribution in the open Universe [16),17),18),19)]: the observed sky $\Delta T/T$ - pattern is a random superposition of the uncorrelated spots. It means that for $\Omega < 1$ ($h \neq 0$) a statistical distribution of $\frac{\Delta T}{T}(\alpha)$ with an angular correlation scale $\sim \Omega$ takes place (see Figure). The correlation function also depens on the primordial perturbation spectrum q_k. So, the large-scale

$\frac{\Delta T}{T}(\alpha)$ - distribution allows for determining both the spectrum q_k and the overall matter density Ω .

7. THE MAIN CONCLUSIONS

We emphasize here two conclusions concerning the large-scale $\Delta T/T$ - fluctuations.

(i) If the quadrupole amplitude $(\frac{\Delta T}{T})_Q < 3\cdot 10^{-5}$ then a standard model of the inflationary universe with the missing mass in the form of massive neutrinos, is incorrect.

(ii) The detection of $\Delta T/T$ on scales from 5° and up to 180° would enable one to draw the conclusions on the primordial spectrum $\delta\rho/\rho$ and the total matter density $\Omega = \rho/\rho_{cr}$.

REFERENCES

1. Lukash, V.N., Naselskij, P.D. and Novikov, I.D., "Anisotropy of the Relic Radiation and the Global Structure of the Universe", Preprint Space Res. Inst. Пр - 921, Moscow (1984).

2. Parijskij, Yu.N., Petrov, Z.N. and Chernov, A.N., Astr. Zh. Lett., 3, 483 (1977).

3. Partridge, R.B., Phys. Scripta 21, 624 (1980).

4. Uson, J.M. and Wilkinson, D.T., Ap. J. Lett. 277, L1 (1984).

5. Fabbri, R., Guidi, J., Melchiorri, F. and Natale, V., Phys. Rev. Lett. 44, 1563 (1980).

6. Lubin, P., Epstein, G. and Smoot, G., Phys. Rev. Lett. 50, 616 (1983).

Fixen, D., Cheng, E. and Wilkinson, D., Phys. Rev. Lett. 50, 620 (1983).

7. Strukov, I.A., Sagdeev, R.Z., Kardashev, N.S., Skulachev, D. and Eysmont, N., In: Advances in Space Research (Pergamon Press), Proc. COSPAR, June 27-July 7 (1984).

8. Lukash, V.N. , JETP Lett. 31, 631; JETP 79, 160 (1980).

9. Kompaneets, D.A. , Lukash, V.N. and Novikov, I.D., Astr. Zh. 59, 424 (1982).

10. Lukash, V.N. and Novikov, I.D., In: Contr. papers 10-Int. Conf. GRG, Padova, ed. Bertotti B. et al.,844, July 4-9 (1983).

11. Guth, A.H., Phys. Rev. D23, 347 (1981).

12. Linde, A.D., Phys. Lett. 108B, 389; 114B, 431; 116B, 340 (1982).

13. Linde, A.D., JETP Lett. 38, 149; Phys. Lett. 129B, 177 (1983).

14. Sachs, R.K., and Wolfe, A.M., Ap. J. 147, 73 (1967).

15. Novikov, I.D., Astr. Zh. 45, 538 (1968).

16. Lukash, V.N., In: Contr. papers 8-Int. Conf. GRG, Waterloo, 237, Aug. 7-12 (1977).

17. Bisnovatyi-Kogan, G.S., Lukash, V.N., and Novikov, I.D.,In: Variability in Stars and Galaxies (IAU/EPS), Proc. 5-Regional Meeting in Astr., Liege, G.1.1, July 28-Aug. 1 (1980).

18. Lukash, V.N., In: Early Evolution of the Universe and its Present Structure (D.Reidel Publ. Comp.), Proc. Symp. 104 IAU, Crete, ed. Abell G.O. and Chincarini G., 149, Aug. 30 - Sept. 2 (1982); Doctoral thesis, Space Res. Inst., Moscow (1983).

19. Lukash, V.N. and Novikov I.D., "The Effect of 'Spottiness' in Large-Scale Structure of the Microwave Background", Preprint Space Res. Inst. Пр - 954, Moscow (1984).

20. Zeldovich, Ya.B. and Sunyaev, R.A.,Ap. Space Sci., 6, 358 (1970).

21. Peebles, P.J.E. and Yu, I.T., Ap.J. 162, 815 (1970).

22. Doroshkevich, A.G., Zeldovich, Ya.B. and Sunyaev, R.A.,Astr. Zh. 55, 913 (1978).

23. Shandarin, S.F., Doroshkevich, A.G. and Zeldovich, Ya.B., Usp. Fiz. Nauk 139, 83 (1983).

24. Silk, J.,Ap. J. 151, 459 (1968).

25. Zabotin, N.A. and Naselskij, P.D. , Astr. Zh. 59, 447 (1982); 60, 467 (1983).

26. Peebles, P.J.E. , Ap. J. Lett. 263, L1 (1982).

27. Starobinskij, A.A., Astr. Zh. Lett. 9, 579 (1983).

PRIMORDIAL BLACK HOLES AND OBSERVATIONAL RESTRICTIONS ON QUANTUM GRAVITY

M.Yu. Khlopov
Keldysh Institute, of Applied Mathematics, Moscow, USSR

P.D. Nasel'skij
Rostov State University, Rostov, USSR

A.G. Polnarev
Space Research Institute, Moscow, USSR

ABSTRACT

Astrophysical upper limits on the primordial black hole (PBH) spectrum restrict the parameters of quantum gravity, cf supergravity. Possible contribution of PBHs into the dark matter of the modern Universe is discussed.

1. INTRODUCTION

Primordial black holes (PBHs) were first discussed by Zeldovich and Novikov[1] (see also[2]). The effect of PBH evaporation was theoretically discovered by Hawking[3]. There are restrictions on the PBH spectrum only. However, even such restrictions provide nontrivial (and may be unique) information on the physical conditions in the very early Universe.

Calculation of the probability of PBH formation on $p=\varepsilon/3$ stage was given by Carr[4]. If primordial spectrum of metric perturbations is characterized by small amplitude $\delta < 1$, PBH formation may take place due to Gaussian "tails" of statistical fluctuations and the probability of PBH formation is $\propto \exp(-1/18\,\delta^2)$.

Absence of PBHs or effects of their evaporation was considered as a proof of small inhomogenity of the very early Universe.

However, the development of GUT models made us revise the simple picture of the early Universe. Cosmological evolution had to be considered in the light of superhigh energy physics, predicting new particles and fields in the very early Universe.

It turned out, that early cosmological evolution must not be "smooth", as believed previously. We are almost convinced now in the neccessity of inflation[5] (See review[6]). Various possibilities of early dust-like stages, realized either as the domination of superheavy metastable particles in the Universe[7], or as the stages of classical scalar field oscillations[8], may arise. Rather complicated successions of such stages and short inflational stages may take place.

The probability of PBH formation depends crucially on the parameters of all these stages. So restrictions on PBH spectrum provide nontrivial restrictions on these parameters.

The predicted PBH spectrum is determined, in general, by two independent factors[4]: by the spectrum of primordial metric perturbations and by the character of the equation of state variations in the early Universe. The development of quantum field theory makes possible to establish the relationship between these two factors in the frame work of unified theoretical description. Pure gravitational and pure field-theoretical aspects are unified in the framework of the theory of all the four interactions: in the framework of supergravity, Caluzza-Clein theory etc (See review[6] and corresponding articles in this volume).

Looking forward, it is reasonable to await that the

future theory of quantum gravity will determine the properties of scalar field, used in the modern inflational models, that it will determine the parameters of scalar field oscillations, resulting in the early dust-like stages, that it will predict the spectrum of primordial perturbations. On the other hand, one may expect, that the future theory will predict new typesof metastable particles.

Retained on subsequent stages of cosmological evolution, new particles and fields, PBHs, new types of astrophysical objects (domain walls[9]), strings[10]) etc) may influence the physical processes in the Universe. Analysis of such influence may result in new interpretation of the observational data. Such interpretation must take into account the new relationship between the physics of the Universe before 1s with the astrophysical processes in the Universe after 1s (See cf review[11]). However, even reinterpreting the data, the consequences of the theory must be compatible with the observations. So the observational data may restrict or even exclude certain variants of the theory. In this sense one may consider the Universe as a laboratory of quantum gravity.

2. ASTROPHYSICAL RESTRICTIONS ON NEW PARTICLES

One may expect, that some of the particles, arising in the theory of quantum gravity unified with particle theory, will be metastable. In the framework of a given model one may estimate the frozen concentration of such particles. Based on these estimations and predicted properties of concrete species of particles, one may estimate the effect of such particles (their decay or annihilation) on the processes, taking place in the Universe after 1s of expansion.

Various restrictions obtained by numerous authors[12)-14)] are put together on fig.1, taken from the review[14)].

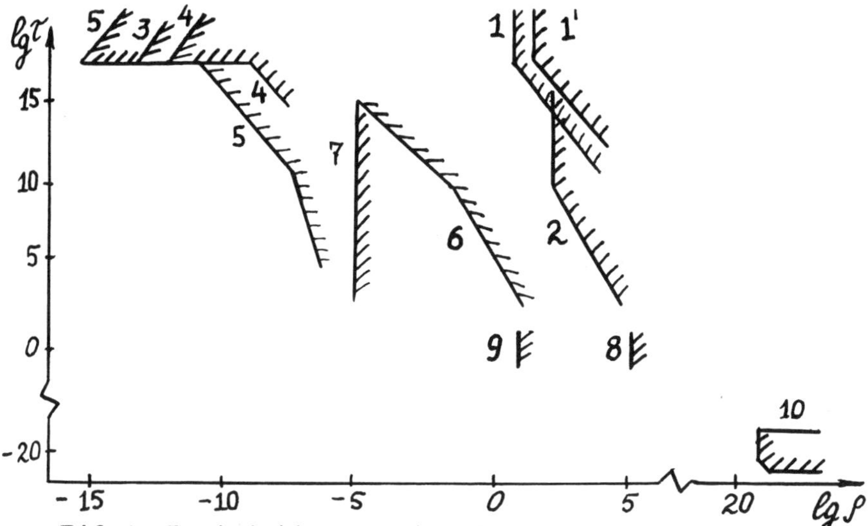

Fif.1. Restrictions on the density, ρ of new particles (in the units of the baryon density) and on their lifetime τ. These restrictions are valid for the density ρ of any type of new astrophysical object with the value of τ specified for concrete type of object. Concerning restrictions on the PBH density ρ_{PBH} τ denotes the time of evaporation. Shaded are the values of ρ and τ, excluded from analysis of influence of any such object (or its products) on the dynamics of the Universe as a whole - on its "flatness" or its age - regions, confined by lines 1 and 1', respectively; on the formation of large scale structure (line 2); on the spectrum of cosmic rays (line 3); on the γ back-ground (line 4); on the spectrum of ultrahigh energy neutrino (line 5); on the formation of thermal background spectrum (line 6); on the concentration of D and ^3He (line 7); on the concentration of ^4He (line 8); on the kinetics of freezing of neutron to proton ratio (line 9). Bounded by line 10 is the region of ρ and τ excluded by restrictions on PBH spectrum (See Sect.5).

Let us pay attention to possible relationship between the properties of nonrelativistic particles and the development of inhomogenities of the Universe.

The modern theory of large scale structure[15] considers stable and metastable particles as possible form of dark matter of the Universe[16]. Astrophysical restrictions, accounting for inhomogeneous distribution of new particles, are several orders of magnitude stronger than in the assumption of their homogeneous distribution.(See fig.1).

However, most of the particles, predicted by the theory, would be metastable with a lifetime much smaller, than 1s. Moreover, many of such particles may be so massive, that their production after 1s of expansion is forbidden. Such particles may exist in the very early Universe only.

In the period, when the thermal energy of particles kT exceeded their rest energy mc^2, these particles were relativistic.

If the lifetime of particles exceeded the cosmological timescale, when $kT < mc^2$, such "longlived" particles may play the role in the development of inhomogenity in the early Universe similar to the role of stable nonrelativistic particles in the development of inhomogenities in the modern Universe.

It means, that in the course of cosmological expansion early dust-like stage must take place. Dust-like stage may also appear as a stage of scalar field oscillations after inflation.

Growth of small primordial inhomogenities on early dust-like stages may result in the formation of primordial black holes (PBHs). PBHs, being formed in the very early Universe, retain on successive stages of expansion. Analysis of their effects on the physical processes in the Universe can be used as a tool to check the predictions concerning superheavy metastable particles.

3. ASTROPHYSICAL RESTRICTIONS ON THE PBH SPECTRUM

According to Hawking[3)] PBHs with the mass, smaller than 10^{15} g, must evaporate to the present time due to quantum processes.

If $\alpha(M)$ is the fractional density of PBHs with mass M at the moment of their evaporation ($\alpha(M)$ =

$= \rho_{PBH}(M)/\rho_{tot}$, where ρ_{tot} is the total cosmological density at that moment), then analysis of the influence of PBH evaporation on different astrophysical processes provide upper limits on $\alpha(M)$.

Correct treatment of PBH evaporation is possible for PBH masses, exceeding the Plank mass 10^{-5}g, only, since for smaller PBH masses their width relative to evaporation is comparable with their mass (On stable PBHs with Plank mass-maximons see[17]. On PBHs with mass, smaller than 10^{-5}g, see [18]).

PBHs with $M < 10^9$g evaporate before 1s. They may produce both additional baryon charge and additional entropy[19]. Upper limits on additional entropy restrict $\alpha(M < 10^9$g).

PBHs with the mass in the interval 10^9-10^{10}g evaporate in time interval 1-10^3s. Evaporating, such PBHs may effect on the abundance of primordial helium[20,21]. It provide upper limits on α ($10^9 < M < 10^{10}$g).

Observed deuterium abundance provides restrictions on the magnitude α (M) for PBHs with masses 10^{10}-10^{13}g, evaporating in the period 10^3-10^{12}s[21].

Analysing the dynamics of recombination[22] and distorsions of the thermal background spectrum[19] one can restrict α (M) for PBHs with $M \sim 10^{13}$-10^{14}g, evaporating during the matter and radiation decoupling.

The most stringent limits on α(M) were obtained for PBHs with masses $M \sim 10^{14}$-10^{15}g from observation of γ background[23] and γ bursts[23,24], as well as from analysis of sinchrotron radioemission[25] and observed fluxes of antiprotons in cosmic rays[14].

PBHs with $M > 10^{15}$g survive to the present time and restrictions on their modern concentration may be obtained from the estimation of their allowed contribution into the cosmological density[1] or of their effect on the dynamics of superclusters (for $M > 10^{15}$ $M_\odot$) and on the large scale structure of the Universe [19]. Review of observational restrictions on the magnitude $\alpha(M)$ see in[4,19,26].

Astrophysical upper limits on the PBH contribution into cosmological density at the moment of their evaporation ($M < 10^{15}$g) or into the modern density ($M > 10^{15}$g) result in restrictions on the magnitude $\beta(M)$, defined as the fraction of the matter, collapsed into PBHs with mass M at the moment of their formation. If the evolution of the early Universe is "smooth", the relationship between α and β would be reduced to multiplication of β by the factor, accounting for the enchancement of relative contribution of PBHs into the cosmological density at the $p = \mathcal{E}/3$ stage. In the presence of early dust-like stages (p=0) this relationship depends significantly on the moment t_k, when the early dust-like stage ends. The magnitude t_k may depend both on particle properties and on their evolution within nonlinear structures, formed on the early dust-like stage. As it was shown in[27] one may give crude but rather reliable upper limit on the magnitude t_k: any early dust-like stage must end up to 1s of cosmological expansion (t_k < 1s). Otherwise they would spoil the observed light element abundances and the thermal spectrum of background radiation.

4. THE MECHANISMS OF PBH FORMATION IN THE VERY EARLY UNIVERSE AND THE EQUATION OF STATE.

The value, which is to be compared with β is the probability W of PBH formation in the early Universe.

For each mass M there exists a moment t(M), at which the mass M is equal to M_h- the mass contained within the cosmological horizon. This moment puts lower limit on the time of PBH formation, since the notion "black hole" is senseless for $M > M_h$.

On the other hand, PBH mass in the moment of its formation must exceed so called Jeans mass $M_J = \rho(c_s t)^3$, where c_s is the speed of sound. On the $p=\mathcal{E}/3$ stage the magnitude M_J is smaller but of the same order as M_h: $M_J = 1/3^{3/2} M_h$. So the formation of PBH with given mass is allowed within a small time interval near t(M).

Remind that PBH formation requires metric perturbation of order 1. It is known also, that metric perturbation of any scale remains constant at $p=\mathcal{E}/3$ stage[28].

If the dispersion of metric perturbation $\delta < 1$, the probability W is of order[4]

$$W \sim 1/\delta \quad \exp(-h^2/18\delta^2) \ll 1 \qquad (1)$$

which corresponds to "Gaussian tails" in the amplitude distribution of metric perturbations. (Detailed statistic analysis see in[29]. Hydrodynamics of PBH formation for $h \sim 1$ at the $p=\mathcal{E}/3$ stage see in[30].

This is not the case for p=0 stage. According to the theory of gravitational instability in the expanding Universe small initial perturbations grow on the dust-like stage as[28]

$$\delta\rho/\rho \propto (t/t(M))^{2/3}, \qquad (2)$$

where t(M) is defined above. At $t_1 = t(M)\,\delta^{-3/2}$ the magnitude $\delta\rho/\rho$ grows up to 1. At this moment inhomogenities of mass scale M separate from cosmological expansion and start to contract[31]. The bulk of contracting matter forms inhomogenities with small gravitational potential (cf pancakelike configurations). However, with small (but nonzero) probability spherically symmetric homogeneous contraction is possible, resulting in black hole formation[7].

The probability of PBH formation depends on the amplitude of initial density perturbation. The smaller is the amplitude, the more fine tuning of density and velocity distributions of particles within the configuration is needed. Estimations[7] give for the probability of PBH formation in the mass interval $M_o < M < M_k$ (where $M_o = m_{Pl} t_o / t_{Pl}$, $M_k = m_{Pl} t_k / t_{Pl} \delta^{-3/2}$, t_o and t_k are the moments of the beginning and of the end of the dust-like stage):

$$W \sim 2 \cdot 10^{-2} \delta^{13/2} \quad (3)$$

As it was shown in[32] the same estimation is valid for the case of scalar field dominance in the Universe.

5. RESTRICTIONS ON EARLY DUST-LIKE STAGES AND INITIAL INHOMOGENITIES

With the use of (3) it is possible to obtain restrictions on the parameters of early dust-like stages, t_o and $\tau = t_k$, which depend on concrete particle and field properties.

Any variant of the theory contradicts to the astrophysical upper limits on the PBH spectrum, if $W > \beta$ at least for one value of M from the considered interval $M_o < M < M_k$ (See fig.2).

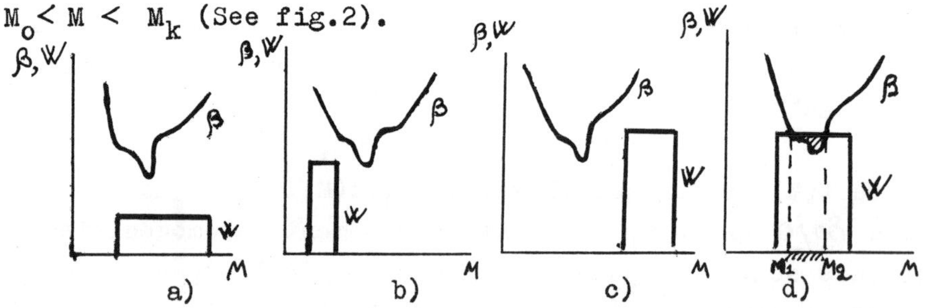

Fig. 2. The situations, allowed (a,b,c) and forbidden (d) from observations.

In the case of "flat" spectrum of initial metric perturbation it is easy to obtain the system of inequalities for t_o, t_k and δ. On fig.3 restrictions on t_o and t_k in dependence on δ are given.

Fig. 3 demonstrates the principal possibility to constrain the parameters of quantum gravity basing on astrophysical restrictions on PBH spectrum.

As was shown in[33] general restrictions, represented on fig.3, can be applied to concrete species of particles, predicted by various models. In particular these restrictions set nontrivial constraints on the mass of superheavy gravitino, arising in the models of N > 1 Supergravity (see fig.4).

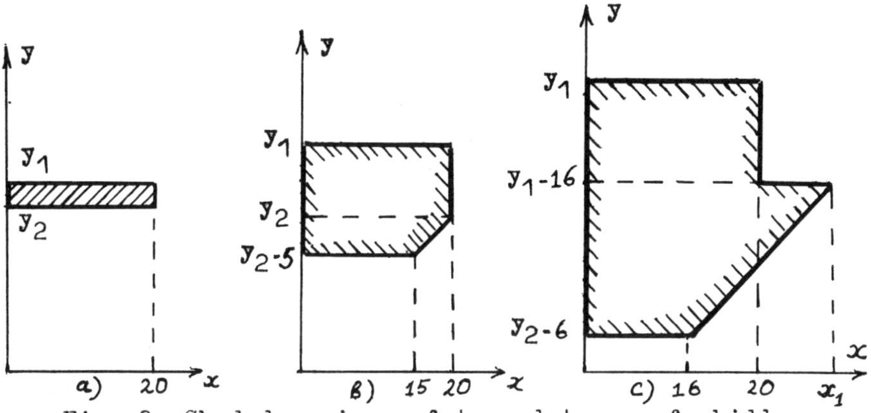

Fig. 3. Shaded regions of t_o and t_k are forbidden from observations. The different figures correspond to different δ : a) $6\cdot 10^{-4}<\delta<1.4\cdot 10^{-3}$; b) $1.4\cdot 10^{-3}<\delta<7.8\cdot 10^{-3}$; c) $\delta > 7.8\cdot 10^{-3}$. Here x= log(to/tpl), y= log(tk/tpl), $x_1 = 20+29/2 \log(\delta/7.8\cdot 10^{-3})$; $y_1 = 25+13 \log(\delta/6\cdot 10^{-4})$; $y_2 = 25-3/2 \log(\delta/6\cdot 10-4)$.

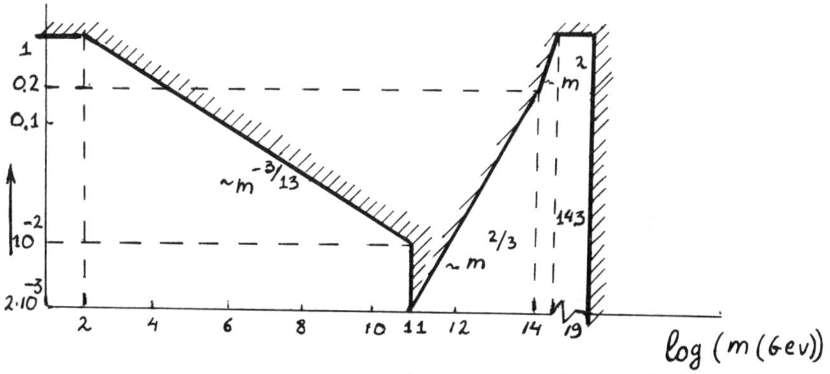

Fig. 4. Restrictions on the mass of gravitino, in dependence on δ .

6. THE PARAMETERS OF INFLATIONARY STAGES AND THE PBH SPECTRUM

There are various models of inflation[5),6)]. According to Starobinsky (see[5)]) the inflation, i.e. exponential growth of the scale factor, can be induced by quantum-gravitational effects near the singularity.

According to another approaches (see[5),6)]) de Sitter stage of expansion can be induced by the domination of the "false" vacuum of Higgs fields.

Haotic inflation[6)] is also considered.

As a rule inflationary cosmology predicts "scale free" ("flat") spectrum of metric perturbations, first suggested by Zeldovich[34)]. It means that small scale amplitudes do not exceed large scale amplitudes of metric perturbations, which are severely restricted by the observed isotropy of the thermal electromagnetic background (see cf[35)]): $\delta < 10^{-4}$. So the models, pretending to explain the observed large scale structure of the Universe and predicting scale free spectrum imply extremely small probability of PBH formation.

However, the only reason to consider single inflational stage in the very early Universe is the reason of simplicity. That is why let us assume a little bit more sophisticated scenario, involving two (or more) inflationary stages. Assume also that the earlier inflation is long enough to provide such global preperties of the modern Universe as flatness, homogenity, isotropy and large scale structure. For example, it may be quantum gravitational inflation mentioned above. Assume at last that the later inflationary stage is short enough and does not affect directly these global properties, but results in high amplitude perturbations at small scales. For example, it may be GUT inflationary stage with "natural" Higgs self-couplings[6)]. So multiple inflationary models may predict essentially non-flat spectrum with "small large-scale amplitude" and relatively "large small-scale amplitude".

It is interesting to note, that alternative way to

obtain the spectrum of similar shape is to involve more than one scalar field in the frame of haotic inflation (see[36]).

It is important to emphasize that the duration of any inflationary stage, induced by any scalar field, is determined by the field self coupling constant λ: the greater is λ, the shorter is the stage. On the other hand, the amplitude of metric perturbations $\delta \propto \lambda^{1/2}$; so the greater is λ, the higher is the amplitude. Thus for "natural" values of $\lambda \sim 10^{-2} - 10^{-4}$, high amplitude small scale part of the spectrum seems to be natural.

In such scenario with nonuniform spectrum, large amount of PBHs may be produced independently on the equation of state (even for $p=\mathcal{E}/3$) after inflation[37].

High probability of PBH formation after the short inflationary stages (due to high δ at small scales) provides restrictions on the parameters of these stages.

To obtain such restrictions one should take into account the following. The spectrum of small scale perturbations is almost flat up to the mass scale M_H of the cosmological horizon at the end of short inflationary stage (t_H). However there is the logarithmic decrease of amplitude to smaller scales[6]. This fact is of great importance when one estimates the probability of PBH formation: the exponential dependence on δ in the eq. (1) results in PBH spectrum peaking at $M_{BH} \sim M_H$. One can obtain[37] the relationship between fundamental parameters of the secondary inflation and the mass of PBH at the peak of the probability

$$M_{BH} \sim M_H \sim \xi m_{pl}^2/M_x^2 , \qquad (6)$$

where $\xi = \exp(2\varkappa)/(\alpha\varkappa)$, α is the GUT gauge constant and $M_x \sim 10^{15}$ Gev is the mass of x boson.

This variant of two-stage inflation scenario is compatible with the observations, if 1) the probability of black hole formation, estimated with the use of Eq.(1) for $M = 10^{15}$g, satisfies $W(M=10^{15}g) < 10^{-25}$; 2) PBHs are formed with the masses $M \ll 10^{15} M_{\odot}$ what is necessary to avoid contradictions with isotropy of microwave background radiation, 3) the modern mass density of PBHs with $M > 10^{15}$g does not exceed the critical density. These rather obvious restrictions reduce to the following inequalities[37]:

$$\varkappa \gtrsim \ln \left[\frac{\alpha M_{BH}}{m_{pl}} \left(\frac{M_x}{m_{pl}} \right)^2 \right], \qquad (5)$$

$$\delta \lesssim 3 \cdot 10^{-2} (1 - 8 \cdot 10^{-3} \ln(M_{BH}/m_{pl}))^{-1/2} . \qquad (6)$$

At $M_{BH} \sim 10^{15}$g, $\varkappa \sim 13-14$ and for $M_{BH} \sim M_{\odot}$, $\varkappa \sim 34-35$. In other words, if $\varkappa \sim 13-35$, the mass of PBH is constrained by $10^{15}\text{g} < M < 10^{33}$g.

7. THE PROBLEM OF DARK MATTER AND PBHs

In the framework of two-stage inflational scenario the modern dark matter may be naturally ascribed to PBHs.

As it follows from eqs.(5,6) this hypothesis may be compatible with astrophysical restrictions on the PBH spectrum. (if $\delta \sim (4-6)10^{-2}$).

Considering PBHs as supporters of dark matter it should be noted that PBH dominance in the period of large scale structure formation is hardly compatible with the existence of voids, and with parameters of large scale structure[15]. However in the unstable neutrino model[38] large scale structure is formed at the stage of unstable neutrino dominance, and the modern dark matter may be determined by PBHs, dominating the density of inhomogenities after neutrino decays. The analysis of PBHs as the supporters of the dark matter is of great importance for cosmology.

8. CONCLUSION

This contribution should be regarded as description

of new method, putting nontrivial restrictions on the physical conditions in the very early Universe. Metastable particles and PBHs may be viewed as some kind of bridge between microphysics of the very early Universe, involving quantum gravity, and macrophysics of the Universe in its present state including its global properties. May be at the first sight this bridge looks to be rather "narrow". But, when direct experiment is impossible and the observational data are so rare, any indirect information about the very early Universe and any restriction on the theory are desirable.

REFERENCES

1) Zeldovich, Ya.B. and Novikov, I.D., Astron.Zh. 43, 758 (1966).

2) Hawking, S.W., Mon.Not. RAS, 152, 75 (1971).

3) Hawking, S.W., Nature 248, 30 (1974); Commun. Math. Phys. 43, 199 (1975).

4) Carr, B.J., Astrophys. J. 201, 1 (1975).

5) Guth, A.H., Phys.Rev. D23, 347 (1981); Linde, A.D., Phys. Lett. B108, 389 (1982); Albrecht, A. and Steinhardt, P.J., Phys.Rev.Lett., 48, 1220 (1982).

6) Linde, A.D., Uspekhi Fiz. Nauk 144, 177 (1984).

7) Khlopov, M.Yu. and Polnarev, A.G., Phys. Lett. B97, 383 (1980); Astron. Zh. 58, 706 (1981); Ibid. 59, 639 (1982).

8) Turner, M.S., Phys.Rev. D28, 1243 (1983);Starobinsky, A.A., Phys.Lett. B91, 99(1980).

9) Zeldovich, Ya.B., Kobsarev, I.Yu. and Okun, L.B., ZhETP 67, 3 (1974).

10) Kibble, T.W.B., J.Phys. A9, 1387 (1976); Shafi,Q., in: The Very Early Universe", Eds. Gibbons, G.W., Hawking S.W., Siclos S.T.C., Cambridge (1982)p.147; Vilenkin A., Ibid p.163.

11) Chechetkin, V.M., Khlopov, M.Yu. and Sapozhnikov,M.G., Rivista Nuovo Cim. 5, N 10 (1982).

12) Gershtein, S.S. and Zeldovich, Ya.B., IETP Lett. 4, 120 (1966); Schwartzman, V.F., IETP Lett. 9, 184 (1969); Cowsik, R. and McClelland, J., Phys. Rev.Lett.

29, 609 (1972) etc.
See detailed references in reviews [13,14].

13) Dolgov, A.D. and Zeldovich, Ya.B., Rev.Mod.Phys. 53, 1 (1981); Schramm, D.N., Ann. N.Y. Acad.Sci. 375, 54 (1981); Polnarev, A.G. and Khlopov, M.Yu., Uspekhi Fiz. Nauk 145, N 3 (1985).

14) Khlopov M.Yu., Chechetkin V.M. EChAYa (1985).

15) Shandarin, S.F., Doroshkevich, A.G. and Zeldovich,Ya. B., Uspekhi Fiz.Nauk 139, 83 (1983).

16) Peebles P.J.E. "The Large Scale Structure of the Universe". Princeton. N.Y. (1980).

17) Markov, M.A., Progr. Theor. Phys. Sup. Ext. p.85 (1965); ZhETP 51, 878 (1966).

18) Zeldovich, Ya.B., in: The Proceed. of the Second Seminar "Quantum Gravity", Moscow 1981, p.146 (1982).

19) Novikov, I.D., Polnarev, A.G., Starobinsky, A.A. and Zeldovich, Ya.B. Astron. & Astrophys. 80, 104 (1979).

20) Vainer, B.V., Nassel'skij,P.D., Pis'ma Astr. Zh. 3, 147 (1977).

21) Zeldovich Ya.B., Starobinsky A.A., Khlopov M.Yu. and Chechetkin, V.M. Pis'ma Astr. Zh. 3, 208 (1977).

22) Nassel'skij,P.D., Pis'ma Astr. Zh. 4, 387 (1978).

23) Page, D.N. and Hawking, S.W., Astrophys. J. 206, 1 (1975).

24) Rees, M.J., Nature, 268, 333 (1977); Blanford, R.D., Mon. Not RAS. 181, 489 (1977).

25) Nassel'skij,P.D. and Pelikhov, N.V., Astron. Zh. 56, 714 (1979).

26) Carr, B.J., in: The Proceed. of the Second Seminar "quantum Gravity", Moscow 1981, p.195 (1982).

27) Polnarev, A.G. and Khlopov, M.Yu., Astron. Zh. 59, 15 (1982).

28) Lifshitz, E.M., ZhETF 16, 587 (1946).

29) Zabbotin, N.A., Marochnik, L.S. and Nassel'skij,P.D., Astrofizika 18, 161 (1982).

30) Nadyozhin, D.K., Novikov, I.D. and Polnarev, A.G., Astron. Zh. 55, 216 (1978); Novikov, I.D. and Polnarev, A.G., Astron. Zh. 57, 250 (1980).

31) Zeldovich, Ya.B. and Novikov, I.D., "Structure and Evolution of the Universe". Moscow, Nauka (1975).